Fish and Fisheries in the Brazilian Amazon

Renato A. M. Silvano
Editor

Fish and Fisheries in the Brazilian Amazon

People, Ecology and Conservation in Black and Clear Water Rivers

Springer

Editor
Renato A. M. Silvano
Department of Ecology
Federal University of Rio Grande do Sul
Porto Alegre, Rio Grande do Sul
Brazil

ISBN 978-3-030-49145-1 ISBN 978-3-030-49146-8 (eBook)
https://doi.org/10.1007/978-3-030-49146-8

This Springer imprint is published by the registered company Springer Nature Switzerland AG
The registered company address is: Gewerbestrasse 11, 6330 Cham, Switzerland

We dedicate this book to the Amazonian people who live off the rivers and forests, especially to fishers who provided an invaluable contribution to our research.

Fisherman in the Tapajos River. (Photo: Renato A. M. Silvano)

Foreword

When hearing about the Amazon, most people living in other regions of the world probably harbor thoughts of lush tropical forests, mysterious rivers, and animals such as jaguars, toucans, anacondas and piranhas. When thinking about the Amazon's human inhabitants, their minds likely conjure images of indigenous tribes living a primitive lifestyle in remote jungles with little contact with modern society. Those who have visited the Amazon or have spent time reading literature or watching educational videos quickly gain a deeper understanding of this vast tropical region. Indeed, Amazonia contains vast expanses of rainforest, diverse indigenous cultures, the world's greatest network of rivers, and the world's greatest concentration of biodiversity. But the region also is home to millions of citizens who are part of a global economy, including over 2 million people living in the city of Manaus where the giant Rio Negro joins the Amazon within the very center of the Amazon Basin.

The Amazon contains a wealth of natural resources that have long attracted explorers, homesteaders and others seeking opportunity. These natural resources include rubber trees that supported economic empires during the later 1880s and early 1900s, Brazil nuts that remain a valuable commodity to this day, as well as other plants, insects and microbes that contain chemical compounds useful in modern medicine. Amazonia's diverse and abundant fishes, the subject of this book, have supported livelihoods and food security for as long as people have lived in the region. Today, other Amazonian resources are attracting those seeking economic opportunity. The land and water of Amazonia are the targets for a new wave of resource exploitation. Land is being cleared of forest at an alarming rate for conversion to agriculture, including cattle ranching and farming of soybean, oil palm, and other crops. This conversion has been underway for centuries, and although the rate of forest loss has varied over time, it appears to be undergoing significant increase in recent years. Amazonia's rivers are being exploited for energy to support Brazil's growing population and economy. In the past several years, the pace of hydropower development in the Amazon Basin has increased tremendously, with plans drawn to construct dozens more hydroelectric dams in every country of the basin.

The rapid pace of human alteration of land and rivers has direct and major effects on aquatic ecology and fishes in the Amazon. It is now well established that seasonal flood pulses are essential for maintaining lateral connectivity of aquatic habitats, which allows fishes to swim from river channels and floodplain lakes into the flooded forest to exploit food resources. Dams not only change this essential pattern of seasonal flooding, but they also block migration routes for some of the most import fish species in the region's commercial and subsistence catches. Therefore, the loss of forests to agriculture and the loss of lateral and longitudinal connectivity of rivers are bound to have negative impacts on fishes and fisheries in the Amazon. The magnitude of these impacts is difficult to forecast with precision, and this is because of our lack of data and understanding about the interactions among multiple anthropogenic stressors as well as the spatial and temporal scales of impacts to land and water. Another major limitation to our ability to make projections about environmental impacts of land conversion and dams is our limited knowledge about the ecology of most Amazonian fish species. Despite intense interest and investigations by hundreds of scientists over the past three decades, most South American fish species remain poorly documented in terms of geographic distribution, habitat use, feeding and reproductive biology. Effective fisheries management requires a firm foundation of not only biological data and knowledge, but also socioeconomic information. In this regard, the book by Renato Silvano and colleagues fills some major information gaps. The authors state that this book is intended to provide much-needed information on fish ecology relevant to local people of the Rio Negro and Rio Tapajos, two major tributaries of the mighty Amazon. They explain that the book does not contain a complete catalogue of Amazon fish biodiversity, which is something that is not even possible at this time given that many dozens of new fish species continue to be discovered each year from this vast river basin. The authors state that their main goal was to expand the biological database for Amazon fishes in hopes of promoting fisheries sustainability and rural livelihoods in the Brazilian Amazon. This admirable goal was pursued during their research project 'Linking sustainability of small-scale fisheries, fishers' knowledge, conservation and co-management of biodiversity in large rivers of the Brazilian Amazon'. That project was funded by the United States Agency for International Development and the US National Academy of Sciences via the Partnerships for Enhanced Engagement in Research (PEER) program. It was my privilege to serve as an advisor for their project, and I am gratified that the labors of this dedicated group of researchers has resulted in several important journal articles and this book. Although many information gaps remain and we desperately need more data from many more fish species and local fishing communities, this book contains a wealth of data and knowledge to advance fisheries sustainability in the Amazon.

Kirk O. Winemiller, PhD
University Distinguished Professor and Regents Professor
Texas A&M University

Preface

There is still much to be known about Amazonian fish and fisheries, considering the large geographic area and the overwhelming diversity of fish species. The riverine population rely heavily on fish for food and income, and some river basins have been threatened by deforestation and development projects, such as dams for hydropower generation or mining. In such context, compilations of existing data about Amazon fish and fisheries could indicate future directions and priorities to research, management, and decision-making, besides generating awareness about the rich fish diversity and the importance of protecting such biodiversity and associated aquatic ecosystems. My first and premature contact with the Amazon and its fish fauna was nearly 25 years ago, when I went to sample fish as part of an interdisciplinary research project in the Jurua River, Western Amazon. Since that time, I was and continue to be amazed by the rich diversity of freshwater fishes found in the Amazon, their varied uses by local people and the detailed knowledge held by local fishers about fish ecology and aquatic ecosystems. I and my research group continued our scientific investigations on fish and people in the Brazilian Amazon along all these years in the rivers Araguaia, Tocantins, Solimões, Negro, and Tapajos. These research projects were funded by the Brazilian institutions: Fundação de Amparo a Pesquisa do Estado de São Paulo (FAPESP/SP), Coordenação de Aperfeiçoamento de Pessoal de Nível Superior (CAPES), Instituto de Desenvolvimento Sustentável Mamirauá (IDSM/CNPq), and Eletronorte. We currently continue our research in Amazonian rivers through funding from an international research project lead by the University of Alberta, Canada. My first experience resulted in a modest book on fish from Jurua River and it is a great pleasure now to present this new and much more comprehensive book on fish and fisheries in the Negro and Tapajos rivers. All data included in this book were gathered and analyzed during the research project "Linking sustainability of small-scale fisheries, fishers' knowledge, conservation and co-management of biodiversity in large rivers of the Brazilian Amazon," which was funded by USAID/NAS. Gathering of data was done through a combination of interdisciplinary methods, such as fish sampling with gillnets, interviews with fishers, and participatory monitoring of fish landings. The book's main goal is providing comprehensive information on fisheries and ecological and cultural aspects of Brazilian Amazonian fish, including illustrated atlas with color photographs of 183 fish species from the two rivers. The book includes nearly all the main fish groups currently exploited (used and marketed) by the riverine people in these clear- and blackwater rivers, thus adding to the existing body of knowledge on fish catalogs for the Amazon Basin, many of which address the more productive and better-known whitewater rivers. This book also provides detailed data on abundance, distribution, and sizes of each fish species, besides patterns of fish use by small-scale fisheries in both rivers. We included distribution maps for all fish species included in chapters, weight–length graphs for those species with 10 or more individuals sampled, and fisheries data (distribution of biomass, seasonal patterns on catches, sizes) for each exploited species or species groups. Some fish species from the same genus were hard to distinguish during sampling, as we could not take all collected fish to be identified in the laboratory. In such cases, data for these species were grouped (abundance and distribution), even though each species is presented in photographs. Fisheries data were usually grouped at the level of fish genus or even families, as this was how fishers recognized these fishes. The intent of the book is to provide useful information on a portion of the fish diversity from the Brazilian Amazon, which is relevant to local people and to ecological processes in aquatic ecosystems. Therefore, this book is not a complete catalogue of

IX

Amazon fish biodiversity nor a detailed taxonomic inventory, as we address only a small part of the Amazonian fish diversity. We hope that this book and the information within it can contribute to raise public and scientific awareness of the relationships among fish conservation, economic development, fishers' livelihoods, and fisheries sustainability in the Brazilian Amazon, a biodiversity hotspot that sustains people's livelihoods.

Renato A. M. Silvano
Porto Alegre, Rio Grande do Sul, Brazil

Financial Support to the Research Project

National Academy of Sciences (NAS) and the U.S. Agency for International Development (USAID), through the Partnerships for Enhanced Engagement in Research (PEER) program (Cycle 4), Grant Award Number: AID-OAA-A-11-00012. https://sites.nationalacademies.org/PGA/PEER/PEERscience/PGA_168062.

We wish to thank the people and institutions that helped with data collection for this book.

- Acácio Freitas Nogueira
- Andrea Leme da Silva
- Angela Castro de Souza
- Daiana Inês Schneider
- Fernando Gertum Becker
- Franciele Lausch Dos Santos
- Heinrich Hasenack
- Josele Trindade da Silva
- Kirk O. Winemiller
- Leonardo Capitani Zanoli
- Oswaldo Gomes de Souza Junior
- Paula Evelyn Rubira Pereyra
- Priscila F. M. Lopes
- Wanderson Michel de Farias Pantoja
- Instituto Chico Mendes de Conservação da Biodiversidade (ICMbio)
- Universidade Federal do Rio Grande do Sul (UFRGS)

To all fishers from the Negro and Tapajós rivers.

Contents

Contributors

Marcelo C. Andrade Universidade Federal do Pará, Núcleo de Ecologia Aquática e Pesca da Amazônia, Belém, Brazil
andrademarcosta@gmail.com

Alpina Begossi Capesca, Nepa, Universidade Estadual de Campinas (Unicamp), Campinas, São Paulo, Brazil

Graduate Program, Universidade Santa Cecilia, Santos, São Paulo, Brazil

FIFO (Fisheries and Food Institute), Rio de Janeiro, Rio de Janeiro, Brazil
albegossi@gmail.com

Junior A. Chuctaya Universidade Federal do Rio Grande Sul (UFRGS), Departamento de Zoologia, Porto Alegre, Rio Grande do Sul, Brazil
junior.chuctaya@gmail.com

Mariana Clauzet Universidade Santa Cecilia (Unisanta) and PPED-Políticas Públicas, Estratégias e Desenvolvimento, Universidade Federal do Rio de Janeiro (UFRJ), Rio de Janeiro, Brazil
mariana.clauzet@gmail.com

Márcia C. F. Dutra Universidade Federal do Rio Grande Sul (UFRGS), Departamento de Ecologia, Porto Alegre, Rio Grande do Sul, Brazil
carolinefriedldutra@gmail.com

Gustavo Hallwass Universidade Federal do Oeste do Pará (UFOPA), Campus Oriximiná, Oriximiná, Pará, Brazil

FIFO (Fisheries and Food Institute), Rio de Janeiro, Rio de Janeiro, Brazil
gustavo.hallwass@gmail.com

Friedrich W. Keppeler Department of Wildlife and Fisheries Sciences, Texas A&M University, College Station, TX, USA

Center for Limnology, UW-Madison, Madison, WI, USA
fkeppeler@gmail.com

Astrid Tatiana R. Martínez Pontificia Universidad Javeriana, Bogota, Colombia
Astroblepus_trm@hotmail.com

Paula Nagl Universidade Federal do Rio Grande Sul (UFRGS), Departamento de Ecologia, Porto Alegre, Rio Grande do Sul, Brazil

Caraïbes Environnement Développement, Baie Mahault, Guadeloupe
paula.nagl@aon.at

Pedro Peixoto Nitschke Universidade Federal do Rio Grande Sul (UFRGS), Departamento de Ecologia, Porto Alegre, Rio Grande do Sul, Brazil
pedropn91@gmail.com

Henrique N. Oliveira Universidade Federal do Rio Grande Sul (UFRGS), Departamento de Ecologia, Porto Alegre, Rio Grande do Sul, Brazil
hnegrello@gmail.com

Anaís R. P. Rowedder Universidade Federal do Rio Grande Sul (UFRGS), Departamento de Ecologia, Porto Alegre, Rio Grande do Sul, Brazil

National Institute of Amazon Researches – INPA, Manaus, Brazil
anaisrprestes@gmail.com

Luís Henrique Tomazoni da Silva Universidade Federal do Rio Grande Sul (UFRGS), Departamento de Ecologia, Porto Alegre, Rio Grande do Sul, Brazil
luis.tomazzoni@gmail.com

Renato A. M. Silvano Universidade Federal do Rio Grande Sul (UFRGS), Departamento de Ecologia, Porto Alegre, Rio Grande do Sul, Brazil
renato.silvano@ufrgs.br

Kaluan C. Vieira Universidade Federal do Rio Grande Sul (UFRGS), Departamento de Ecologia, Porto Alegre, Rio Grande do Sul, Brazil
kaluancv@gmail.com

1

Introduction: Fishes of the Brazilian Amazon

Friedrich W. Keppeler and Renato A. M. Silvano

Contents

Renato A. M. Silvano (ed.), *Fish and Fisheries in the Brazilian Amazon*, https://doi.org/10.1007/978-3-030-49146-8_1

1.1 Fish Diversity

Fishes were the first vertebrates with origin back to the Cambrian period, 540–485 million of years ago (Nelson 2006; Helfman et al. 2009). During the course of their evolution, fish acquire important innovations that were later passed to other vertebrates, including jaws, teeth, brains, body endoskeleton, appendages and internal organ systems (Helfman et al. 2009). Currently, 35,558 species of fish are known to science, but this number increases every year (Eschmeyer et al. 2019). In the last 20 years, around 7900 new species have been described, an average of almost 400 new species per year (Eschmeyer et al. 2019). These impressive numbers compress more than half of the diversity found in all vertebrates. Fishes originated in the sea and later colonize estuaries and inland waters. Interestingly, around 45% of fish species are restricted to lakes, rivers, and other freshwater ecosystems that make up only ~0.01% of available water on Earth (Lévêque et al. 2008). Why is the number of species per volume of water so much higher in freshwaters than in salt waters? This is an intriguing question, especially if we consider that freshwaters are more transient (i.e., disappear at a faster rate) than marine ecosystems, and therefore should experience higher rates of species extinction (Lévêque et al. 2008). Several hypotheses have been raised, including: (i) higher productivity in freshwaters; (ii) higher net diversification rates in freshwaters mainly due to more barriers limiting population connectivity in freshwaters, greater geographic range size of marine fishes, and more habitat complexity in freshwaters; (iii) higher transition rates of species between marine to freshwater ecosystems than freshwater to marine ecosystems over evolutionary time (Bloom et al. 2013). The importance of each one of these factors is still a matter of debate.

1.2 High-Diverse Neotropical and Amazonian Fish Assemblages

The Neotropical region is richer (~31% of all freshwater fish species) and more functionally diverse than all biogeographic regions in the world (Lévêque et al. 2008; Toussaint et al. 2016). The Amazon alone, which is the largest basin in the world, harbor 2716 species of fish, 1696 (63%) of them endemic (Dagosta and De Pinna 2019). The origin of the high species diversity found in Amazon has puzzled naturalists and ecologists for centuries. The refuge hypothesis proposed in 1969 by Haffer (1969) advocated that during the past glacial ages the Amazon passed through several periods of expansion and retraction, which facilitated speciation events via population geographic isolation (vicariance). This idea was popular for many years until the emergence of palynological and geological studies in the 1990s/2000s that showed that Amazonian lowlands formed a forest continuum through the Tertiary and late Quaternary periods (65 million years ago to present), refuting Haffer's hypothesis of fragmentation (Da Rocha and Kaefer 2019). Studies have shown that the high biological diversity in the Amazon has formed from a much longer period of evolution, involving complex vicariant (such as the Andean uplift, marine transgressions, climate shifts, and riverine barriers) and non-vicariant processes (such as sexual selection and competition; Hoorn et al. 2010; Da Rocha and Kaefer 2019). It is important to notice that some macroecological variables related to the latitudinal gradient of species, including temperature and solar radiation, also contribute to the higher diversity found in the Amazon basin (Brown 2014).

Eighty percent of all Amazonian fish belong to the Otopysi clade, which includes the first (Characiformes – 39.1% of the species), second (Siluriformes – 35.2%), and fifth (Gymnotiformes – 6%) richest orders (Dagosta and De Pinna 2019). Perciformes (10.5%), mainly composed of the family Cichlidae, is the third most diverse order of fish in the Amazon Basin, followed by Cyprinodontiformes (6%; Dagosta and De Pinna 2019). Among the least diverse orders, 10 are from marine lineages that colonize freshwaters secondarily (Dagosta and De Pinna 2019). The most specious families in the Amazon are dominated by small body species, such as Characidae (23.4%), Loricariidae (13.3%), Cichlidae (9.5%), Cynolebiidae (5.2%), and Callichthyidae (4.9%; Dagosta and De Pinna 2019). The only endemic fish family in the Amazon is Tarumaniidae. Biogeographical studies have suggested that the distribution of Amazonian fishes is little explained by the river basins, which suggest high levels of species exchange among basins and may reflect the complex geomorphological history of the region (Dagosta and De Pinna 2017, 2019). Lowland regions are more diverse and have taxa with a wider geographical range than uplands regions. This seems to happen due to higher susceptibility of lowlands to hydrogeological changes and a higher level of interconnection among lowland rivers (Lundberg et al. 1988; Dagosta and De Pinna 2017, 2019).

1.3 Ecology of Amazon Fishes

The structure of local fish assemblage in the Amazon is usually strongly associated with habitat type and abiotic conditions. For instance, macrophytes are important nursery habitats for many species as roots, steam, and leaves function as physical and visual barriers against predators (Petry et al. 2003). Surface dwellers and epibenthic fishes with high maneuverability are often associated with structurally complex microhabitats in forested areas (Arantes et al. 2017; Montag et al. 2019). On the other hand, water velocity and dissolved oxygen have been shown to be important abiotic factors limiting the occurrence of certain species. For instance, depth-body fishes are usually incapable of inhabiting fast flow waters in Amazonian streams (Bower and Winemiller 2019). Hypoxic waters in floodplain lakes often limit the occurrence of large-body fishes, especially predators without specific adaptations to breathe atmospheric air (Dos Anjos et al. 2008). Associations between environmental condition and fish assemblage structure are stronger during low-water periods than high-water periods. This happens because, during high-water periods, the flood pulse increases the connectivity between different habitats, homogenizing their environmental conditions and fish composition (Thomaz et al. 2007).

A huge effort has been done in the last decades to understand the interrelationship between Amazonian fish species, particularly in terms of trophic interactions. Species interactions tend to be more intense during low-water periods when fishes are more densely packed (Lowe-Mcconnell 1999; Arrington et al. 2005). This is especially true in isolated lakes and pools, where predation rates increase exponentially as the water goes down and the frequency of encounters between predatory fishes and their prey increases (Matthews and Marsh-Matthews 2003; Peterson et al. 2017). Some studies have also hypothesized that during the low-water period, dietary breadth, and dietary overlap among species decline as a response of fishes to reduce interspecific competition; tests of this hypothesis generated mixed results (Mérona and Rankin-de-Mérona 2004; Peterson et al. 2017). During high-water periods, resources are abundant as streams, rivers, and lakes flood to the surrounding forest. The proportion of allochthonous resources, such as seeds and ants, in the diet of fishes increase considerably (Mérona and Rankin-de-Mérona 2004; Peterson et al. 2017), an exception being fishes in small forested streams that rely on allochthonous material along the whole year (Lowe-Mcconnell 1999). Seasonal changes in water level are also associated with the migration of several periodic species from nutritious-rich white-water rivers to nutritious-poor black- and clear-water rivers (Barthem and Goulding, 1997; Winemiller and Jepsen 1998; Benedito-Cecilio and Araujo-Lima 2002). Fishers and also piscivorous fish (e.g., *Cichla* spp.) from oligotrophic rivers, such as the Tapajós and Negro River, are known to depend on this seasonal subsides to meet their energy requirements (Benedito-Cecilio and Araujo-Lima, 2002; Hoeinghaus et al. 2006; Keppeler et al. 2017).

Amazonian fishes have different life history strategies to cope with the environmental variation at different temporal (e.g., season, interannual) and spatial scales (e.g., microhabitats, rivers). Winemiller and Rose (1992) proposed to divide fishes into three main life history strategies: opportunistic, periodic, and equilibrium. Opportunistic species, such as livebearers (Poeciliidae) and killifishes (Rivulidae), have high reproductive effort, short generation time, fast maturation, small body, long reproductive periods, and low investment per offspring. This strategy is favored when there are unpredictable changes in the system, frequent disturbance, and low influence of interspecific interactions. Periodic species, such as jaraquis and curimbas (Prochilodontidae), have moderate reproductive effort, long generation time, large body sizes, few reproductive events per year, large batch sizes, and low investment per offspring. The periodic strategy is favored when there are predictable spatial and temporal variations in the system. Equilibrium species, such as tucunarés (*Cichla* spp.) and pirarucus (*Arapaima* spp.), have moderate reproductive effort, medium-to-long generation times, variable body size, high investment per offspring (often involving parental care), and small batch sizes. The equilibrium strategy is beneficiated in a stable environment with few perturbations and when competition and other interspecific interactions are intense. Rather than just three discrete categories, these life history strategies are endpoints of a three-lateral continuum. Indeed, a recent study conducted by Espírito-Santo et al. (2013) could not assign most of the Amazonian fishes from headwater streams into the three life history strategies proposed by Winemiller and Rose (1992).

1.4 Documenting Fish Diversity and Use in the Brazilian Amazon

Considering the high number of species and the complex ecological interactions, documenting the diversity of Amazonian fishes is a challenging, but urgently needed task. The aquatic ecosystems in the Brazilian Amazon are under increasing pressure from land use and development projects, especially the floodplain forests and rapids in some tributaries of the Amazon River. There are many dams planned to be built at upstream reaches of Amazonian tributaries, including the Tapajos River, which may disrupt hydrological flow and fish migrations and affect both fish diversity and people who rely on fish for food and income (Winemiller et al. 2016). The existing protected areas may not always include aquatic habitats and the commercial fisheries and deforestation may have had already decreased the size and abundance of some commercial fish species near large cities (Tregidgo et al. 2017; Keppeler et al. 2018). Therefore, in face of all these environmental changes and competing proposals of using the Amazonian aquatic habitats for economic development, there is an urgent need to improve our knowledge on the diversity and use of fishes in the

Brazilian Amazon. There are books documenting the fish diversity with photographs and ecological data for some Amazonian rivers, such as the Juruá (Silvano et al. 2001), Tocantins (Santos et al. 2004), Amazon (Ferreira et al. 1998), Madeira (Ribeiro et al. 2016), besides a comprehensive book on Amazon fishes, including Orinoco and Guianas (van der Sleen and Albert 2017), among others.

This book brings detailed data from a research project addressing fish and small-scale fisheries in the Negro and Tapajos rivers, in the Brazilian Amazon. This research had the major goal of understanding the diversity, abundance, and use of fish by riverine people in less productive and poorly known clear and black water rivers. The primary focus of this book is not taxonomical, but just to record a fraction of the fish diversity in these two rivers and the relevance of these fishes to local people, including those inhabiting protected areas. During the research, we sampled 224 fish species in these two rivers and provided photos and more detailed data on abundance, distribution, fish sizes, and fisheries for 183 fish species distributed in three chapters (◘ Table 1.1.). We thus aim to contribute to improve the existing knowledge on Amazon fishes, small-scale fisheries, and the aquatic biodiversity, thus highlighting the relevance of this complex, unique, and imperiled tropical socio-ecological system.

◘ Table 1.1 Complete list of all 224 fish species sampled during the research in 16 fishing communities in the Negro and Tapajos rivers, Brazilian Amazon. Chapter indicates the chapter showing photos and more data of the species, but not all species could be included in chapters. The number 1 indicates the occurrence of a given species in each studied river (Negro and Tapajos)

Order and family	Fish species	Chapter	Negro	Tapajos
Beloniformes				
Belonidae	*Pseudotylosurus microps*	6		1
Characiformes				
Acestrorhynchidae	*Acestrorhynchus falcirostris*	4	1	1
	Acestrorhynchus microlepis	4	1	1
Anostomidae	*Anostomoides atrianalis*	4	1	
	Laemolyta proxima	4	1	1
	Laemolyta taeniata	4	1	1
	Leporinus cf. *fasciatus*	4	1	1
	Leporinus fasciatus	4	1	1
	Rhytiodus argenteofuscus	4		1
	Schizodon fasciatus	No		1
	Schizodon vittatus	4		1
Characidae	*Agoniates halecinus*[a]	4	1	1
	Brycon melanopterus	No		1
	Brycon pesu[b]	4	1	
	Brycon sp.[b]	No		1
	Bryconops alburnoides[c]	4	1	1
	Bryconops caudomaculatus[c]	4	1	1
	Bryconops gracilis[c]	4		1
	Bryconops melanurus[c]	4	1	1
	Chalceus macrolepidotus[d]	4	1	1
	Charax delimai	4	1	
	Charax gibbosus	4	1	
	Gymnocorymbus thayeri	No	1	

Table 1.1 (continued)

Order and family	Fish species	Chapter	Negro	Tapajos
	Moenkhausia lepidura	4	1	
	Moenkhausia megalops	4		1
	Poptella compressa	No	1	
	Tetragonopterus manaos	4	1	
	Triportheus albus[a]	4	1	
	Triportheus angulatus[a]	4		1
	Triportheus auritus[a]	4		1
	Triportheus rotundatus[a]	4		1
Chilodontidae	*Caenotropus labyrinthicus*	4		1
Ctenoluciidae	*Boulengerella cuvieri*	4	1	1
	Boulengerella maculata	4	1	1
Curimatidae	*Curimata* cf. *knerii*	No	1	
	Curimata knerii	4		1
	Curimata sp.	4		1
	Curimata ocellata	4	1	1
	Curimata roseni	4	1	
	Curimata vittata	4	1	
	Curimatella immaculata	4		1
	Curimatella meyeri	4	1	1
	Cyphocharax abramoides	4	1	1
	Cyphocharax leucostictus	4	1	
	Potamorhina latior	4	1	
	Potamorhina pristigaster	4	1	
	Psectrogaster cf. *falcata*	4		1
	Psectrogaster rutiloides	4	1	
Cynodontidae	*Cynodon gibbus*	4	1	1
	Hydrolycus tatauaia	No		1
	Hydrolycus wallacei	4	1	
	Rhaphiodon vulpinus	4	1	1
	Roestes molossus[e]	4	1	
Erythrinidae	*Hoplias malabaricus*	4	1	1
Hemiodontidae	*Anodus elongatus*	4	1	1
	Anodus orinocensis	4	1	
	Argonectes longiceps	4	1	1
	Bivibranchia fowleri	4		1
	Hemiodus argenteus	4		1
	Hemiodus atranalis	4	1	1

(continued)

1

Table 1.1 (continued)

Order and family	Fish species	Chapter	Negro	Tapajos
	Hemiodus goeldii	No	1	
	Hemiodus gracilis	4	1	1
	Hemiodus immaculatus	4	1	1
	Hemiodus microlepis	No	1	1
	Hemiodus sp.	4	1	1
	Hemiodus tocantinensis	No		1
	Hemiodus unimaculatus	4	1	1
	Micromischodus sugillatus	4	1	1
Prochilodontidae	*Semaprochilodus insignis*	4	1	1
	Semaprochilodus taeniurus	4	1	1
Serrasalmidae	*Catoprion absconditus*	4	1	1
	Colossoma macropomum	4		1
	Metynnis altidorsalis	4		1
	Metynnis hypsauchen	4	1	1
	Metynnis lippincottianus	4	1	1
	Metynnis luna	No	1	
	Metynnis melanogrammus	4	1	
	Myloplus lucienae	4	1	
	Myloplus asterias	4		1
	Myloplus lobatus	4	1	1
	Myloplus schomburgkii	4	1	
	Myloplus nigrolineatus	4	1	
	Pristobrycon cf. *scapularis*	4		1
	Pristobrycon maculipinnis	4	1	
	Pygocentrus nattereri	4		1
	Pygopristis denticulata	4	1	1
	Serrasalmus altuvei	No	1	
	Serrasalmus eigenmanni	4	1	1
	Serrasalmus elongatus	4	1	
	Serrasalmus gouldingi	4	1	
	Serrasalmus hollandi	4		1
	Serrasalmus maculatus	4		1
	Serrasalmus manueli	4	1	1
	Serrasalmus medinai	4	1	
	Serrasalmus odyssei	4	1	
	Serrasalmus rhombeus	4	1	1
	Serrasalmus serrulatus	4	1	

Table 1.1 (continued)

Order and family	Fish species	Chapter	Negro	Tapajos
Clupeiformes				
Engraulidae	*Anchovia surinamensis*	6	1	1
	Anchoviella guianensis	No		1
	Lycengraulis batesii	6	1	1
Pristigasteridae	*Ilisha amazonica*	No		1
	Pellona castelnaeana	6	1	1
	Pellona flavipinnis	6	1	1
Gymnotiformes				
Rhamphichthyidae	*Rhamphichthys pantherinus*	5	1	1
	Rhamphichthys rostratus	5	1	1
	Steatogenys aff. *ocellatus*	No		1
Sternopygidae	*Eigenmannia macrops*	5		1
	Rhabdolichops troscheli	5	1	1
	Sternopygus macrurus	No		1
Myliobatiformes				
Potamotrygonidae	*Potamotrygon constellata*	No		1
	Potamotrygon humerosa	6		1
	Potamotrygon motoro	6		1
	Potamotrygon orbignyi	6		1
Osteoglossiformes				
Osteoglossidae	*Osteoglossum bicirrhosum*	6	1	1
Perciformes[f]				
Cichlidae	*Acarichthys heckelii*	6	1	1
	Acaronia nassa	6	1	1
	Astronotus ocellatus	6	1	1
	Biotodoma cupido	No	1	
	Caquetaia spectabilis	6		1
	Chaetobranchus flavescens	6	1	
	Cichla aff. pinima	6		1
	Cichla nigromaculata	6	1	
	Cichla ocellaris	6	1	1
	Cichla pinima	6		1
	Cichla temensis	6	1	
	Crenicichla lenticulata	6	1	
	Crenicichla lugubris	6	1	
	Crenicichla marmorata	No		1
	Geophagus altifrons	6		1

(continued)

1

Table 1.1 (continued)

Order and family	Fish species	Chapter	Negro	Tapajos
	Geophagus cf. *altifrons*	No		1
	Geophagus sp. Red head	6		1
	Geophagus winemilleri	6	1	
	Heros efasciatus	No		1
	Hoplarchus psittacus	6	1	
	Mesonauta festivus	No	1	
	Satanoperca acuticeps	6	1	1
	Satanoperca jurupari	6	1	
	Satanoperca lilith	6	1	1
	Symphysodon aequifasciatus	No		1
	Uaru amphiacanthoides	6	1	1
Sciaenidae	*Pachypops fourcroi*	6	1	1
	Pachyurus junki	6	1	1
	Plagioscion squamosissimus	6	1	1
Carangiformes				
Achiridae	*Hypoclinemus mentalis*	6		1
Siluriformes				
Auchenipteridae	*Ageneiosus akamai*	5	1	1
	Ageneiosus apiaka	5		1
	Ageneiosus dentatus	5		1
	Ageneiosus inermis	5	1	
	Ageneiosus intrusus	5	1	
	Ageneiosus lineatus	5	1	
	Ageneiosus polystictus	5	1	
	Ageneiosus ucayalensis	5	1	
	Ageneiosus vittatus	No	1	
	Auchenipterichthys coracoideus	5	1	1
	Auchenipterichthys longimanus	5	1	1
	Auchenipterus nuchalis	5	1	1
	Centromochlus existimatus	No	1	
	Centromochlus heckelii	5	1	1
	Tatia cf. *aulopygia*	No	1	
	Tatia cf. *intermedia*	No	1	
	Tatia intermedia	No	1	
	Tatia sp.	5		1
	Trachelyichthys decaradiatus	5		1
	Trachelyopterichthys sp. Pintado	No		1
	Trachelyopterus coriaceus	No		1

Table 1.1 (continued)

Order and family	Fish species	Chapter	Negro	Tapajos
	Trachelyopterus galeatus	5	1	1
	Trachycorystes trachycorystes	5	1	
	Tympanopleura brevis	No	1	
Callichthyidae	*Hoplosternum littorale*	5		1
Cetopsidae	*Cetopsis coecutiens*	5	1	1
Doradidae	*Acanthodoras* cf. *cataphractus*	5	1	
	Amblydoras affinis	5		1
	Anadoras weddellii	5	1	1
	Anduzedoras oxyrhynchus	5	1	
	Astrodoras asterifrons	5	1	
	Hassar orestis	5		1
	Megalodoras uranoscopus	5	1	1
	Tenellus trimaculatus	5	1	1
	Oxydoras niger	5		1
	Platydoras armatulus	5		1
	Platydoras hancockii	5	1	
	Tenellus ternetzi	5	1	
	Trachydoras brevis	5	1	
	Trachydoras sp.	No	1	
Heptapteridae	*Goeldiella eques*	5	1	
	Pimelodella cristata	5	1	
	Pimelodella sp.	5		1
Loricariidae	*Ancistrus hoplogenys*	No		1
	Ancistrus sp.	5	1	1
	Aphanotorulus emarginatus	5		1
	Dekeyseria amazonica	5	1	
	Dekeyseria picta	No	1	
	Hemiodontichthys acipenserinus	5	1	1
	Hypoptopoma elongatum	No	1	
	Hypostomus weberi	5	1	
	Leporacanthicus galaxias	No		1
	Limatulichthys griseus	5	1	
	Loricaria cataphracta	5	1	1
	Loricariichthys acutus	5	1	1
	Peckoltia cf. *cavatica*	No		1
	Peckoltia vittata	No		1
	Pseudoloricaria laeviuscula	5		1

(continued)

1

Table 1.1 (continued)

Order and family	Fish species	Chapter	Negro	Tapajos
	Pterygoplichthys gibbiceps	5	1	1
	Rineloricaria daraha	5	1	
	Scobinancistrus pariolispos	No		1
Pimelodidae	*Brachyplatystoma capapretum*	5	1	
	Brachyplatystoma filamentosum	5	1	1
	Calophysus macropterus	5	1	
	Cheirocerus goeldii	No		1
	Hypophthalmus edentatus	5	1	
	Hypophthalmus fimbriatus	5	1	1
	Hypophthalmus marginatus	5	1	1
	Pimelodina flavipinnis	5	1	
	Pimelodus blochii	5	1	
	Pimelodus ornatus	5	1	
	Pimelodus tetramerus	5		1
	Pinirampus pinirampu	5	1	1
	Platynematichthys notatus	5		1
	Platystomatichthys sturio	5		1
	Pseudoplatystoma punctifer	5	1	1
	Pseudoplatystoma tigrinum	5	1	1
	Sorubim elongatus	5		1
Tetraodontiformes				
Tetraodontidae	*Colomesus tocantinensis*	6		1

[a]This species now belongs to the family Triportheidae
[b]This species now belongs to the family Bryconidae
[c]This species now belongs to the family Iguanodectidae
[d]This species now belongs to the family Chalceidae
[e]This species now belongs to the family Acestrorhynchidae
[f]The family Cichlidae now belongs to the Order Cichliformes

References

Arantes C, Winemiller KO, Petrere M, Castello L, Freitas C, Hess LL (2017) Relationships between forest cover and fish diversity in the Amazon River floodplain. J Appl Ecol 55:386–395

Arrington DA, Winemiller KO, Layman CA (2005) Community assembly at the patch scale in a species-rich tropical river. Oecologia 144:157–167

Barthem R, Goulding M (1997) The catfish connection: ecology, migration, and conservation of Amazon predators. Columbia University Press, New York

Benedito-Cecilio E, Araujo-Lima CARM (2002) Variation in the carbon isotope composition of *Semaprochilodus insignis*, a detritivorous fish associated with oligotrophic and eutrophic Amazonian rivers. J Fish Biol 60:1603–1607

Bloom DD, Weir JT, Piller KR, Lovejoy NR (2013) Do freshwater fishes diversify faster than marine fishes? A test using state-dependent diversification analyses and molecular phylogenetics of new world silversides (Atherinopsidae). Evolution 67:2040–2057

Bower LM, Winemiller KO (2019) Fish assemblage convergence along stream environmental gradients: an intercontinental analysis. Ecography 42:1–13

Brown JH (2014) Why are there so many species in the tropics? J Biogeogr 41:8–22

Dagosta FCP, De Pinna M (2017) Biogeography of Amazonian fishes: deconstructing river basins as biogeographic units. Neotropical Ichthyology 15:e170034

Dagosta FCP, De Pinna M (2019) Distribution and biogeographical patterns, with a comprehensive list of species. Bull Am Mus Nat Hist 431:1–163

Da Rocha DG, Kaefer IL (2019) What has become of the refugia hypothesis to explain biological diversity in Amazonia? Ecol Evol 9:4302–4309

Dos Anjos MB, De Oliveira RR, Zuanon JAS (2008) Hypoxic environments as a refuge against predatory fish in the Amazonian floodplains. Braz J Biol 68:45–50

Eschmeyer WN, Fricke R, Van der Laan, R (2019) Catalog of fishes. Available at: http://researcharchive.calacademy.org/research/ichthyology/catalog/SpeciesByFamily.asp. Accessed 10 July 2019

Espírito-Santo HMV, Rodríguez MA, Zuanon J (2013) Reproductive strategies of Amazonian stream fishes and their fine-scale use of habitat are ordered along a hydrological gradient. Freshw Biol 58:2494–2504

Ferreira EJG, Zuanon JAS, Santos GM (1998) Peixes Comerciais do Médio Amazonas: região de Santarém. Edições IBAMA, Brasília. 211 pp.

Haffer J (1969) Speciation in Amazonian forest birds. Science 165:131–137

Helfman GS, Collette B, Facey DE, Bowen BW (2009) The diversity of fishes: biology, evolution and ecology. Wiley-Blackwell, West Sussex

Hoeinghaus D, Winemiller KO, Layman CA, Arrington DA, Jepsen DB (2006) Effects of seasonality and migratory prey on body condition of *Cichla* species in a tropical floodplain river. Ecol Freshw Fish 15:398–407

Hoorn C, Wesselingh P, Ter Steege H, Bermudez MA, Mora A, Sevink J, Sanmartín I, Sanchez-Meseguer A, Anderson CL, Figueiredo JP, Jaramillo C, Riff D, Negri FR, Hooghiemstra H, Lundberg J, Stadler T, Särkinen T, Antonelli A (2010) Amazonia through time: Andean uplift, climate change, landscape evolution, and biodiversity. Science 330:927–931

Keppeler FW, Hallwass G, Silvano RAM (2017) Influence of protected areas on fish assemblages and fisheries in a large tropical river. Oryx 51:268–279

Keppeler FW, Souza AC, Hallwass G, Begossi A, Almeida MC, Isaac VJ, Silvano RAM (2018) Ecological influences of human population size and distance to urban centres on fish communities in tropical lakes. Aquat Conserv Mar Freshwat Ecosyst 28:1030–1043

Lévêque C, Oberdorff T, Paugy D, Stiassny MLJ, Tedesco PA (2008) Global diversity of fish (Pisces) in freshwater. Hydrobiologia 595:545–567

Lowe-Mcconnell RH (1999) Estudos ecológicos de comunidades de peixes tropicais. EDUSP, São Paulo

Lundberg JG, Linares OJ, Antonio ME, Nass P (1988) *Phractocephalus hemiliopterus* (Pimelodidae, Siluriformes) from the upper Miocene Urumaco formation, Venezuela: a further case of evolutionary stasis and local extinction among south American fishes. J Vertebr Paleontol 8:131–138

Matthews WJ, Marsh-Matthews E (2003) Effects of drought on fish across axes of space, time and ecological complexity. Freshw Biol 48:1232–1253

Mérona B, Rankin-de-Mérona J (2004) Food resource partitioning in a fish community of the Central Amazon floodplain. Neotropical Ichthyology 2:75–84

Montag LFA, Winemiller KO, Keppeler FW, Leão H, Benone NL, Torres NR, Prudente BS, Begot TO, Bower LM, Saenz DE, Lopez-Delgado EO, Quintana YM, Hoeinghaus DJ, Juen L (2019) Land cover, riparian zones and instream habitat influence stream fish assemblages in the eastern Amazon. Ecol Freshw Fish 28:317–329

Nelson JS (2006) Fishes of the world. In: Ohn Wiley & Sons, Inc. USA, Hoboken

Peterson CC, Keppeler FW, Saenz DE, Bower LM, Winemiller KO (2017) Seasonal variation in fish trophic networks in two clearwater streams in the central llanos region, Venezuela. Neotropical Ichthyology 15:e160125

Petry P, Bayley PB, Markle DF (2003) Relationships between fish assemblages, macrophytes and environmental gradients in the Amazon River floodplain. J Fish Biol 63:547–579

Ribeiro AC, Vilara GT, Filho JAL, Doria CRC (2016) Ecologia e Biologia de Peixes do rio Madeira. Editora EDUFRO, Porto Velho, Brasil

Santos GM, de Mérona B, Juras AA, Jégu M (2004) Peixes do baixo rio Tocantins: 20 anos depois da Usina Hidrelétrica Tucuruí. Eletronorte, Brasília. 216 pp

Silvano RAM, Oyakawa OT, Amaral BD, Begossi A (2001) Peixes do Alto Rio Juruá (Amazônia, Brasil). Editora da USP, São Paulo

Thomaz SM, Bini LM, Bozelli RL (2007) Floods increase similarity among aquatic habitats in river-floodplain systems. Hydrobiologia 579:1–13

Toussaint A, Charpin N, Brosse S, Villéger S (2016) Global functional diversity of freshwater fish is concentrated in the Neotropics while functional vulnerability is widespread. Sci Rep 6:22125

Tregidgo DJ, Barlow J, Pompeu PS, de Almeida RM, Parry L (2017) Rainforest metropolis casts 1,000-km defaunation shadow. Proc Natl Acad Sci 114:8655–8659

van der Sleen P, Albert JS (2017) Field guide to the fishes of the Amazon, Orinoco, and Guianas (Princeton field guides). Princeton University Press, Princeton

Winemiller KO, Jepsen DB (1998) Effects of seasonality and fish movement on tropical river food webs. J Fish Biol 53:267–296

Winemiller KO, Rose KA (1992) Patterns of life-history diversification in north American fishes: implications for population regulation. Canadian Journal of Fisheries and Aquatic Science 49:2196–2218

Winemiller KO, McIntyre PB, Castello L, Fluet-Chouinard E, Giarrizzo T, Nam S, Baird IG, Darwall W, Lujan NK, Harrison I, Stiassny MLJ, Silvano RAM, Fitzgerald DB, Pelicice FM, Agostinho AA, Gomes LC, Albert JS, Baran E, Petrere M Jr, Zarfl C, Mulligan M, Sullivan JP, Arantes CC, Sousa LM, Koning AA, Hoeinghaus DJ, Sabaj M, Lundberg JG, Armbruster J, Thieme ML, Petry P, Zuanon J, Torrente-Vilara G, Snoeks J, Ou C, Rainboth W, Pavanelli CS, Akama A, van Soesbergen A, Sáenz L (2016) Balancing hydropower and biodiversity in the Amazon, Congo and Mekong. Science 351:128–129

Studied Regions and Methodology

Gustavo Hallwass, Pedro Peixoto Nitschke, and Friedrich W. Keppeler

Contents

Renato A. M. Silvano (ed.), *Fish and Fisheries in the Brazilian Amazon*, https://doi.org/10.1007/978-3-030-49146-8_2

2.1 Studied Rivers: Ecological Characteristics

This research addressed two large and important tributaries of the Amazon River, the Tapajós and Negro rivers (Fig. 2.1). The average annual precipitation in the Lower Tapajós is 1820 mm, while in the Negro river it is 2504 mm. In both cases, the rain is distributed unevenly along the year, which makes the water levels in each river to vary up to 10 m from the dry (Tapajós: November/December, Negro: October/February) to the wet (Tapajós: May/June, Negro: June/July) seasons (Goulding et al. 2003; ANA 2019). The high discharge during the wet season connects the main river channel with a complex landscape, composed of channels, small streams (*igarapés*), flooded forests (*igapó*), and lakes. The climate is tropical and humid with mean temperatures around 26 °C. The highest temperatures usually occur during the dry season and the lowest during the wet season, but the temperature variation tends to be greater within a single day than between seasons (Sioli 1984).

The Tapajós river has an extension of 2700 km encompassing the Brazilian states of Rondônia, Amazonas, Mato Grosso, and Pará, the last two representing more than 95% of the river basin area. The Tapajós basin drains 489,000 km^2 of a region located in the old rock formations of the Pre-Cambrian; its headwaters rise in the central Brazilian *cerrado*, a savanna biome (Goulding et al. 2003), and the Tapajós mouth flows into the Lower Amazon river near the city of Santarém, in Pará State. The Tapajós has clear and oligotrophic waters with low sediments in suspension and dissolved solids and high visibility (1–5 m). The waters of the Tapajós basin have conductivity ranging around 10–40 μS cm^{-1} and pH that varies between 6 and 7 (Sioli 1984; Junk et al. 2011). This research was conducted between the coordinates 2° 45′00″ S–3° 25′00″ S and 55°19′00″ W–54°59′00″ W in the Lower Tapajós river within the municipalities of Santarém and Aveiros, Pará State (Fig. 2.1). The main channel of this river section is wide, reaching more than 15 km between margins in some areas. The floodplain area of the Lower Tapajós is small when compared to other large tributaries of the Amazon river and with the Lower Amazon.

The Negro river is one of the most important tributaries of the Amazon river, having 2250 km of length and draining an area of 691,000 km^2, mostly located in the old Precambrian Guiana shield. Different from the Tapajós, the Negro river has blackwaters, which are characterized by pH values lower than 5.0 (acid) and conductivity lower than 25 μS cm^{-1}. The water of the Negro river is transparent with low quantities of nutrients and suspended matter, but with high amounts of humic acids and tannins from forest materials dissolved in the water, which are responsible for its blackish coloration (Junk et al. 2011). The study area is located between the coordinates 1° 35′34″ S–2° 24′39″ S and 62°36′20″ W–61°02′11″ W in the Middle Negro river and Unini river (one of the largest tributaries of the Negro river) within the municipalities of Novo Airão and Barcelos (Fig. 2.1). The floodplain forest surrounding the Middle Negro river has slow growth and low litter production because of the low fertility and low pH values (acid) of blackwaters, but still plays an important function sustaining aquatic productivity, including fisheries and a rich fish diversity (Goulding et al. 1988; Junk et al. 2011).

2.2 Federal Protected Areas

The Brazilian legislation that delimits the categories and objectives of protected areas is the National System of Nature Conservation Units (SNUC in Portuguese), established through the Federal Law # 9.985/2000. All federally protected areas are managed by the Chico Mendes Institute of Biodiversity Conservation (ICMBio in Portuguese). Both the Lower Tapajós river and the Middle Negro river have protected areas of sustainable use in the form of Extractive Reserves (RESEX in Portuguese): RESEX Tapajós-Arapiuns and RESEX Unini; and National Forests (FLONA in Portuguese): FLONA Tapajós. The Extractive Reserves and the National Forests allow local communities to live inside their boundaries as long as they follow some specific management rules. The goal of these protected areas is to ensure both biodiversity conservation and sustainable exploitation of natural resources, through the implementation of local rules in agreement with local populations, protected areas' managers, and the Brazilian legislation. In both protected areas, the human population relies on a diversified system of subsistence, including small-scale agriculture, extractive forest production, livestock farming (mainly chickens), fishing, and hunting (Fig. 2.2). Besides protected areas of sustainable use (e.g., RESEX), the Middle Negro river also has two protected areas of integral protection, the National Park of Jaú and the National Park of Anavilhanas, which have aims that involve the environmental preservation, scientific research, education, and ecological tourism without the presence of local populations. There are also other types of protected areas near the studied region, including Federal Indigenous Lands in both rivers, the State Park of Negro river, Environmental Protected Area (EPA) of Rio Negro Right Bank and EPA of Rio Negro Left Bank, and the Municipal EPA of Alter do Chão, in the Tapajós river. EPA is a less restrictive type of protected area, which was not considered as a strict protected area for the purposes of this study.

In the Tapajós river, we studied the RESEX Tapajós-Arapiuns (677,513.24 ha). This reserve was officially designated in 1998 (Decree s/n° of 1998), in a bottom-up fashion, after an almost 20-year struggle by locals against

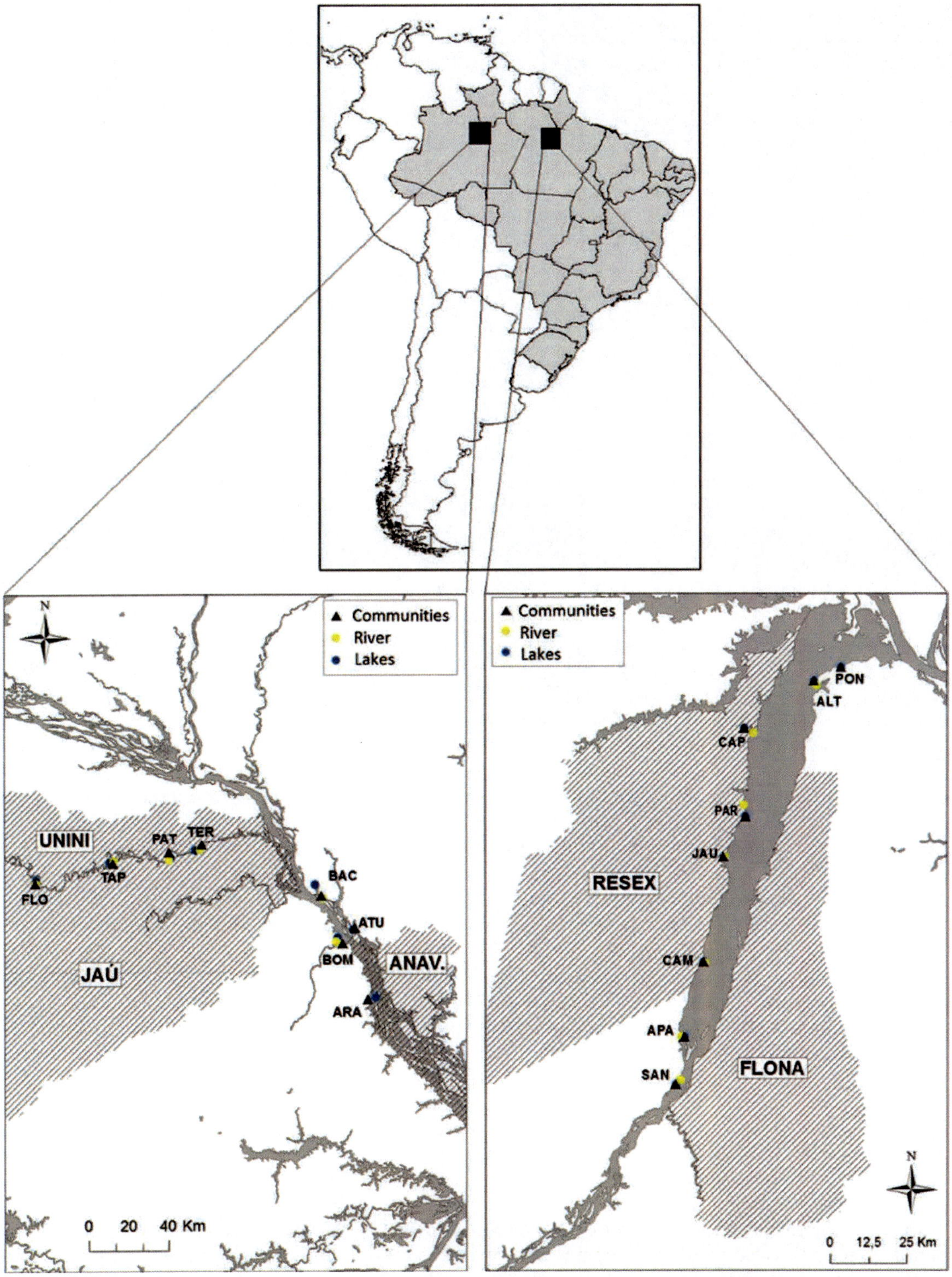

Fig. 2.1 South America with Brazil in grey (top map) and the localization of the two studied rivers in the Brazilian Amazon. The map on the left shows the Negro river, the National Park of Anavilhanas (ANAV.), and the Unini river between the Extractive Reserve of Unini (UNINI) and the National Park of Jaú (JAÚ). The map on the right shows the Tapajós river, the Extractive Reserve of Tapajós-Arapiuns (RESEX), and the National Forest of Tapajós (FLONA). Triangles represent the communities and circles represent sites where experimental fisheries were conducted (yellow = river, blue = lake). Codes for the names of riverine communities are in Table 2.1

Fig. 2.2 Processing of manioc flour in a RESEX Unini's community, in the Unini river, Negro river basin (Photo: Renato Silvano)

illegal logging in the area. Its buffer zone covers 10 km around the area of the reserve in terrestrial and aquatic environments. The management plan of this Extractive Reserve was first published in 2008 with little attention to fishing and fish. However, an updated version of the management plan was published in 2014, including 99 fish species recorded from several habitats (rivers, lakes, and streams) and identifying approximately 5200 artisanal small-scale fishers. Initiatives in the establishment of community fisheries rules were identified in this updated management plan (ICMBio 2014).

In the Lower and Middle Negro river, there is a mosaic of more than 10 federal and state protected areas, from which three federal reserves were studied: two National Parks, Anavilhanas and Jaú, and the Extractive Reserve of Unini. The Anavilhanas National Park (350,469.8 ha) was created in 1981 (Decree n° 86.061/1981) and is one of the largest river archipelagos in the world with approximately 400 islands, 60 lakes, and several river channels. According to the last management plan, published in 2017, some possible conflicting activities in the protected area are fishing, logging, hunting, mining, and pollution. The National Park of Jaú (2,367,333.44 ha), located in the Jaú, Negro and Unini rivers, was created in 1980 (Decree n° 85.200/1980) and is one of the most important protected areas in the region. The demand for the park establishment was made by researchers in the 1970s aiming for the protection of the regional biodiversity, while the management plan was published in 1998. This category of protected area (National Park) does not allow people living inside the park, but some traditional population of afro-descendants who lived there before the park creation acquired the right to remain in the area.

The establishment of the Jaú Park, which includes the right margin of the Unini River, was one of the drivers to create the RESEX Unini. For several decades, the local communities located in the left margin of the Unini River fought against commercial fishing from urban centers, such as Manaus and Novo Airão. In 2002, locals created the Unini River Residents Association to stablish an extractive reserve in the left margin of the river and ensure better conditions of living. Thus, in 2006, following the demands of local communities (bottom-up fashion), the RESEX Unini (849,684.79 ha) was established by the Brazilian government (Decree s/n° of 2006), implementing the ban of commercial and sport fisheries, as well as limited access to outsiders. Environmental peculiarities, including the presence of rapids and the construction of an enforcement base of the ICMBio, favor the control of commercial fishing vessels. The management plan published in 2014 describe the management of several natural resources by local communities in this protected area, including a large and valuable commercial fish species (the pirarucu - *Arapaima gigas*), chelonians eggs, woody vine, Brazilian nuts, and small-scale agriculture; all these activities are supervised by the ICMBio and local associations. More recent local demands include the opening of artisanal commercial fishing, similar to what is done in the RESEX Tapajós-Arapiuns, and sport fisheries. These demands are coming from the local riverine communities and have the aim to improve social-economic aspects in the region.

2.3 Fish Sampling

The scientific sampling of fish was conducted in four communities inside and four outside extractive reserves (Tapajós-Arapiuns and Unini) in each river, totaling 16 communities. In each community, a lake (lentic environment) and the river (lotic environment) were sampled; two sampling units per community, during two seasons (low and high-water), amounting to 32 sampling sites and 64 samples in total in both rivers. The communities were selected according to similarities in terms of socio-economic aspects, including dedication to fishing activities by most inhabitants, and considering a minimum distance of 10 km between studied communities, aiming to cover a wide spatial scale and avoiding the overlap between sampling units (◘ Fig. 2.1). Before sampling, research aims and objectives were explained to each community leader and permission to conduct the study was requested. Community leaders were asked to indicate river sites and floodplain lakes where the community usually perform their fishing activities and which would be suitable for fish sampling.

Each community (river and lake) was sampled twice, once in the high-water season (Tapajós: June, Negro: July) and another in the low-water season (Tapajós: October, Negro: November). Fishes were collected using gillnets for 24 hours in each environment and each net was checked every 4 hours to avoid fish damage by predation (◘ Fig. 2.3). Fishes were collected using two sets of gillnets (420 m^2 each) with different mesh sizes (15,

◘ **Fig. 2.3** Fisherman checking a gillnet in the Unini river near the Floresta community during the dry season (Photo: Renato Silvano)

Fig. 2.4 Setting gillnets during fish sampling in the Tapajós river during the dry season (Photo: Renato Silvano)

25, 35, 50, 60, 70, 80 mm between opposite knots): one set was placed along the shoreline and the other in the center of the river/lake (Fig. 2.4). During the high-water season, another gillnet (43 m^3) with three different mesh sizes (50, 60, 70 mm between opposite knots) was placed inside the floodplain forest targeting frugivorous fishes. One type of gillnet named "feiticeira" (54 m^2) with three mesh sizes (25, 35, and 50 mm) was also used to increase the diversity of fishes sampled. All fishes sampled were weighed and measured to standard length (SL) (Fig. 2.5). Some fishes were anesthetized with clove oil, preserved in a 10% formalin solution, and identified to the species level. A total of 13,624 individuals of 224 fish species (~859 kg) were sampled (Table 2.1).

2.4 Interviews

A total of 281 fishers of 16 communities from both rivers (the Tapajós and Negro) were individually interviewed (Fig. 2.6) using a standardized semi-structured questionnaire focused on fishers' socio-economic profile as well as fisheries dynamics, the ecology of target fish species, and changes in fisheries. Community leaders were contacted and their permission to conduct the research on the community was requested. After agreeing with the research, community leaders were asked to nominate other fishers to be interviewed. The study's goals were explained to each fisherman nominated and their permission to

Fig. 2.5 Measuring and identifying fish species collected during the sampling in the Negro River (Photo: Renato Silvano)

participate in the study was requested. After the interview, nominated fishers were asked to indicate others, according to a snowball sampling procedure, as have been performed in other surveys (Silvano et al. 2006; Hallwass et al. 2013). The interview sampling finished when the indicated fishers had already been interviewed or when the indicated fishers were not in the community. This method of sampling did not include the whole community, as some fishers were not available to be interviewed at the time of sampling. The snowball method applied in this study directs the interviews towards the more experienced fishers, who were recognized as such by other community members.

2.5 Participatory Monitoring of Fish Landings

At the end of each interview, those fishers who had at least 5 years of elementary education and fished at least three times per week were invited to voluntarily record their own fish landings during one year. Those who accepted the invitation were individually trained and each received a pencil, a watch, a scale, and paper forms to record their first five fish landings on each month (Fig. 2.7). The monitoring of fish landings started in July 2016 and finished in June 2017 in the Tapajós river, while in the Negro river the monitoring started in August 2016 and finished in July 2017.

2

Table 2.1 Riverine communities studied in both rivers, inside and outside of Extractive Reserves, showing the number of families in each community, the number of interviewed fishers, the number of fishers who recorded their own fish landings, the number of fish landings recorded, and the number of fish sampled (individuals). Location of the studied communities is shown in Fig. 2.1

River and protected area	Communities	N. of families	N. of interviewed fishers	N. of fishers who recorded landings	N. of fish landings recorded	N. of fish sampled (individuals)[a]	Map codes
Tapajós river							
Tapajós-Arapiuns extractive reserve	Capichauã	23	19	9	167	284	CAP
	Parauá	200	24	15	416	475	SUR
	Jauarituba	49	20	12	258	595	JAU
	Cametá	120	25	15	372	1249	CAM
Outside reserve	Ponta Pedras	56	14	10	247	174	PON
	Alter do Chão	>200	14	10	373	583	ALT
	Apacê	130	20	12	348	1437	APA
	Santa Cruz	180	25	9	173	1117	SAN
Negro river							
Unini extractive reserve	Terra Nova	33	20	14	225	533	TER
	Patauá	23	17	7	69	1243	PAT
	Tapiira	36	18	15	329	384	TAP
	Floresta	28	19	10	365	1001	FLO
Outside reserve	Aracari	10	9	5	88	1205	ARA
	Bom Jesus	49	18	12	252	794	BOM
	Aturiá	5	5	5	67	1297	ATU
	Bacaba	26	14	11	81	1253	BAC
Total			*281*	*171*	*3830*	*13,624*	

[a]The total number of fish individuals sampled, including individuals preyed upon, were accounted for in the total individuals; thus the final number may vary among locations because only species with more than 10 individuals sampled were selected for establishing length-weight relations

For each fish landing, fishers were requested to record the composition and weight of the catch (per fish species or group of species), the fishing spot (name and environment), the fishing gear employed, time spent fishing, and the number of fishers who participated in the fishing trip. Every 30 days, phone calls were made to the fishers, whenever possible (only in the Tapajós river), to discuss and resolve any problems. The fish landing forms were collected every 4 months. One hundred and seventy-one fishers were individually trained, 92 in the Tapajós river and 79 in the Negro river (Table 2.1). After detailed conference and exclusion of inaccurate and dubious data, a total of 3830 fish landings were recorded and analyzed, 2354 in the Tapajós river and 1476 in the Negro river (Table 2.1). Data from these fish landings are shown in the chapters about fisheries (► Chap. 3) and in the chapters about fish (► Chaps. 4, 5 and 6) to provide fisheries data of target fish species.

2.6 Data Analysis

Recorded data were analyzed according to the goals of each chapter. In ► Chap. 3 "Small-scale fisheries, livelihoods, and food security of riverine people," average, standard deviation, and proportion (%) of citations were used to summarize socioeconomic data

Fig. 2.6 Fisherman showing the size of fish he regularly catches in the Negro river during one of our interviews (Photo: Pedro Nitschke)

from interviews. Analysis of variance with a posteriori test were run to compare production per fishing trip, productivity (Catch Per Unit Effort, CPUE), and effort (fishing time and time to fishing spot). Analyses of variance were based on the Kruskal-Wallis test and a posteriori Dunn tests because data did not present normal distribution, even after transformations. Composition data from fish landings were analyzed based on a non-metric multi-dimensional scaling (NMDS) ordination, using the Bray-Curtis similarity index and randomization test (1000 permutations) to check differences on catch composition among the 16 fishing communities sampled in four hydrological seasons (dry, flooding, wet, and receding). The data matrix consisted of biomass proportion (%) of 46 fish species caught in all studied communities (Table 3.3 in ▶ Chap. 3). Analysis of similarities (ANOSIM) were carried out to compare the composition of the most caught fish species among the rivers (the Tapajós and Negro), inside and outside Protected Areas, seasons, and communities. The main fish species caught inside and outside Extractive Reserves in the Tapajós and Negro rivers were plotted in figures. NMDS and ANOSIM analyses were performed in the software program PRIMER 6 (Clarke and Gorley 2006).

In ▶ Chaps. 4 to 6 about fish, length–weight relations of each fish species with more than 10 individuals sampled were shown. Length–weight relationships are useful for ecological and fisheries science to compare the growth and welfare of species between different regions (Froese 1998). Sampled fishes were measured by standard length (cm) and total weight (g). Length–weight relationship was defined by linear regressions of log-transformed data where: log (Weight) = log(a) + b*log (Length). Parameters *a* and *b*, which are related to body shape and growth (respec-

Fig. 2.7 Fisherman in the Tapajós River being trained to record his fish landing. In detail, the kit delivered to each fisherman to record the fish landings is shown (Photo: Pedro Nitschke)

tively) and estimated by the linear regression, were provided. Outliers of each fish species were verified using diagnostic plots and were removed before the analysis (Froese and Binohlan 2000). The contribution of individual species to small-scale fisheries was estimated by the fish landings data. For each community, the total biomass of fish caught registered in fish landings was calculated and used as a measure of fishing pressure. The seasonal variation in fish biomass and fishing gear used was also accessed using fish landing data.

References

Agência Nacional de Águas (ANA). HidroWeb. Available at: http://www.snirh.gov.br/hidroweb/publico/apresentacao.jsf. Accessed 9 July 2019

Clarke KR, Gorley RN (2006) PRIMER v6: user manual/tutorial. PRIMER-E, Plymouth

Froese R (1998) Length-weight relationships for 18 less-studied fish species. J Applied Ichthyology 14:117–118

Froese R, Binohlan C (2000) Empirical relationships to estimate asymptotic length, length at first maturity and length at maximum yield per recruit in fishes, with a simple method to evaluate length frequency data. J Fish Biol 56:758–773

Goulding M, Carvalho ML, Ferreira EG (1988) Rio Negro, rich life in poor water. SPB Academic Publishing, The Hague

Goulding M, Barthem R, Ferreira E (2003) The Smithsonian atlas of the Amazon. Smithsonian Books, Washington

Hallwass G, Lopes PFM, Juras AA, Silvano RAM (2013) Fishers' knowledge identifies environmental changes and fish abundance trends in impounded tropical rivers. Ecol Appl 23:392–407

ICMBio (Instituto Chico Mendes de Conservação da Biodiversidade) (2014) Plano de manejo - Reserva ExtrativistacTapajós-Arapiuns. Ministério do Meio Ambiente, Brasília

Junk EJ, Piedade MTF, Schöngart J, Cohn-Haft M, Adeney JM, Wittmann F (2011) A classification of major naturally-occurring Amazonian lowland wetlands. Wetlands 31:623–640

Silvano RAM, MacCord PFL, Lima RV, Begossi A (2006) When does this fish spawn? Fishermen's local knowledge of migration and reproduction of Brazilian coastal fishes. Environ Biol Fishes 76:371–386

Sioli H (1984) The Amazon - limnology and landscape ecology of a mighty tropical river and its basin, Monographiae biologicae. Junk Publishers, Dordrecht

Small-scale Fisheries, Livelihoods, and Food Security of Riverine People

Gustavo Hallwass, Luís Henrique Tomazoni da Silva, Paula Nagl, Mariana Clauzet, and Alpina Begossi

Contents

Renato A. M. Silvano (ed.), *Fish and Fisheries in the Brazilian Amazon*, https://doi.org/10.1007/978-3-030-49146-8_3

3.1 Small-Scale Fisheries: A Global Context

Small-scale fisheries (SSF) have a fishery production similar to large-scale fisheries while employing a much larger number of fishers and usually with a reduced amount of fish discards (Pauly 2006). Therefore, SSF optimize the use of natural resources and play an essential social role through the generation of food and income for millions of people (Béné et al. 2009; Fluet-Chouinard et al. 2018). SSF encompass the majority of fisheries conducted in developing countries (FAO 2014). However, SSF are usually underevaluated or neglected regarding statistics and scientific studies, generating political and economic invisibility of this important activity (Lynch et al. 2017).

Freshwater environments can be considered as "natural banks", whose capital (fish) provides food and income for riverine populations and, consequently, plays an important role in alleviating poverty and promoting food security in many developing and low-income countries (Béné et al. 2009). The fishery benefits directly from the extraction of natural capital provided by aquatic ecosystems (Costanza et al. 1997). Therefore, a better understanding of the dynamics of the freshwater fisheries is fundamental for the development of public policies aiming at the conservation of the aquatic ecosystems and the livelihoods of the riverine populations that rely on these resources (Bayley and Petrere 1989; Welcomme et al. 2010; De Graaf et al. 2015). A recent study on household consumption of freshwater fish in low-income countries indicates that underreported catches or hidden harvests reached around 65% of the total fish catches that are officially reported (Fluet-Chouinard et al. 2018). The authors calculated that freshwater fish is responsible for animal protein intake of 119.1 million people in 36 low income countries (Fluet-Chouinard et al. 2018). Therefore, the SSF ensure food security for poor populations, despite receiving little attention from scientific studies and official statistics.

On the other hand, even SSF have potential to reduce biodiversity and deplete fishing resources (Pinnegar and Engelhard 2007). Historical and archaeological research has demonstrated that even ancient and less technological fisheries caused the extinction of large species hundreds and thousands of years ago (Jackson et al. 2001; Pinnegar and Engelhard 2007). In the Central Brazilian Amazon, one archaeological study indicated that fish fauna accounted for more than 75% of the vertebrate species consumed and recovered in an ancient settlement of indigenous populations from 750 to 1020 A.D. (Prestes-Carneiro et al. 2015). There are, however, big challenges in aquatic ecosystems associated with Amazonian fisheries, such as the dilemma of survival and poverty of local inhabitants, associated with the conservation of biodiversity. For example, as local inhabitants depend upon local protein for their consumption and marketing of fish for their survival, and considering that many of the exploited fish species are classified as "unknown" (Begossi et al. 2018), one of the biggest challenges is to integrate (or conciliate) biodiversity conservation and livelihood survival. Borja (2014) reviews the ecology of marine ecosystems and summarizes eight "challenges," which may also apply to Amazonian riverine ecosystems: (1) understanding the role of biodiversity in ecosystems; (2) understanding human pressures on ecosystems; (3) understanding the impacts of global change; (4) assessing ecosystems' health; (5) delivering ecosystem services by protecting waters; (6) recovering ecosystem structure; (7) uses of the ecosystem approach and spatial planning; (7) modeling ecosystems for better management. In the context of Amazonian riverine fisheries, the data in this chapter could contribute to understand the challenges related to human pressures, the role of biodiversity, and delivering ecosystem services for conservation and people well-being.

3.2 Amazon Freshwater Fisheries

The Amazon is the largest river basin of the world and has the world's richest freshwater fish diversity (Welcomme 1985; Junk et al. 2007). In the Amazon basin, the fisheries are essentially small-scale (Bayley and Petrere 1989). Despite the high richness of fish species, the Amazon fisheries are selective, especially the commercial fisheries (Hallwass and Silvano 2016). The fleet from small-scale fisheries in the Amazon is usually composed of paddled or small motorized canoes with low autonomy and storage capacity, and the crew is usually composed of one or two fishers. The most employed fishing gear is the gillnet (including several mesh sizes and techniques of use); however, handline, longline, castnet, and harpoon are also used according to habitat explored, hydrological season, and the decision of fishers (Hallwass et al. 2013a; Silvano et al. 2017).

Although the inland fisheries in the Amazon are mostly small-scale, there are evidences of overexploitation of some target fish species (Smith 1985; Petrere et al. 2004; Isaac et al. 2008; Garcia et al. 2009). Indeed, recent studies in the Amazon indicate an influence of the proximity to urban centers, which negatively affects the size of fish, as the largest fish are found in sites distant from larger cities (Tregidgo et al. 2017; Keppeler et al. 2018). Thus, conservation and management measures are necessary to maintain fish stocks and ensure food security of riverine populations who rely on fish as their main source of animal protein (Isaac et al. 2015a).

However, the Amazon region has recently been suffering several kinds of environmental pressures and impacts. For example, the increase of the illegal deforestation in the Amazon is driven by the agribusiness interest for cattle and soybean production (Nepstad et al. 2002; Torres 2005; Fearnside 2006). Furthermore, several hydroelectric projects proposed and already under construction disregard technical advice, being authorized based on political pressures to support private interests of large companies (Fearnside 2001, 2014, 2015). The future projects of hydroelectric and dam development in the Amazon Basin can harm fish biodiversity and the associated fisheries (Winemiller et al. 2016). To make things worse, recent projects in the National Congress aim to change the status of already established Protected Areas and indigenous lands in the Amazon (reduction of area, change of category and reclassification) to facilitate the construction of hydroelectric facilities (including large dams) and concessions for mineral exploitation (Bernard et al. 2014; Ferreira et al. 2014). All of these changes would directly or indirectly affect the fish stocks exploited by riverine populations.

Another environmental pressure is the gold mining and the mercury (Hg) contamination, especially in Tapajós river (Roulet et al. 1998, 1999; Akagi and Naganuma 2000; Cynamon et al. 2001). The mercury is released in the environment during gold exploration, besides small amounts that are released by the soil erosion. Some cities of the Upper Tapajós river have their economic growth based on the exploration of gold that affects directly the river (Roulet et al. 1998, 1999; Nevado et al. 2010). The Hg is a heavy metal that accumulates in the trophic chain and is a long-term problem for the aquatic environment and for fishing (Nevado et al. 2010). Studies have indicated high levels of Hg in riverine communities along the Tapajós river, indicating the impact of gold mining in the population health (Akagi and Naganuma 2000; Nevado et al. 2010). Therefore, the uncontrolled gold mining can directly affect the food security and health of riverine communities which rely on fish for food supply.

Management and conservation efforts in the Amazon basin related to fish have been implemented at local and regional scales. Fishing agreements regulating the access through fish quotas and gear restrictions in floodplain lakes are examples of management at the local level (Castro and McGrath 2003; Almeida et al. 2009; Silvano et al. 2014). On the other hand, management at the regional level could be the closure of fishing during the fish breeding season by government laws applied to the whole country or to river basins (Hallwass et al. 2013a; Corrêa et al. 2014). Other regional conservation measure is the establishment of Protected Areas, which were created mainly to conserve forests and terrestrial fauna, but could also protect fish communities. Recent studies have shown positive influences of Protected Areas on conservation of fish assemblages and to increase fishing productivity in less productive Amazonian rivers, such as in the black water Negro river (Sobreiro et al. 2010) and in the clear water Tapajós and Tocantins rivers (Silvano et al. 2014; Keppeler et al. 2017).

The goal of this chapter is to describe the dynamics of SSF inside and outside Extractive Reserves in Tapajós and Negro rivers, Brazilian Amazon. Extractive Reserves are a kind of Protected Area, which are described in more detail in ▶ Chap. 2 (Methods). We described the socioeconomic profile of fishers (fishing experience and frequency, literacy level, income, and fish destination) and fishing characteristics, such as the type of vessels, exploited habitats, and fishing gears used. Fishing productivity, fishing effort, and fish composition were also analyzed, discussing the most caught fish species. All the above comparisons considered potential differences between the two studied rivers.

3.3 Small-Scale Fisheries in the Tapajós and Negro Rivers

Clear and black waters, such as Tapajós and Negro rivers, respectively, are oligotrophic environments, which have lower fishery production if compared to white water rivers, and have been less studied. However, these rivers support a large number of small-scale fishers and ensure food security and income for the majority of the riverine population. During this study, 281 fishers were interviewed and 3830 fish landings were recorded by fishers in both rivers, inside and outside Extractive Reserves (ER); see ◘ Table 2.1 (▶ Chap. 2) for more details.

3.3.1 Socioeconomic Characteristics

The age of interviewed fishers ranged from 18 to 82 years old. The oldest fishers were interviewed in Tapajós river, outside ER, and the younger fishers in Negro river, outside ER (◘ Table 3.1). The fishing experience was related to the age of fishers; in average fishers started to fish at the age of 13–14 years. Thus, the oldest interviewed fishers are the most experienced ones (◘ Table 3.1). Besides that, fishers' experience refers to the studied region, since residence time in the region was similar to the time when fishing started (◘ Table 3.1). Fishers started to fish very young, usually accompanying a relative, such as parents, uncles, or, sometimes, a neighbor close to the family. This pattern is very common in SSF in the Amazon basin (Hallwass et al. 2013b). Kinship relationships are very strong in the use of natural resources, especially in fisheries, promoting the transmission of knowledge from older generations to new ones (Begossi 1996, 2008, 2014; Castello et al. 2009; Diamond 2001; Johnson 2006).

3

Table 3.1 Socioeconomic profile of 281 interviewed fishers from 16 fishing communities in Tapajós river (8 communities) and in Negro river (8 communities). In Tapajós river 88 fishers were interviewed inside Extractive Reserve and 73 outside. In Negro river 74 fishers were interviewed inside Extractive Reserve and 46 outside. Data shown are average and standard deviation, or in proportions (%), due to different numbers of samples in each analyzed region

Fisher's profile	Tapajós river		Negro river	
	Extractive reserve	Outside	Extractive reserve	Outside
Age (years old)	45.8 ± 12.8	51 ± 12.3	40.9 ± 13.6	37.2 ± 12.9
Fishing experience (years)	31.7 ± 14.3	36.5 ± 13.8	28.1 ± 12.9	23.7 ± 13.3
Living time (years)	36.6 ± 16.6	37.9 ± 18.2	27 ± 16	23.3 ± 16.5
Study time (%)				
Illiterate	3.5	5.5	24.3	11.1
1–3 years	16.5	20.5	29.7	20.0
4–8 years	63.5	58.9	31.1	51.1
More than 8 years	16.5	15.1	14.9	17.8
Fishing frequency (week) (%)				
Daily	35.2	20.5	35.1	39.1
3 times or more	36.4	57.5	36.5	36.9
1–2 times	21.6	19.2	20.3	23.9
Fish consumption (%)	55.4 ± 28	49 ± 29.4	99.1 ± 6.2	47.7 ± 37.8
Fish sale (%)	43.6 ± 28	51 ± 29.4	0.9 ± 6.2	52.3 ± 37.8
Sale destination (%)				
Community	100	76.3	–	23
Middleman	17	18.6	–	67
Tourists	–	5.1	–	–
Monthly income (US$[a])				
0.0–156.4	56.8	38.4	50.8	38.7
156.5–312.5	30.7	24.6	26.1	32.3
312.6–468.8	9.1	10.9	15.4	9.7
468.9–625	3.4	15.1	4.6	12.9
More than 625	–	10.9	–	–

[a]US dollar quotation calculated (1.0 US$ = R$ 3.2) as the average of purchase values for the year 2017

More than half of the interviewed fishers studied from 4 to 8 years, except in ER of Negro river (Table 3.1), where the number of illiterate fishers was higher, as 24% of fishers do not know how to read and write. The number of fishers who studied more than 8 years ranged from 15% to 18%. The level of formal education is low among Amazon fishers, due to the lack of access and logistical difficulties to arrive at schools. For instance, the ER in Negro river, which showed a higher proportion of illiterate fishers, is also the most distant region from urban centers, resulting in a restricted logistical access.

Most of the interviewed fishers cited to fish daily or 3 times or more per week (Table 3.1), indicating a high dependence on fishing, either for food or income. In this sense, according to the fishers, similar amounts of fish caught are destined for consumption and sale in Tapajós river (inside and outside ER) and outside ER in Negro river (Table 3.1). In the ER of the Negro river, the sale is not significant because the commercialization to outsiders is forbidden. The fish is usually sold within communities in the Tapajós river, while in the Negro river, fishers from outside ER sell the fish to middlemen.

Nevertheless, the income obtained from fishing is low, especially inside the ER in both rivers (◘ Table 3.1). The mixed destination of fish caught for sale or consumption is a typical pattern of SSF, aiming to ensure food security and livelihoods.

Fishers from outside ER in Tapajós river, near to Santarém city, especially in *Ponta de Pedras* and *Alter do Chão* communities, have higher economic benefits from fish sales if compared with fishers from other communities and regions. These communities receive lots of tourists from Santarém, other Brazilian regions, and even other countries. Because of tourism and the consequent high demand for fish, the prices paid for fish are higher in these communities in the Tapajós river (Hallwass et al. 2017). On the other hand, fishers from inside ER in Negro river manage the pirarucu (*Arapaima gigas*), the only fish that is allowed a quota for sale once a year. However, fishers there request the release of the commercial exploitation of large catfish and the permission for conducting sport fishing inside the ER to attract tourism.

3.3.2 Fishing Characteristics: Vessels, Exploited Habitats, and Fishing Gears

All the following results are based on participatory monitoring, for which each fisher recorded his own fish landings for one year (see ► Chap. 2 on methods for more details). Fishers from inside ER, in both rivers, used more frequently paddle canoes during fishing trips, 55.5% of records in Tapajós and 51.8% in Negro river, while motorized canoes with small outboard motors, locally called *rabeta*, represented 41.2% of fish landings in Tapajós river inside ER, and 44.4% in Negro river, inside ER. On the other hand, outside ER in both rivers, the results are the opposite, *rabetas* (motorized canoes with small outboard motors) were more commonly used, 58.6% of fishing landings in Tapajós river and 60.3% in Negro river, while paddle canoes were used in 38.8% of fishing trips in the Tapajós river and 36.1% in the Negro river, outside ERs. Boats with higher autonomy of travel and storage capacity were less used, ranging from 2.6% of fishing trips outside ER in Tapajós river to 3.8% inside ER in Negro river. However, most of the fish production is provided by fishing trips using *rabetas*, 78.5% of production inside ER and 86.4% outside ER in Tapajós river, and 81.2% outside ER in Negro river, while inside ER in Negro river *rabetas* represent 55.5% of fish production and paddle canoes 40.2%. The higher frequency of use of *rabetas* outside ERs in both rivers indicated a wider range of exploitation (distance), potential higher fishing pressure, and increased costs with fuel for fisheries outside Protected Areas (ERs). Therefore, the higher frequency of paddled canoe used inside the ER could contribute to reduce the fishing pressure inside these Protected Areas and maintain the fishing productivity over time (Keppeler et al. 2017). However, the *rabeta* presents a relatively high proportional fish production considering its frequency of use. Thus, fishers can go further and, despite the cost of fuel, they may optimize their catches using *rabetas* by increasing the amount of fish caught, in agreement to the Optimal Foraging Theory applied to fishers' decision (Begossi and Richerson 1992; Oliveira and Begossi 2011). *Rabetas* are the most common type of vessels used in small-scale fisheries in the Amazon (Cardoso and Freitas 2008; Gonçalves and Batista 2008; Hallwass et al. 2011; Isaac et al. 2015b), which may be due to the cost-benefit of this kind of vessel.

The most employed fishing gear was the gillnet (◘ Table 3.2). Gillnets are more frequently used and responsible for the most of catches in Tapajós river (inside and outside ER) and outside ER in Negro river, considering only gillnet and "mixed 1," on which gillnets were used with another gear (◘ Table 3.2). The gillnets are widely employed in Amazon fisheries, being the most used fishing gear in the Amazon (Fernandes et al. 2009; Hallwass et al. 2011, 2013a; Isaac et al. 2015b; Hallwass and Silvano 2016; Silvano et al. 2017). Gillnets were the least selective gear recorded in the fish landings, as fishes are selected according to the mesh size used. Handlines were widely used in all analyzed regions, mainly inside the ER in Negro river, being, together with harpoon, the main fishing gear employed by fishers there. The harpoon was also important to fishing outside ER in Negro river. Handlines are selective fishing gear, selecting fish species and sizes according to the size of the hook and kind of bait used. Thus, handlines select the target fish based on mouth opening (related to the size of the fish) and diet. In the Lower Tocantins river, which is also a clear water river, fishers achieve high fishing productivity using handlines during the high-water season, due to the catches of pescada (*Plagioscion squamosissimus*) (Hallwass et al. 2011). The harpoon also is an efficient method in the Negro river, possibly due to ecological characteristics of the water, such as low level of sediments and transparency of the water (Sioli 1984). Begossi and others (2005) show a high frequency of use (usually at night) and fishing productivity of harpoons in the Negro river.

Castnets were mainly used outside ER in Tapajós river, aiming to catch fish shoals moving along the river, such as aracus (family Anostomidae), jaraqui (*Semaprochilodus* spp.), and charuto (*Hemiodus* spp.), according to Hallwass and others (2017). The efficiency of this fishing method is closely linked to handling skills and the ability to recognize and identify fish shoals by fishers. Due to the catch of shoals, castnets could be considered as selective. Nevertheless, the use of castnet is seasonal, being mostly used when

3

Table 3.2 Frequency of records from participatory monitoring of fish landings in total number (*N*) and relative values (proportion, %), and fish production in amount (kg) and relative values (proportion, %), of fishing gears employed and fishing habitats exploited in the small-scale fisheries inside and outside Extractive Reserves (ER) in Tapajós and Negro rivers

	Tapajós river				Negro river			
	Extractive reserve		Outside		Extractive reserve		Outside	
	N (%)	kg (%)	*N* (%)	kg (%)	*N* (%)	kg (%)	*N* (%)	kg (%)
Fishing gears								
Gillnets	623 (50.9)	9019.9 (53.2)	291 (29.1)	5515.6 (30.4)	123 (12.9)	884.1 (15.6)	233 (47)	3905.2 (52.6)
Handline	197 (16.1)	1734.6 (10.2)	185 (18.5)	1725.5 (9.5)	382 (40)	1688.9 (29.8)	69 (13.9)	373.4 (5)
Longline	26 (2.1)	716.7 (4.2)	36 (3.6)	716.5 (3.9)	46 (4.8)	124.8 (2.2)	–	–
Harpoon	36 (2.9)	159.8 (0.9)	22 (2.2)	393.5 (2.2)	319 (33.4)	2004.2 (35.3)	86 (17.3)	1255 (16.9)
Castnet	109 (8.9)	761.5 (4.5)	221 (22.1)	3784.7 (20.9)	–	–	–	–
Mixed 1[a]	171 (14)	3012.7 (17.8)	174 (17.4)	4213.5 (23.2)	36 (3.8)	266.7 (4.7)	81 (16.3)	1488.5 (20)
Mixed 2[a]	54 (4.4)	1492.5 (8.8)	45 (4.5)	1318.5 (7.3)	9 (0.9)	290 (5.1)	27 (5.4)	407.3 (5.5)
Mixed 3[a]	8 (0.6)	59.4 (0.3)	27 (2.7)	458.6 (2.5)	40 (4.2)	413.3 (7.3)	–	–
Total	**1224**	**16,957.1**	**1001**	**18,126.4**	**955**	**5672.05**	**496**	**7429.4**
Fishing habitats								
River	933 (74.5)	22882.5 (89.8)	758 (74.7)	12708.8 (69.7)	345 (35.9)	1881.9 (33.2)	120 (24.3)	2290.6 (30.8)
Lake	146 (11.7)	758.1 (2.9)	88 (8.7)	2660.1 (14.6)	138 (14.4)	854.3 (15.1)	132 (26.7)	2071.8 (27.9)
Streams	20 (1.6)	122 (0.5)	25 (2.5)	811 (4.4)	62 (6.4)	387.1 (6.8)	75 (15.2)	616.5 (8.3)
Floodplain	19 (1.5)	226 (0.9)	15 (1.5)	165.2 (0.9)	160 (16.6)	781.5 (13.8)	77 (15.6)	414.4 (5.6)
River & lake	108 (8.6)	1086.3 (4.3)	117 (11.5)	1650.7 (9)	197 (20.5)	1374.3 (24.3)	80 (16.2)	1887.2 (25.4)
River & streams	26 (2.1)	410.9 (1.6)	11 (1.1)	244.5 (1.3)	59 (6.1)	383.2 (6.8)	10 (2)	160 (2.2)
Total	**1252**	**25,485.8**	**1014**	**18,240.3**	**961**	**5662.3**	**494**	**7440.5**

[a]Mixed 1: Gillnets and another gear; Mixed 2: Gillnets and two another gears; Mixed 3: Handline and another gear

the shoals of fish are moving in the river, during migrations for breeding or feeding (Welcomme 1985; Fernandes 1997).

The river channel was the main fishing habitat exploited in the Tapajós river (inside and outside ER) (Table 3.2). Floodplain lakes were also exploited by fishers in the Tapajós river, but the river channel concentrates most of the fish catches because this stretch of the Tapajós river has a small flooded area with lakes and other habitats. This characteristic of to fish mainly in river channel is peculiar of this stretch of the Tapajós river and agrees with a recent study in other Amazonian clear water river, Xingu (Isaac et al. 2015b). On the other hand, in the Tocantins river, fishing grounds are more homogeneously distributed among different available habitats (Hallwass et al. 2011, 2013a). In the Negro river (inside and outside ER), the fishing trips were more evenly distributed among the river channel, floodplain lakes, flooded area, and a mixed of river and lake, although the river is slightly more exploited in Negro river, too (Table 3.2). The studied region of the Negro river is more heterogeneous, thus offering several habitats to be exploited. The alternate exploitation of several kinds of fishing habitats reflects the adaptation of fishers to the seasonality and availability of habitats, as well as the ecological characteristics of target species (Hallwass et al. 2013a; Huntington et al. 2017). A high diversity of habitats and fishing gears may be used (chosen) by fishers daily, in order to catch the largest amount and the preferred fish. Thus, fishers make decisions on a daily basis regarding how, where and which fishes to fish according to their goals and local ecological knowledge (Salas and Gaertner 2004; Fulton et al. 2011; Hallwass et al. 2013a).

3.3.3 Fishing Productivity and Effort

In total, 44,369.63 kg of fish were caught in 3830 fish landings recorded by fishers, in both rivers (Tapajós and Negro) inside and outside ERs. On average, fishers caught 14 kg per fishing trip; however, this amount differed among the regions analyzed ($H = 385.4$; $p < 0.0001$). The catch per fishing trip was higher outside ERs in both rivers, and the lower values of catch were recorded inside ER in Negro river (Fig. 3.1). On the other hand, the fishing productivity calculated through the catch per unit of effort (CPUE) in kg per fisher per hour showed similar values among communities located outside and inside the ER in Tapajós river ($H = 34.5$; $p < 0.0001$), while communities inside ER in Negro river showed lowest values of CPUE (Fig. 3.2). These results indicated a higher fishing pressure outside Protected Areas (ERs) where we observed larger fish catches.

Although the average number of fishers per fishing trip was similar among the analyzed regions (1.6 ± 0.9 fishers), other variables related to fishing effort differed among regions, such as total time fishing ($H = 311.4$; $p < 0.0001$), as well as variables related to fishing costs, such as time to fishing spot ($H = 122.7$; $p < 0.0001$). Communities outside of ER spend more time fishing (Fig. 3.3) and need to travel longer distances to fish (calculated here as time to fishing spot) (Fig. 3.4). Fishers inside ERs employed less effort, spent less time fishing and traveled shorter distances to fishing spots. This difference in the effort made the CPUE of fishers inside ER of Tapajós river similar to outside ER. However, fishers from ER of Negro river had a lower CPUE, even considering that they spent less time fishing. Inside ER of Negro river, fishers used more handlines and harpoons than gillnets (Table 3.2). These methods are more selective and demand that the fisher be constantly and actively managing the fishing gear, whereas gillnets are less selective and can cover a large area without constant supervision by fishers. Thus, the ER could be reducing the fishing effort and hence the fishing pressure. Protected Areas have an important role in maintaining the fish production and productivity, according to studies in clear water Amazonian rivers

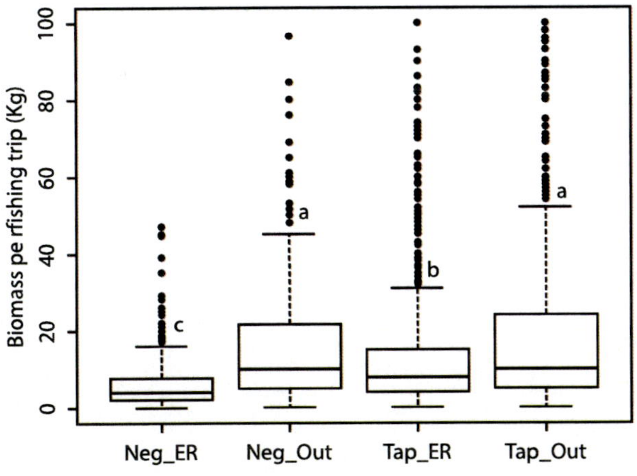

Fig. 3.1 Comparison of biomass caught per fishing trip (Kg) among communities inside (ER) and outside (Out) the Extractive Reserve in Tapajós (Tap) and Negro (Neg) rivers. Median (darker line in the box plot), minimum and maximum values (vertical lines), and outer lines of the box plots showing the quartiles (25% and 75%), Dunn test: $a > b > c$, $p < 0.05$. Circles are outliers

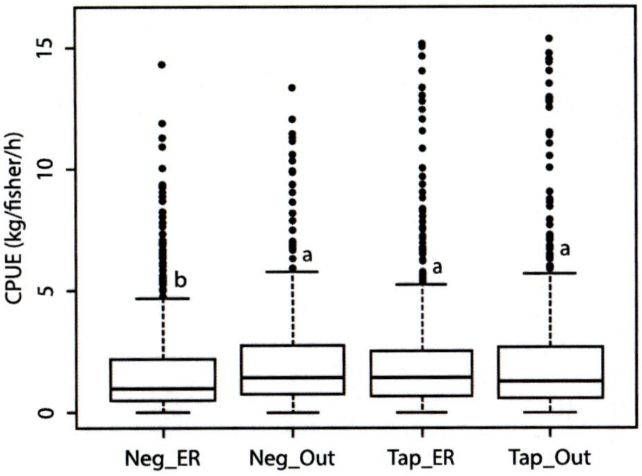

Fig. 3.2 Comparison of the CPUE (kg/fisher/hour) per fishing trip among communities inside (ER) and outside (Out) the Extractive Reserve in Tapajós (Tap) and Negro (Neg) rivers. Median (darker line in the box plot), minimum and maximum values (vertical lines), and outer lines of the box plots showing the quartiles (25% and 75%), Dunn test: $a > b$, $p < 0.05$. Circles are outliers

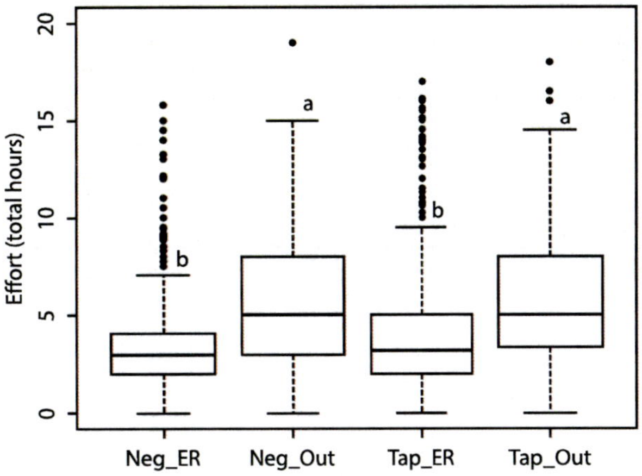

Fig. 3.3 Comparison of effort (total fishing time, in hours) among communities inside (ER) and outside (Out) the Extractive Reserve in Tapajós (Tap) and Negro (Neg) rivers. Median (darker line in the box plot), minimum and maximum values (vertical lines), and outer lines of the box plots showing the quartiles (25% and 75%), Dunn test: $a > b$, $p < 0.05$. Circles are outliers

(Silvano et al. 2014; Keppeler et al. 2017). Furthermore, studied communities outside Protected Areas (ERs) are nearer to large cities than communities inside ERs (see map on ► Chap. 2, on Methods), and this factor could have an additive effect increasing fishing effort to maintain catches outside ERs.

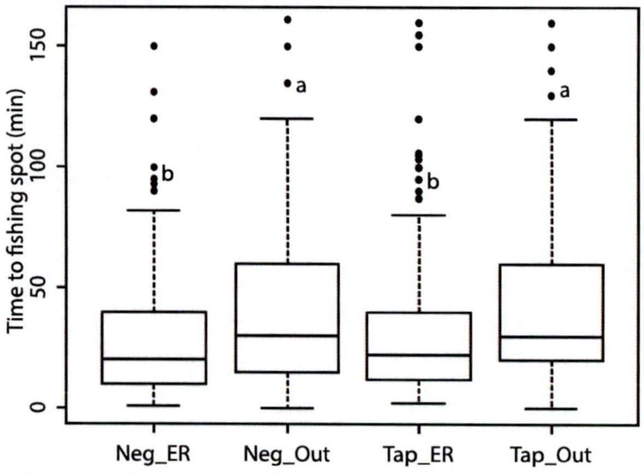

Fig. 3.4 Comparison among communities inside (ER) and outside (Out) the Extractive Reserve in Tapajós (Tap) and Negro (Neg) rivers, considering fishing costs measured as time to fishing spot (minutes). Median (darker line in the box plot), minimum and maximum values (vertical lines), and outer lines of the box plots showing the quartiles (25% and 75%), Dunn test: $a > b$, $p < 0.05$. Circles are outliers

In relation to seasonality, the fishing productivity (CPUE) differed among seasons in both rivers, Tapajós ($H = 24.9$; $p = 0.0007$) and Negro ($H = 87.7$; $p < 0.0001$). The lowest CPUE were observed during high-water and flooding seasons in the Tapajós river, outside ER, while the CPUE did not differ among seasons inside ER (Fig. 3.5). On the other hand, the seasonal hydrological changes had a strong influence on fishing productivity (CPUE) in the Negro river (Fig. 3.6). Low and receding water seasons showed higher CPUE values than high and flooding seasons in both regions, inside and outside ER in the Negro river. The observed lower fishing productivity during flooding and high-water seasons is a common pattern in the Amazon fisheries (Fernandes 1997; Cardoso and Freitas 2008; Keppeler et al. 2017), due to the increase in the water volume and the consequent dispersal of fish over a wide area, which makes it harder to catch fish.

While the fisheries in the Negro river showed a clear seasonal pattern, the seasonal influence was weaker in the Tapajós river. These differences could be due to the large floodplain area, several channels, and islands, such as Anavilhana's archipelago, in the Negro river, which are explored by fish shoals during flooding and high-water seasons, reducing the catchability of fisheries and, consequently, fishing productivity. On the other hand, the lower stretch of the Tapajós river has a wide river channel, reaching more than 10 km between mar-

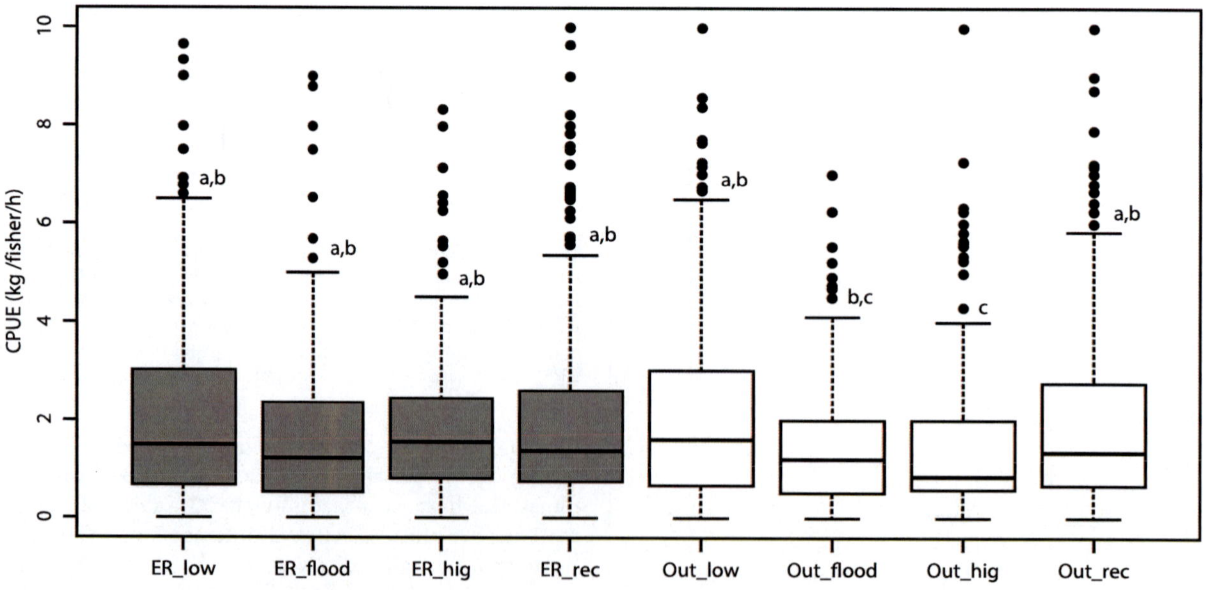

Fig. 3.5 Comparison of CPUE (kg/fisher/hour) per fishing trip among seasons inside (ER, gray) and outside (Out, white) the Extractive Reserve in the Tapajós river. Hydrological seasons: low water, flooding, high-water, and receding. Median (darker line in the box plot), minimum and maximum values (vertical lines), and outer lines of the box plots showing the quartiles (25% and 75%), Dunn test: $a > b > c$, $p < 0.05$. Circles are outliers

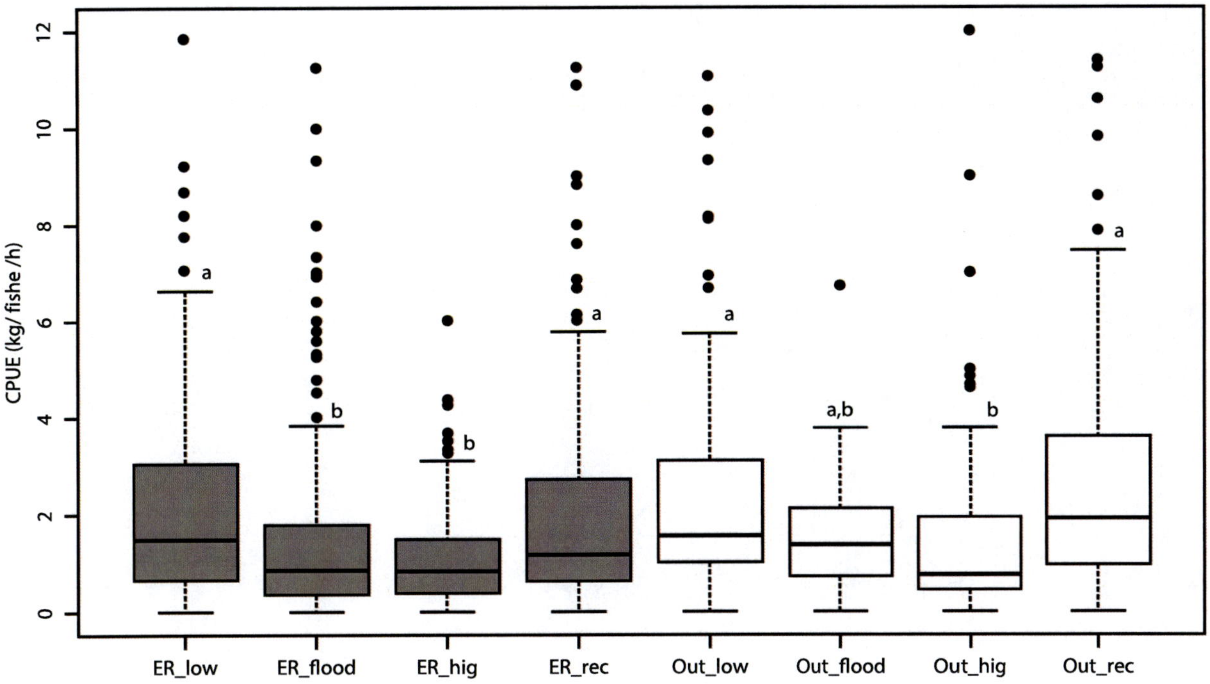

Fig. 3.6 Comparison of CPUE (kg/fisher/hour) per fishing trip among seasons inside (ER, gray) and outside (Out, white) the Extractive Reserve in the Negro river. Hydrological seasons: low water, flooding, high-water and receding. Median (darker line in the box plot), minimum and maximum values (vertical lines), and outer lines of the box plots showing the quartiles (25% and 75%), Dunn test: $a > b$, $p < 0.05$. Circles are outliers

gins and usually with reduced floodplain area, channels, and islands that could provide refuge for exploited fish. Indeed, fishers from Tapajós river fish mainly in the river channel (Table 3.2), which has lower environmental heterogeneity that could provide refuges for fish.

3.3.4 Fish Species Caught

Ecological and local variables are the main factors influencing fish catch composition in both rivers. The fish catch composition differed between the two rivers (Global $R = 0.69$; $p = 0.001$, see data matrix in Table 3.3). These differences may be partially due to ecological characteristics of each river, such as pH, transparency, electrical conductivity (Sioli 1984; Junk et al. 2011, see ▶ Chap. 2 for more details about ecology of studied rivers and statistical analyses), which may influence the fish communities available in the environment and consequently fish catches. The composition of fish catches differed also among the studied communities according to local characteristics related to fish availability and demand (Global $R = 0.65$; $p = 0.001$). Fishers from each community catch fish species according to individual decisions in relation to fishing gear used, fishing spots and habitats explored during each fishing trip. While the composition of caught fish differed among communities located inside and outside the Extractive Reserve in both rivers, the explanation of this factor was weak due to low Global R-value (Global $R = 0.1$; $p = 0.006$), and seasonality did not influence the composition of catches (Global $R = 0.0$; $p = 0.42$). These results indicate that Protected Areas (ERs) did not influence catch composition. Therefore, based on these results, we can suggest two non-mutually exclusive hypotheses: (1) the exploitation level of fish stocks is moderate, not causing a drastic decline of preferred target species (usually the largest) and replacement of these by smaller ones, as suggested by *fishing down process* in multi-species fisheries (Welcomme 1999); (2) the reduced fishing pressure inside ERs could help to maintain the integrity of fish stocks inside ERs and to maintain fish communities outside ERs. Previous studies in floodplain lakes in the Brazilian Amazon indicate that managed or unfished lakes can potentially contribute to the maintenance of fishing stocks of fished lakes (MacCord et al. 2007; Silvano et al. 2009, 2014; Campos-Silva and Peres 2016), which could be similar to spillover effects in marine protected areas (McClanahan and Mangi 2000; Gell and Roberts 2003; Russ et al. 2004). However, only long-term fisheries data could confirm these hypotheses.

3

Table 3.3 Data matrix based on biomass of 46 fish species (or groups of species) caught inside and outside Extractive Reserve (ER) in Tapajós and Negro rivers

Common name	Scientific name	Tapajós river		Negro river		Total (kg)
		Inside ER (kg)	Outside ER (kg)	Inside ER (kg)	Outside ER (kg)	
Acará	*Satanoperca* spp. and other species of Cichlidae	29.8	47.3	212.75	342.13	**631.98**
Acará-Açú	*Astronotus ocellatus*	3.6	33	33.95	121	**191.55**
Acaratinga	*Geophagus* spp.	2295.87	1203.55	–	–	**3499.42**
Acari	Loricariidae	61.4	26	–	6	**93.4**
Aracú	*Leporinus* spp., *Schizodon* spp.	196.38	2863.57	145.34	296.2	**3501.49**
Arari	*Chalceus macrolepidotus*	7	3.5	14.41	10	**34.91**
Arraia	*Potamotrygon* spp.	7	156	–	–	**163**
Aruanã	*Osteoglossum bicirrhosum*	–	38	95.5	42.5	**176**
Babão	*Goslinia platynema*	–	–	–	2.5	**2.5**
Bacu	*Lithodoras dorsalis, Megalodoras uranoscopus*	6	–	–	–	**6**
Bararuá	*Uaru amphiacanthoides*	9.1	3.1	–	–	**12.2**
Barbado	*Pinirampus pirinampu*	8	17	44	14	**83**
Baru	*Uaru amphiacanthoides*	–	–	3.8	72.3	**76.1**
Branquinha	*Potamorhina* spp.,*Curimata inornata, Psectrogaster amazonica*	110.15	82	14	39.5	**245.65**
Cara-de-gato	*Platynematichthys notatus*	11.5	–	–	–	**11.5**
Charuto	*Hemiodus* spp., *Argonectes* spp.	432.65	1402.7	3	–	**1838.35**
Corvina	*Pachypops fourcroi*	–	2	–	–	**2**
Cuiú	*Oxydoras niger*	6.75	245.5	7	–	**259.25**
Curimatá	*Prochilodus nigricans*	–	276	–	–	**276**
Dourada	*Brachyplatystoma rousseauxi*	775.5	455	–	183	**1413.5**
Filhote	*Brachyplatystoma filamentosum*	1831.8	806	–	281	**2918.8**
Fura-calça	*Pimelodella* spp., *Pimelodina* spp.	55	93	–	–	**148**
Jacunda	*Crenicichla* spp.	16.95	5	11.25	–	**33.2**
Jandiá	Pimelodidae	–	–	3	10	**13**
Jaraqui	*Semaprochilodus* spp.	372.95	1137	198.25	1918.3	**3626.5**
Mandi	*Pimelodus* spp.	–	–	1.08	–	**1.1**
Mapará	*Hypophthalmus* spp.	2220.7	373.5	1	2	**2597.2**
Matrinchã	*Brycon* spp.	–	98	35.75	207.7	**341.45**
Mistura	Small fish, several families	39.1	7	6.5	1.5	**54.1**

Table 3.3 (continued)

Common name	Scientific name	Tapajós river		Negro river		Total (kg)
		Inside ER (kg)	Outside ER (kg)	Inside ER (kg)	Outside ER (kg)	
Pacu	*Myleus* spp., *Myloplus* spp.	904.25	1581.5	829.17	705.75	**4020.67**
Peixe-cachorro	*Rhaphiodon vulpinus*	1.5	7	0.3	–	**8.8**
Pescada	*Plagioscion squamosissimus*	4289.35	1360.5	11.8	23.5	**5685.15**
Piracatinga	*Calophysus macropterus*	11.5	–	–	–	**11.5**
Piramutaba	*Brachyplatystoma vaillantii*	–	3	–	–	**3**
Piranambu	*Pinirampus pirinampu*	4.5	–	–	–	**4.5**
Piranha	*Serrasalmus* spp.	23.9	96.1	1115.43	464.5	**1699.92**
Pirapitinga	*Piaractus brachypomus*	55.3	137	–	25	**217.3**
Pirarara	*Phractocephalus hemioliopterus*	–	31	78.25	53	**162.25**
Pirarucu	*Arapaima gigas*	105	–	–	14	**119**
Sarda	*Pellona* spp.	380.9	234.5	–	3	**618.4**
Sardinha	*Triportheus* spp.	5	36	2.3	13	**56.3**
Surubim	*Pseudoplatystoma* spp.	202.8	515	18	137	**872.8**
Tambaqui	*Colossoma macropomum*	47.1	789.4	–	44	**880.5**
Tamoatá	*Hoplosternum littorale*	30	20	–	–	**50**
Traíra	*Hoplias malabaricus*	21.5	37.1	174.24	36.75	**269.59**
Tucunaré	*Cichla* spp.	1433	2812	1438.25	1755.58	**7438.83**
Total		**16,012.79**	**17,033.82**	**4498.31**	**6824.71**	**44,369.63**

This database was employed to run the non-metric multi-dimensional scaling (NMDS) ordination and analysis of similarities (ANOSIM)

In order to exclude ecological effects related to the type of river, the same analyses were run separately for each river. In both rivers, specific communities had a higher influence on fish catch composition than location inside or outside Extractive Reserves and seasonality did not influence the catches (Global $R = 0.03$; $p = 0.26$ in Tapajór river; and Global $R = -0.0$; $p = 0.48$ in Negro river). In the Tapajós river, the composition of catches was 15% more related to communities than to Extractive Reserve (Global $R = 0.42$; $p = 0.001$ related to communities; and Global $R = 0.27$; $p = 0.001$ related to ER), Fig. 3.7a, b). Similarly, in the Negro river, riverine communities explained more the composition of fish catches (Global $R = 0.51$; $p = 0.001$) in comparison to ER (Global $R = 0.28$; $p = 0.001$). However, the difference between Community and ER factors was higher in the Negro river (23%) and the stress of ordination was lower, indicating a stronger influence on catches (Fig. 3.8a, b). Therefore, factors related to riverine communities, such as local heterogeneity, ecological, and socioeconomic characteristics, had a higher influence on the composition of fish caught than Protected Areas (ERs). Each community has its own characteristics, and fishers have to deal with daily choices in relation to exploited habitats and fishing spots, gear, availability, demand, preferences, and ability to catch the several fish species (see ► Chaps. 4, 5, and 6 on fish species). These factors and choices seem to influence the composition of catches, which differed among fishing communities. Thus, local initiatives aimed at the local level tend to be more effective and have faster results in programs for management and conservation of fish stocks (Begossi 2008; Danielsen et al. 2010).

In the Tapajós river inside the ER, a total of 16,012.8 kg was caught and recorded by fishers. The most caught fish species was the pescada (*Plagioscion*

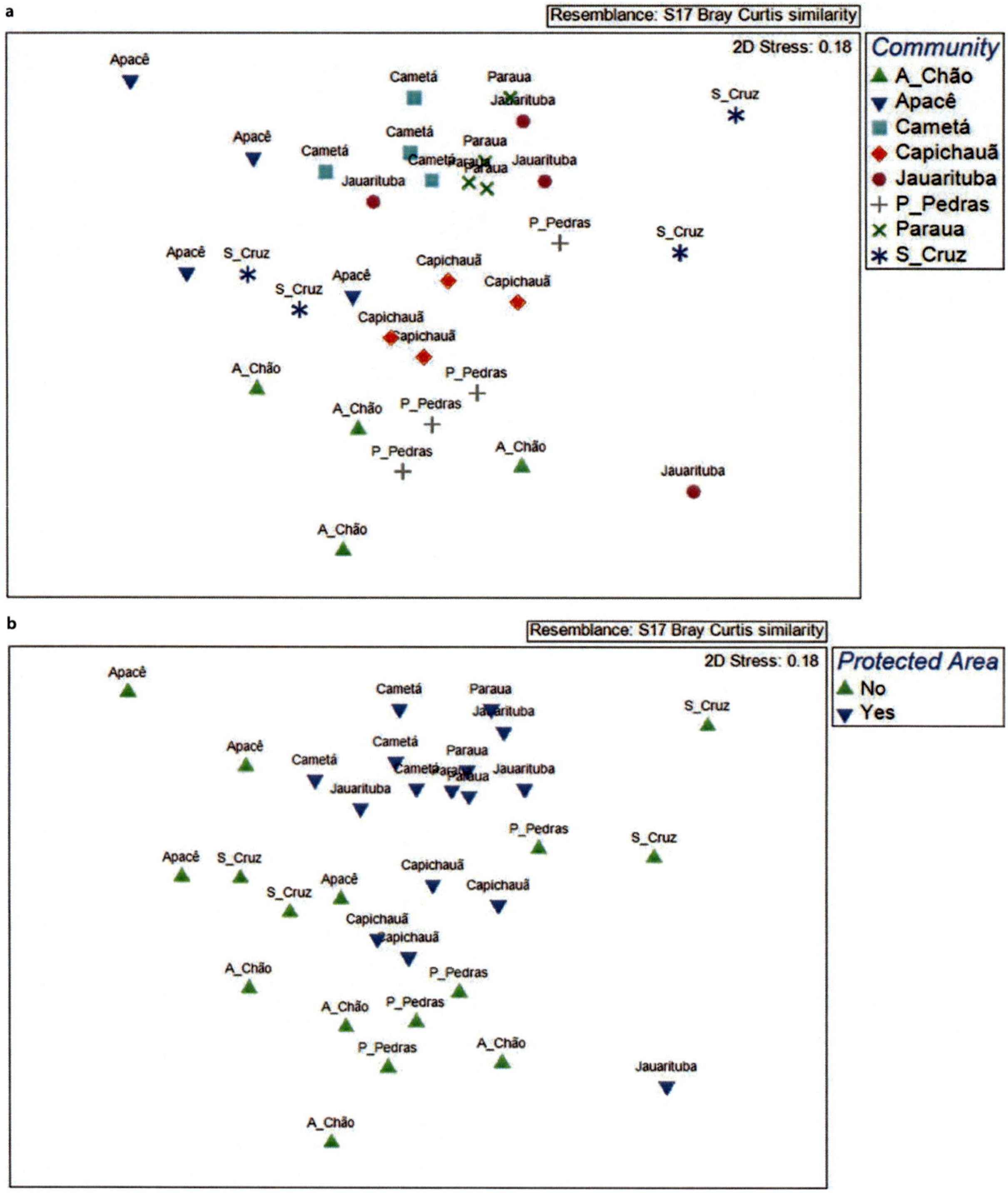

Fig. 3.7 NMDS ordination (stress = 0.18) plot based on the composition (% of biomass) of 46 fish species (or groups of species) caught (Table 3.3) in eight riverine communities during four hydrological seasons (low-water, flooding, high-water and receding) in the Tapajós river (points of plot): **a** plot organized according to studied communities; **b** plot organized according to localization of fishing communities, inside and outside Extractive Reserve (Protected Area)

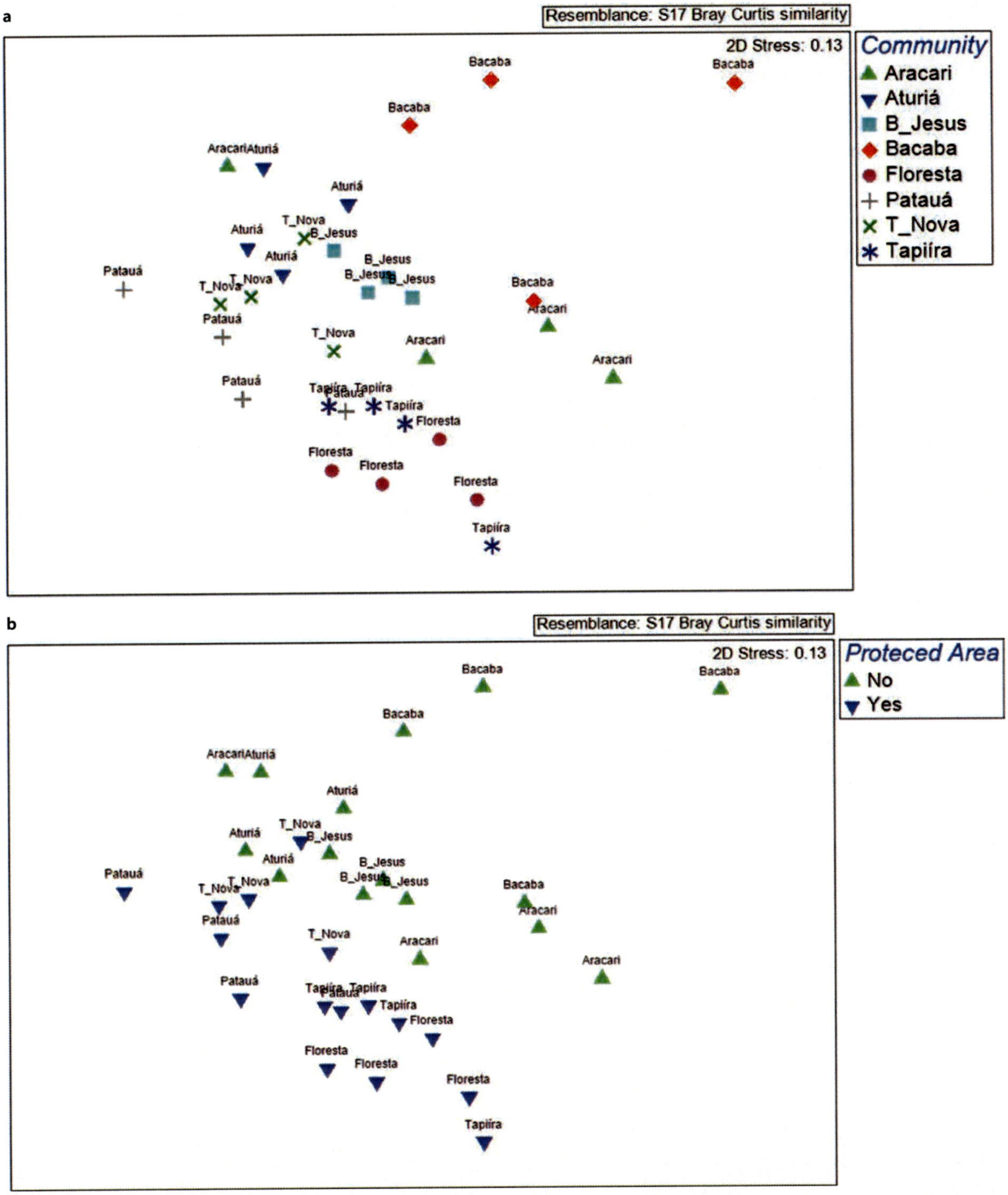

Fig. 3.8 NMDS ordination (stress = 0.13) plot based on the composition (% of biomass) of 46 fish species (or groups of species) caught (Table 3.3) in eight riverine communities during four hydrological seasons (low-water, flooding, high-water and receding), in the Negro river (points of plot): **a** plot organized according to studied communities; **b** plot organized according to localization of fishing communities, inside and outside Extractive Reserve (Protected Area)

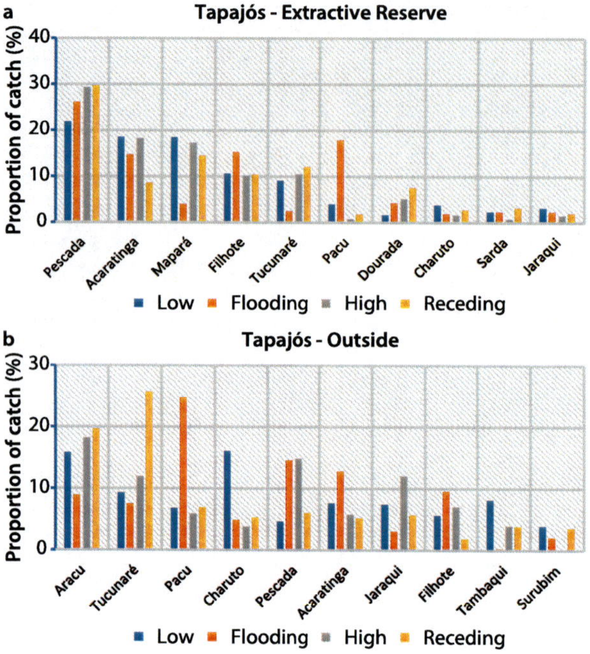

Fig. 3.9 Proportion of biomass caught (%) of the ten most caught fish species landed in the four hydrological seasons (low, flooding, high and receding) in Tapajós river: **a** inside Extractive Reserve (N = 16,012.8 kg of fish); **b** outside Extractive Reserve (N = 17,033.8 kg of fish)

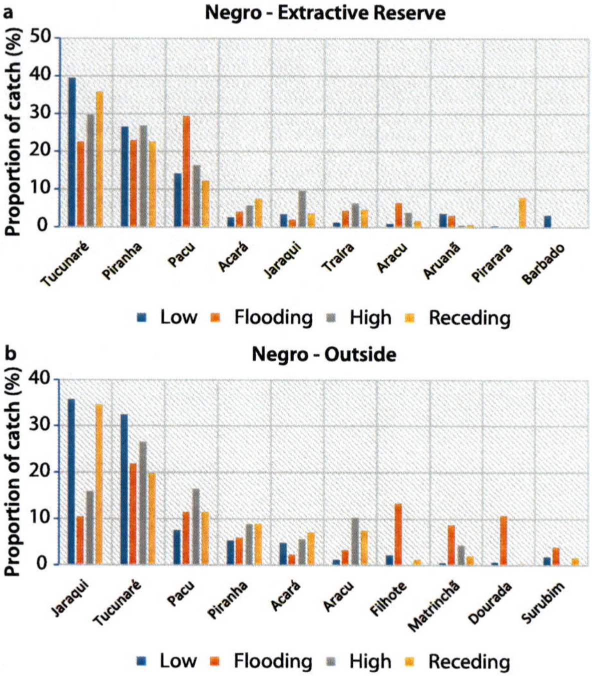

Fig. 3.10 Proportion of biomass caught (%) of the ten most caught fish species landed in the four hydrological seasons (low, flooding, high and receding) in Negro river: **a** inside Extractive Reserve (N = 4498.1 kg of fish); **b** outside Extractive Reserve (N = 6824.7 kg of fish)

squamosissimus) representing more than 20% of total caught in all seasons, followed by acaratinga (*Geophagus* spp.), mapará (*Hypophthalmus marginatus*), filhote (*Brachyplatystoma filamentosum*), tucunaré (*Cichla* spp.), and pacu (several species of genus *Myloplus*, *Metynnis*, and *Myleus*) (Fig. 3.9a). Pescada and acaratinga, the main fish species caught, together represented approximately 40% of all biomass caught in all seasons. Despite their relatively low commercial value, these species are essential for food security of riverine communities in the Tapajós river (Hallwass et al. 2017). Pescada is the most caught fish species in the fisheries of Lower Tocantins river, another clear water Amazonian river (Hallwass et al. 2011), and in the Amazon region as a whole (Hallwass and Silvano 2016), playing an important role in providing food and income for the riverine populations. The flooding season influenced the catch of other most caught fish species; while flooding reduced catches of mapará and tucunaré, it also increased catches of filhote and pacu (Fig. 3.9a). During the rise of the water level (flooding season), some fish species disperse in flooded areas, and others have their trigger for migration activated, such as filhote and pacu, making these fish species more susceptible to catch as they move along the river. These fish species migrate in shoals for breeding, but with different patterns, the filhote migrates along the river channels (longitudinal migration), and the pacu migrates from river channel to floodplain areas (lateral migration) (Welcomme 1985; Barthem and Goulding 1997; Fernandes 1997).

Outside the ER in the Tapajós river fishers recorded 17,033.8 kg of fish landed. The main fish species caught were the aracu (several species of family Anastomidae), tucunaré, pacu, charuto (*Hemiodus* spp.), pescada, and acaratinga (Fig. 3.9b). The relative proportions of most caught fish species were more homogeneous (less selective) outside ER in Tapajós river. The main fish species caught there have higher commercial importance (value). The shoals of aracu are caught with castnets, mainly in the communities of *Alter do Chão* and *Ponta de Pedras*, located near Santarém, which facilitates the commercialization of the fish. On the other hand, the tucunaré is one of the most valuable fish species in the region (Hallwass et al. 2017) and is caught mainly in receding water season. Since the tucunaré has high commercial value and sedentary behavior, this species should be locally managed to ensure the maintenance of fish stocks, income, and food security for riverine populations. The pacu is among the preferred fish for food among riverine communities, and it is mainly caught in the flooding season. On the other hand, the charuto provides the demands of local bars and restaurants, which sell this fish as a snack on the beach. Therefore, the charuto is mainly caught in the dry season, when the beaches of the Tapajos river appear and there is a greater flow of tourists in the region. Besides, there is a festival for this fish (*Festival do Charuto*) in *Ponta de Pedras* community, outside the ER, showing the economic and cultural importance of this fish species. Thus, the charuto should be considered in future fishery management plans in the region or as an economic alternative to diversify fish catches.

On the other hand, inside the ER in Negro river, in total, 4498.1 kg of fish were recorded in fish landings. The most caught fish species were tucunaré, piranha (several species of Family Serrasalmidae), and pacu (◘ Fig. 3.10a). Similarly, as observed in the Tapajós river, pacu was mainly caught during flooding season, when the shoals of this species enter in the flooded area to feed. Tucunaré was the most caught fish species inside ER, ranging from 20% to 40% of total caught, according to the season. This species is highly valuable, but sales of fish to the outside of ER are forbidden. In this sense, ER residents request the opening of the sport fishing in the area to generate an extra source of income. The piranha was the second most caught fish species and is among the preferred food fish of inhabitants of ER, according to interviews with local fishers. The two most caught fish, with high proportional abundance, are predators, which may indicate relatively low levels of fishing pressure inside the ER, and the integrity of the fish community.

In the Negro river, outside ER were recorded 6825.7 kg of fish caught. The most caught fish species were the jaraqui (*Semaprochilodus* spp.) caught mainly in receding and low water seasons, tucunaré, pacu, and piranha (◘ Fig. 3.10b). The jaraqui has its seasonal catch possibly linked to the migratory movements of this species (Ribeiro and Petrere 1990). The jaraqui is widely caught in the Amazon region (Hallwass and Silvano 2016) and plays an important role in the fish market and in the regional economy. The second fish species most caught was the tucunaré, which has high commercial value in the Amazonian region (Hallwass et al. 2017). Therefore, tucunaré may be well managed in less productive rivers, such as Tapajós and Negro. Pacu was more caught than piranha outside the ER in the Negro River. Fishing communities outside ER did not exhibit a feeding preference for piranha as observed among the communities from inside the ER.

3.4 Concluding Remarks

Interviewed fishers have a fishing experience in the studied region, and knowledge on their fisheries, but they have a low level of formal education, especially inside the ER in Negro river, due to the lack of access and logistical difficulties to arrive at schools. Fishers have a high dependence on fish, for income and food, and they show a high frequency of fishing. Although *rabetas* are responsible for most of the fishery production, paddled canoes are most often used inside ERs in both rivers, Tapajós and Negro. Gillnets were the most important fishing gear (regarding the frequency of use and fish production) in Tapajós river (inside and outside ER) and outside ER in Negro river. Inside ER in Negro river, selective fishing gears were more used, such as handlines and harpoons. The river channel was the main fishing ground in the Tapajós river (inside and outside ER), while in Negro river (inside and outside ER), the fishing spots are more varied, including river channel, floodplain lakes, and flooded area.

Extractive reserves have an important role in maintaining the fish production and productivity through the reduction of fishing pressure, as fishers there spend less time fishing and less time traveling to fishing spots, and use more selective fishing gears. However, ERs did not influence the composition of fish catches, which was more influenced by ecological, local, and socioeconomic factors, such as the type of rivers and studied fishing communities. Fishing communities in the Negro river are more selective than communities from Tapajós river, and the fish stocks appear to be healthier inside the ER in Negro river. Tucunaré is a valuable fish for the income generation and food security of fishing communities and, because of this, receives a higher fishing pressure. The sedentary behavior and the natural abundance of tucunaré in clear and black waters indicate this species as a promising target for management programs in less productive rivers, such as Tapajós and Negro.

In order to conserve the biodiversity and fish stocks in ERs, it is important to ensure the generation of income for the residents aiming to conciliate the conservation of biodiversity and the livelihood of stakeholders. Incentives to sustainable tourism and added value on products from Protected Areas could be ways to improve the income of their inhabitants and to ensure the conservation of biodiversity.

3

References

Akagi H, Naganuma A (2000) Human exposure to mercury and the accumulation of methylmercury that is associated with gold mining in the Amazon Basin, Brazil. J Health Sci 46:323–328

Almeida OT, Lorenzen K, McGrath DG (2009) Fishing agreements in the lower Amazon: for gain and restraint. Fish Manag Ecol 16:61–67

Barthem R, Goulding M (1997) The catfish connection: ecology, migration, and conservation of Amazon predators. Columbia University Press, New York

Bayley PB, Petrere M (1989) Amazon fisheries: assessment methods, current status and management points. In: Dodge DP (ed) Proceedings of the international large river symposium. Canadian special publication of fisheries and aquatic sciences 106, pp. 385–398

Begossi A (1996) The fishers and buyers from Buzios Island (Brazil): kin ties and production. Ciência e Cultura (SBPC) 48(3):142–147

Begossi A (2008) Local knowledge and training towards management. Environ Dev Sustain 10:591–603

Begossi A (2014) Ecological, cultural, and economic approaches to managing artisanal fisheries. Environ Dev Sustain 16:5–34

Begossi A, Richerson PJ (1992) The animal diet of families from Buzios Island (Brasil): an optimal foraging approach. J Hum Ecol 3:433–458

Begossi A, Silvano RAM, Ramos RM (2005) Foraging behavior among fishers from the Negro and Piracicaba Rivers, Brazil: implications for management. In: Brebbia CA, Antunes do Carmo JS (eds) River management III. WIT Press, Southampton, pp 503–513

Begossi A, Salivonchyk SV, Hallwass G, Hanazaki N, Lopes PFM, Silvano RAM, Dumaresq D, Pittock J (2018) Fish consumption on the Amazon: a review of biodiversity, hydropower and food security issues. Braz J Biol 79:345–357

Béné C, Steel E, Luadia BK, Gordon A (2009) Fish as the "bank in the water" – evidence from chronic-poor communities in Congo. Food Policy 34:108–118

Bernard E, Penna LAO, Araújo E (2014) Downgrading, downsizing, degazettement, and reclassification of protected areas in Brazil. Conserv Biol 28:939–950

Borja A (2014) Grand challenges in marine ecosystem ecology. Front Mar Sci 1:1–6

Castello L, Viana JP, Watkins G, Pinedo-Vasquez M, Luzadis VA (2009) Lessons from integrating fishers of arapaima in small-scale fisheries management at the Mamirauá Reserve, Amazon. Environ Manag 43:197–209

Campos-Silva JV, Peres CA (2016) Community-based management induces rapid recovery of a high-value tropical freshwater fishery. Sci Rep 6:34745

Cardoso RS, Freitas CEC (2008) A pesca de pequena escala no rio Madeira pelos desembarques ocorridos em Manicoré (Estado do Amazonas), Brasil. Acta Amazon 38:781–788

Castro F, McGrath DG (2003) Moving toward sustainability in the local management of floodplain lake fisheries in the Brazilian Amazon. Hum Organ 62:123–133

Corrêa MAA, Kahn JR, Freitas CEC (2014) Perverse incentives in fishery management: the case of the defeso in the Brazilian Amazon. Ecol Econ 106:186–194

Costanza R, d'Arge R, de Groot R, Farber S, Grasso M, Hannon B, Limburg K, Naeem S, O'Neil R, Paruelo J, Raskin RG, Sutton P, van den Belt M (1997) The value of the world's ecosystem services and natural capital. Nature 387:253–260

Cynamon D, Lebre E, Costa MA (2001) Management challenges on small-scale gold mining activities in Brazil. Environ Res 87: 181–198

Danielsen F, Burgess ND, Jensen PM, Pirhofer-Walzl K (2010) Environmental monitoring: the scale and speed of implementation varies according to the degree of people's involvement. J Appl Ecol 47:1166–1168

De Graaf G, Bartley D, Jorgensen J, Marmulla G (2015) The scale of inland fisheries, can we do better? Alternative approaches for assessment. Fish Manag Ecol 22:64–70

Diamond J (2001) Unwritten knowledge. Nature 410:521

FAO (2014) The state of world fisheries and aquiculture. FAO, Rome

Fearnside PM (2001) Environmental impacts of Brazil's Tucuruí dam: unlearned lessons for hydroelectric development in Amazonia. Environ Manag 27:377–396

Fearnside PM (2006) Desmatamento na Amazônia: dinâmica, impactos e controle. Acta Amazon 36:395–400

Fearnside PM (2014) Impacts of Brazil's Madeira River dams: unlearned lessons for hydroelectric development in Amazonia. Environ Sci Pol 38:164–172

Fearnside PM (2015) Amazon dams and waterways: Brazil's Tapajós Basin plans. Ambio 44:426–439

Fernandes CC (1997) Lateral migration of fishes in Amazon floodplains. Ecol Freshw Fish 6:36–44

Fernandes VLA, Vicentini RN, Batista VS (2009) Caracterização do uso de malhadeiras pela frota pesqueira que desembarca em Manaus e Manacapuru, Amazonas. Acta Amazon 39:405–414

Ferreira J, Aragão LEOC, Barlow J, Barreto P, Berenguer E, Bustamante M, Gardner TA, Lees AC, Lima A, Louzada J, Parry L, Peres CA, Pardini R, Pompeu PS, Tabarelli M, Zuanon J (2014) Brazil's environmental leadership at risk, mining and dams threaten protected areas. Science 346:706–707

Fluet-Chouinard E, Funge-Smith S, McIntyre PB (2018) Global hidden harvest of freshwater fish revealed by household surveys. Proc Natl Acad Sci 115:7623–7628

Fulton EA, Smith ADM, Smith DC, van Putten IE (2011) Human behaviour: the key source of uncertainty in fisheries management. Fish 12:2–17

Garcia A, Tello S, Vargas G, Duponchelle F (2009) Patterns of commercial fish landings in the Loreto region (Peruvian Amazon) between 1984 and 2006. Fish Physiol Biochem 35:53–67

Gell FR, Roberts CM (2003) Benefits beyond boundaries: the fishery effects of marine reserves. Trends Ecol Evol 18(9):448–455

Gonçalves C, Batista VS (2008) Avaliação do desembarque pesqueiro efetuado em Manacapuru, Amazonas, Brasil. Acta Amazon 38:135–144

Hallwass G, Silvano RAM (2016) Patterns of selectiveness in the Amazonian freshwater fisheries: implications for management. J Environ Plan Manag 59:1537–1559

Hallwass G, Lopes PF, Juras AA, Silvano RAM (2011) Fishing effort and catch composition of urban market and rural villages in Brazilian Amazon. Environ Manag 47:188–200

Hallwass G, Lopes PF, Juras AA, Silvano RAM (2013a) Behavioral and environmental influences on fishing rewards and the outcomes of alternative management scenarios for large tropical rivers. J Environ Manag 128:274–282

Hallwass G, Lopes PF, Juras AA, Silvano RAM (2013b) Fishers' knowledge identifies environmental changes and fish abundance trends in impounded tropical rivers. Ecol Appl 23: 392–407

Hallwass G, Nunes MUS, Lopes PFM (2017) Pessoas e peixes no Rio Tapajós: pesca e impactos ambientais. In: Silvano RAM (ed) Conservação, Pesca e Ecologia de Peixes do Baixo Rio Tapajós, Amazônia Brasileira. Editora Rima, São Carlos-SP, Brasil, pp. 23–44

Huntington H, Begossi A, Gearheard SF, Kersey B, Loring PA, Mustonen T, Paudel PK, Silvano RAM, Vave R (2017) How small communities respond to environmental change: patterns from tropical to polar ecosystems. Ecol Soc 22:9

Isaac V, Almeida MC, Giarizzo T, Deus CP, Vale R, Klein G, Begossi A (2015a) Food consumption as an indicator of the conservation of natural resources in riverine communities of the Brazilian Amazon. An Acad Bras Cienc 87:2229–2242
Isaac VJ, Almeida MC, Cruz REA, Nunes LG (2015b) Artisanal fisheries of the Xingu River basin in Brazilian Amazon. Braz J Biol 75(3):125–137
Isaac VJ, Silva CO, Ruffino ML (2008) The artisanal fishery fleet of the lower Amazon. Fish Manag Ecol 15:179–187
Jackson JBC, Kirby MX, Berger WH, Bjorndal KA, Botsford LW, Bourque BJ, Bradbury RH, Cooke R, Erlandson J, Estes JA, Hughes TP, Kidwell S, Lange CB, Lenihan HS, Pandolfi JM, Peterson CH, Steneck RS, Tegner MJ, Warner RR (2001) Historical Overfishing and the Recent Collapse of Coastal Ecosystems. Science 293:629–638
Johnson DS (2006) Category, narrative, and value of governance for small-scale fisheries. Mar Policy 30:747–756
Junk WJ, Soares MGM, Bayley PB (2007) Freshwater fishes of the Amazon River basin: their biodiversity, fisheries, and habitats. Aquat Ecosyst Health Manage 10:153–173
Junk WJ, Piedade MTF, Schöngart J, Cohn-Haft M, Adeney JM, Wittmann F (2011) A Classification of Major Naturally-Occurring Amazonian Lowland Wetlands. Wetlands 31:623–640
Keppeler FW, Hallwass G, Silvano RAM (2017) Influence of protected areas on fish assemblages and fisheries in a large tropical river. Oryx 51:268–279
Keppeler FW, Souza AC, Hallwass G, Begossi A, Almeida MC, Isaac VJ, Silvano RAM (2018) Ecological influences of human population size and distance to urban centres on fish communities in tropical lakes. Aquat Conserv Mar Freshwat Ecosyst 28:1030–1043
Lynch AJ, Cowx IG, Fluet-Chouinard E, Glaser SM, Phang SC, Beard TD, Bower SD, Brooks JL, Bunnell DB, Claussen JE, Cooke SJ, Kao YC, Lorenzen K, Myers BJE, Reid AJ, Taylor JJ, Youn S (2017) Inland fisheries – invisible but integral to the UN sustainable development agenda for ending poverty by 2030. Glob Environ Chang 47:167–173
MacCord PFL, Silvano RAM, Ramires MS, Clauzet M, Begossi A (2007) Dynamics of artisanal fisheries in two Brazilian Amazonian reserves: implications to co-management. Hydrobiologia 583:365–376
McClanahan TR, Mangi S (2000) Spillover of exploitable fishes from a marine park and its effect on the adjacent fishery. Ecol Appl 10(6):1792–1805
Nepstad D, McGrath D, Alencar A, Barros AC, Carvalho G, Santilli M, Diaz MCV (2002) Frontier Governance in Amazonia. Science 295:629–631
Nevado JJB, Martín-Doimeadios RCR, Bernardo FJG, Moreno MJ, Herculano AM, Nascimento JLM, Crespo-López ME (2010) Mercury in the Tapajós River basin, Brazilian Amazon: a review. Environ Int 36:593–608
Oliveira LEC, Begossi A (2011) Last trip return rate influence patch choice decisions of small-scale shrimp trawlers: optimal foraging in São Francisco, Coastal Brazil. Hum Ecol 39:323–332
Pauly D (2006) Major trends in small-scale marine fisheries, with emphasis on developing countries, and some implications for the social sciences. Maritime Stud 4:7–22
Petrere MJr, Barthem RB, Córdoba EA, Gómez BC (2004) Review of the large catfish fisheries in the upper Amazon and the stock depletion of piraíba (*Brachyplatystoma filamentosum* Lichtenstein). Rev Fish Biol Fish 14:403–414
Pinnegar JK, Engelhard GH (2007) The 'shifting baseline' phenomenon: a global perspective. Rev Fish Biol Fish 18:1–16
Prestes-Carneiro G, Béarez P, Bailon S, PyDaniel AR, Neves EG (2015) Subsistence fishery at Hatahara (750-1230 CE), a pre-Columbian central Amazonian village. J Archaeol Sci Rep: Reports 8:454–462
Ribeiro MCLB, Petrere M Jr (1990) Fisheries ecology and management of the Jaraqui (*Semaprochilodus taeniurus, S. insignis*) in Central Amazonia. Regul Rivers Res Manag 5:195–215
Roulet M, Lucotte M, Canuel R, Rheualt I, Tran S, Goch YGD (1998) Distribution and partition of total mercury in waters of the Tapajós River Basin, Brazilian Amazon. Sci Total Environ 213:203–211
Roulet M, Lucotte M, Fafarella N, Serique G, Coelho H, Passos CJD (1999) Effects of recent human colonization on the presence of mercury in Amazonian ecosystems. Water Air Soil Pollut 112:297–313
Russ GR, Alcala AC, Maypa AP, Calumpong HP, White AT (2004) Marine reserve benefits local fisheries. Ecol Appl 14(2): 597–606
Salas S, Gaertner D (2004) The behavioral dynamics of fishers: management implications. Fish Fish 5:53–167
Silvano RAM, Ramires M, Zuanon J (2009) Effects of fisheries management on fish communities in the floodplain lakes of a Brazilian Amazonian reserve. Ecol Freshw Fish 18:156–166
Silvano RAM, Hallwass G, Lopes PF, Ribeiro AR, Lima RP, Hasenack H, Juras AA, Begossi A (2014) Co-management and spatial features contribute to secure fish abundance and fishing yields in tropical floodplain lakes. Ecosystems 17:271–285
Silvano RAM, Hallwass G, Juras AA, Lopes PF (2017) Assessment of efficiency and impacts of gillnets on fish conservation in a tropical freshwater fishery. Aquat Conserv Mar Freshwat Ecosyst 27:521–533
Sioli H (1984) The Amazon: limnology and landscape ecology of a mighty tropical river and its basin. Dr. W. Junk Publishers, Dordrecht
Smith NJH (1985) The Impact of Cultural and Ecological Change on Amazonian Fisheries. Biol Conserv 32:355–373
Sobreiro T, Freitas CEC, Prado KL, Nascimento FA, Vicentini R, Moraes AM (2010) An evaluation of fishery co-management experience in an Amazonian black-water river (Unini River, Amazon, Brazil). Environ Dev Sustain 12:1013–1024
Torres M (2005) Amazônia revelada: Os descaminhos ao longo da BR-163. Conselho Nacional de Desenvolvimento Científico e Tecnológico (CNPq), Brasília, Brasil
Tregidgo DJ, Barlow J, Pompeu PS, de Almeida RM, Parry L (2017) Rainforest metropolis casts 1,000-km defaunation shadow. Proc Natl Acad Sci 114:8655–8659
Welcomme R (1985) River fisheries. FAO, Rome
Welcomme RL (1999) A review of a model for qualitative evaluation of exploitation levels in multi-species fisheries. Fish Manag Ecol 6:1–19
Welcomme RL, Cowx IG, Coates D, Béné C, Funge-Smith S, Halls A, Lorenzen K (2010) Inland capture fisheries. Philos Trans R Soc B 365:2881–2896
Winemiller KO, McIntyre PB, Castello L, Fluet-Chouinard E, Giarrizzo T, Nam S, Baird IG, Darwall W, Lujan NK, Harrison I, Stiassny MLJ, Silvano RAM, Fitzgerald DB, Pelicice FM, Agostinho AA, Gomes LC, Albert JS, Baran E, Petrere M Jr, Zarfl C, Mulligan M, Sullivan JP, Arantes CC, Sousa LM, Koning AA, Hoeinghaus DJ, Sabaj M, Lundberg JG, Armbruster J, Thieme ML, Petry P, Zuanon J, Torrente-Vilara G, Snoeks J, Ou C, Rainboth W, Pavanelli CS, Akama A, van Soesbergen A, Sáenz L (2016) Balancing hydropower and biodiversity in the Amazon, Congo and Mekong. Science 351:128–129

Atlas of Fish of Tapajós and Negro Rivers I: Characiformes

Renato A. M. Silvano, Pedro P. Nitschke, Kaluan C. Vieira, Paula Nagl, Astrid T. R. Martínez, Junior A. Chuctaya, Márcia C. F. Dutra, Friedrich W. Keppeler, Anaís R. P. Rowedder, and Marcelo C. Andrade

Contents

Renato A. M. Silvano (ed.), *Fish and Fisheries in the Brazilian Amazon*, https://doi.org/10.1007/978-3-030-49146-8_4

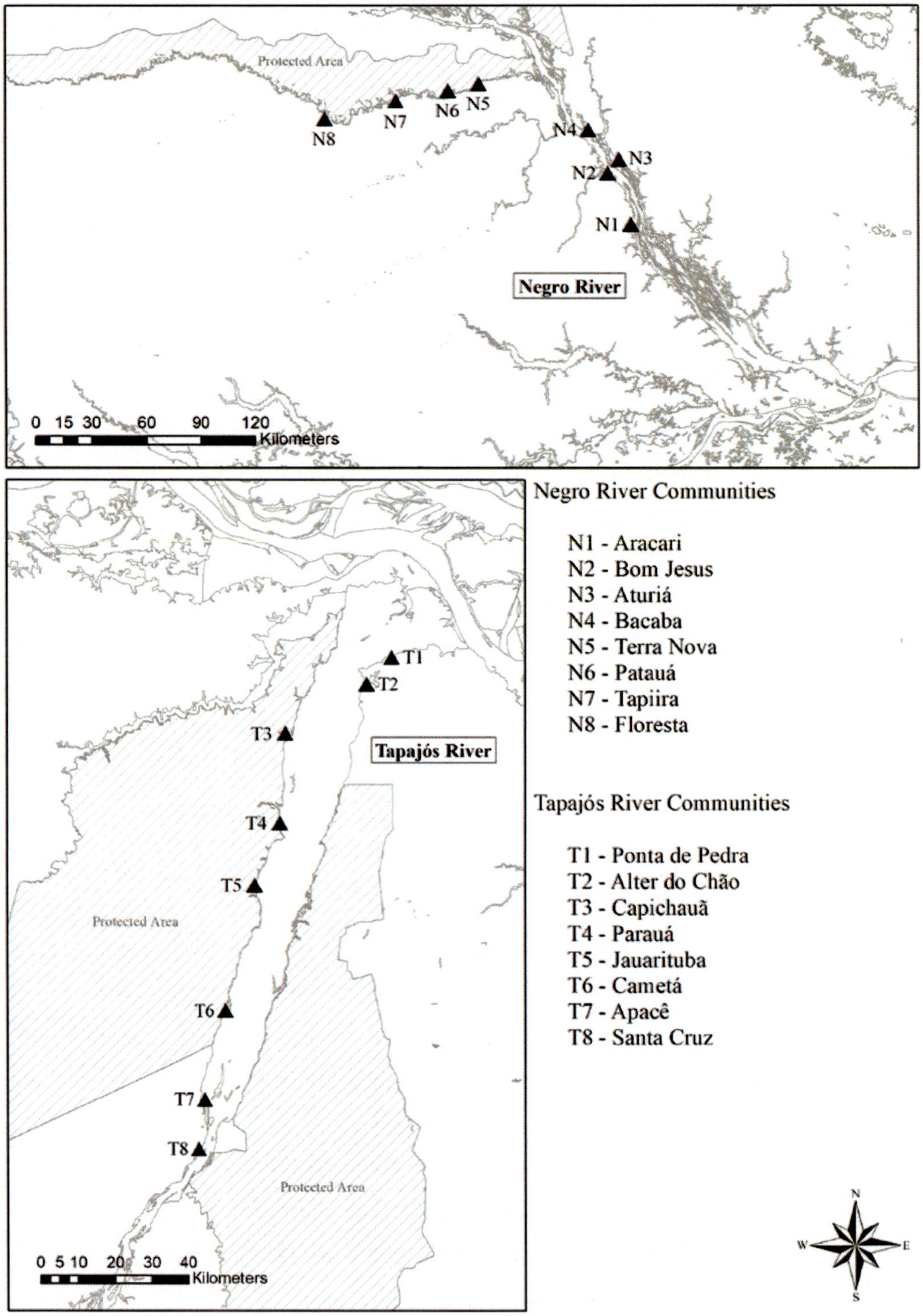

Fig. 4.1 Map of the studied region, showing the 16 studied communities with codes for community localization in maps showing the occurrence of fish species

4.1 Family Acestrorhynchidae

4.1.1 Summary of Biology and Ecology (Literature)

Diversity: family composed of seven genera, including *Acestrorhynchus* with 14 species.

Habitat: fishes from this family can be found in a variety of different habitats, but have a preference for lentic environments, such as lakes, igarapés (small Amazonian streams), and margins of large rivers with oxygenated water. Their spatial distribution is restricted to South America, where peaks of diversity can be found in the Amazon and Orinoco basin and drainages of the São Francisco, Paraná, Paraguay, and La Plata rivers.

Ecology and feeding: all species in this family are piscivorous and avid predators that actively pursue their prey (e.g., mostly small characids) in the water column. They usually forage alone or in small groups of two to three individuals.

Morphology: species of this family vary from small to medium body sizes. The larger species can reach to a standard length of up to 400 mm. These fishes possess a fusiform body shape with small and numerous scales, long snout, conical teeth, and dorsal fin located posteriorly to the middle of the body.

Reproduction: there are pieces of evidence suggesting that these fish have multiple spawning per year.

Economic importance: species of this family are usually non-important to artisanal and commercial fisheries. Small species may have economic importance for the aquarium trade.

References: Mago-Leccia (1970), Britski et al. (1986), Menezes (1992), Saccol-Pereira et al. (2006), Zuanon et al. (2015), Toledo-Piza (2017), Fricke et al. (2020).

Fish species: *Acestrorhynchus falcirostris* (Cuvier 1819)

- **Common name(s):** peixe-cachorro, cangoia

Fig. 4.2 *Acestrorhynchus falcirostris* sampled in the Negro River. (Photo: Renato Silvano)

Fish species: *Acestrorhynchus microlepis* (Jardine 1841)

Common name(s): peixe-cachorro, cangoia

Fig. 4.3 *Acestrorhynchus microlepis* sampled in the Tapajós River. (Photo: Renato Silvano)

Observation: abundance and size data were grouped for these two species, due to the difficulty in distinguishing them during sampling: both species were collected in both rivers; see Table 1.1 (► Chap. 1).

Size of fish sampled (standard length, both rivers) for *Acestrorhynchus* spp.: min. = 5.6 cm, max. = 44 cm, mean = 16 cm, median = 15.5 cm

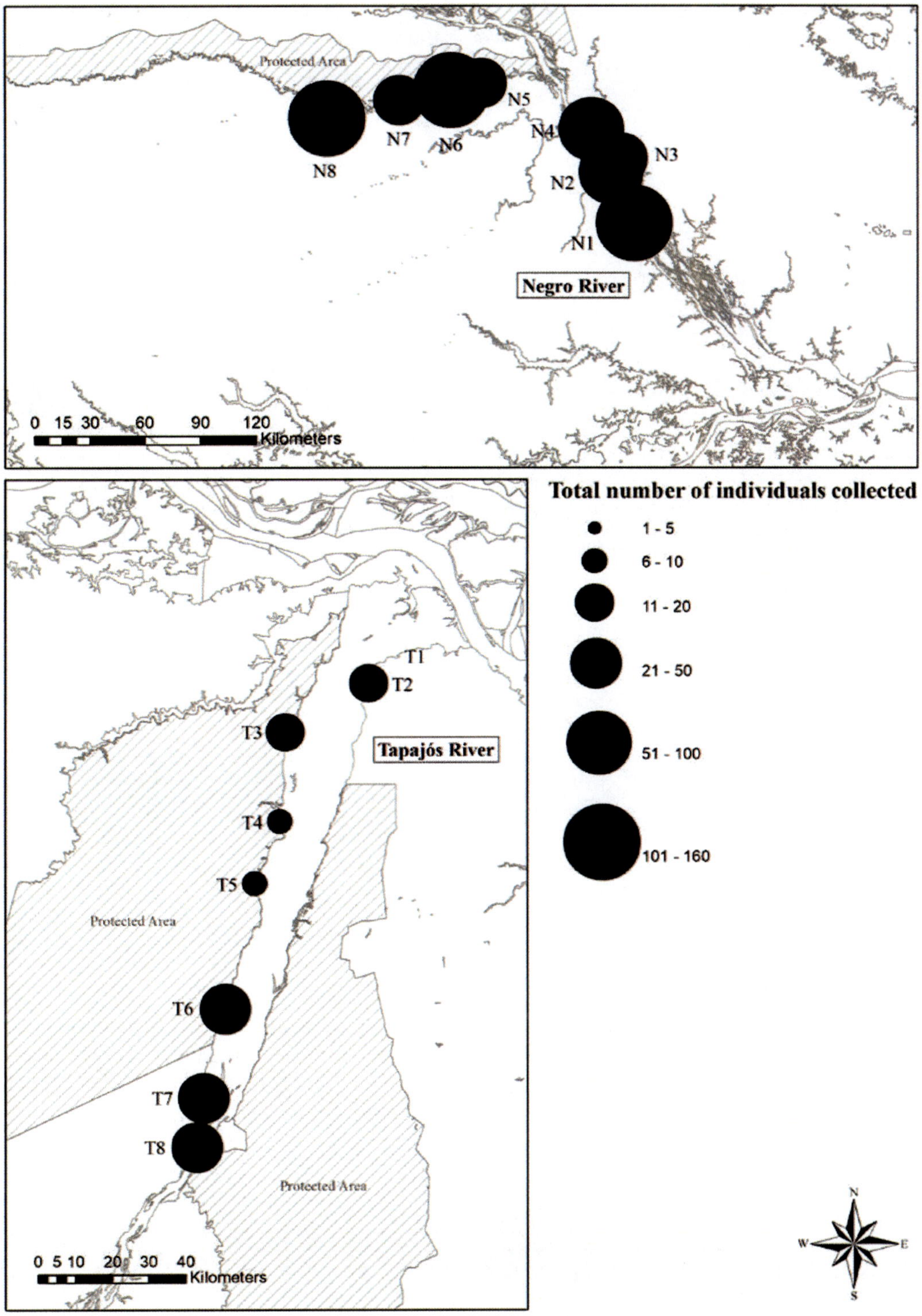

Fig. 4.4 Distribution of individuals of *Acestrorhynchus* spp. sampled in the Negro and Tapajós rivers

Fisheries: peixe cachorro (*Acestrorhynchus* spp. and larger species of Cynodontidae) is not important to fisheries and has few or no commercial value, being only occasionally caught in the Tapajós and Negro rivers. Biomass caught is in ◘ Table 3.3 (▶ Chap. 3).

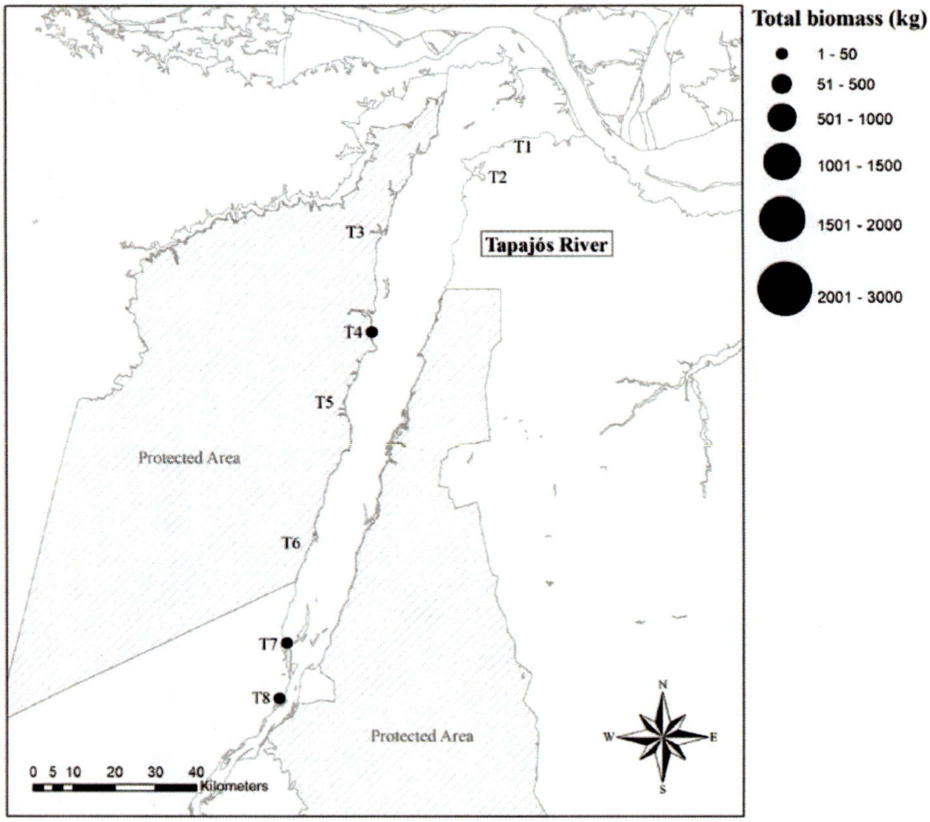

◘ **Fig. 4.5** Distribution of biomass of peixe cachorro (Acestrorhynchidae and Cynodontidae) caught in small-scale fisheries in the Tapajós River

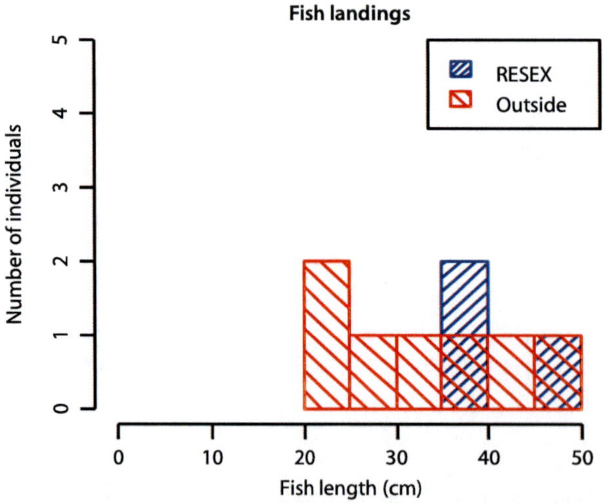

◘ **Fig. 4.6** Size distribution of peixe cachorro (Acestrorhynchidae and Cynodontidae) caught in small-scale fisheries in eight communities inside and outside the Extractive Reserves (RESEX) in the Tapajós River, N Resex = 3, N outside = 7

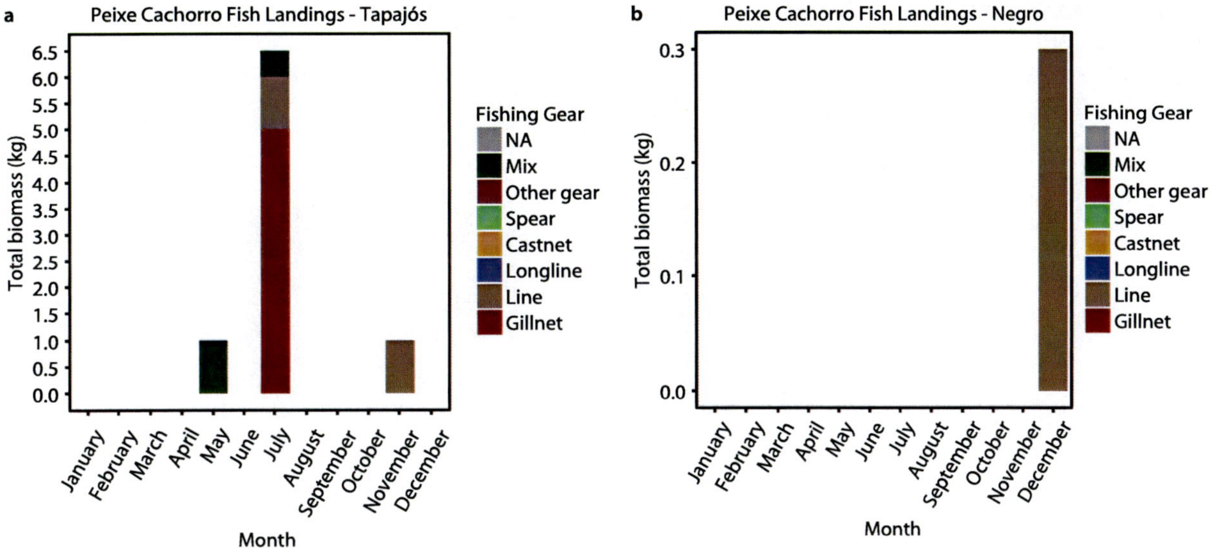

Fig. 4.7 Seasonal variation of biomass of peixe cachorro (Acestrorhynchidae and Cynodontidae) caught per fishing gear in small-scale fisheries in 16 communities in the Tapajós **a** and Negro **b** Rivers. NA: fishing gear not informed; Mix: more than one gear was used

4.2 Family Anostomidae

4.2.1 Summary of Biology and Ecology (Literature)

Diversity: family composed of 15 genera and 149 species with high morphological diversity. *Leporinus* (82 species) is the genus with the highest amount of species in the family and in the order Characiformes. Anostomids are restricted to the South and Central America and can be found in all Brazilian river basins. Most of the species in this family are endemic to the Amazon region.

Habitat: anostomids can be found in a variety of environments, including large rivers of white, black, and clear waters, lakes and streams.

Ecology and feeding: fishes of this family are herbivores and omnivores, feeding basically on fruits, seeds, roots, sponges, insects, and other aquatic invertebrates. Studies suggest that anostomids are opportunistic foragers. Most species form schools that undergo trophic and reproductive migrations. Given their trophic interactions with plants and fruits and their high mobility, anostomids may exert an ecological role in the seed dispersion of riparian plants.

Morphology: small- to medium-sized fishes with some species reaching a length of up to 40cm. Anostomids possess only three to four teeth on their superior and inferior jaws. The premaxilla (especially the ascending process) is enlarged when compared to the maxilla. Premaxilla teeth are also enlarged when compared to teeth on the maxilla. Body is elongated with a short anal fin composed of less than 11 ramified rays.

Reproduction: fishes from the genus *Leporinus* usually reproduce from November to February, but this may vary among river basins. During this reproductive period, fish migrate upstream to spawn. The first maturation of males occurs during their second year, while in females during the third year.

Economic importance: some larger species, such as *Leporinus* spp. and *Schizodon* spp., are important for commercial and artisanal fisheries. Although not among the most valuable for commercial purposes, fishes of this family may be important as food source for riverine people. Small and colorful species of this family may be important in the aquarium trade.

References: Knöeppel (1972), Winterbottom (1974), Godoy (1975), Saul (1975), Géry (1961, 1977), Santos (1980, 1981), Taphorn (1992), Nelson (1994), Santos and Jegu (1996), Santos and Rosa (1998), Garavello and Britski (2003), Balassa et al. (2004), Santos et al. (2006), Ribeiro et al. (2016), Hallwass et al. (2020), Fricke et al. (2020).

4

Fish species: *Anostomoides atrianalis* Pellegrin 1909

Common name(s): aracu-cata-barro, aracu

Size of fish sampled (standard length, both rivers): 12 cm

Fisheries: not relevant to fisheries, no commercial value.

Fig. 4.8 *Anostomoides atrianalis* sampled in the Negro River. (Photo: Renato Silvano)

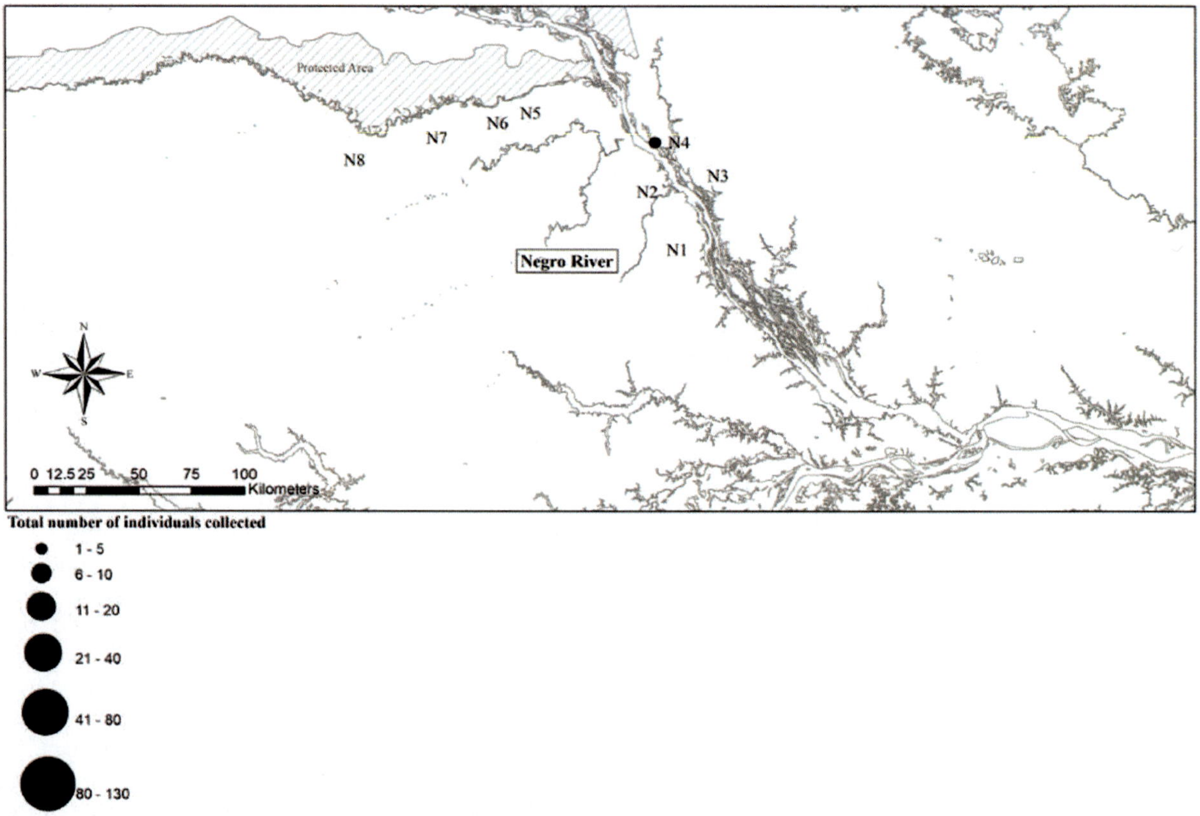

Fig. 4.9 Distribution of individuals of *Anostomoides atrianalis* sampled in the Negro River

Fish species: *Laemolyta proxima* (Garman 1890)

Common name(s): piau, aracu-piau, aracu, aracu-camunaru

Size of fish sampled (standard length, both rivers): mean and median = 15.9 cm

Fisheries: not very important, may be occasionally caught and consumed with aracus; see *Leporinus fasciatus* for aracu fisheries data.

Fig. 4.10 *Laemolyta proxima* sampled in the Negro River. (Photo: Renato Silvano)

4

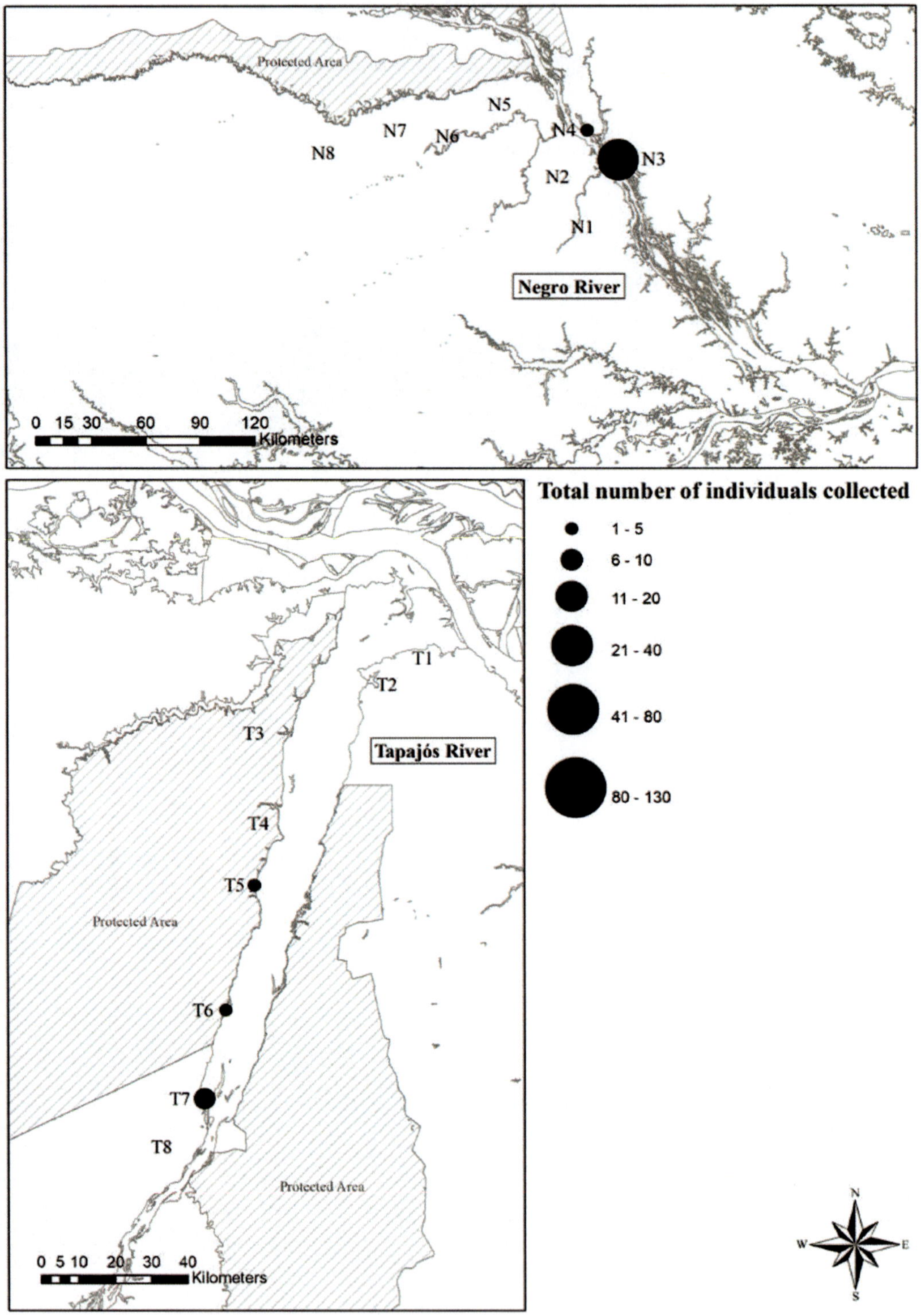

Fig. 4.11 Distribution of individuals of *Laemolyta proxima* sampled in the Negro and Tapajós rivers

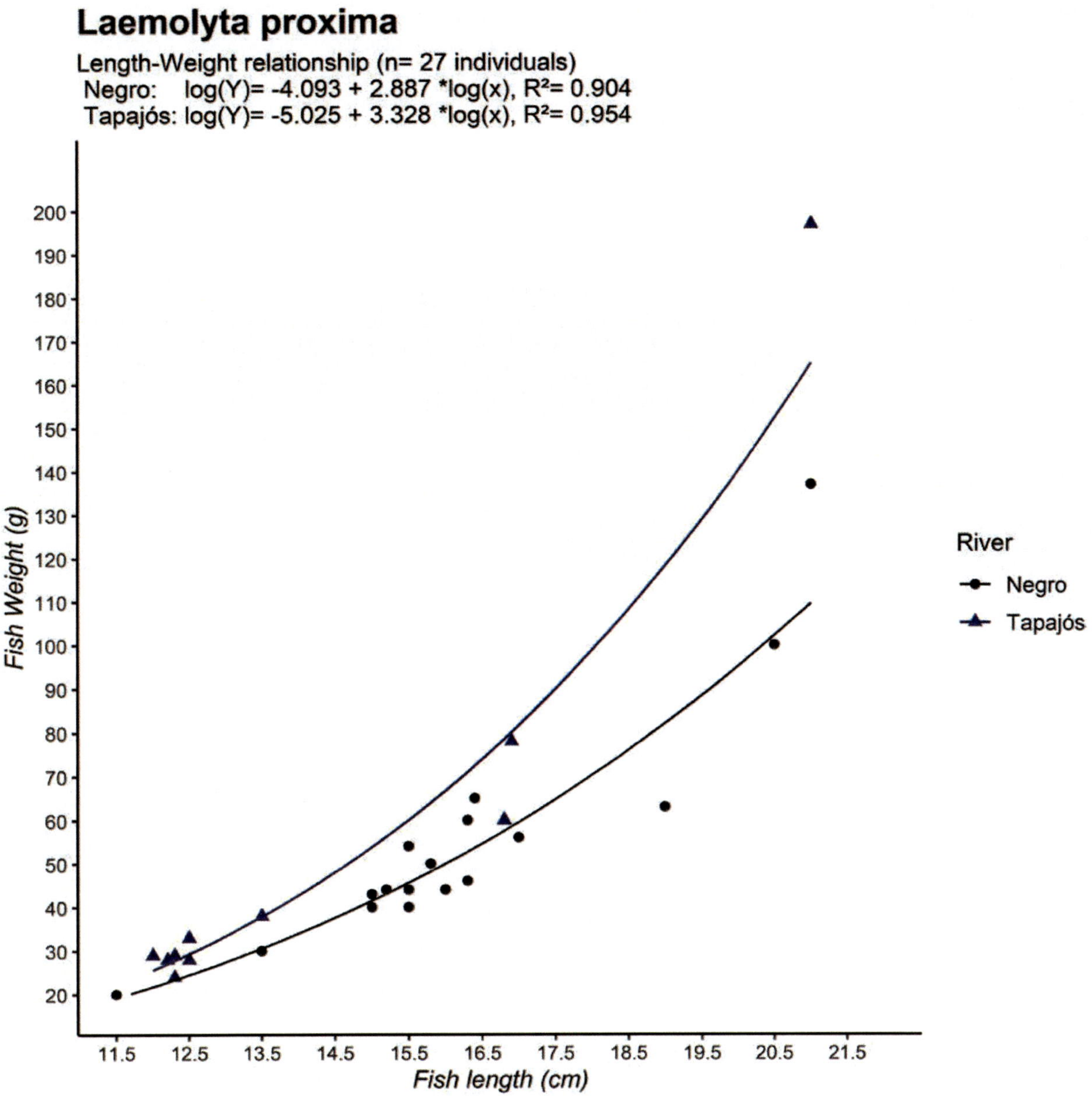

Fig. 4.12 Length-weight relationship for *Laemolyta proxima* sampled in Tapajós and Negro rivers

Fish species: *Laemolyta taeniata* (Kner 1858)

Common name(s): piau, aracu, aracu-biri, lápis

Size of fish sampled (standard length, both rivers): min. = 10 cm, max. = 26 cm, mean and median = 16.7 cm

Fisheries: not very important, may be occasionally caught and consumed with aracus; see *Leporinus fasciatus* for aracu fisheries data.

4

Fig. 4.13 *Laemolyta taeniata* sampled in the Negro River. (Photo: Renato Silvano)

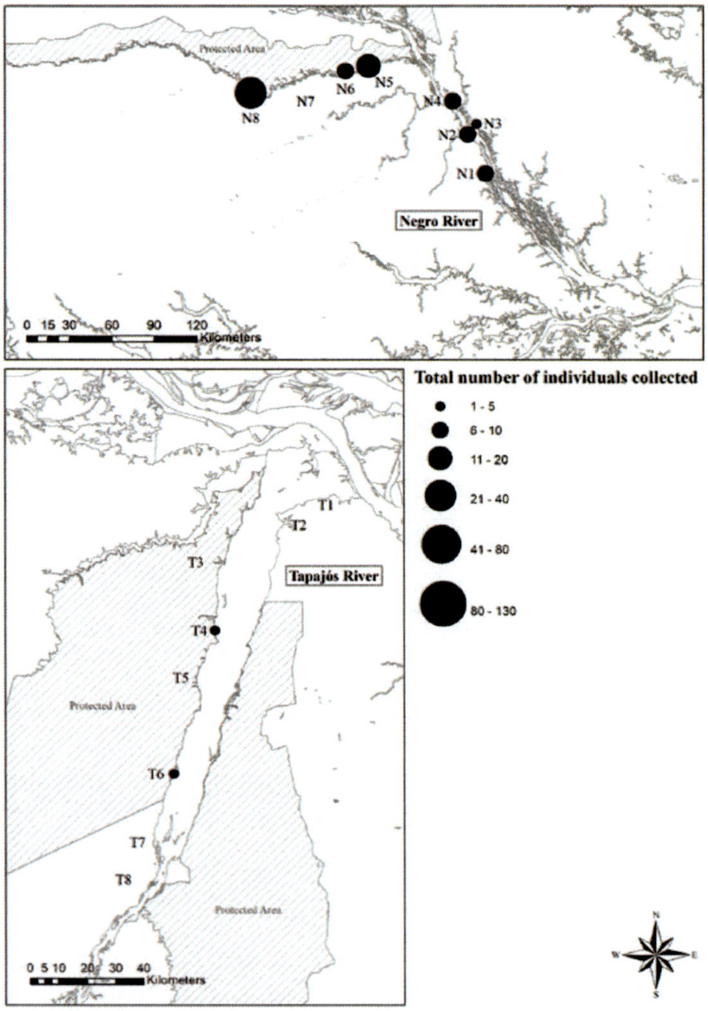

Fig. 4.14 Distribution of individuals of *Laemolyta taeniata* sampled in the Negro and Tapajós rivers

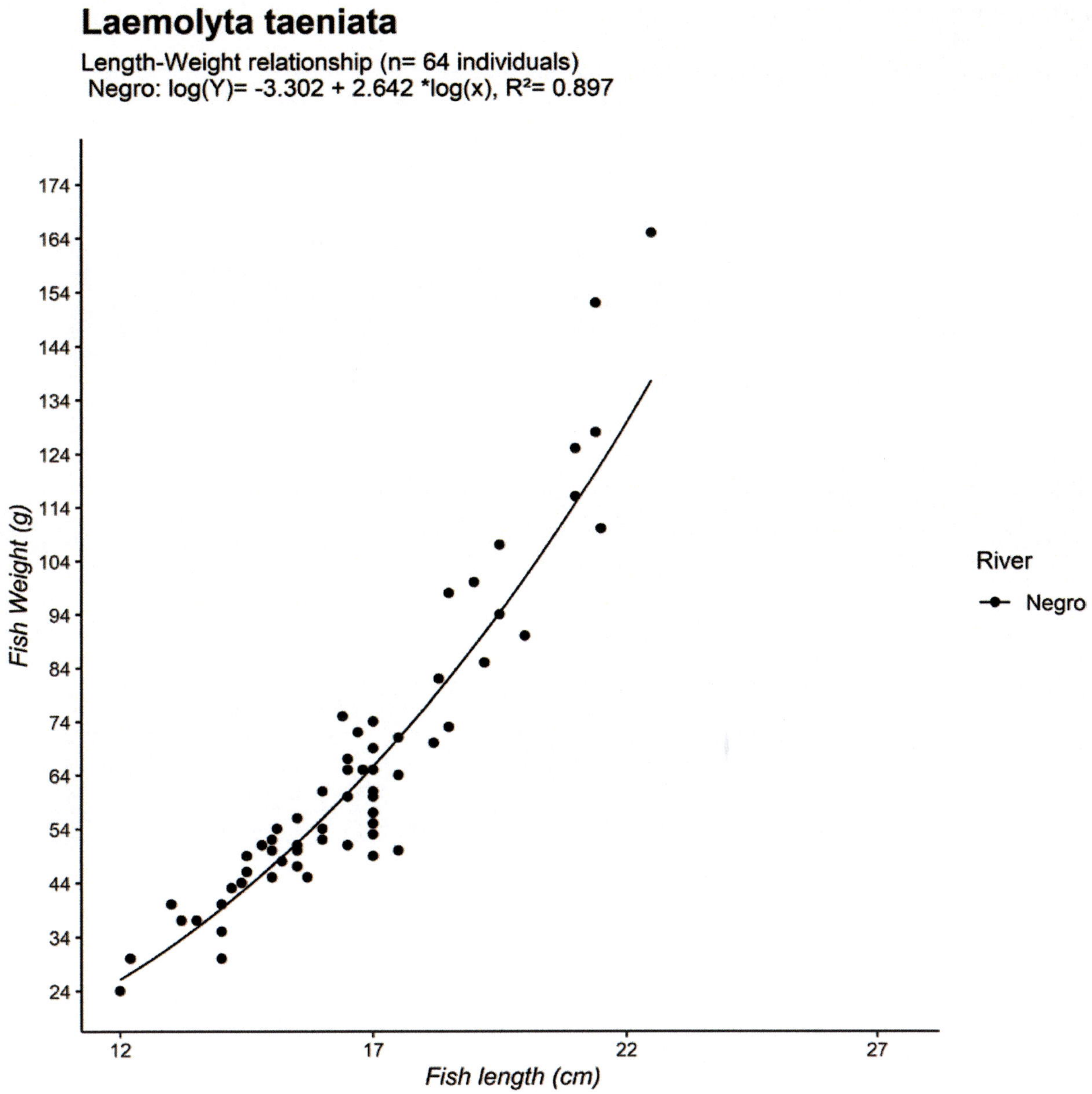

Fig. 4.15 Length-weight relationship for *Laemolyta taeniata* sampled in the Negro river

Fish species: *Leporinus fasciatus* (Bloch 1794)

Common name(s): aracu, aracu-pinima, aracu-flamengo, aracu-bicho

Fig. 4.16 *Leporinus fasciatus* sampled in the Negro River. (Photo: Renato Silvano)

Fish species: *Leporinus* cf. *fasciatus*

Common name(s): aracu, aracu-tesoura

Fig. 4.17 *Leporinus* cf. *fasciatus* sampled in the Negro River. (Photo: Renato Silvano)

Observation: abundance and size data were grouped for these two species, due to the difficulty in distinguishing them during sampling: both species were collected in both rivers; see Table 1.1 (▶ Chap. 1).

Size of fish sampled (standard length, both rivers): min. = 10.5 cm, max. = 27 cm, mean = 17.3 cm, median = 15.5 cm

Fisheries: aracus (*Leporinus* spp. and *Schizodon* spp.) are regularly exploited by fisheries in both rivers, mainly for consumption but also for sale. Biomass caught is in Table 3.3 (▶ Chap. 3).

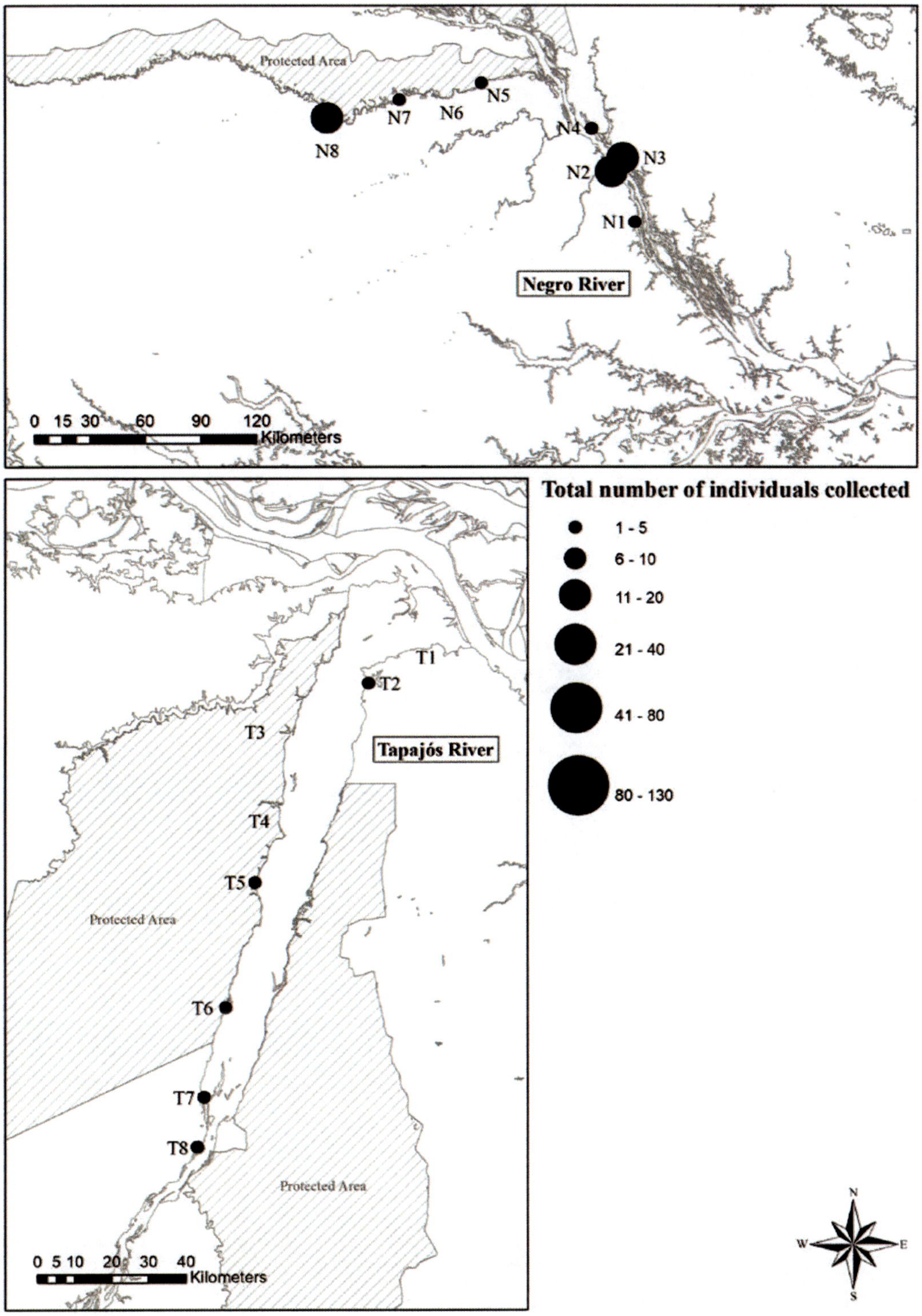

Fig. 4.18 Distribution of individuals of *Leporinus* spp. sampled in the Negro and Tapajós rivers

4

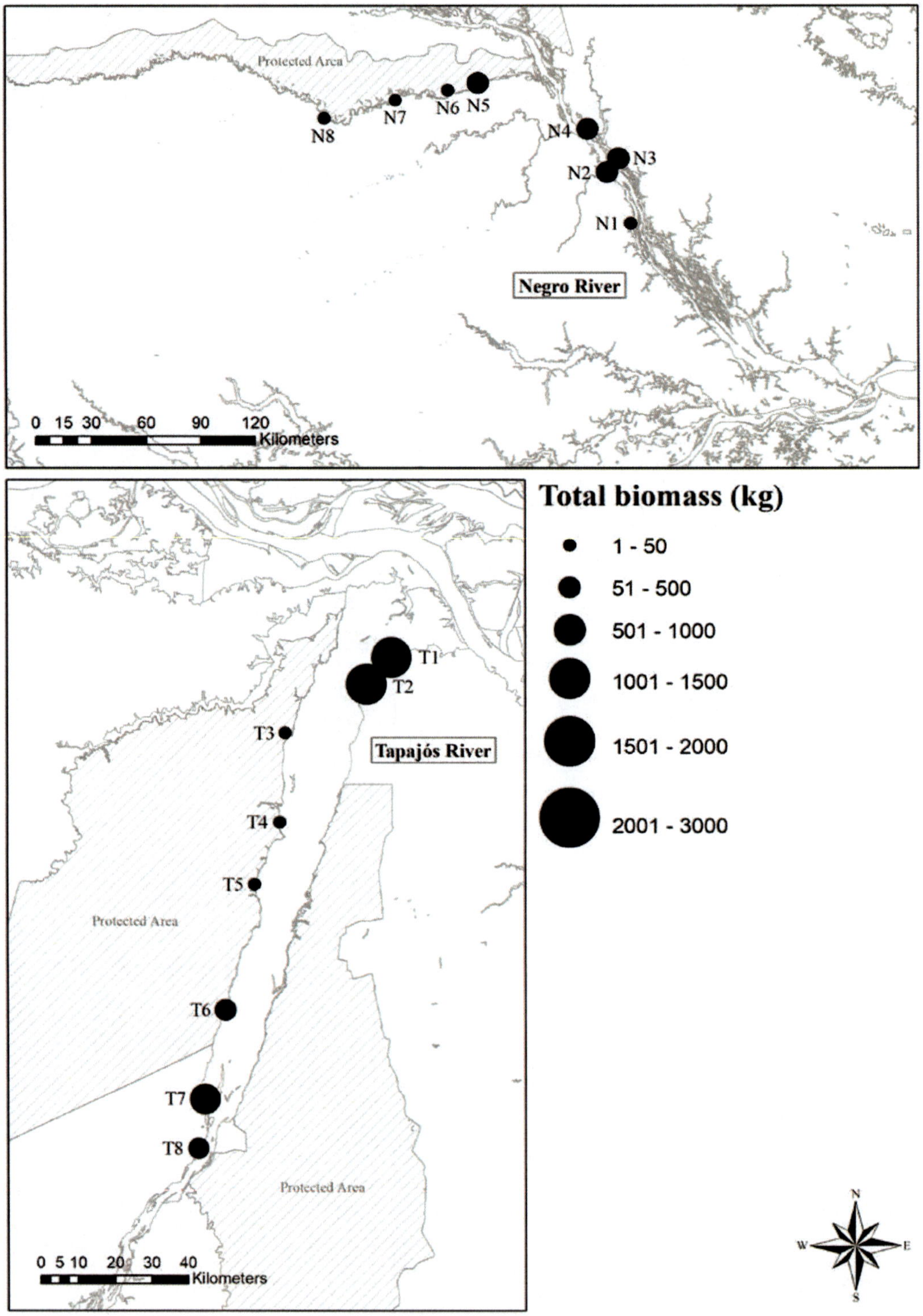

Fig. 4.19 Distribution of biomass of aracus (*Leporinus* spp. and *Schizodon* spp.) caught in small-scale fisheries in the Negro and Tapajós rivers

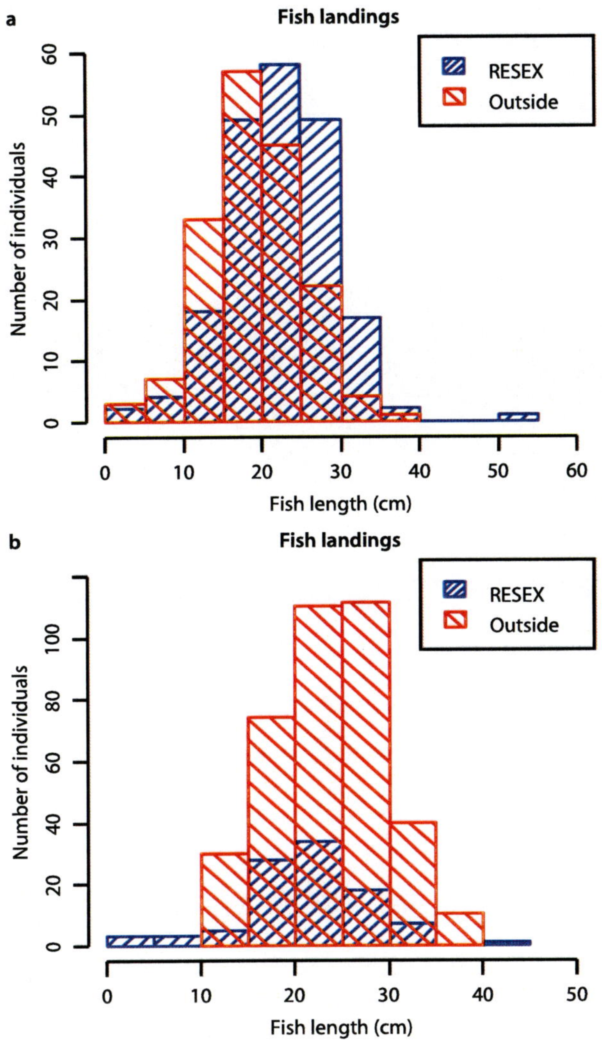

Fig. 4.20 Size distribution of aracus (*Leporinus* spp. and *Schizodon* spp.) caught in small-scale fisheries in 16 communities inside and outside the Extractive Reserves (RESEX) in the rivers **a** Negro, N Resex = 200, N outside = 172 and **b** Tapajós, N Resex = 99, N outside = 375

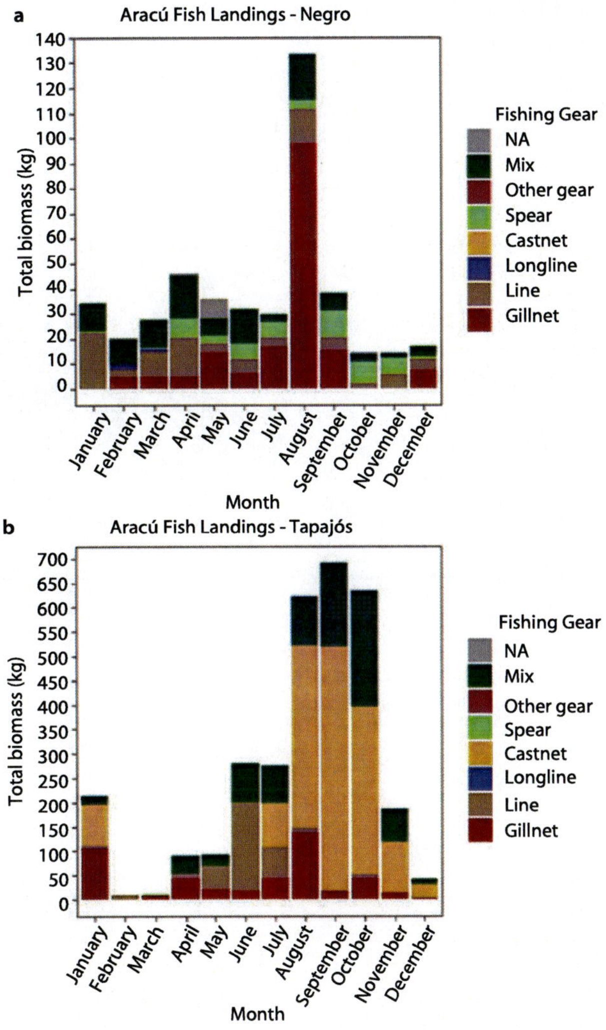

Fig. 4.21 Seasonal variation of biomass of aracus (*Leporinus* spp. and *Schizodon* spp.) caught per fishing gear in small-scale fisheries in16 communities in the Negro **a** and Tapajós **b** rivers. NA: fishing gear not informed; Mix: more than one gear was used

Fish species: *Rhytiodus argenteofuscus* **Kner 1858**

Common name(s): aracu, aracu-pau-de-vaqueiro

Size of fish sampled (standard length, both rivers): min. = 23 cm, max. = 23.2 cm, mean and median = 23.1 cm

Fisheries: this species may be occasionally caught and consumed with other aracus; see *Leporinus fasciatus* for aracu fisheries data.

Fig. 4.22 *Rhytiodus argenteofuscus* sampled in the Tapajós River. (Photo: Renato Silvano)

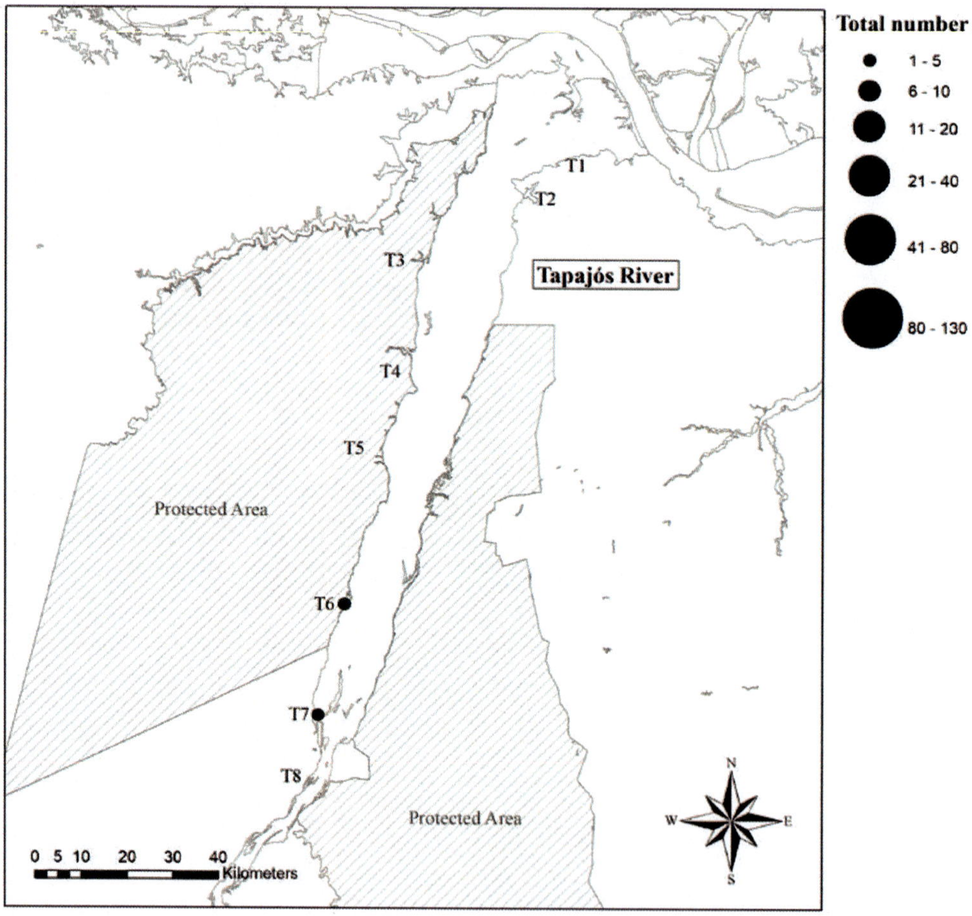

Fig. 4.23 Distribution of individuals of *Rhytiodus argenteofuscus* sampled in the Tapajós River

Fish species: *Schizodon vittatus* (Valenciennes 1850)

Common name(s): aracu, aracu-branco

Fig. 4.24 *Schizodon vittatus* sampled in the Tapajós River. (Photo: Renato Silvano)

Observation: abundance and size data were grouped for the two species (*Schizodon vittatus* and *S. fasciatus*), due to the difficulty in distinguishing them during sampling: both species were collected only in Tapajós River; see Table 1.1 (▶ Chap. 1).

Size of fish sampled (standard length, both rivers): min. = 11.5 cm, max. = 26.4 cm, mean = 17.3 cm, median = 18 cm

Fisheries: aracus (*Leporinus* spp. and *Schizodon* spp.) are regularly exploited by fisheries in both rivers, mainly for consumption but also for sale. Biomass caught is in Table 3.3 (▶ Chap. 3); see *Leporinus fasciatus* for aracu fisheries data.

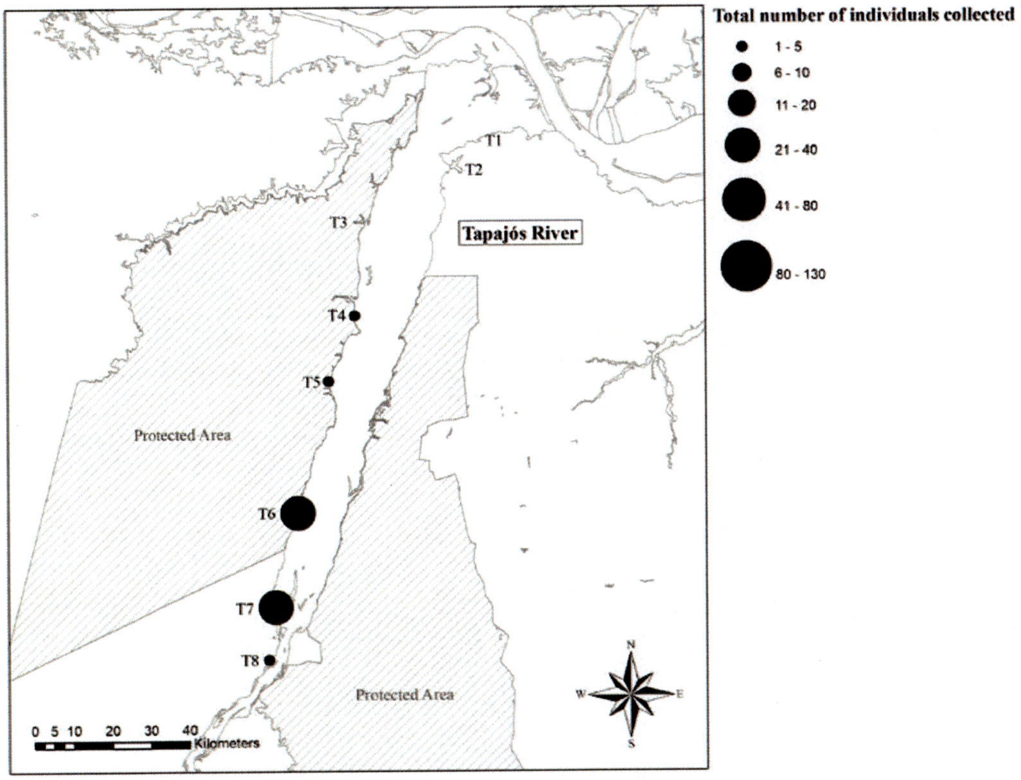

Fig. 4.25 Distribution of individuals of *Schizodon* spp. sampled in the Tapajós River

4

4.3 Family Characidae

4.3.1 Summary of Biology and Ecology (Literature)

Diversity: the Characidae family is a numerous and diverse family of fishes composed of 9 subfamilies and around 162 genera and 1213 species. Members of this family can be found from the south of Texas to the south of South America. After this book was organized some changes had occurred and some genera from this family were assigned to other families, which are Bryconidae (*Brycon*), Iguanodectidae (*Bryconops*), Triportheidae (*Agoniates, Triportheus*), Chalceidae (*Chalceus*).

Habitat: characids inhabit most of the freshwater systems in the Neotropics and are dominant in the Amazonian igarapés (small streams).

Ecology and feeding: their feeding behavior varies from species to species. Some are carnivores (invertivores/piscivores), while others are omnivores, herbivores (including frugivores), and detritivores. Some species, such as those from the genus *Roeboides*, are adapted to eat other fish scales. Some species, such as *Brycon* spp., are frugivores and important seed dispersers in tropical floodplains.

Morphology: body size and shape is variable in the fishes of this family. Characids possess a complete set of fins, including a small adipose fin on the caudal peduncle and body completely covered with cycloid scales.

Reproduction: multiple types of reproduction strategies can be found in characids. Some species have a partial spawning, while others spawn once a year in a large synchronized reproductive event.

Economic importance: some species of this family are important in artisanal and commercial fisheries (e.g., *Triportheus* spp. and *Brycon* spp.). Many small and colorful species are also very important for the aquarium trade.

References: Nelson (1994), Santos et al. (2006), Baumgartner et al. (2012), Ribeiro et al. (2016), Ohara et al. (2017), Zuanon et al. (2015), Lima et al. (2013), van der Sleen et al. (2017), Fricke et al. (2020).

Fish species: ***Agoniates halecinus*** **Müller & Troschel 1845**

Common name(s): sardinha-dentuda, sardinha, xaréu

Size of fish sampled (standard length, both rivers): mean = 14.5 cm, median = 14.8 cm

Fisheries: not relevant to fisheries, no commercial value.

Fig. 4.26 *Agoniates halecinus* sampled in the Tapajós River. This species now belongs to the family Triportheidae. (Photo: Renato Silvano)

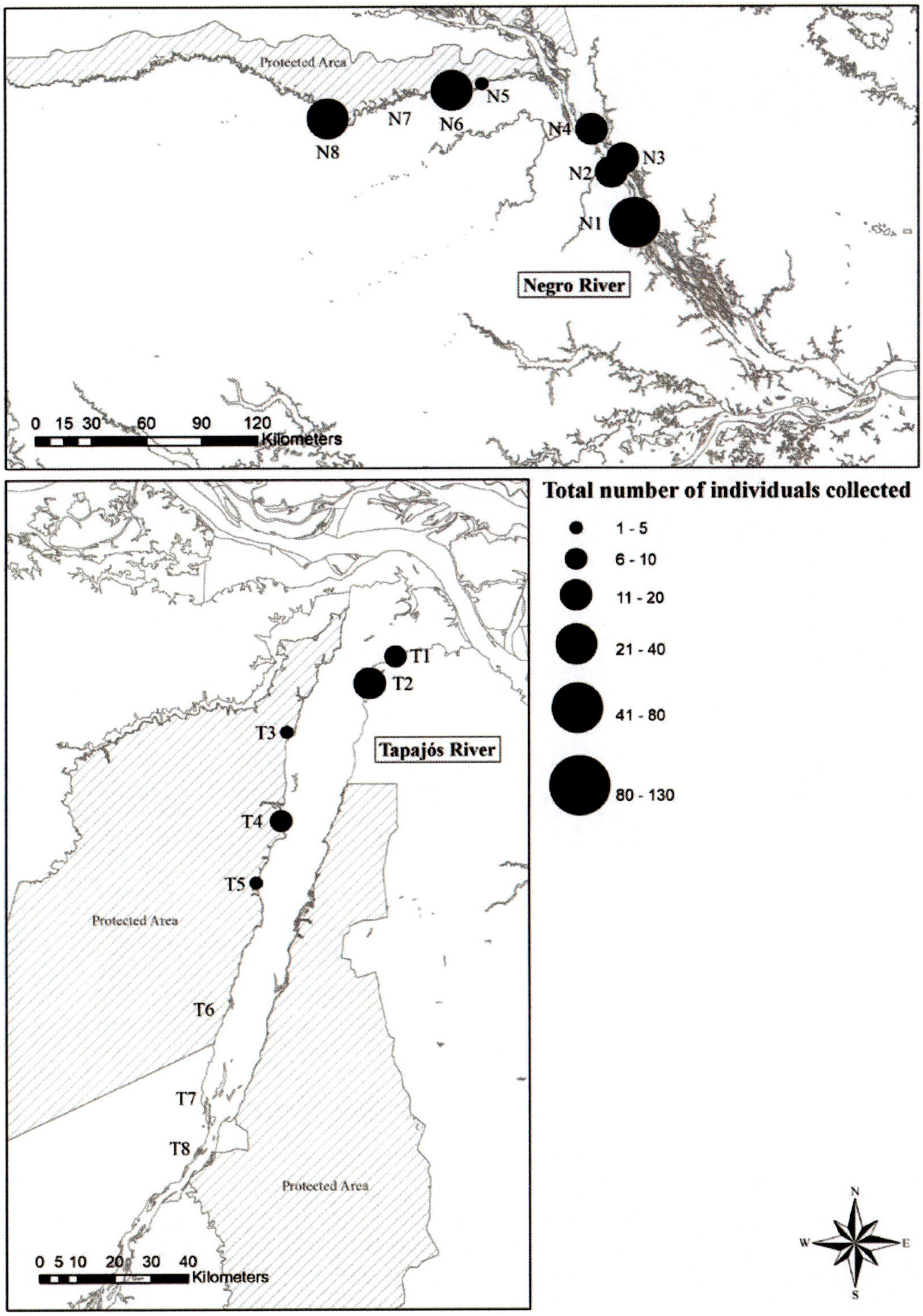

Fig. 4.27 Distribution of individuals of *Agoniates halecinus* sampled in the Negro and Tapajós rivers

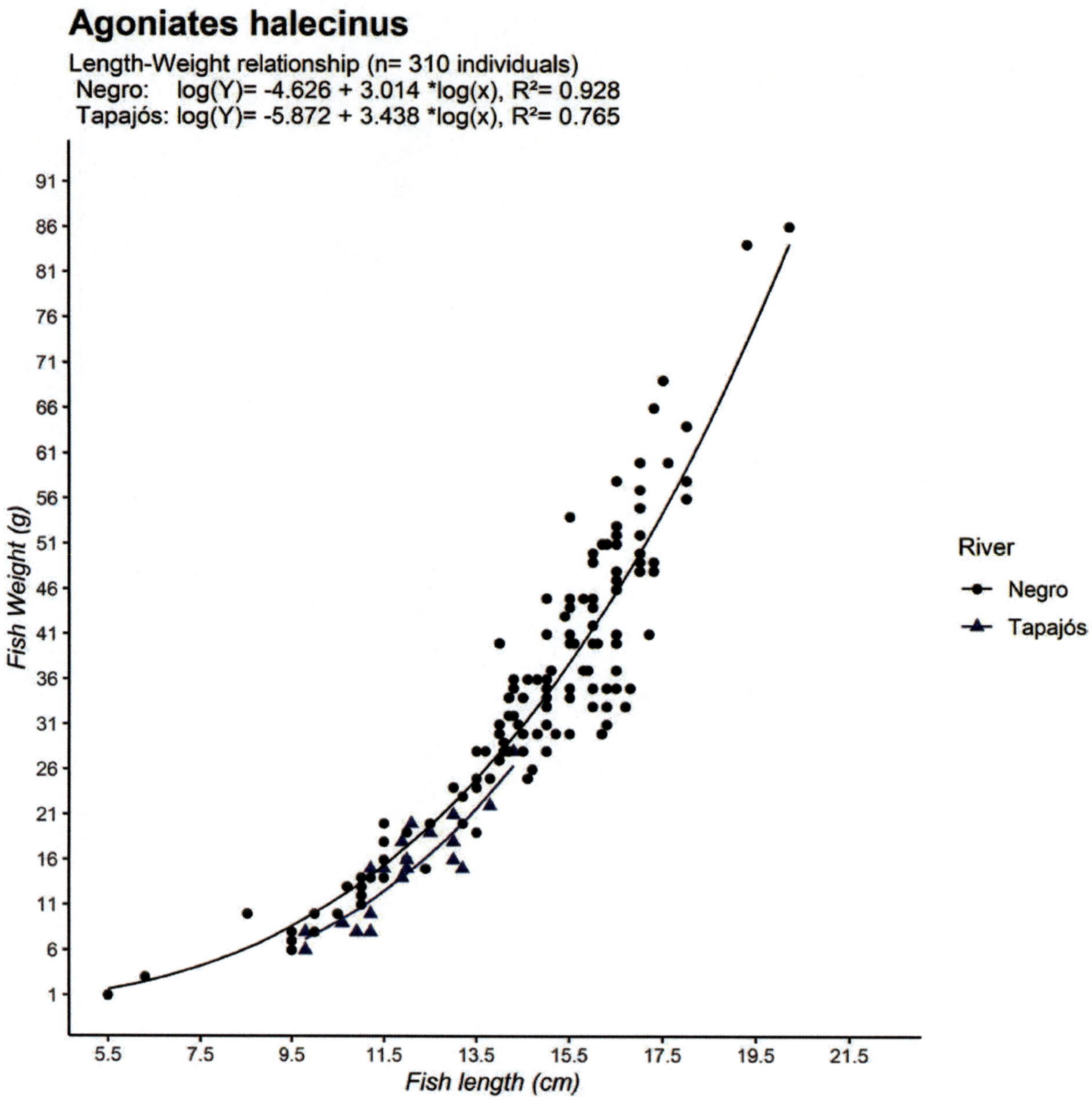

Fig. 4.28 Length-weight relationship for *Agoniates halecinus* sampled in Tapajós and Negro rivers

Fish species: *Brycon pesu* **Müller & Troschel 1845**

Common name(s): jatuarana

Size of fish sampled (standard length, both rivers): min. = 11 cm, max. = 13.5 cm, mean = 12.1 cm, median = 11.8 cm

Fisheries: matrinchã and jatuarana (*Brycon* spp.) are important for food and commerce to fisheries in both rivers, although not being among the most caught fishes. Biomass caught is in Table 3.3 (▶ Chap. 3).

Fig. 4.29 *Brycon pesu* sampled in the Negro River. This species now belongs to the family Bryconidae. (Photo: Renato Silvano)

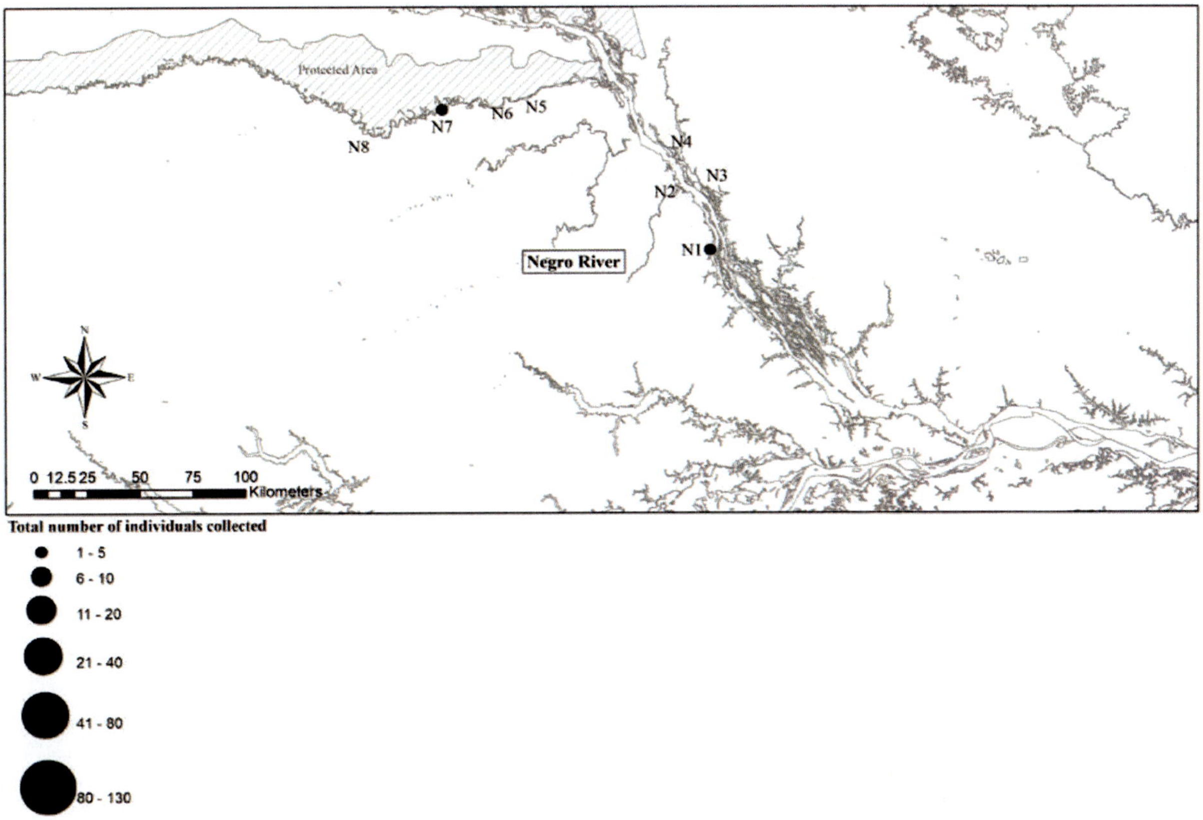

Fig. 4.30 Distribution of individuals of *Brycon pesu* sampled in the Negro river

4

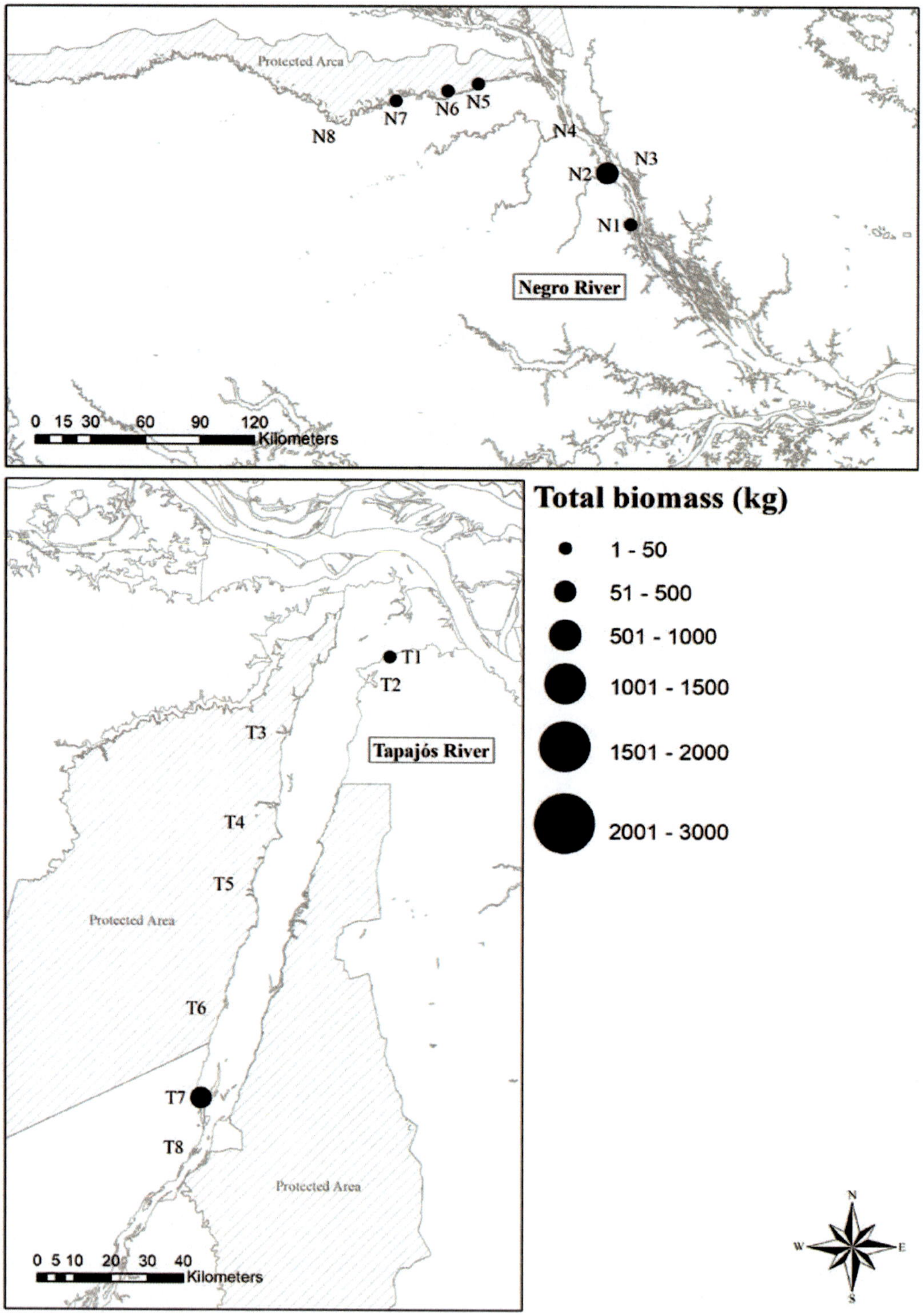

Fig. 4.31 Distribution of biomass of matrinchã (*Brycon* spp.) caught in small-scale fisheries in the Negro and Tapajós rivers

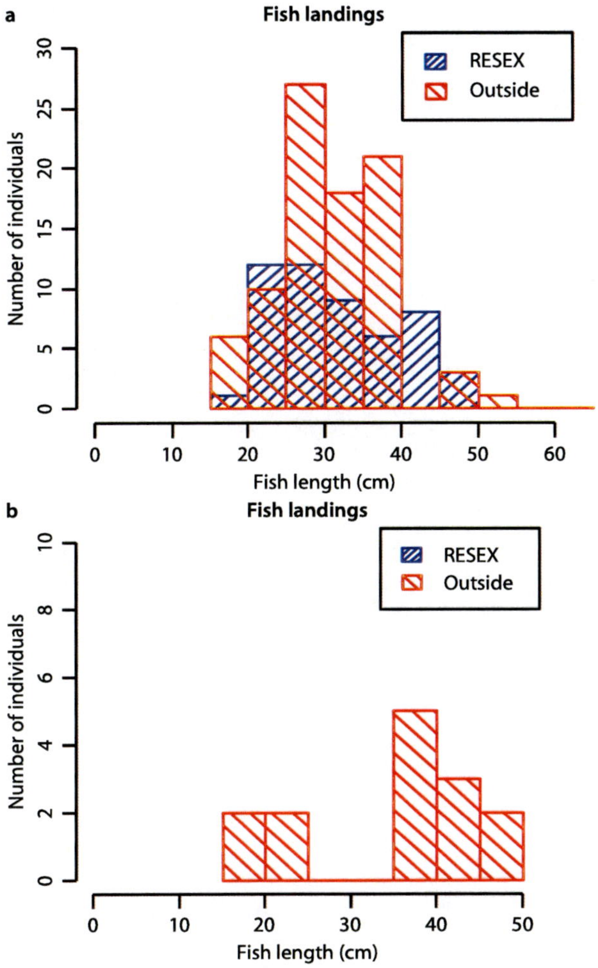

Fig. 4.32 Size distribution of matrinchã (*Brycon* spp.) caught in small-scale fisheries in 16 communities inside and outside the Extractive Reserves (RESEX) in the rivers **a** Negro, N Resex = 51, N outside = 87 and **b** Tapajós, N Resex = 0, N outside = 14

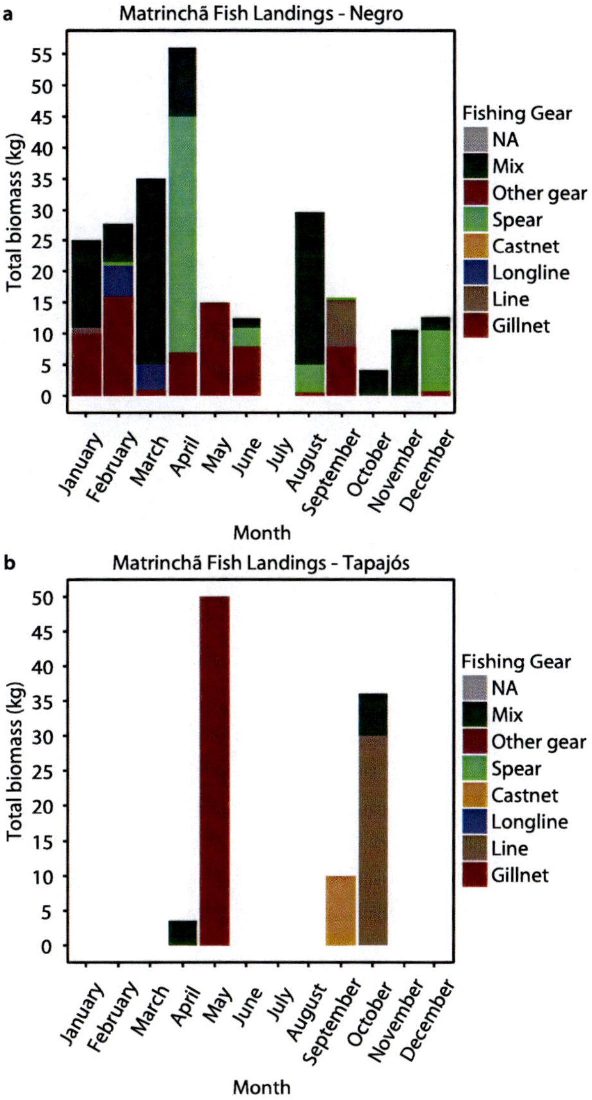

Fig. 4.33 Seasonal variation of biomass of matrinchã (*Brycon* spp.) caught per fishing gear in small-scale fisheries in 16 communities in the rivers **a** Negro and **b** Tapajós. NA: fishing gear not informed; Mix: more than one gear was used

Fish species: *Bryconops alburnoides* Kner 1858

Common name(s): piabão, catrapola, peixe-duro

Size of fish sampled (standard length, both rivers): mean = 11.5 cm, median = 12.1 cm

Fisheries: not relevant to fisheries, due to its small size.

Fig. 4.34 *Bryconops alburnoides* sampled in the Negro River. This species now belongs to the family Iguanodectidae. (Photo: Renato Silvano)

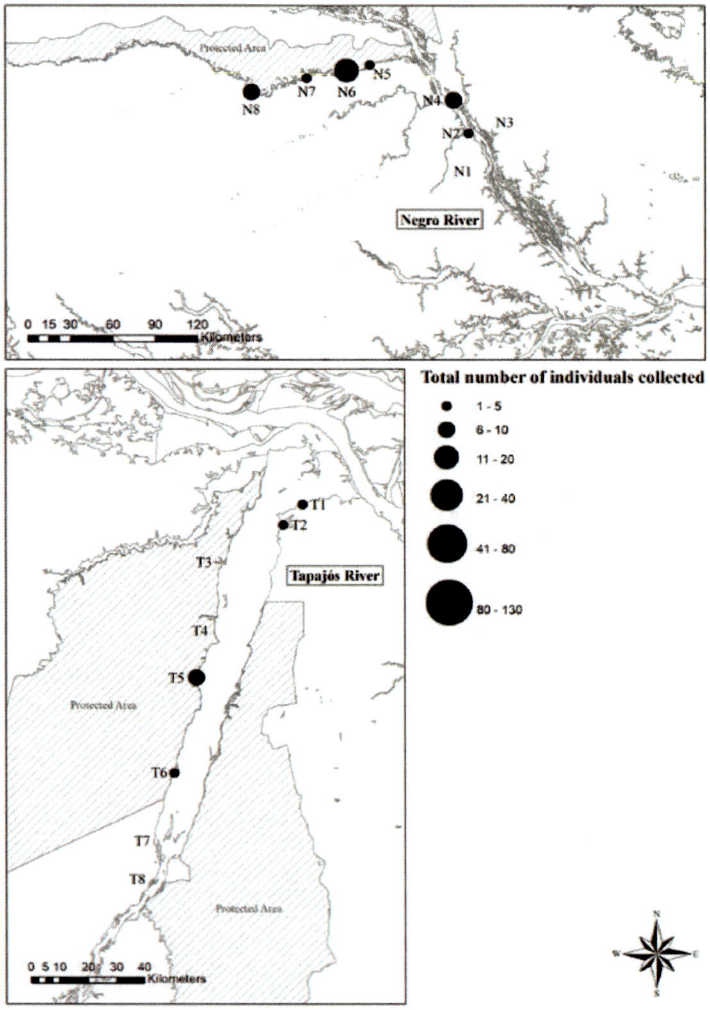

Fig. 4.35 Distribution of individuals of *Bryconops alburnoides* sampled in the Negro and Tapajós rivers

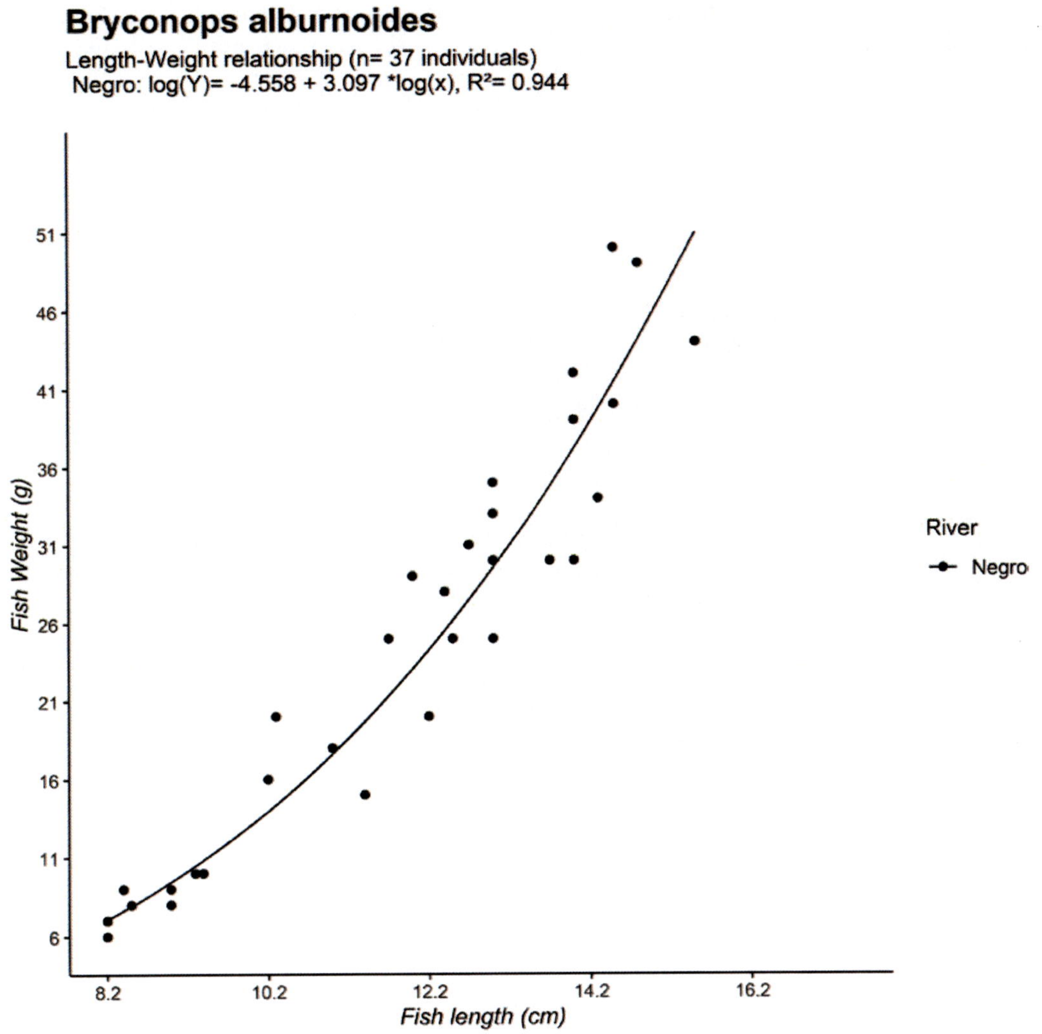

Fig. 4.36 Length-weight relationship for *Bryconops alburnoides* sampled in Negro River

Fish species: *Bryconops caudomaculatus* (Günther 1864)

Common name(s): cataprola, peixe-duro

Size of fish sampled (standard length, both rivers): min. = 9.6 cm, max. = 13.7 cm, mean = 11.8 cm, median = 12 cm

Fisheries: not relevant to fisheries, due to its small size.

4

Fig. 4.37 *Bryconops caudomaculatus* sampled in the Tapajós River. This species now belongs to the family Iguanodectidae. (Photo: Renato Silvano)

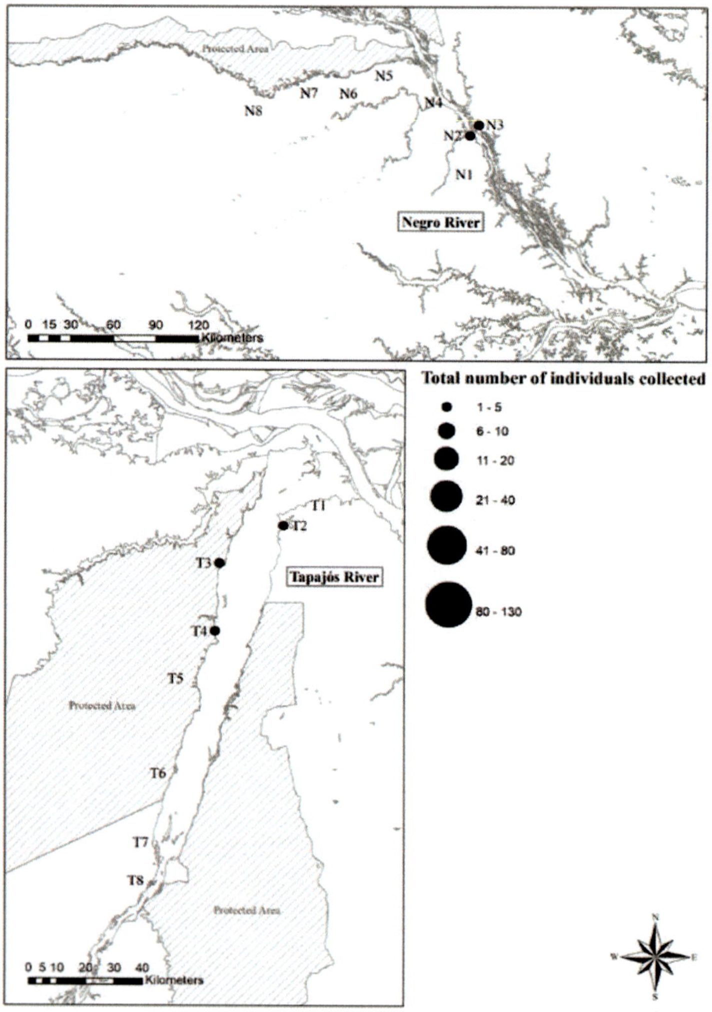

Fig. 4.38 Distribution of individuals of *Bryconops caudomaculatus* sampled in the Negro and Tapajós rivers

Fish species: *Bryconops gracilis* (Eigenmann 1908)

Common name(s): cataprola, peixe-duro, matupiri

Size of fish sampled (standard length, both rivers): min. = 10.7 cm, max. = 14.5 cm, mean = 13.2 cm, median = 13.5 cm

Fisheries: not relevant to fisheries, due to its small size.

Fig. 4.39 *Bryconops gracilis* sampled in the Tapajós River. This species now belongs to the family Iguanodectidae. (Photo: Renato Silvano)

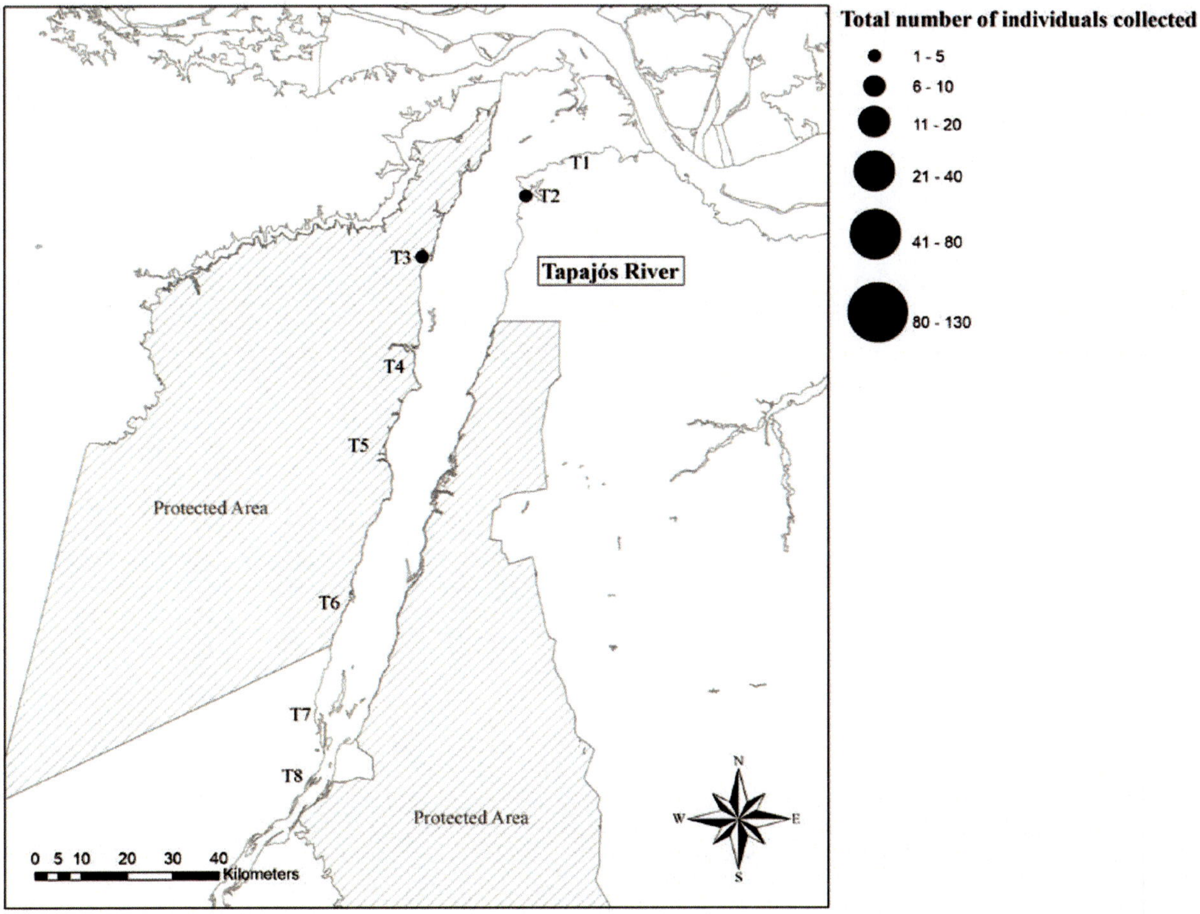

Fig. 4.40 Distribution of individuals of *Bryconops gracilis* sampled in the Tapajós River

Fish species: *Bryconops melanurus* (Bloch 1794)

Common name(s): piabão, cataprola

Size of fish sampled (standard length, both rivers): mean = 9.7 cm, median = 9 cm

Fisheries: not relevant to fisheries, due to its small size.

Fig. 4.41 *Bryconops melanurus* sampled in the Negro River. This species now belongs to the family Iguanodectidae. (Photo: Renato Silvano)

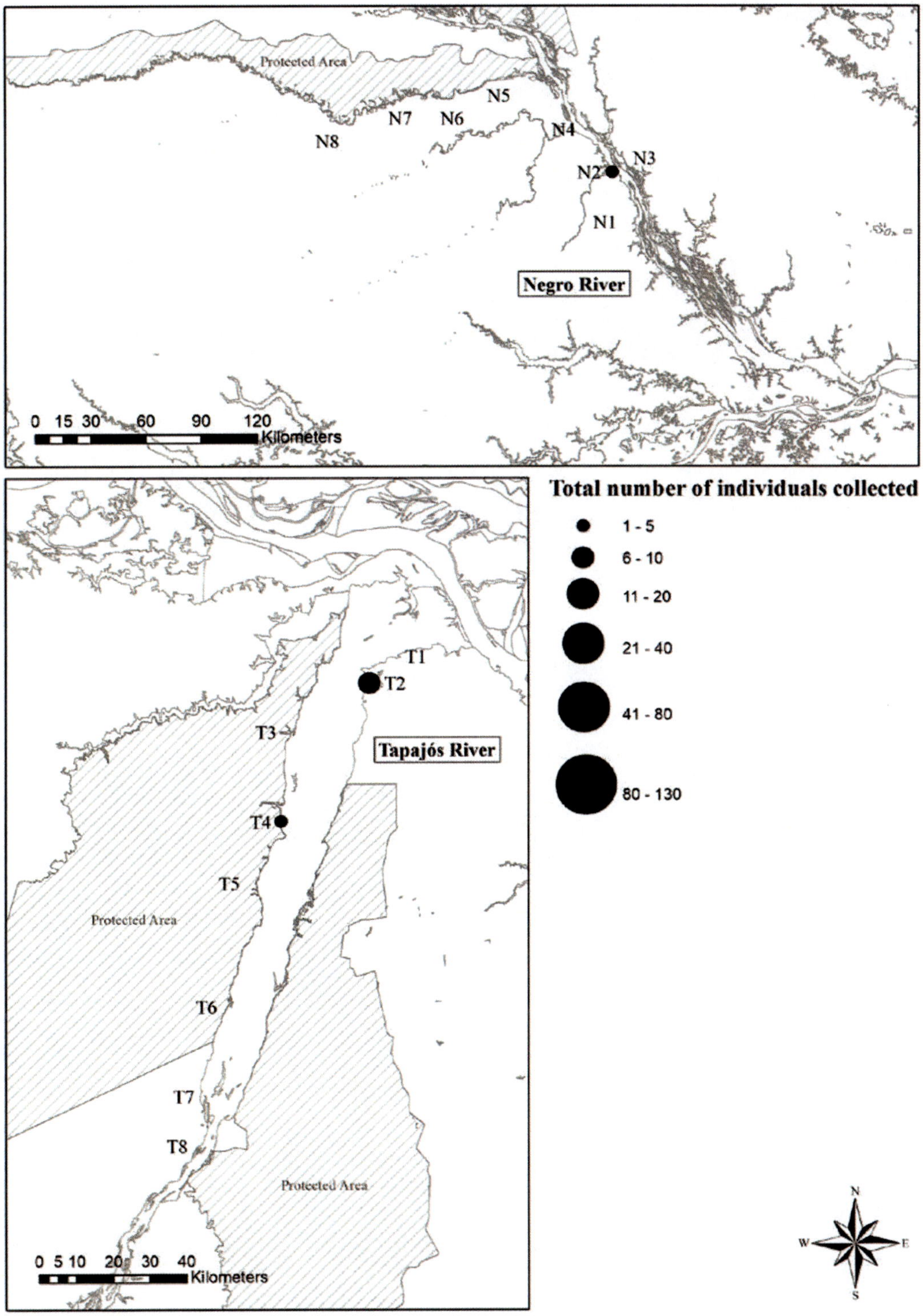

Fig. 4.42 Distribution of individuals of *Bryconops melanurus* sampled in the Negro and Tapajós rivers

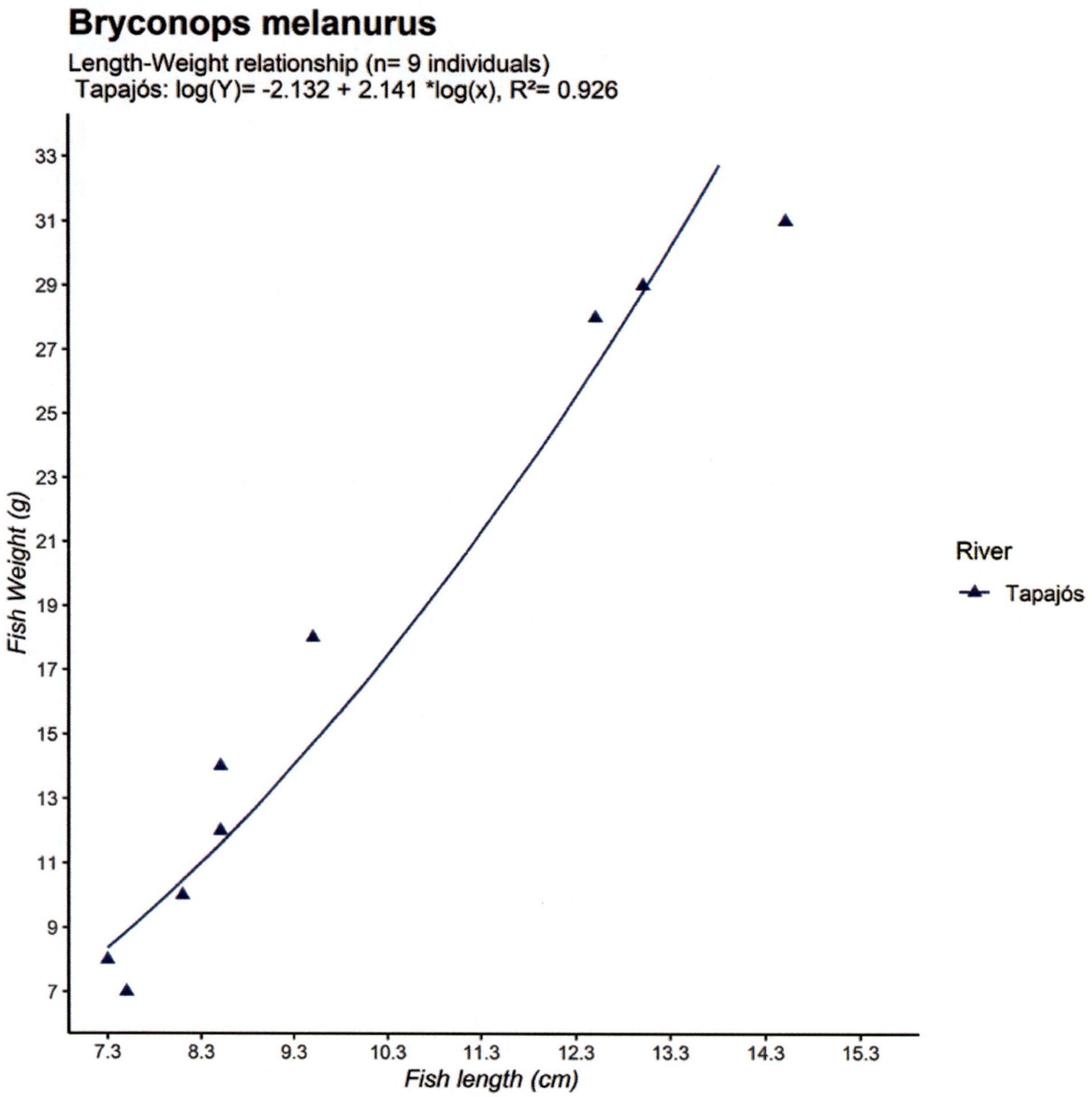

Fig. 4.43 Length-weight relationship for *Bryconops melanurus* sampled in Tapajós River

Fish species: *Chalceus macrolepidotus* Cuvier 1818

Common name(s): arari

Size of fish sampled (standard length, both rivers): mean = 16.6 cm, median = 18 cm

Fig. 4.44 *Chalceus macrolepidotus* sampled in the Negro River. This species now belongs to the family Chalceidae. (Photo: Renato Silvano)

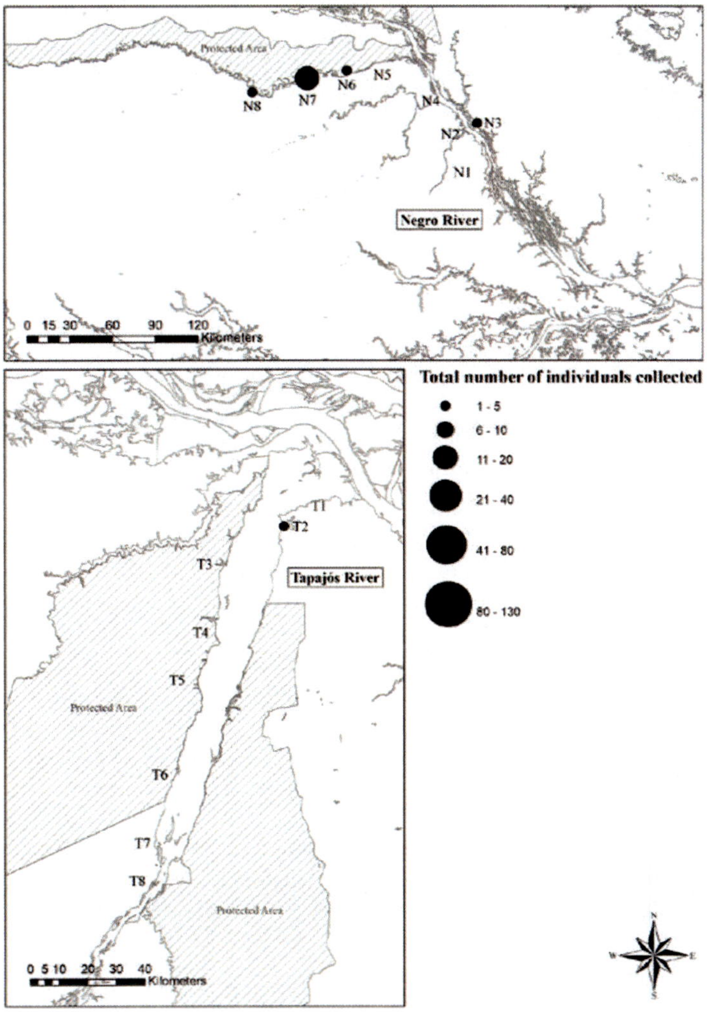

Fig. 4.45 Distribution of individuals of *Chalceus macrolepidotus* sampled in the Negro and Tapajós rivers

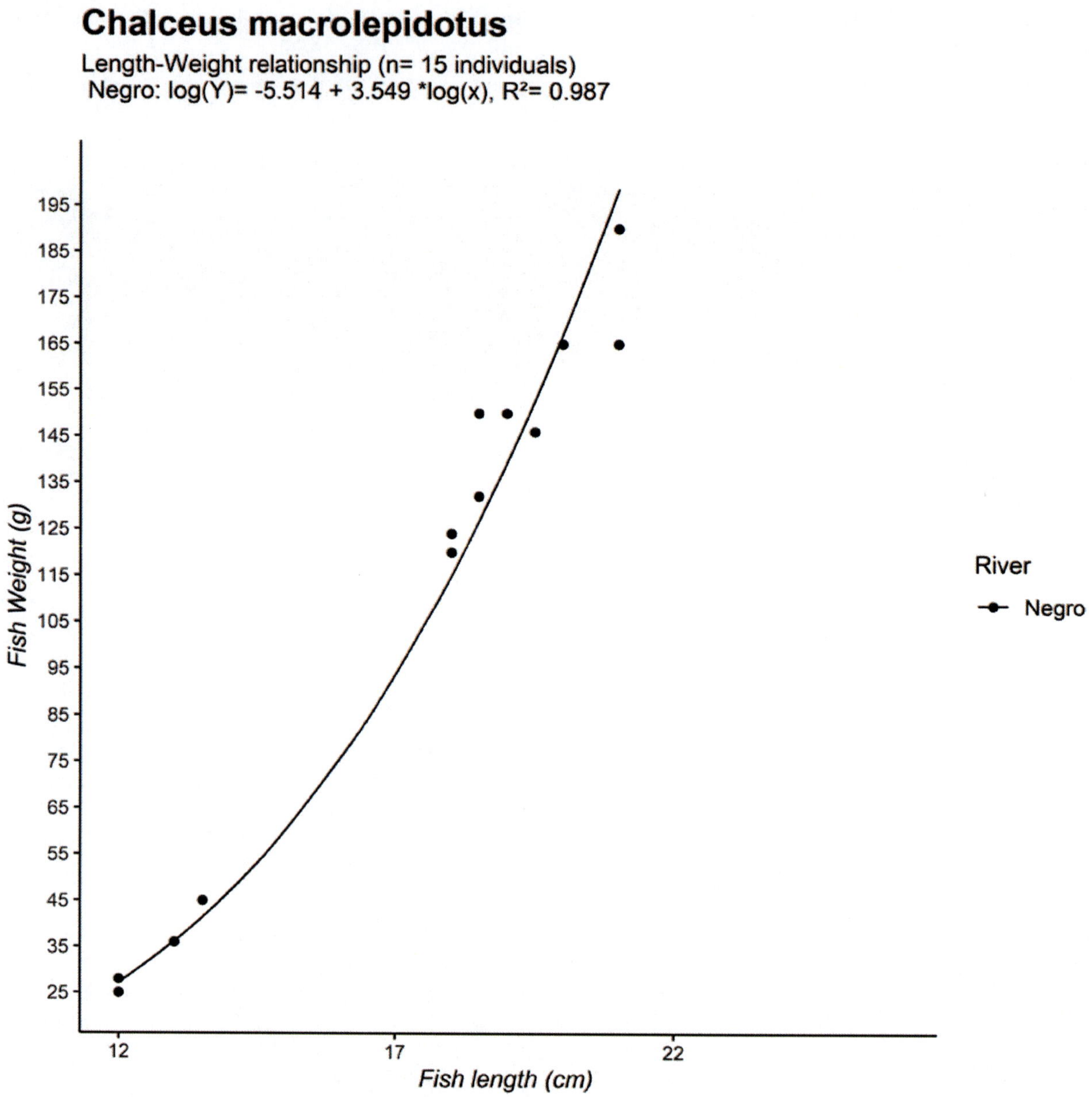

Fig. 4.46 Length-weight relationship for *Chalceus macrolepidotus* sampled in the Negro River

Fisheries: arari (*Chalceus macrolepidotus*) is occasionally exploited in fisheries of both rivers, but has low or no commercial value. Biomass caught is in Table 3.3 (▶ Chap. 3).

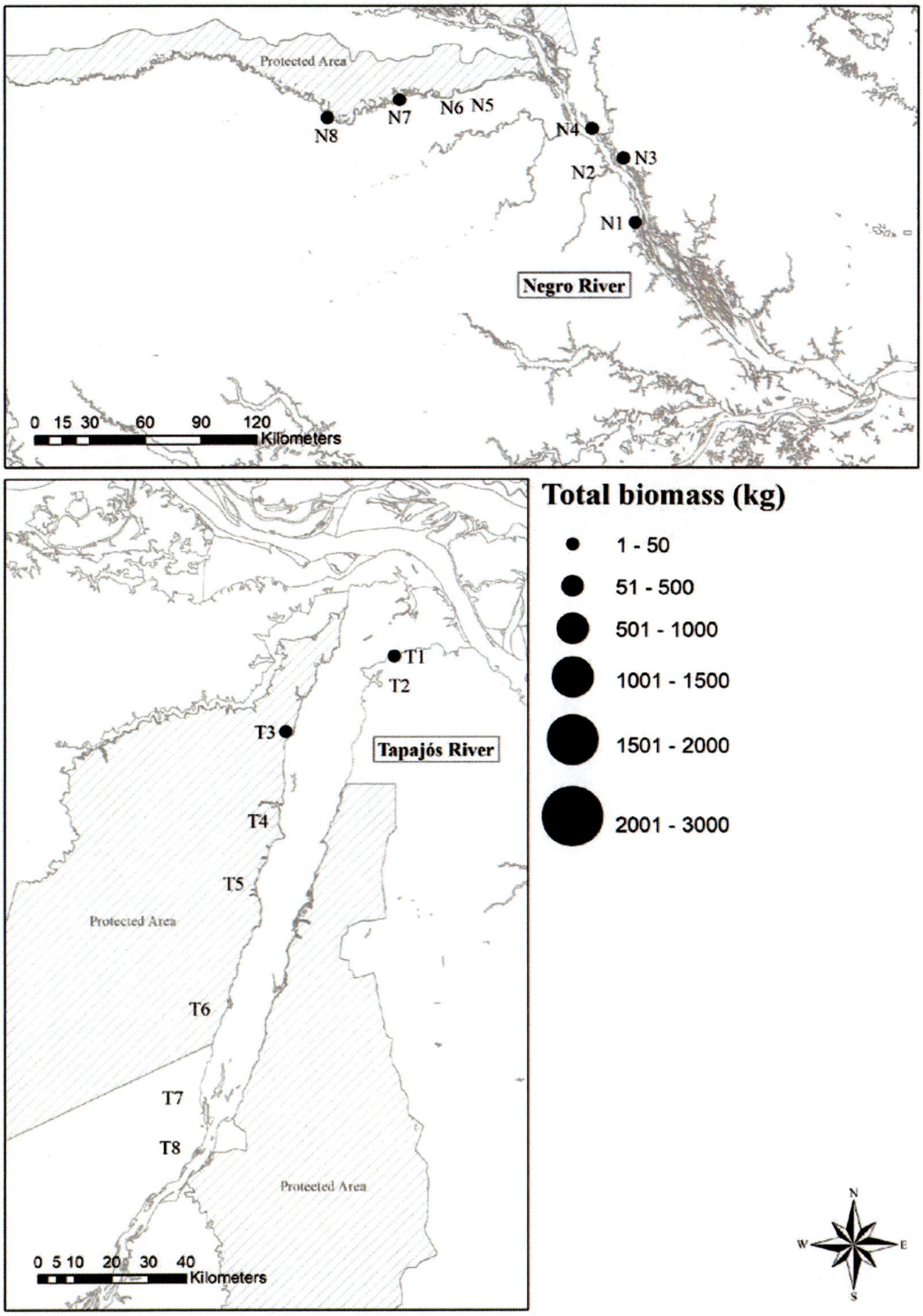

Fig. 4.47 Distribution of biomass of arari (*Chalceus macrolepidotus*) caught in small-scale fisheries in the Negro and Tapajós rivers

4

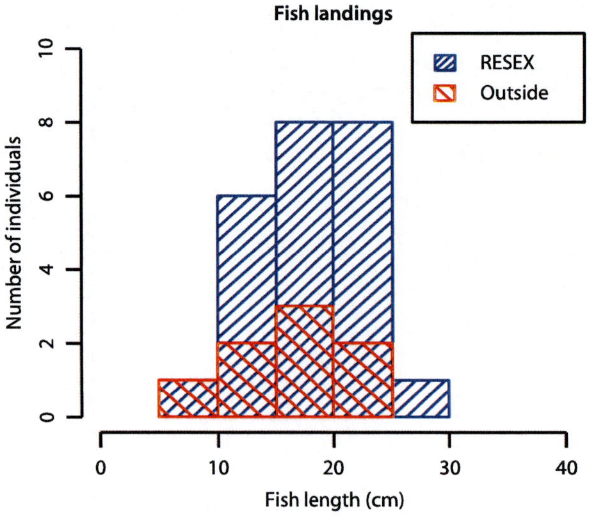

Fig. 4.48 Size distribution of arari (*Chalceus macrolepidotus*) caught in small-scale fisheries in eight communities inside and outside the Extractive Reserves (RESEX) in the Negro River, N Resex = 24, N outside = 8

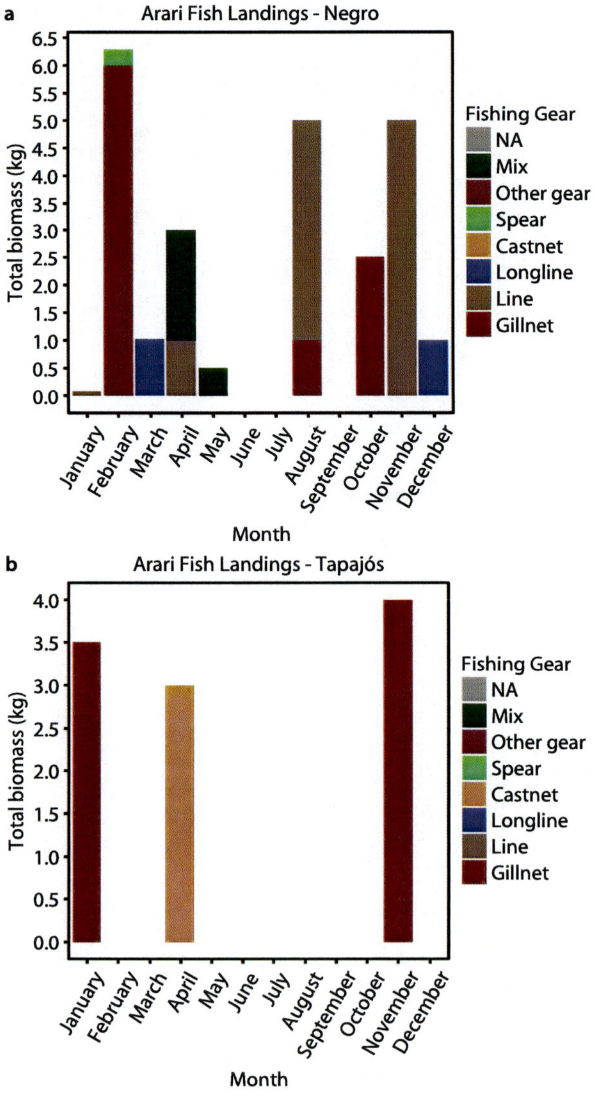

Fig. 4.49 Seasonal variation of biomass of arari (*Chalceus macrolepidotus*) caught per fishing gear in small-scale fisheries in 16 communities in the rivers **a** Negro and **b** Tapajós. NA: fishing gear not informed; Mix: more than one gear was used

Fish species: *Charax delimai* Menezes & Lucena 2014

Common name(s): pirandirá

Fig. 4.50 *Charax delimai* sampled in the Negro River. (Photo: Renato Silvano)

Fish species: *Charax gibbosus* (Linnaeus 1758)

Common name(s): not available

Fig. 4.51 *Charax gibbosus* sampled in the Negro River. (Photo: Renato Silvano)

Observation: abundance and size data were grouped for these species, due to the difficulty in distinguishing them during sampling: both species were collected only in Negro River; see ◻ Table 1.1 (▶ Chap. 1).

Size of fish sampled (standard length, both rivers): min. = 6 cm, max. = 11.5 cm, mean = 8.1 cm, median = 8 cm

Fisheries: not relevant to fisheries, due to its small size.

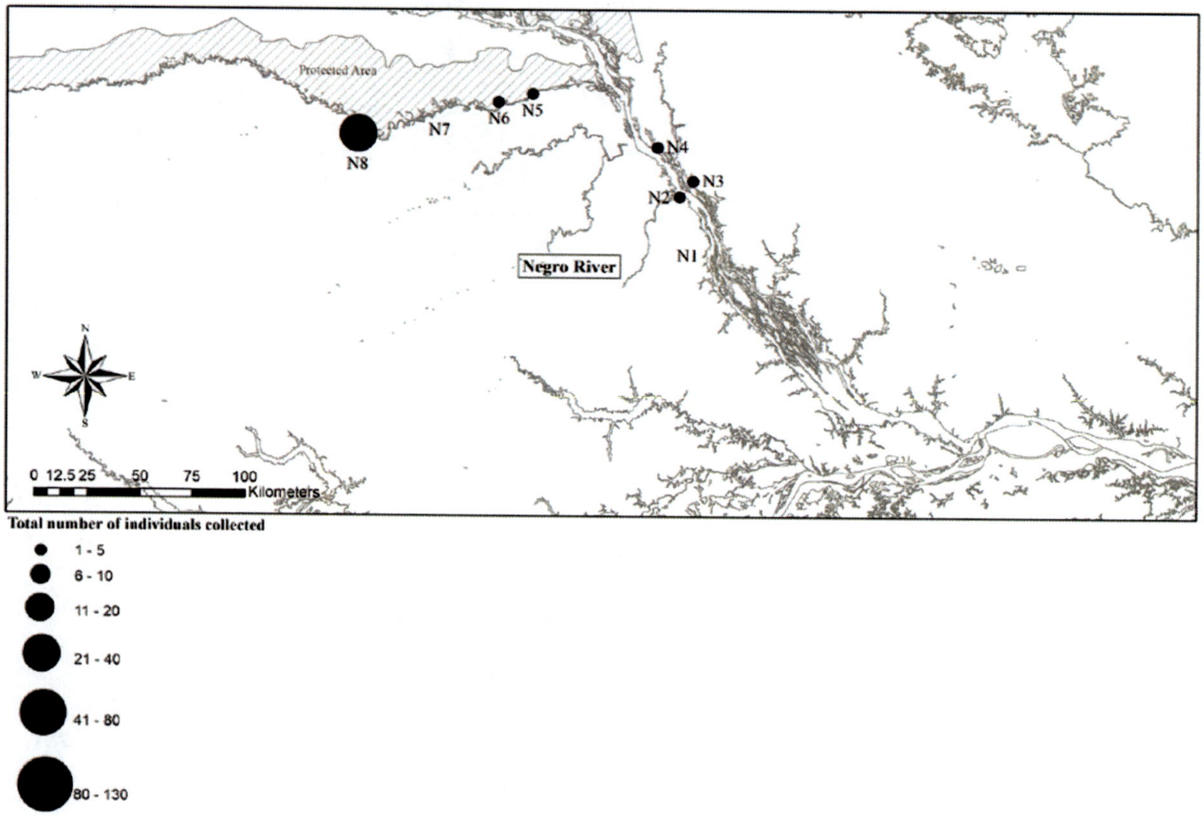

◻ **Fig. 4.52** Distribution of individuals of *Charax* spp. sampled in the Negro River

Fish species: ***Moenkhausia lepidura*** **(Kner 1858)**

Common name(s): piabão

Size of fish sampled (standard length, both rivers): min. = 5.4 cm, max. = 9.5 cm, mean = 8.2 cm, median = 8.5 cm

Fisheries: not relevant to fisheries, due to its small size. Some species from this group of fishes (Tetragonopterinae) may be exploited as ornamental fish in the aquarium trade.

Fig. 4.53 *Moenkhausia lepidura* sampled in the Negro River. (Photo: Renato Silvano)

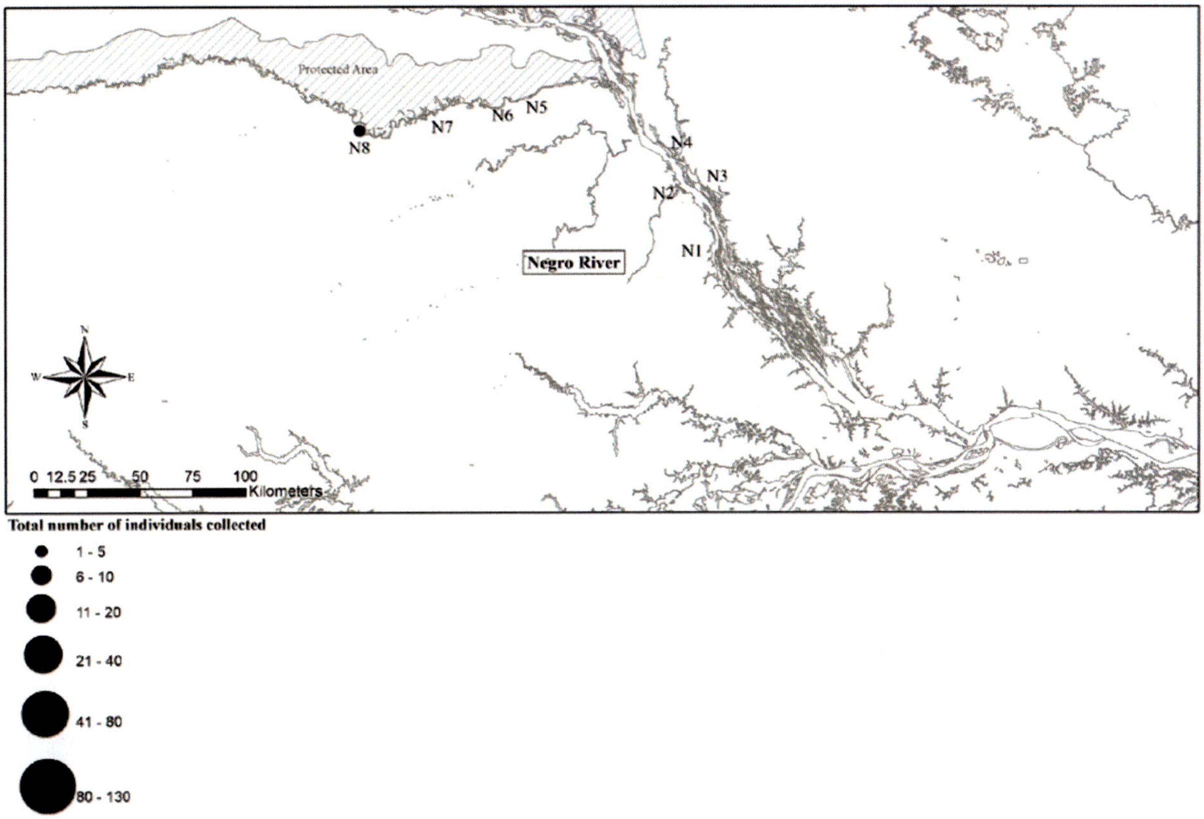

Fig. 4.54 Distribution of individuals of *Moenkhausia lepidura* sampled in the Negro River

Fish species: *Moenkhausia megalops* (Eigenmann 1907)

Common name(s): matupiri

Size of fish sampled (standard length, both rivers): mean = 7.1 cm, median = 6.6 cm

Fisheries: not relevant to fisheries, due to its small size. Some species from this group of fishes (Tetragonopterinae) may be exploited as ornamental fish in the aquarium trade.

Fig. 4.55 *Moenkhausia megalops* sampled in the Tapajós River. (Photo: Renato Silvano)

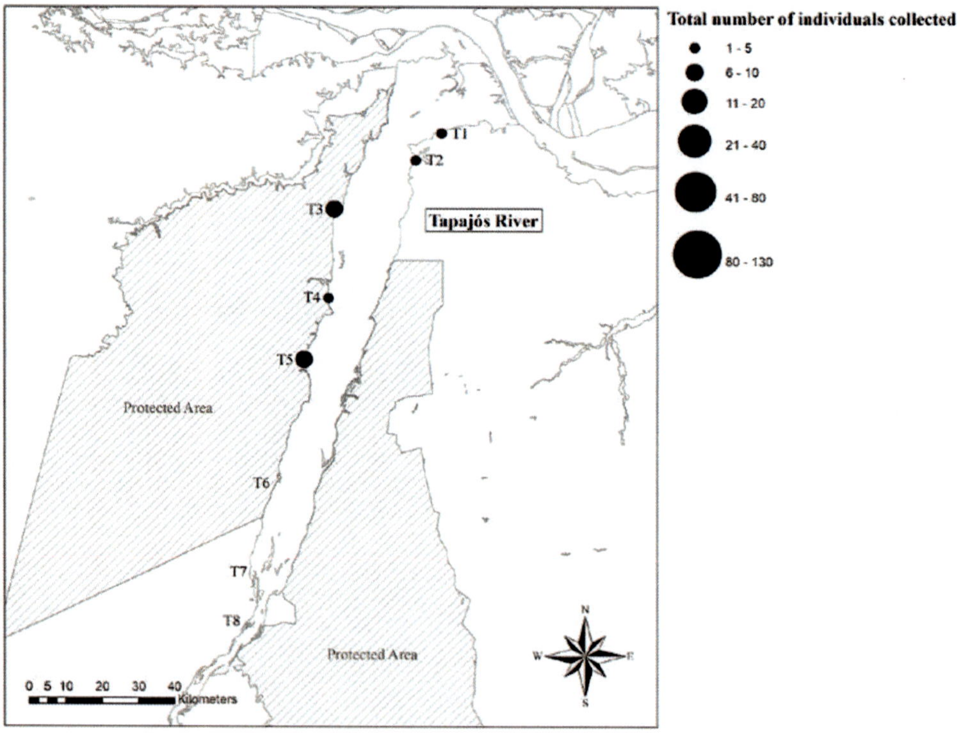

Fig. 4.56 Distribution of individuals of *Moenkhausia megalops* sampled in the Tapajós River

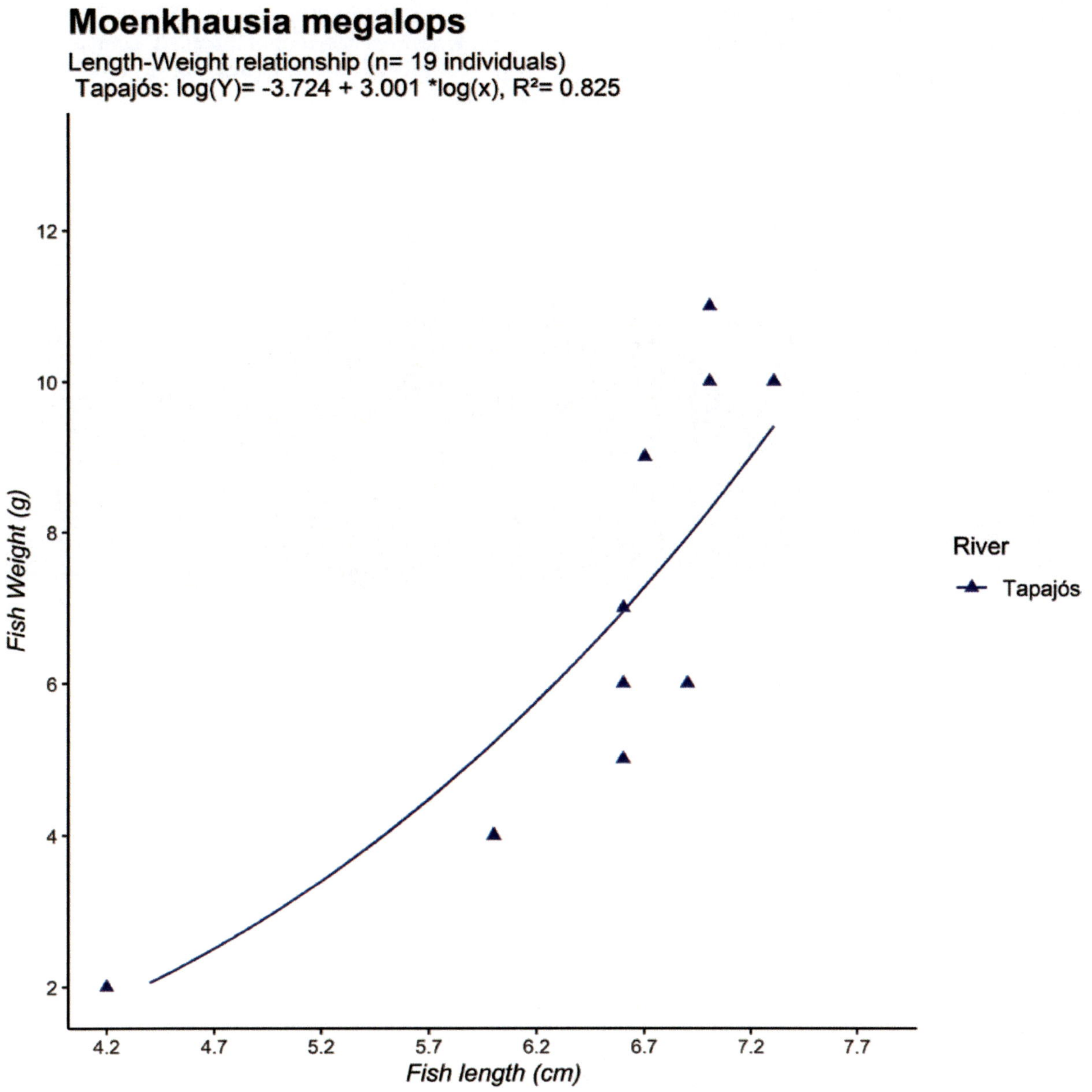

Fig. 4.57 Length-weight relationship for *Moenkhausia megalops* sampled in the Tapajós river

Fish species: *Tetragonopterus manaos* Urbanski, Melo, Silva & Benine 2018

Common name(s): matupiri

Size of fish sampled (standard length, both rivers): mean = 6.7 cm, median = 6.3 cm

Fisheries: not relevant to fisheries, due to its small size. Some species from this group of fishes (Tetragonopterinae) may be exploited as ornamental fish in the aquarium trade.

4

Fig. 4.58 *Tetragonopterus manaos* sampled in the Negro River. (Photo: Renato Silvano)

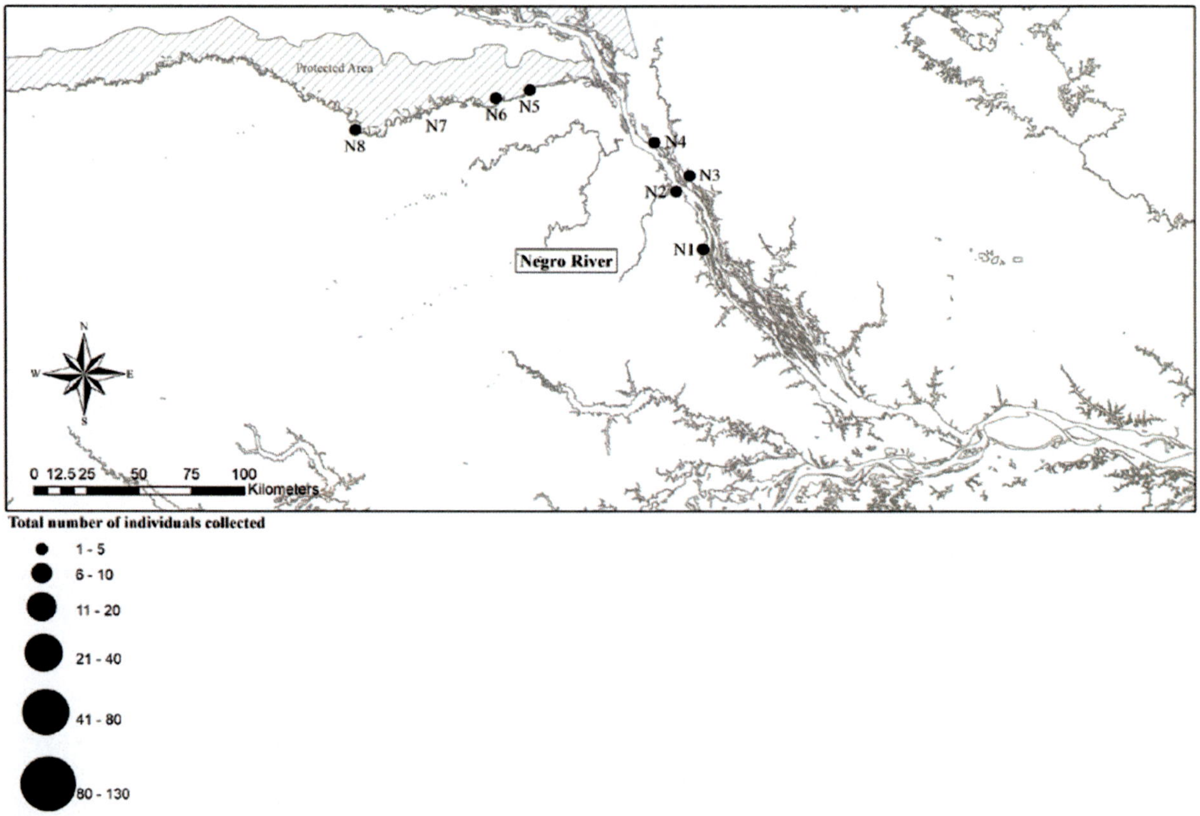

Fig. 4.59 Distribution of individuals of *Tetragonopterus manaos* sampled in the Negro River

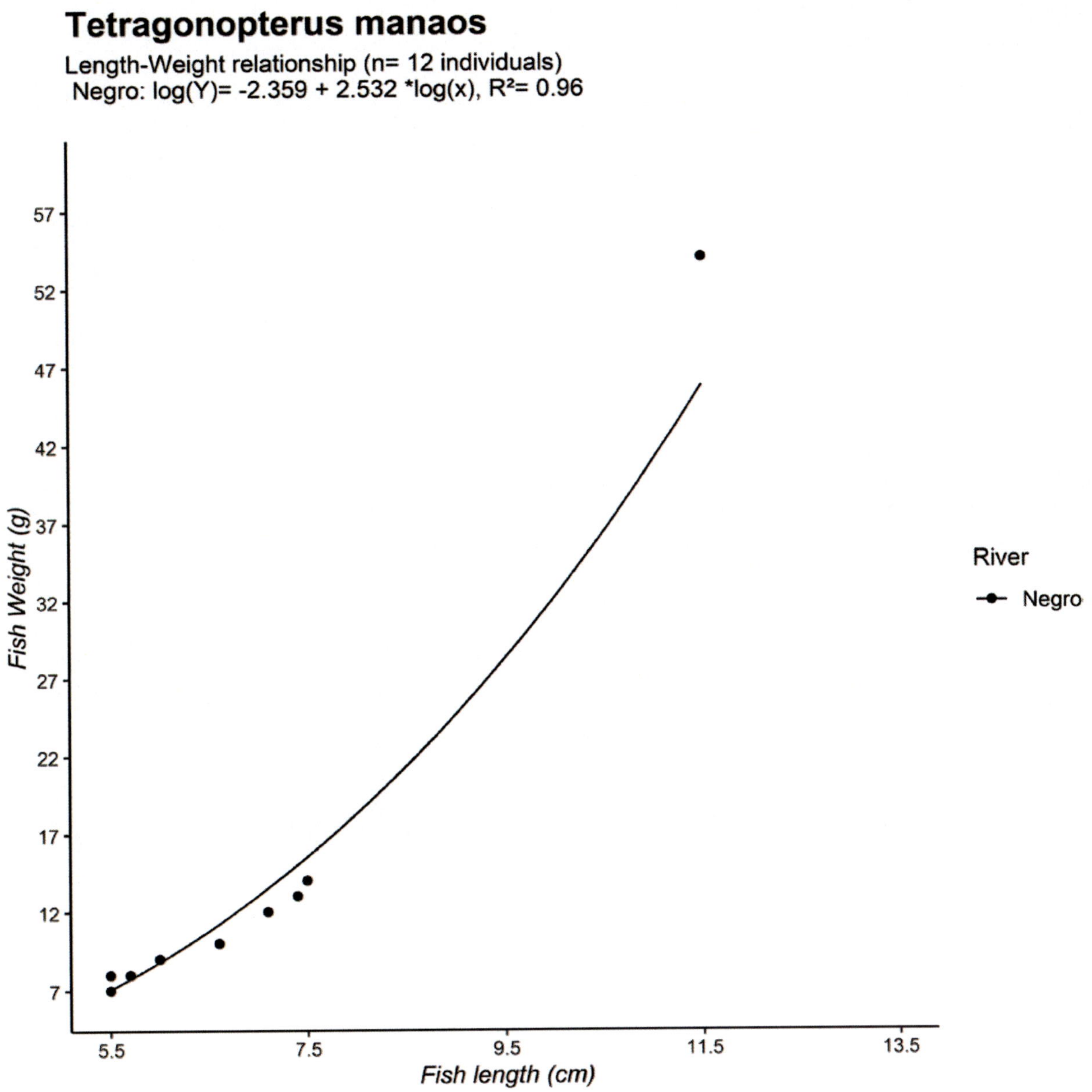

Fig. 4.60 Length-weight relationship for *Tetragonopterus manaos* sampled in Negro River

Fish species: *Triportheus albus* Cope 1872

Common name(s): sardinha

Fig. 4.61 *Triportheus albus* sampled in the Negro River. This species now belongs to the family Triportheidae. (Photo: Renato Silvano)

Fish species: *Triportheus angulatus* (Spix & Agassiz 1829)

Common name(s): sardinha, sardinha-papuda

Fig. 4.62 *Triportheus angulatus* sampled in the Tapajós River. This species now belongs to the family Triportheidae. (Photo: Renato Silvano)

Fish species: *Triportheus auritus* (Valenciennes 1850)

Common name(s): sardinha

Fig. 4.63 *Triportheus auritus* sampled in the Tapajós River. This species now belongs to the family Triportheidae. (Photo: Renato Silvano)

Fish species: *Triportheus rotundatus* (Jardine 1841)

Common name(s): sardinha, sardinha-papuda

Fig. 4.64 *Triportheus rotundatus* sampled in the Tapajós River. This species now belongs to the family Triportheidae. (Photo: Renato Silvano)

Observation: abundance and size data were grouped for all these species, due to the difficulty in distinguishing them during sampling: *Triportheus albus* was collected only in Negro River and the other three species (*T. angulatus, T. auritus, T. rotundatus*) were collected only in Tapajós River; see Table 1.1 (► Chap. 1).

Size of fish sampled (standard length, both rivers): min. = 6.5 cm, max. = 23.9 cm, mean = 12 cm, median = 11.5 cm

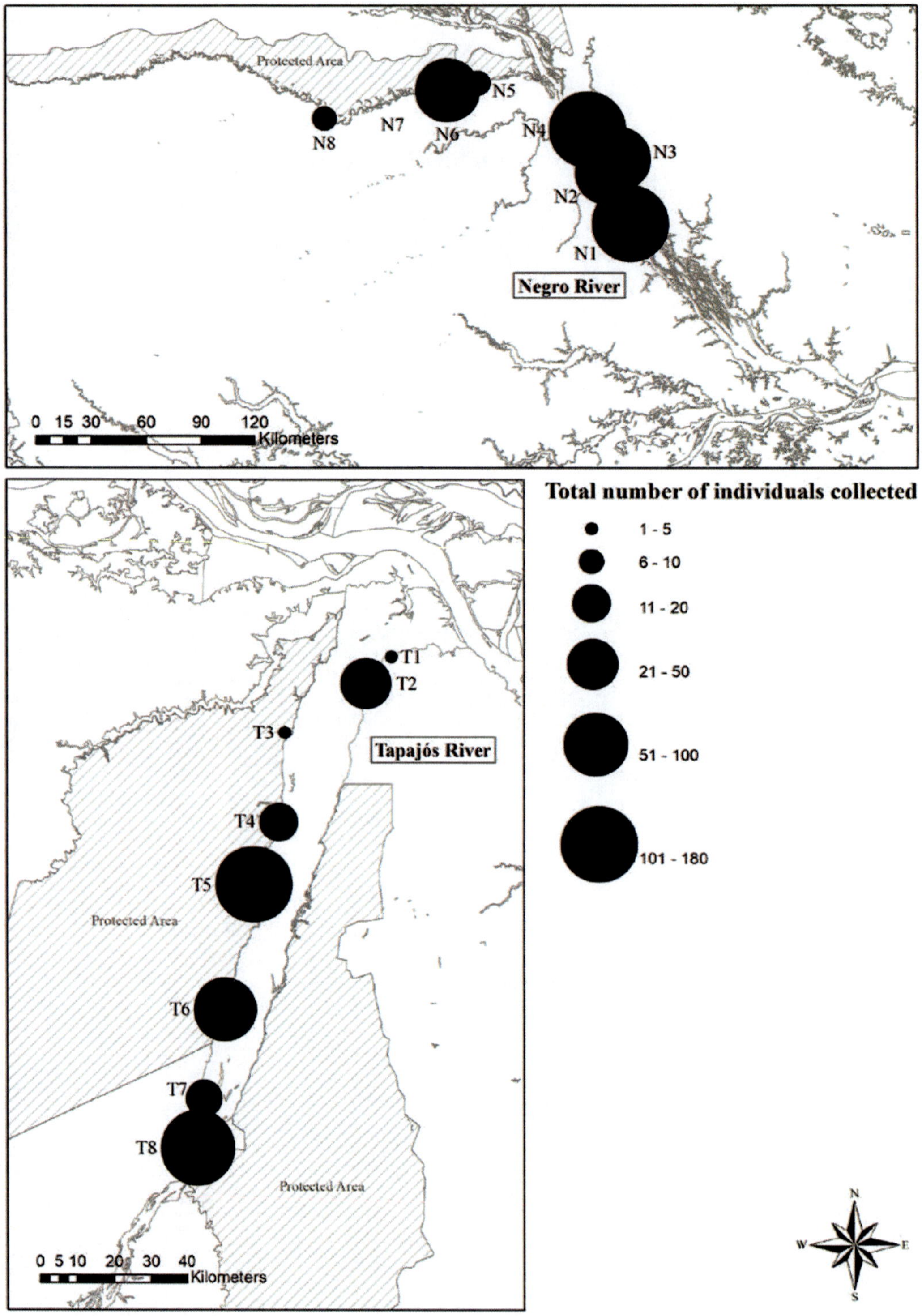

Fig. 4.65 Distribution of individuals of *Triportheus* spp. sampled in the Negro and Tapajós rivers

Fisheries: sardinha (*Triportheus* spp.) is occasionally exploited by fisheries in both rivers mostly for food, as these fishes have low commercial value. Biomass caught is in ◘ Table 3.3 (▶ Chap. 3).

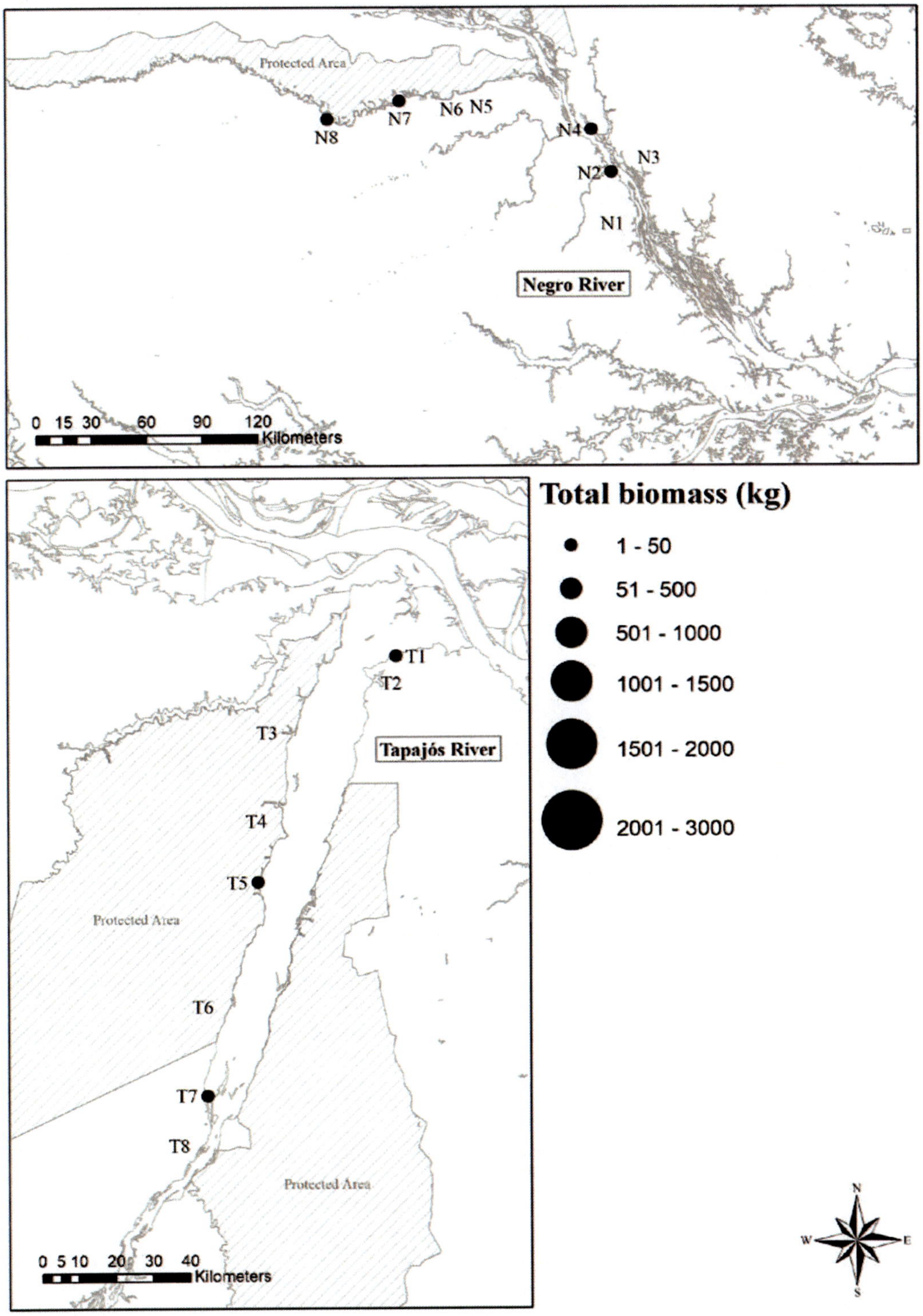

◘ **Fig. 4.66** Distribution of biomass of sardinha (*Triportheus* spp.) caught in small-scale fisheries in the Negro and Tapajós rivers

4

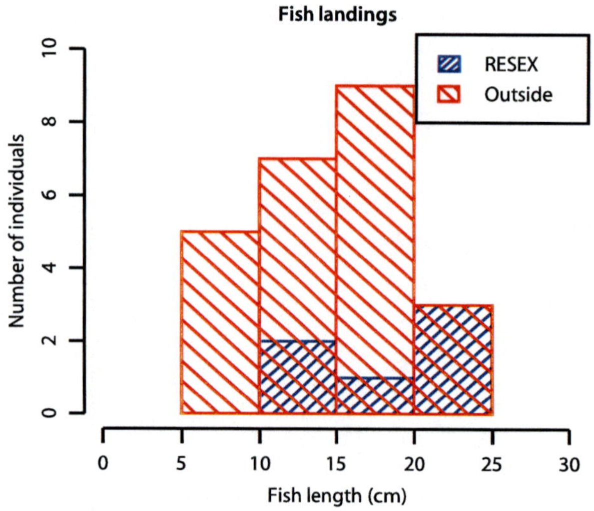

Fig. 4.67 Size distribution of sardinha (*Triportheus* spp.) caught in small-scale fisheries in eight communities inside and outside the Extractive Reserves (RESEX) in the Negro River, N Resex = 6, N outside = 24

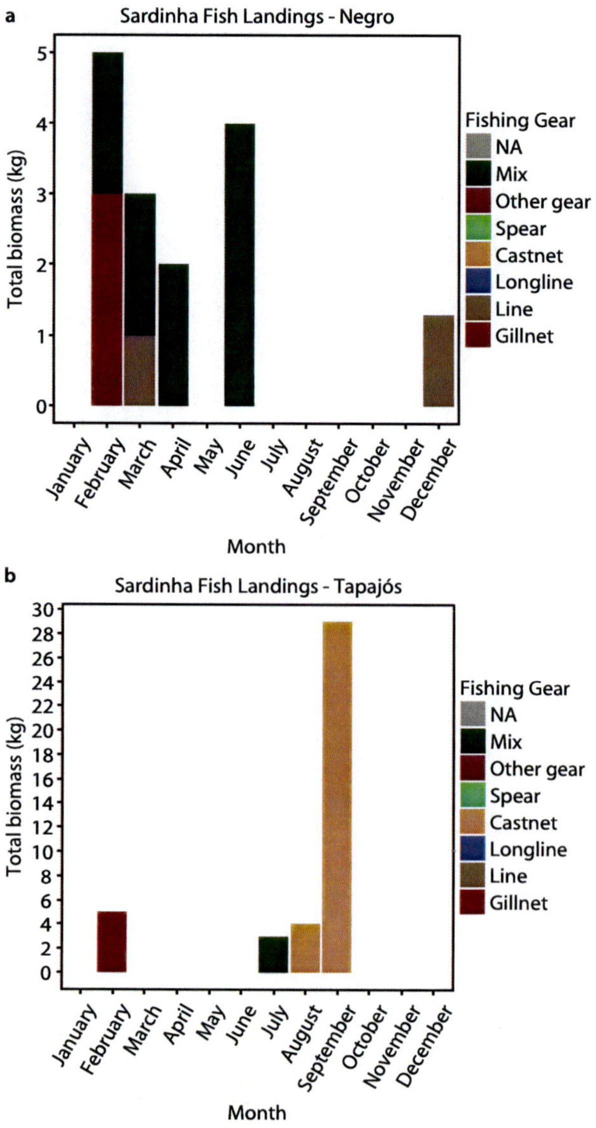

Fig. 4.68 Seasonal variation of biomass of sardinha (*Triportheus* spp.) caught per fishing gear in small-scale fisheries in 16 communities in the rivers **a** Negro and **b** Tapajós. NA: fishing gear not informed; Mix: more than one gear was used

4.4 Family Chilodontidae

4.4.1 Summary of Biology and Ecology (Literature)

Diversity: the Chilodontidae family is composed of two genera, Caenotropus and Chilodus, each one with four species. Species of this family are restricted to the Amazon basin, Orinoco river, and Guianese rivers.

Habitat: chilodontids are present in rivers of white, clear and black waters.

Ecology and feeding: species of this family feed on invertebrates, periphyton, and detritus. Chilodontids are known to move at an oblique angle with the head facing down (headstander). This behavior is similar to what is found in some members of the family Anostomidae.

Morphology: chilodontids are small-bodied fishes (max length of 18 cm) with a robust and plump body (fusiform shape). They possess large and well-attached scales. The sixth lateral-line scale is clearly smaller than other scales of the same series. Their mouths have a unique series of small and flexible teeth attached to the lips of the superior maxilla.

Reproduction: no information found.

Economic importance: although some species of the genus *Chilodus* are relevant for the aquarium trade, the species of *Caenotropus* found in the studied rivers have no relevance for fisheries.

References: Nelson (1994), Vari et al. (1995), Torrente-Vilara et al. (2013), Ohara et al. (2017), Melo e Sidlauskas (2017).

Fish species: *Caenotropus labyrinthicus* (Kner 1858)

Common name(s): cururu, piracururu, cascudinho

Size of fish sampled (standard length, both rivers): mean = 10.3 cm, median = 10 cm

Fisheries: not relevant to fisheries, no commercial value.

Fig. 4.69 *Caenotropus labyrinthicus* sampled in the Tapajós River. (Photo: Renato Silvano)

4

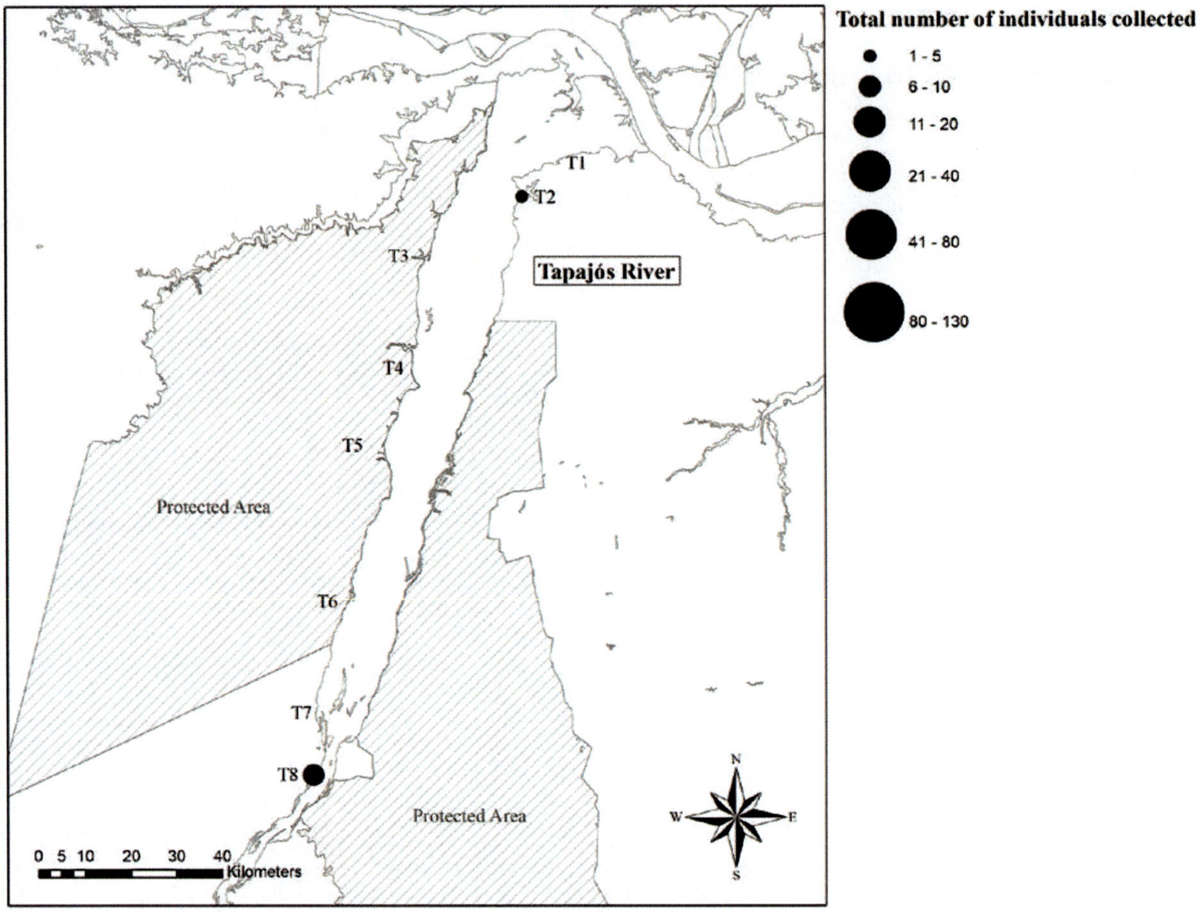

Fig. 4.70 Distribution of individuals of *Caenotropus labyrinthicus* sampled in the Tapajós River

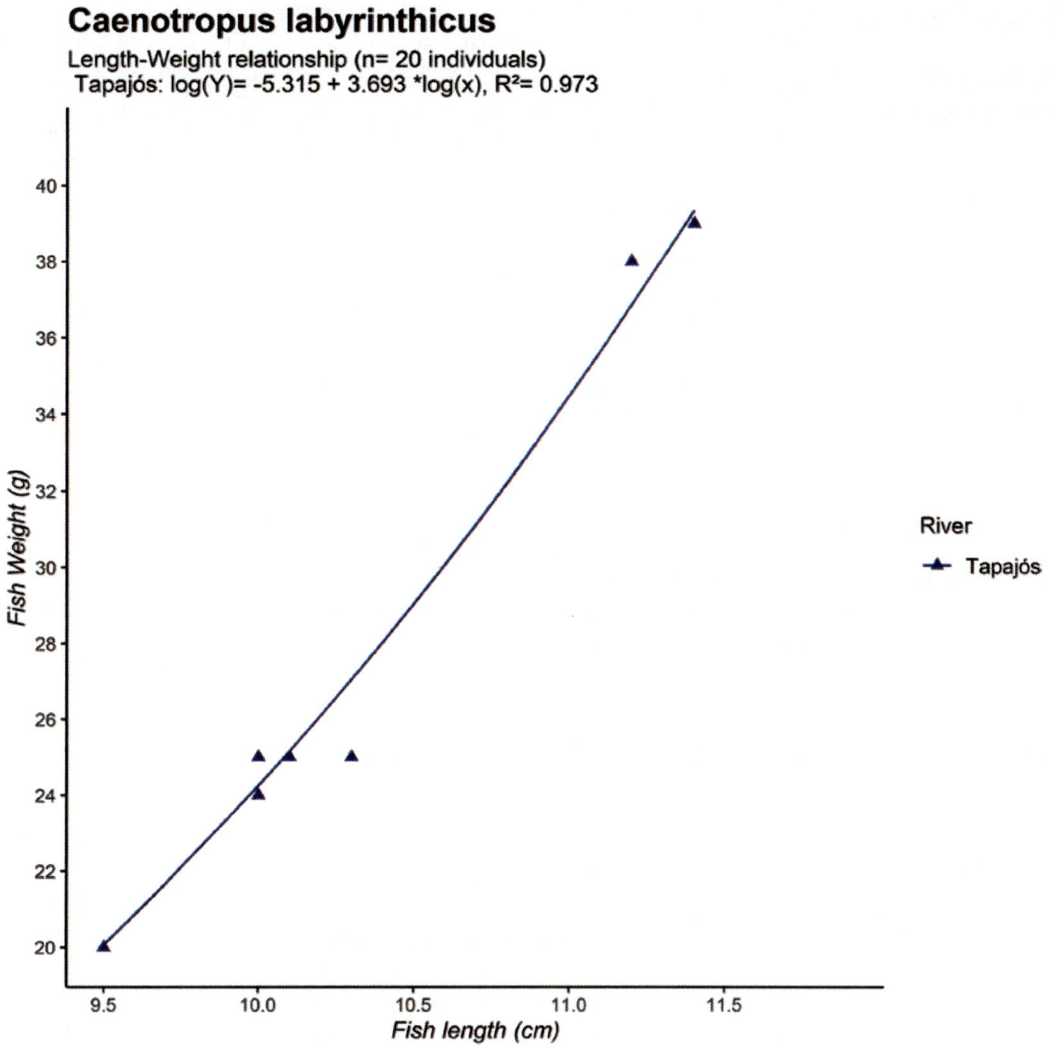

Fig. 4.71 Length-weight relationship for *Caenotropus labyrinthicus* sampled in the Tapajós River

4

4.5 Family Ctenoluciidae

4.5.1 Summary of Biology and Ecology (Literature)

Diversity: family composed of only seven valid species, distributed in two genera: *Boulengerella*, with five species, and *Ctenolucius* with two species. *Ctenolucius* do not occur in Brazil, while *Boulengerella* can be found in the Guianas, Orinoco, Amazon, and Tocantins basins.

Habitat: these species are found mainly on lotic environments with clear water, but can also occur in black and white water rivers.

Ecology and feeding: ctenoluciids, including those from the genus *Boulengerella*, are mainly piscivorous. Ctenoluciids have a preference for lotic environments with transparent waters, which is related to their visual dependency to capture prey. These fishes are ambush predators that capture prey using their fast propulsion.

Morphology: fishes with medium to large body sizes, elongated body shape, and long snout. Their mouth possesses numerous conical teeth.

Reproduction: First sexual maturation in individuals with around 25 cm lenght and the spawning period usually occurs during the flooding period.

Economic importance: ctenoluciids are not commonly caught in commercial and artisanal fisheries, but large individuals of *Boulengerella* may be valuable for recreational fishers.

References: Santos et al. (1984); de Queiroz et al. (2013,b); Ribeiro et al. (2016), Ohara et al. (2017).

Fish species: *Boulengerella cuvieri* (Spix & Agassiz 1829)

Common name(s): bicuda, urumara, peixe-agulha, peixe-agulhão

Size of fish sampled (standard length, both rivers): mean = 26.6 cm, median = 26.5 cm

Fisheries: Not relevant for fisheries, no commercial value.

Fig. 4.72 *Boulengerella cuvieri* sampled in the Negro River. (Photo: Renato Silvano)

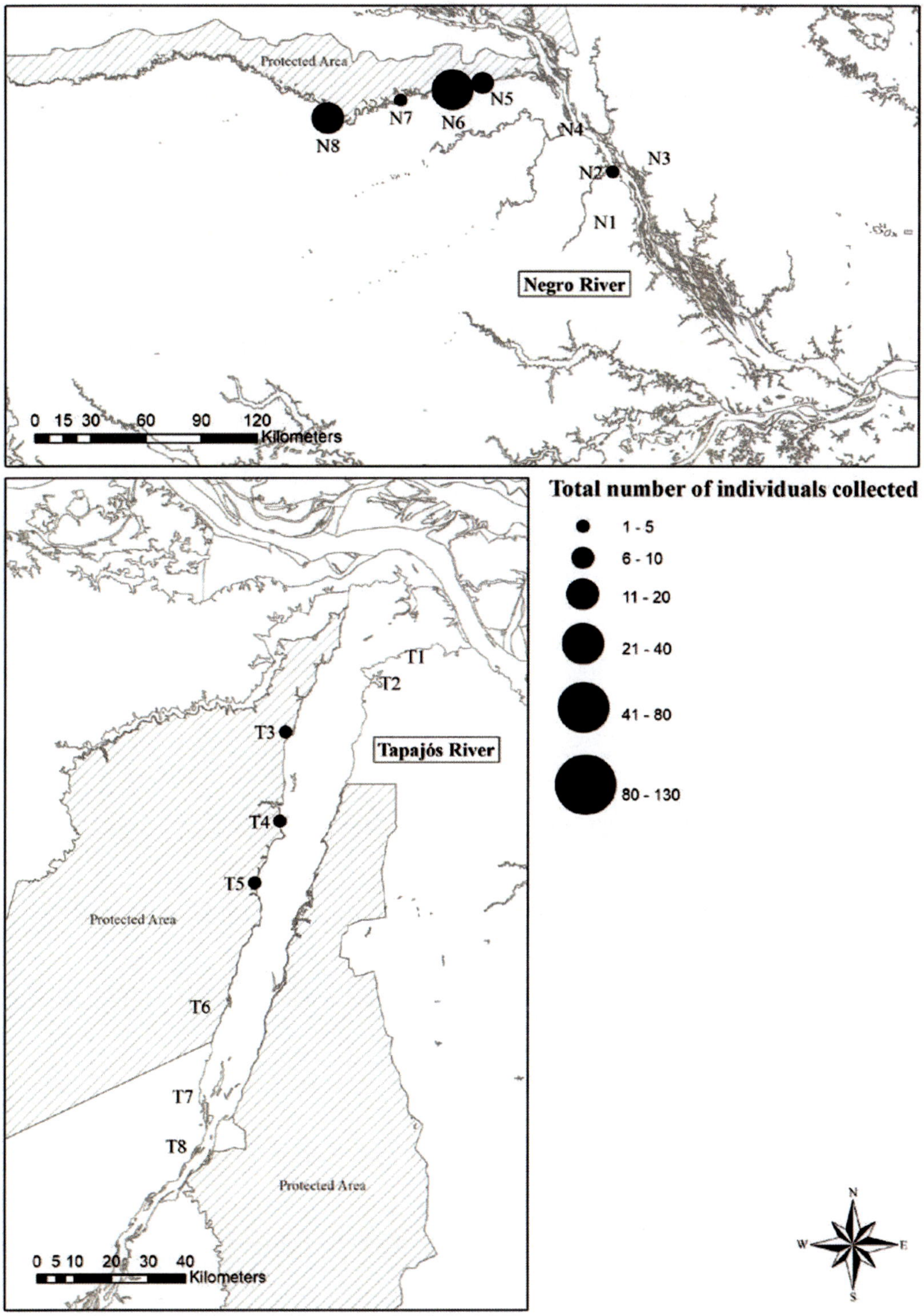

Fig. 4.73 Distribution of individuals of *Boulengerella cuvieri* sampled in the Negro and Tapajós rivers

4

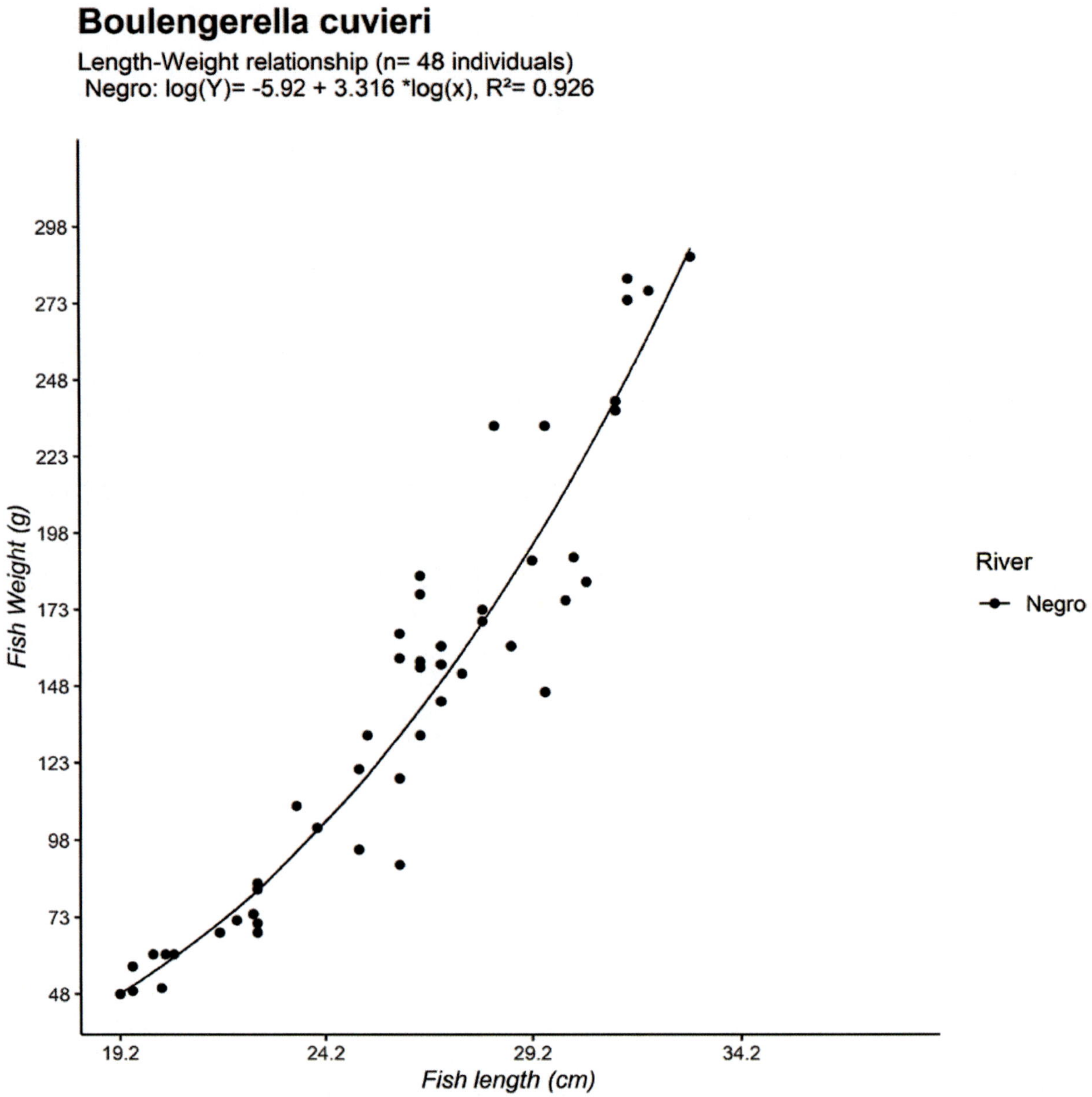

Fig. 4.74 Length-weight relationship for *Boulengerella cuvieri* sampled in Negro River

Fish species: *Boulengerella maculata* (Valenciennes 1850)

Common name(s): bicuda, urumara, peixe-agulha, peixe-agulhão

Size of fish sampled (standard length, both rivers): min. = 11.6 cm, max. = 29.5 cm, mean = 23.6 cm, median = 24 cm

Fisheries: Not relevant for fisheries, no commercial value.

Fig. 4.75 *Boulengerella maculata* sampled in the Tapajós River. (Photo: Renato Silvano)

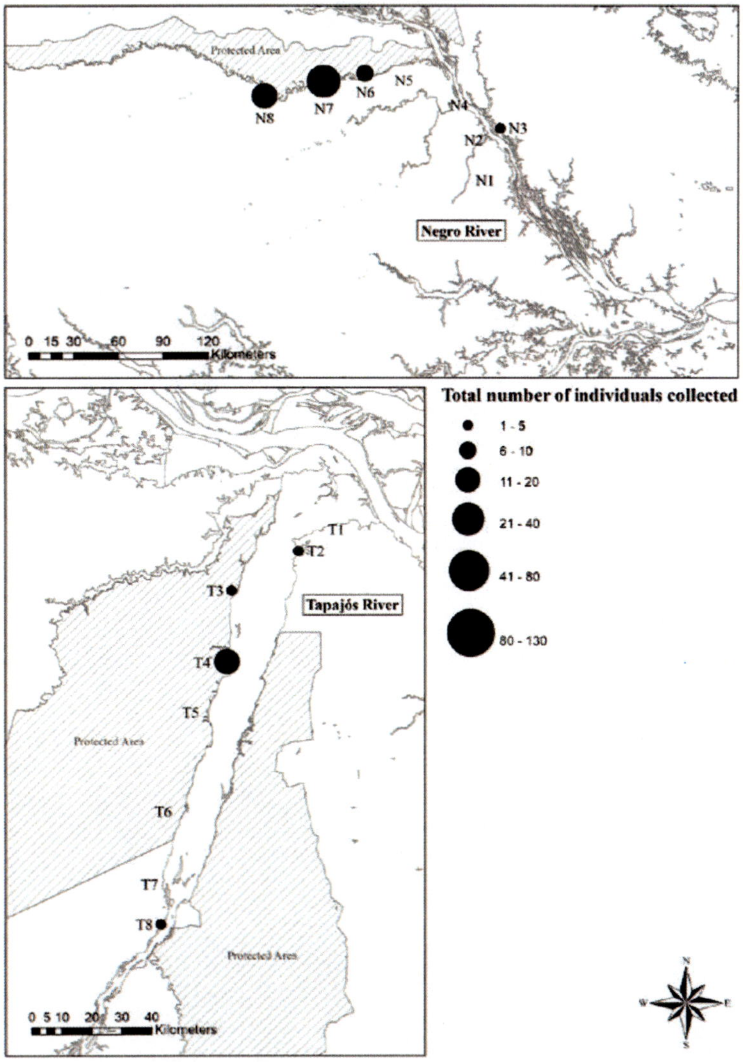

Fig. 4.76 Distribution of individuals of *Boulengerella maculata* sampled in the Negro and Tapajós rivers

4

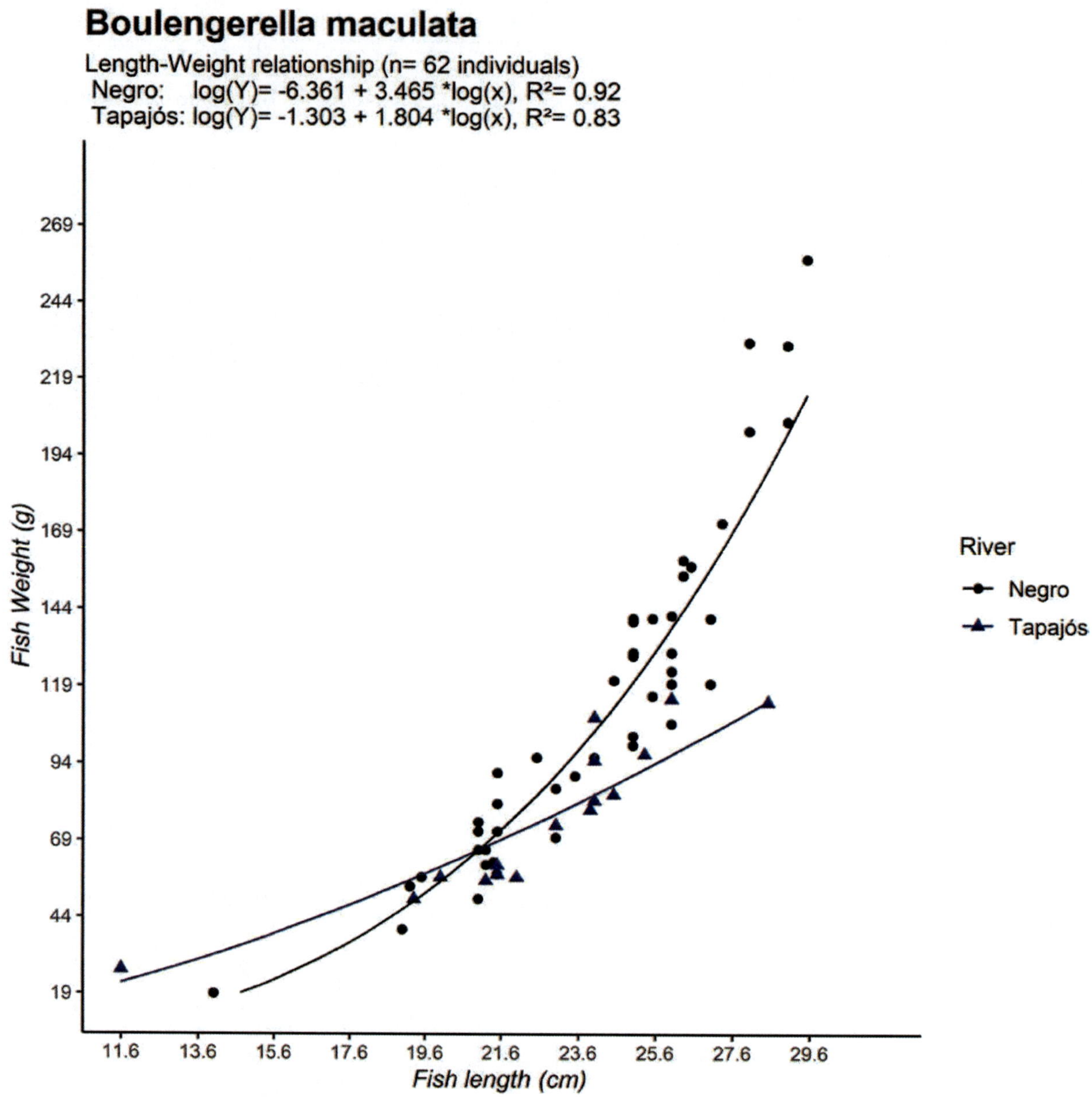

Fig. 4.77 Length-weight relationship for *Boulengerella maculata* sampled in Tapajós and Negro rivers

4.6 Family Curimatidae

4.6.1 Summary of Biology and Ecology (Literature)

Diversity: the Curimatidae family is composed of eight genera and 117 species distributed throughout Central and South America with high occurrence in the Amazon basin.

Habitat: fishes of this family inhabit preferentially lakes and floodplains of clear, black, and white water rivers. However, species of the genera *Cyphocharax* and *Steindachnerina* can also be found in clear water streams. These fishes have a diurnal habit.

Ecology and feeding: curimatids are primarily detritivores, consuming organic flocculated matter, algae, detritus, and associated microorganisms. This group of fish may have an important, although few studied, ecological role in the nutrient cycling process of freshwater ecosystems. Their benthic feeding behavior and usually high abundance also affect the transparency of the water column, which has important consequences for species interactions and phytoplankton productivity in freshwaters.

Morphology: these fishes have a small to medium body size, fusiform shape, and uniform light color, and some species with a single spot on the base of the caudal peduncle. The mouth is terminal or subterminal with absence of teeth in adults.

Reproduction: many species of this family undergo long trophic and reproductive migrations. Reproduction is usually long with complete or partial spawning that generally occurs during the beginning of the raising water period.

Economic importance: some species are very abundant and thus captured in commercial and artisanal fisheries in the Amazon basin, especially those larger species of the genera *Potamorhina* and *Psectrogaster*. However, these fishes have low commercial value and are usually caught to provide food for riverine people. Some smaller species may be also used in the aquarium trade.

References: Santos et al. (2006), Vari and Ropke (2013), Zuanon et al. (2015), Ribeiro et al. (2016), Ohara et al. (2017), Frable (2017).

Fish species: ***Curimata knerii* Steindachner 1876**

Common name(s): branquinha, jaraqui-branco

Fig. 4.78 *Curimata knerii* sampled in the Tapajós River. (Photo: Renato Silvano)

Fish species: *Curimata* sp.

Common name(s): branquinha

Fig. 4.79 *Curimata* sp. sampled in the Tapajós River. (Photo: Renato Silvano)

Fish species: *Curimata ocellata* Eigenmann & Eigenmann 1889

Common name(s): piraruira, charuto, cubiu, orana

Fig. 4.80 *Curimata ocellata* from the Tapajós River (photo from other region of the Tapajós River). (Photo: Renato Silvano)

Fish species: *Curimata roseni* Vari 1989

Common name(s): branquinha, branquinha-cascuda, cascudinha

Fig. 4.81 *Curimata roseni* sampled in the Negro River. (Photo: Renato Silvano)

Fish species: *Curimata vittata* (Kner 1858)

Common name(s): branquinha

Fig. 4.82 *Curimata vittata* sampled in the Negro River. (Photo: Renato Silvano)

Observation: abundance and size data were grouped for all these species, due to the difficulty in distinguishing them during sampling: *Curimata knerii* and *C. macrops* were collected only in Tapajós River, and *C. roseni* and *C. vittata* were collected only in the Negro River, while *C. ocellata* was collected in both rivers; see ◘ Table 1.1 (▶ Chap. 1).

Size of fish sampled (standard length, both rivers): min. = 6 cm, max. = 20.5 cm, mean = 11.8 cm, median = 11.5 cm

Fisheries: branquinha (several species from Curimatidae family) are occasionally caught mostly for food (low or no commercial value) in both rivers. Biomass caught is in ◘ Table 3.3 (▶ Chap. 3).

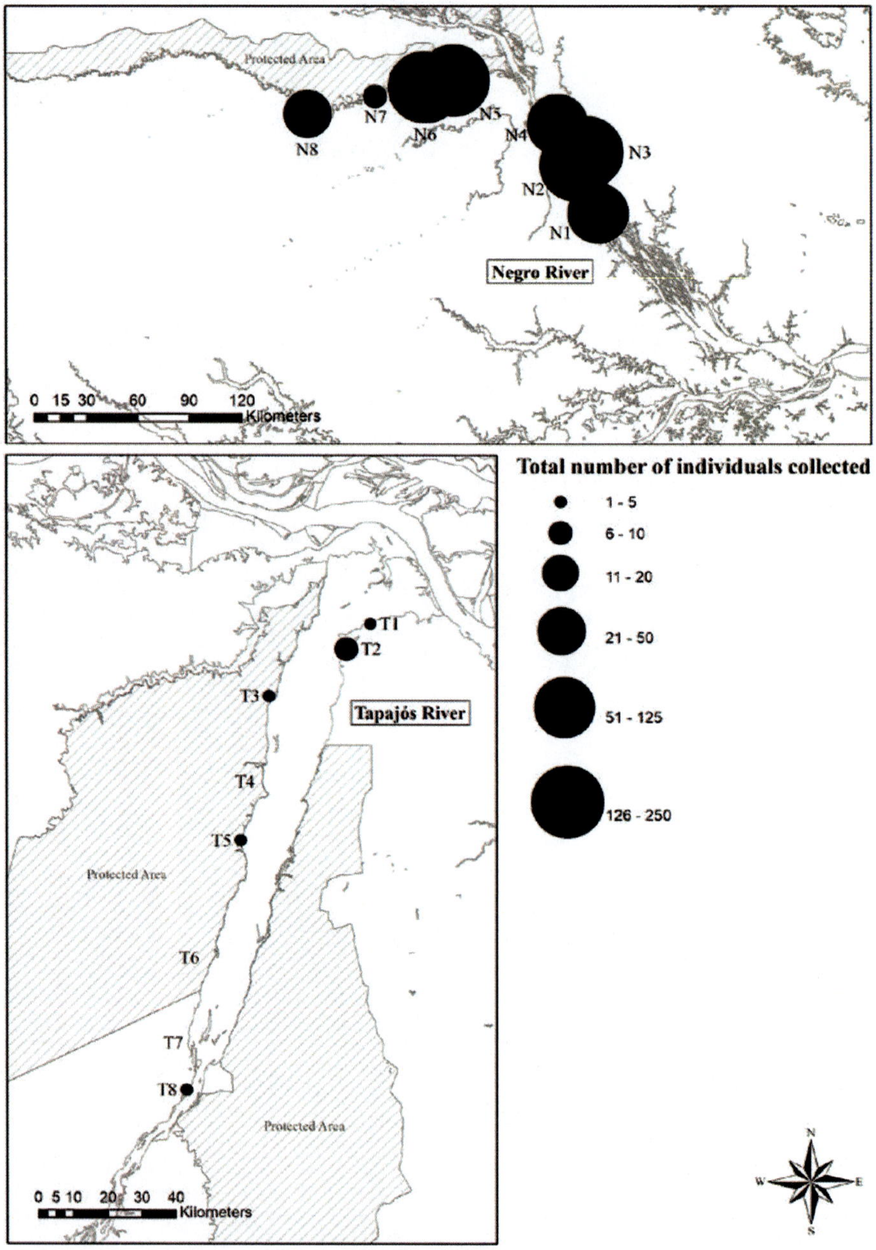

◘ **Fig. 4.83** Distribution of individuals of *Curimata* spp. sampled in the Negro and Tapajós rivers

Fish species: *Curimatella immaculata* (Fernández-Yépez 1948)

Common name(s): branquinha

Size of fish sampled (standard length, both rivers): mean = 8.7 cm, median = 8.5 cm

Fisheries: branquinha (several species from Curimatidae family) are occasionally caught mostly for food (low or no commercial value) in both rivers; see *Curimatella meyeri* for fisheries data on branquinha (Curimatidae). Biomass caught is in ◘ Table 3.3 (► Chap. 3).

◘ **Fig. 4.84** *Curimatella immaculata* sampled in the Tapajós River. (Photo: Renato Silvano)

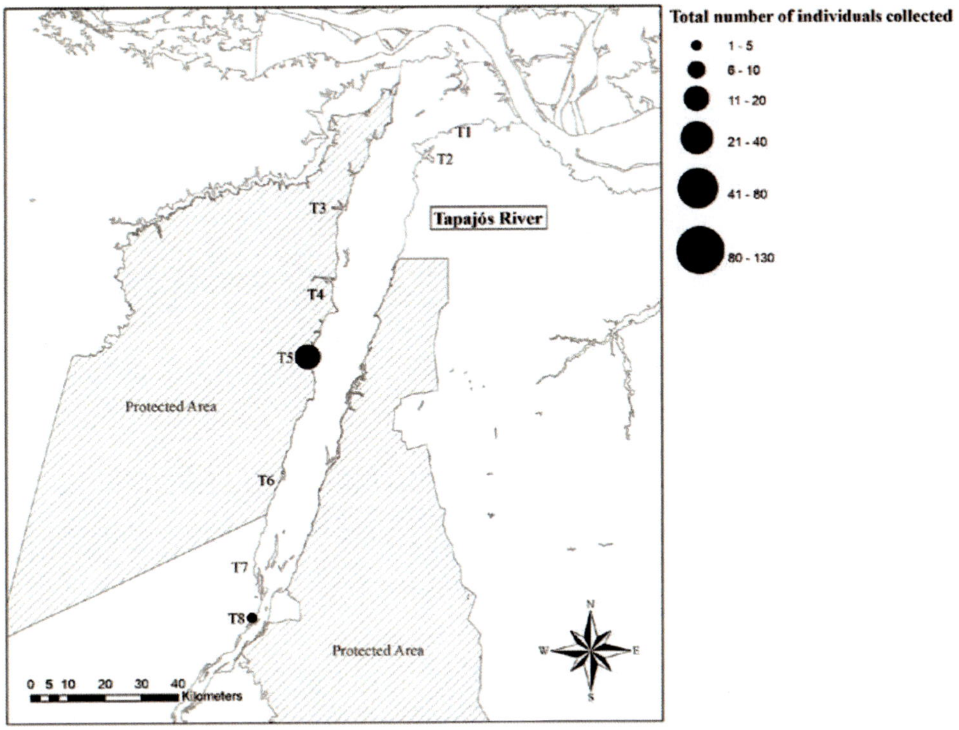

◘ **Fig. 4.85** Distribution of individuals of *Curimatella immaculata* sampled in the Tapajós River

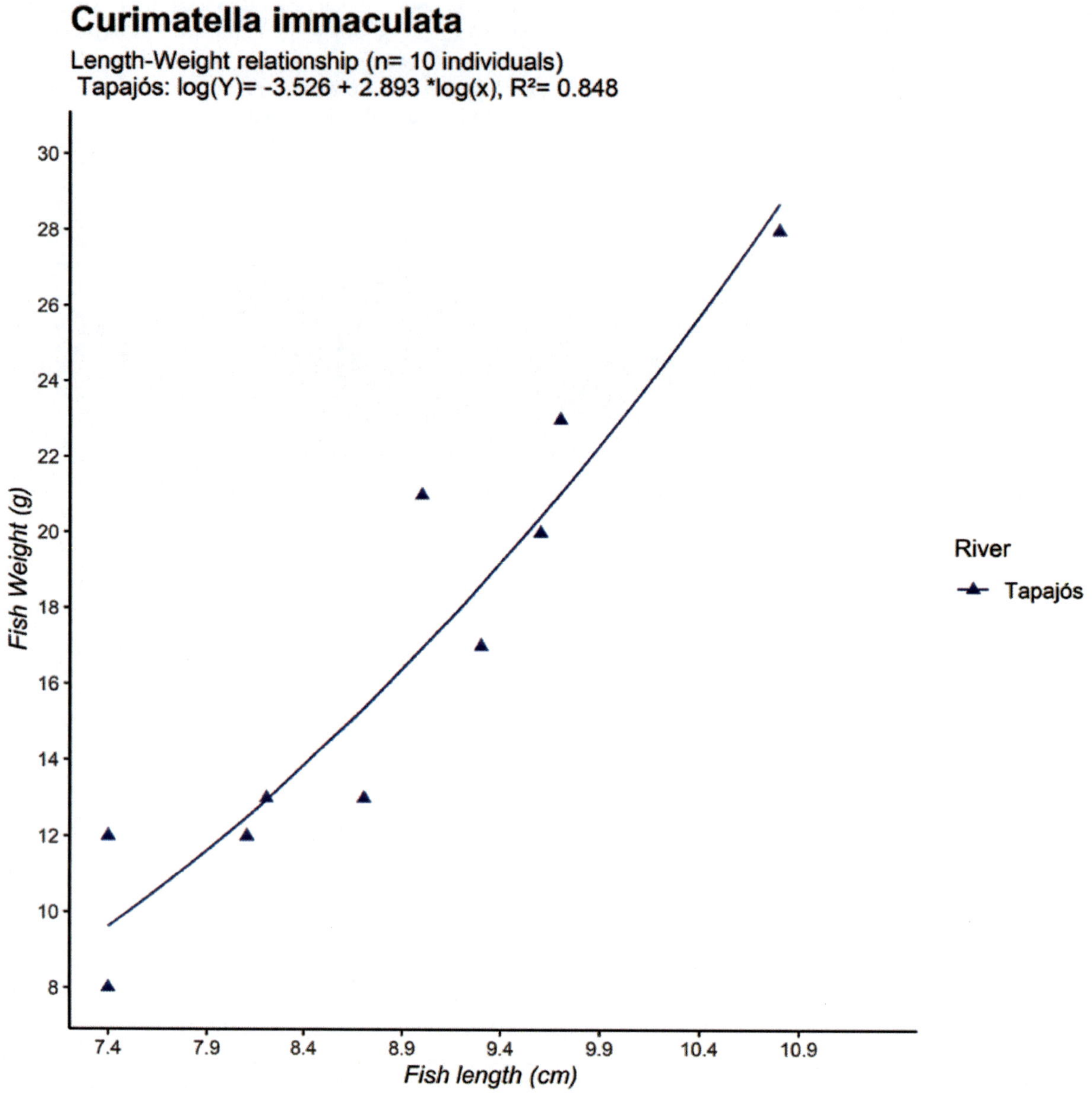

Fig. 4.86 Length-weight relationship for *Curimatella immaculata* sampled in the Tapajós river

Fish species: *Curimatella meyeri* (Steindachner 1882)

Common name(s): branquinha, branquinha-escama-grossa, cururu, piracururu, chorona

Size of fish sampled (standard length, both rivers): mean = 10.9 cm, median = 10 cm

Fig. 4.87 *Curimatella meyeri* sampled in the Negro River. (Photo: Renato Silvano)

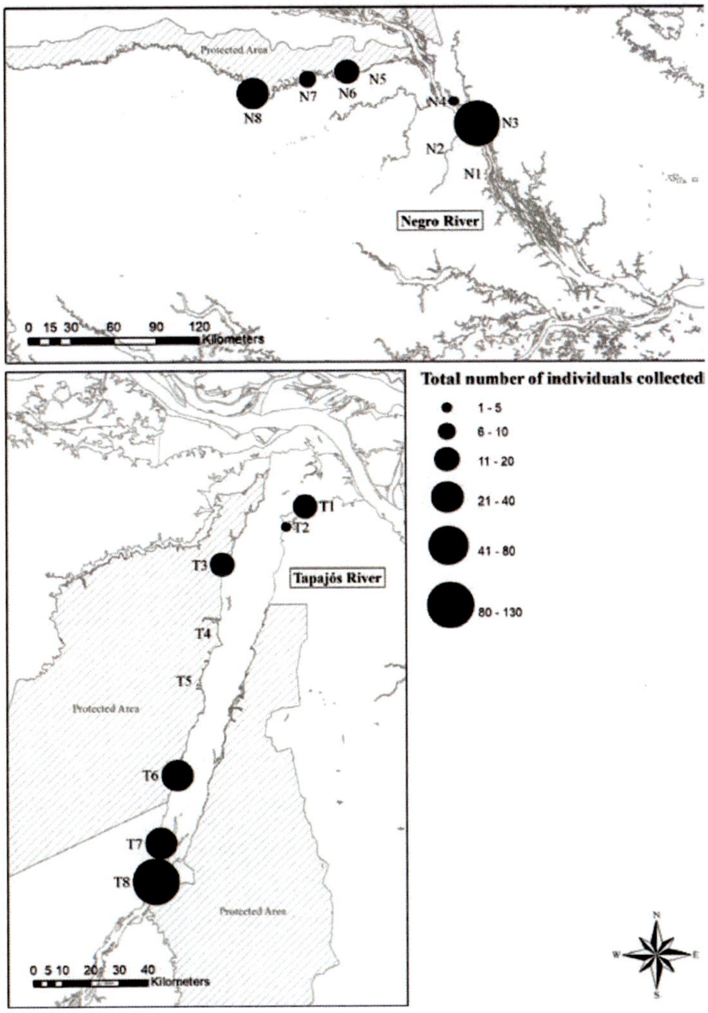

Fig. 4.88 Distribution of individuals of *Curimatella meyeri* sampled in the Negro and Tapajós rivers

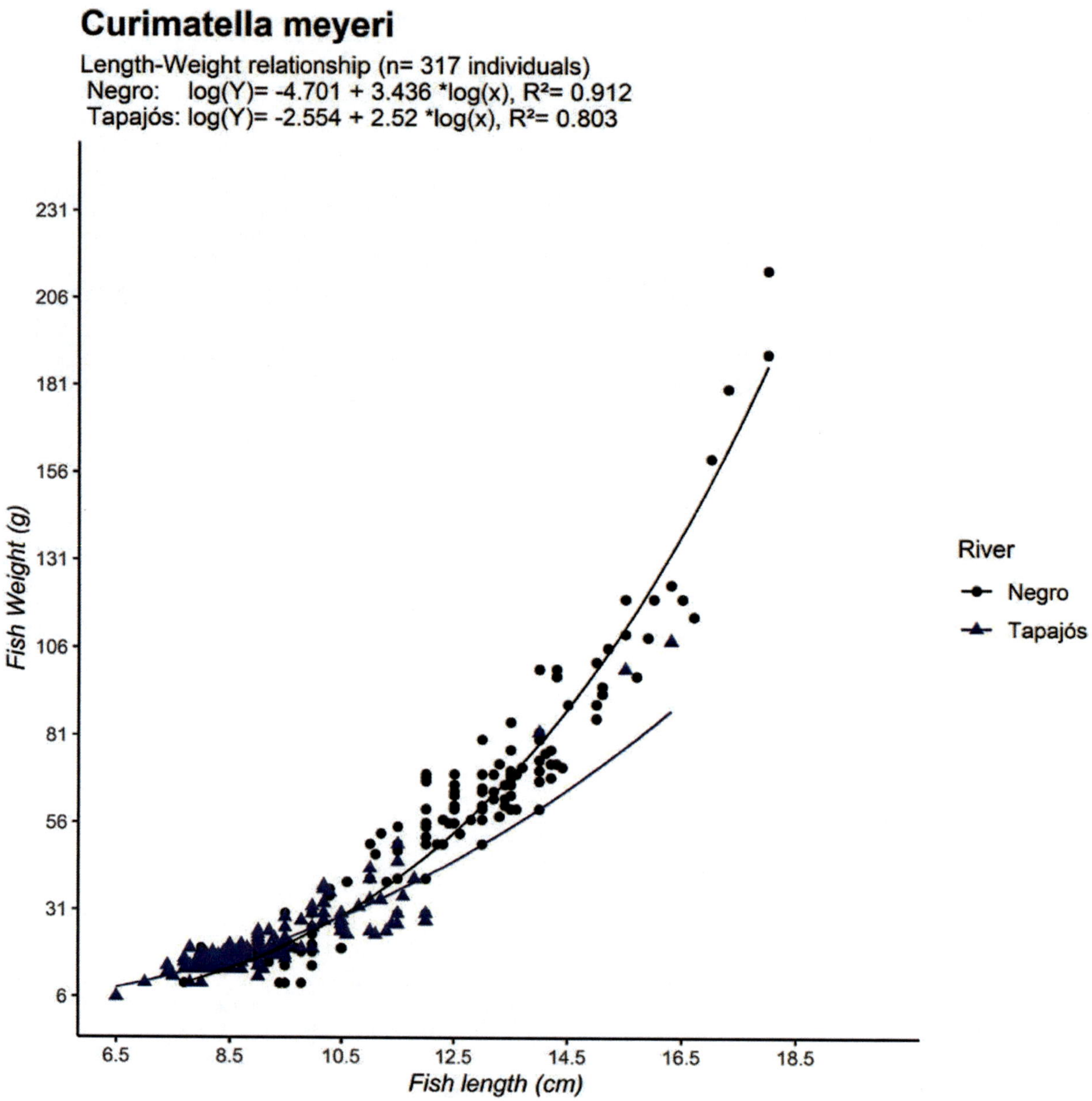

Fig. 4.89 Length-weight relationship for *Curimatella meyeri* sampled in Negro and Tapajós rivers

Fisheries: branquinha (several species from Curimatidae family) are occasionally caught mostly for food (low or no commercial value) in both rivers. Biomass caught is in Table 3.3 (▶ Chap. 3).

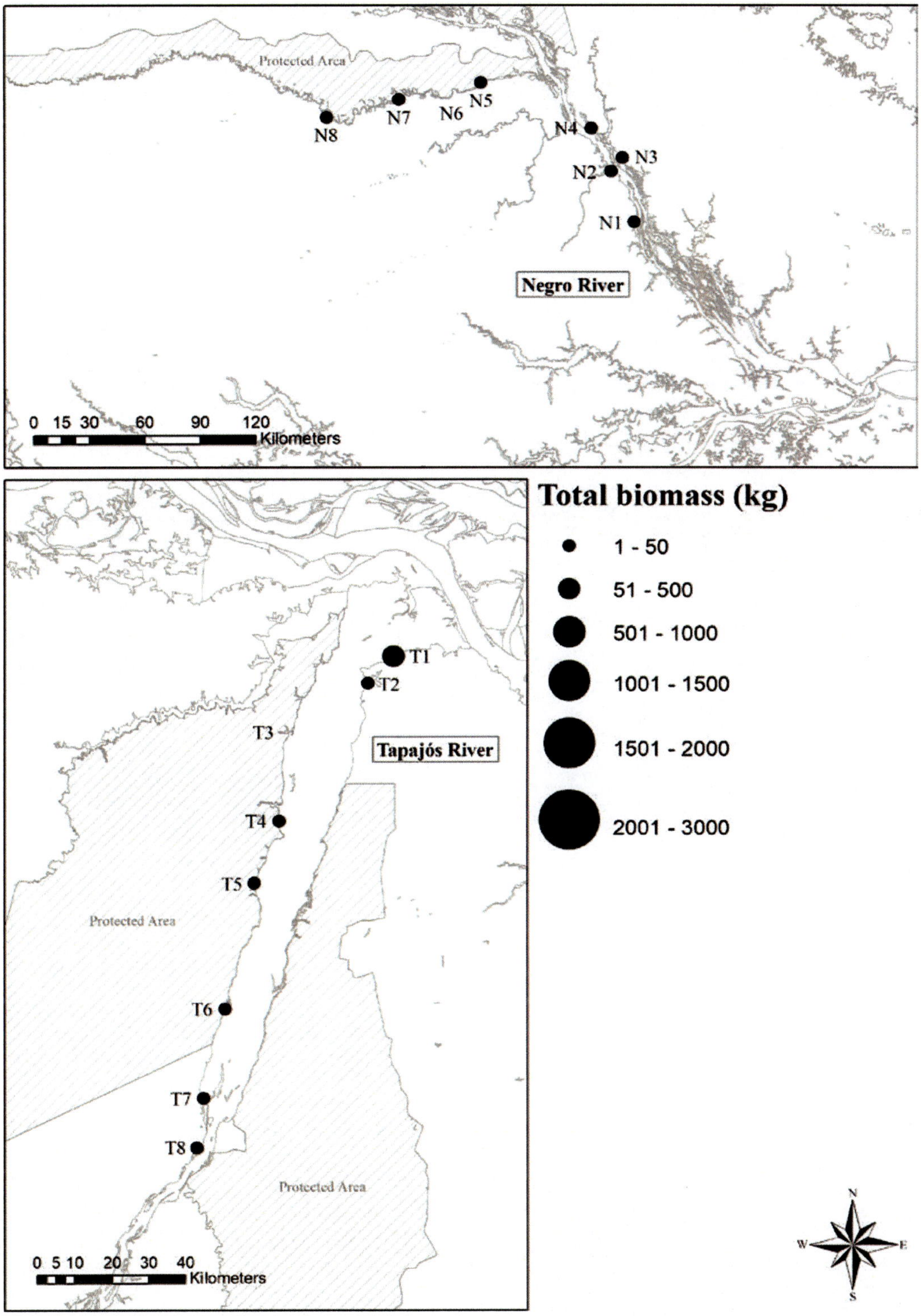

Fig. 4.90 Distribution of biomass of branquinha (Curimatidae) caught in small-scale fisheries in the Negro and Tapajós rivers

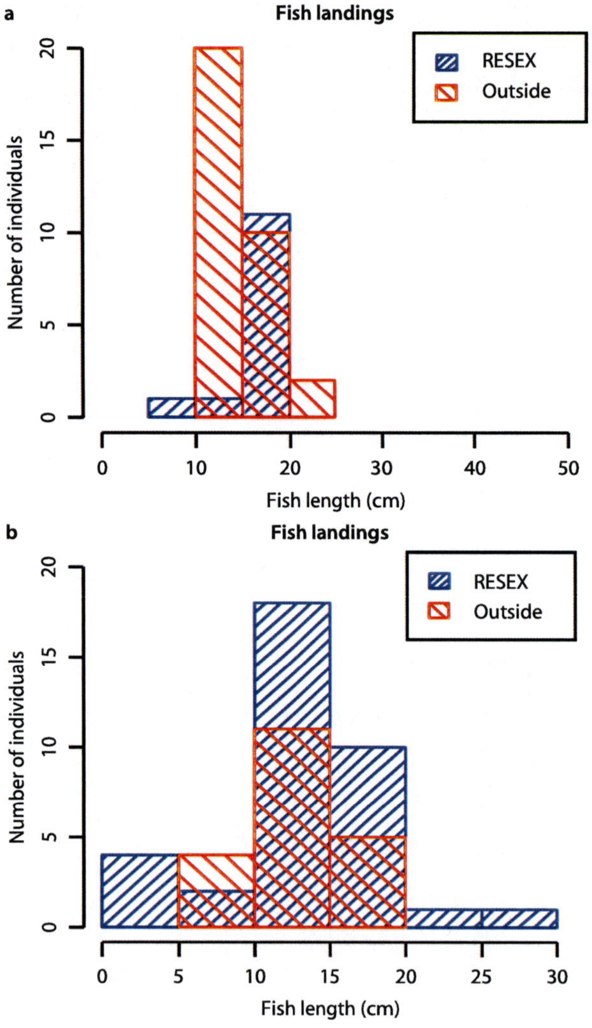

Fig. 4.91 Size distribution of branquinha (Curimatidae) caught in small-scale fisheries in 16 communities inside and outside the Extractive Reserves (RESEX) in the rivers **a** Negro, N Resex = 13, N outside = 32 and **b** Tapajós, N Resex = 36, N outside = 20

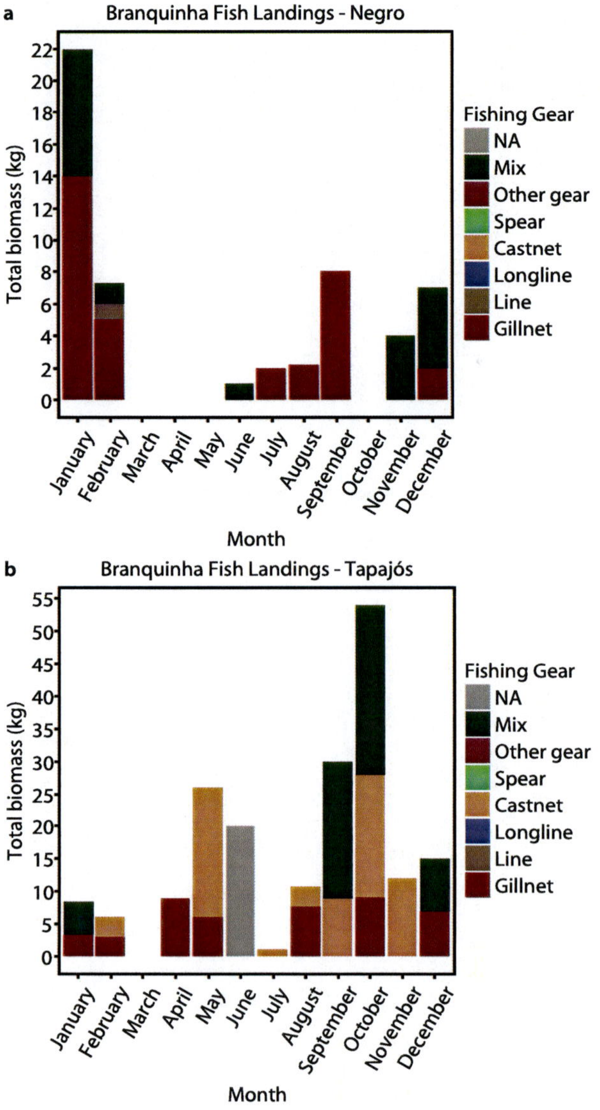

Fig. 4.92 Seasonal variation of biomass of branquinha (Curimatidae) caught per fishing gear in small-scale fisheries in 16 communities in the rivers **a** Negro and **b** Tapajós. NA: fishing gear not informed; Mix: more than one gear was used

Fish species: *Cyphocharax abramoides* (Kner 1858)

Common name(s): branquinha

Fig. 4.93 *Cyphocharax abramoides* sampled in the Negro River. (Photo: Renato Silvano)

Fish species: *Cyphocharax leucostictus* (Eigenmann & Eigenmann 1889)

Common name(s): branquinha

Fig. 4.94 *Cyphocharax leucostictus* sampled in the Negro River. (Photo: Renato Silvano)

Observation: abundance and size data were grouped for these two species: *Cyphocharax abramoides* was collected in both rivers and *C. leucostictus* was collected only in the Negro River; see ◘ Table 1.1 (▶ Chap. 1).

Size of fish sampled (standard length, both rivers): min. = 5.4 cm, max. = 18.5 cm, mean = 9.6 cm, median = 9.5 cm

Fisheries: branquinha (several species from Curimatidae family) are occasionally caught mostly for food (low or no commercial value) in both rivers; see *Curimatella meyeri* for fisheries data on branquinha (Curimatidae). Biomass caught is in ◘ Table 3.3 (▶ Chap. 3).

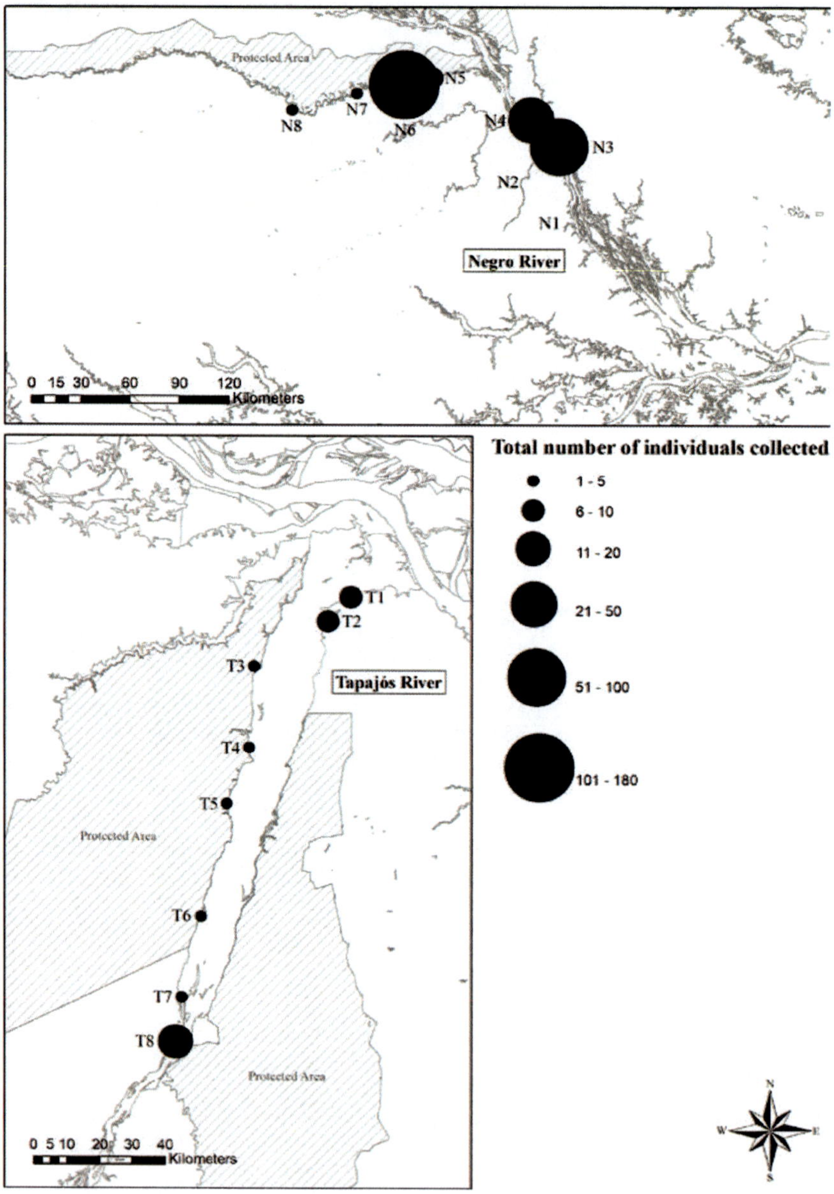

◘ **Fig. 4.95** Distribution of individuals of *Cyphocharax* spp. sampled in the Negro and Tapajós rivers

Fish species: *Potamorhina latior* (Spix & Agassiz 1829)

Common name(s): branquinha, branquinha-peito-de-aço

Fig. 4.96 *Potamorhina latior* sampled in the Negro River. (Photo: Renato Silvano)

Fish species: *Potamorhina pristigaster* (Steindachner 1876)

Common name(s): branquinha, branquinha-peito-de-aço

Size of fish sampled (standard length, both rivers): min. = 6 cm, max. = 22 cm, mean = 12.3 cm, median = 12 cm

Fisheries: branquinha (several species from Curimatidae family) are occasionally caught mostly for food (low or no commercial value) in both rivers; see *Curimatella meyeri* for fisheries data on branquinha (Curimatidae). Biomass caught is in Table 3.3 (► Chap. 3).

Fig. 4.97 *Potamorhina pristigaster* sampled in the Negro River. (Photo: Renato Silvano)

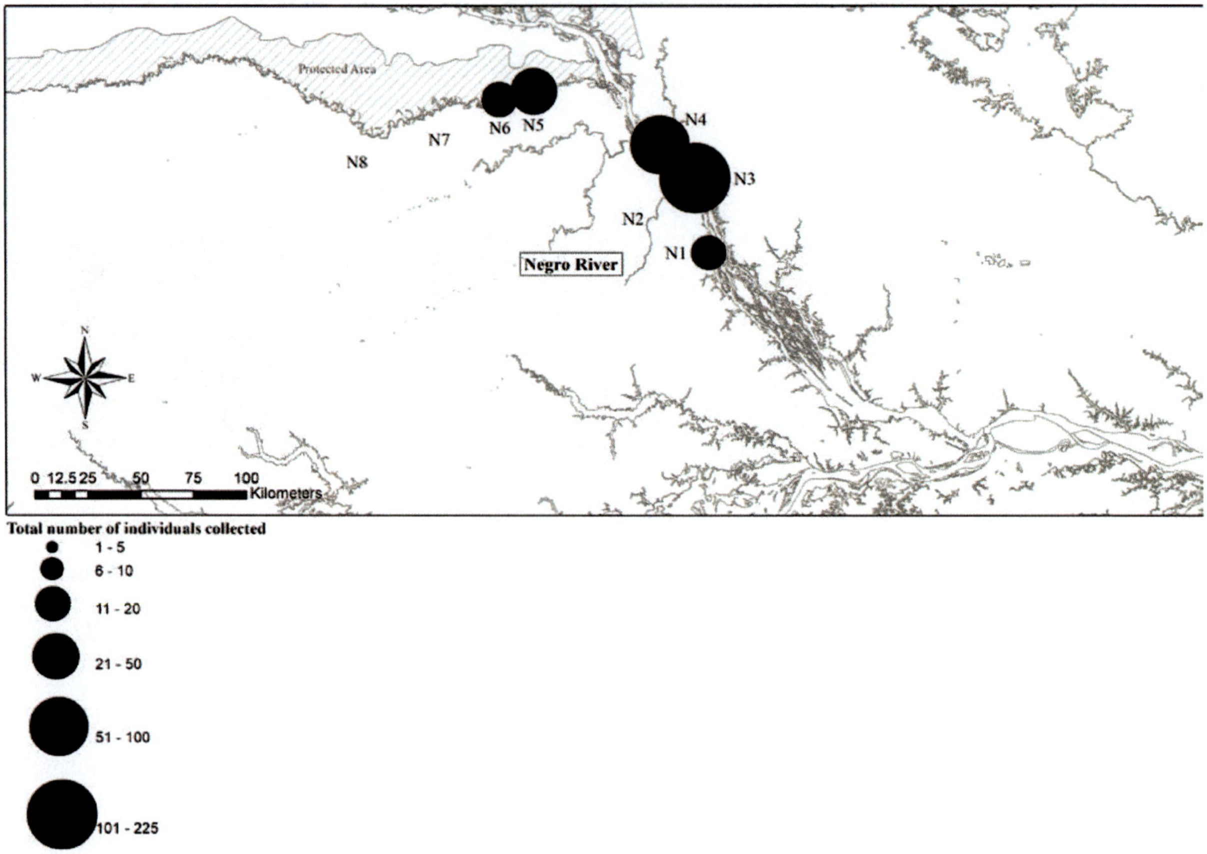

Fig. 4.98 Distribution of individuals of *Potamorhina* spp. sampled in the Negro River

Fish species: *Psectrogaster* cf. *falcata*

Common name(s): branquinha, jaraqui-branco

Size of fish sampled (standard length, both rivers): mean = 9.4 cm, median = 8.5 cm

Fisheries: branquinha (several species from Curimatidae family) are occasionally caught mostly for food (low or no commercial value) in both rivers; see *Curimatella meyeri* for fisheries data on branquinha (Curimatidae). Biomass caught is in Table 3.3 (► Chap. 3).

Fig. 4.99 *Psectrogaster* cf. *falcata* sampled in the Tapajós River. (Photo: Renato Silvano)

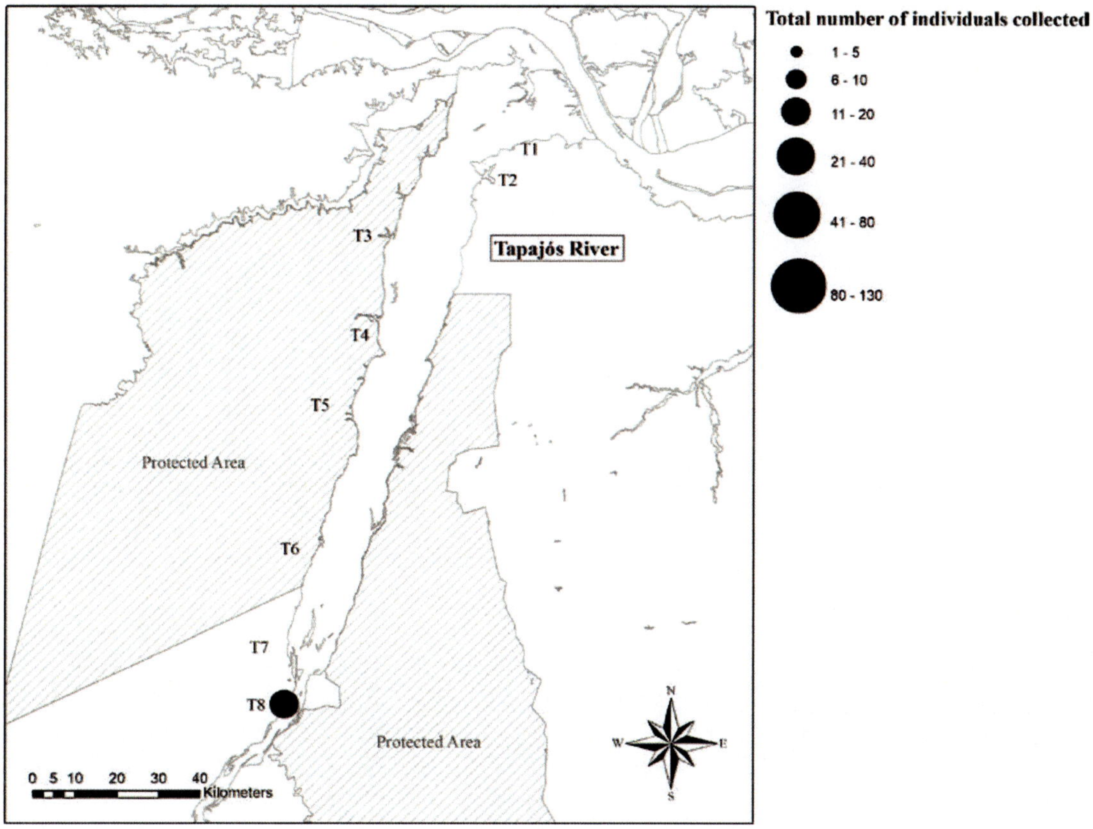

Fig. 4.100 Distribution of individuals of *Psectrogaster* cf. *falcata* sampled in the Tapajós River

4

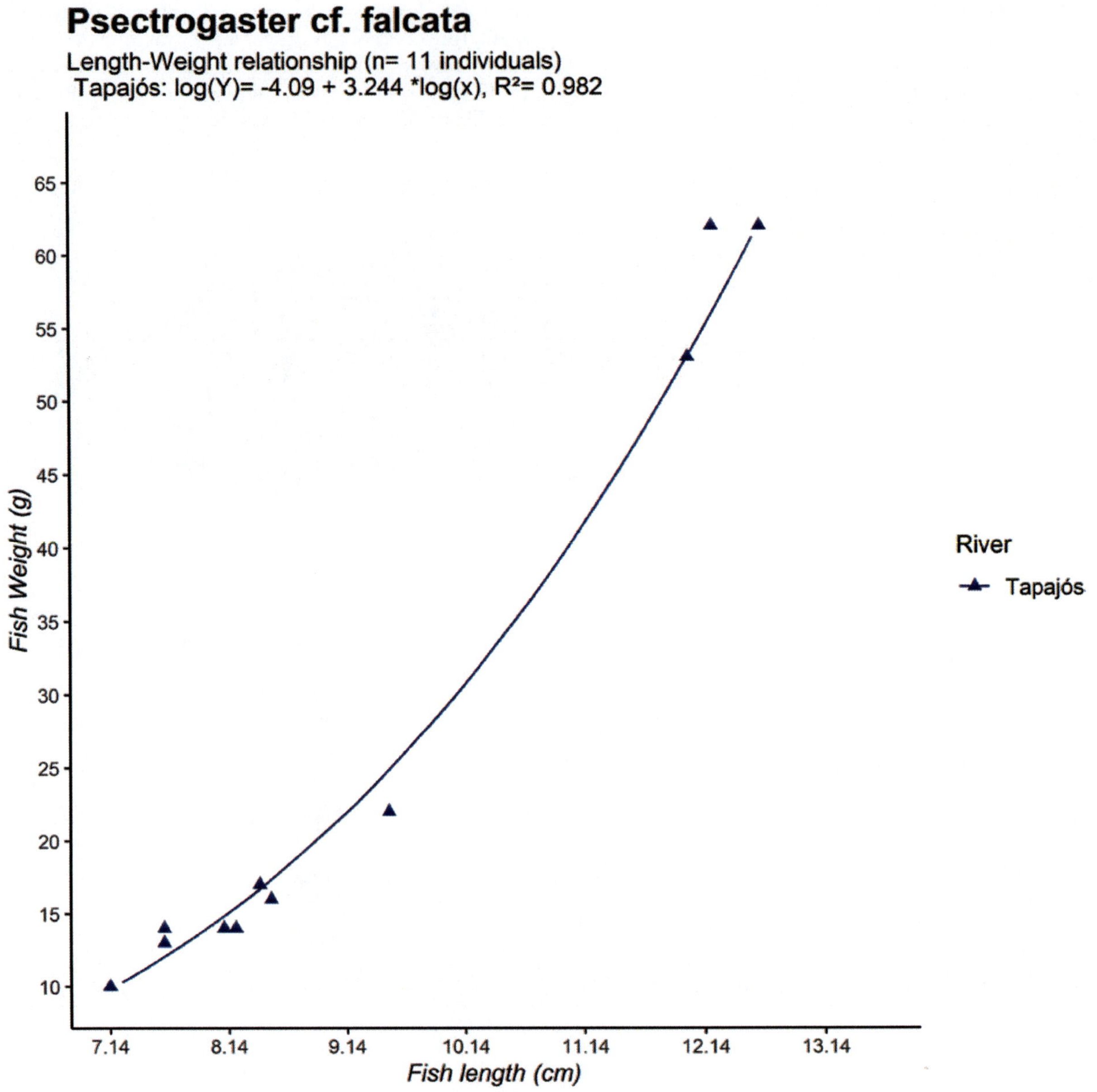

Fig. 4.101 Length-weight relationship for *Psectrogaster* cf. *falcata* sampled in Tapajós River

Fish species: *Psectrogaster rutiloides* (Kner 1858)

Common name(s): branquinha, branquinha-curimatá

Size of fish sampled (standard length, both rivers): min. = 15.5 cm, max. = 16 cm, mean and median = 15.8 cm

Fisheries: branquinha (several species from Curimatidae family) are occasionally caught mostly for food (low or no commercial value) in both rivers; see *Curimatella meyeri* for fisheries data on branquinha (Curimatidae). Biomass caught is in ◘ Table 3.3 (► Chap. 3).

◘ **Fig. 4.102** *Psectrogaster rutiloides* sampled in the Negro River. (Photo: Renato Silvano)

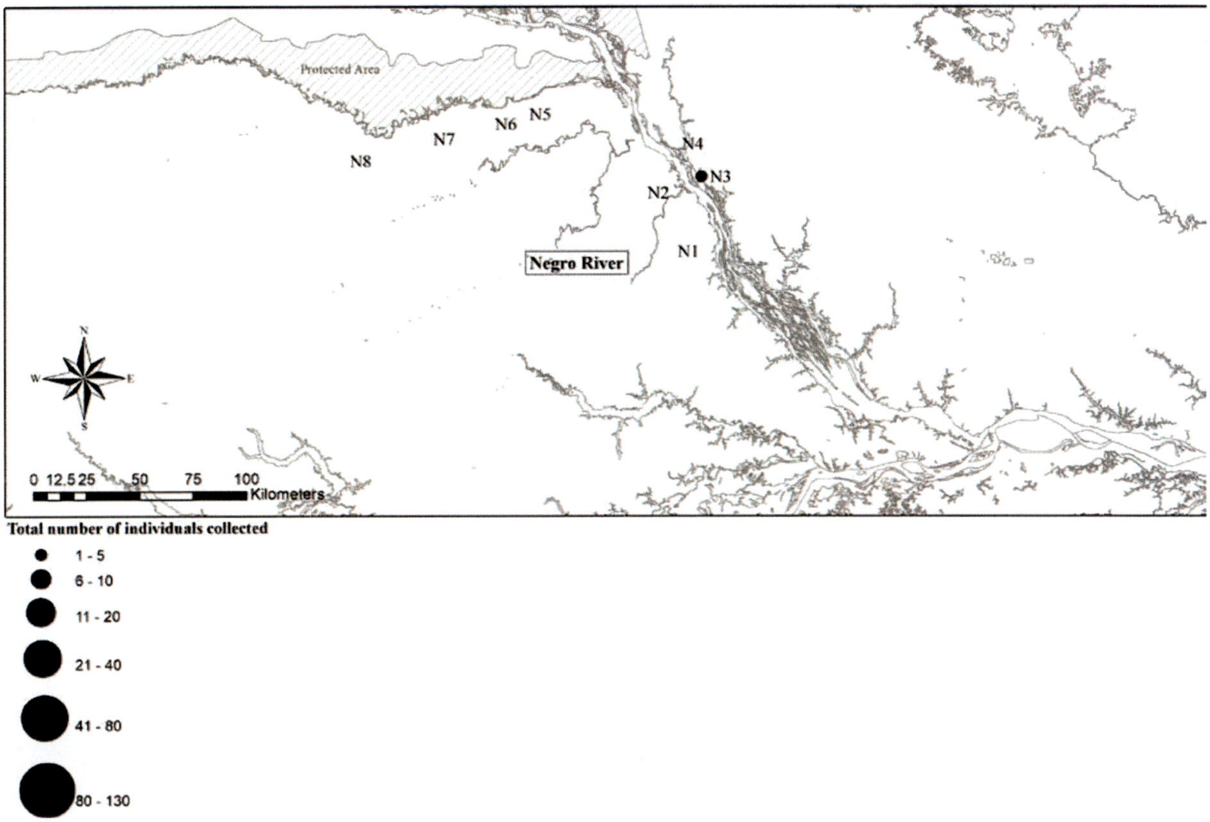

◘ **Fig. 4.103** Distribution of individuals of *Psectrogaster rutiloides* sampled in the Negro River

4

4.7 Family Cynodontidae

4.7.1 Summary of Biology and Ecology (Literature)

Diversity: family composed of eight species divided into four genera. After this book was organized, some changes had occurred and the genus *Roestes* was assigned to other family (Acestrorhynchidae).

Habitat: fishes of this family are mainly distributed in the Amazon and Orinoco basins, except *Rhaphiodon vulpinus*, which also occurs in other South American basins.

Ecology and feeding: members of this family are adapted to the pelagic habitat. Most species are visual predators that feed on other fishes.

Morphology: these fishes are easily recognizable by their long canines and that's why they are popularly known as peixe cachorro (dog-fish). Fishes of this family have medium to large body sizes. Their body shape is elongated and laterally compressed. Pectoral fins are enlarged and located in an upper body position. They possess a pre-ventral keel in the median line of the body and spiny gill rakers; the mouth is oblique and large. One member of this family, *Rhaphiodon vulpinus*, is known for having the capacity to move their head upward, in a 45° degree angle.

Reproduction: some species, such as *Cynodon gibbus*, undergo reproductive migrations.

Economic importance: some larger species of this family are occasionally exploited by artisanal and recreational fisheries, but these fishes have low commercial value. Heads of these fishes can be used by handcrafters, due to their long and sharp canines.

References: Toledo-Piza et al. (2013), Dos Anjos (2009).

Fish species: *Cynodon gibbus* (Spix & Agassiz 1829)

Common name(s): peixe-cachorro, pirandirá, saranha

Size of fish sampled (standard length, both rivers): mean = 19.2 cm, median = 19 cm

Fisheries: peixe cachorro (*Acestrorhynchus* spp. and larger species of Cynodontidae) is not important to fisheries and has few or no commercial value, being only occasionally caught in the Tapajós and Negro rivers. Biomass caught is in ◘ Table 3.3 (▶ Chap. 3).

◘ **Fig. 4.104** *Cynodon gibbus* sampled in the Tapajós River. (Photo: Renato Silvano)

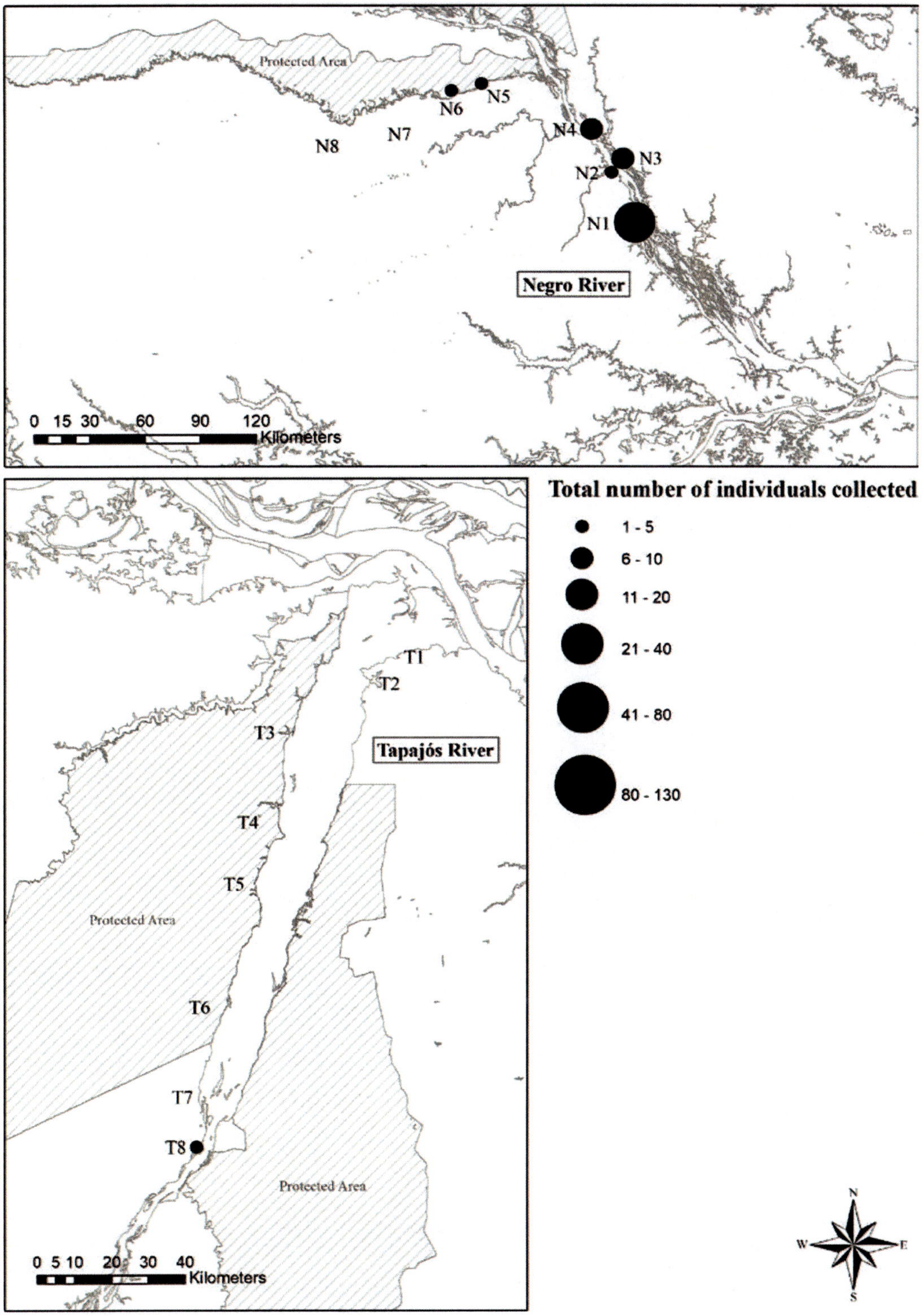

Fig. 4.105 Distribution of individuals of *Cynodon gibbus* sampled in the Negro and Tapajós Rivers

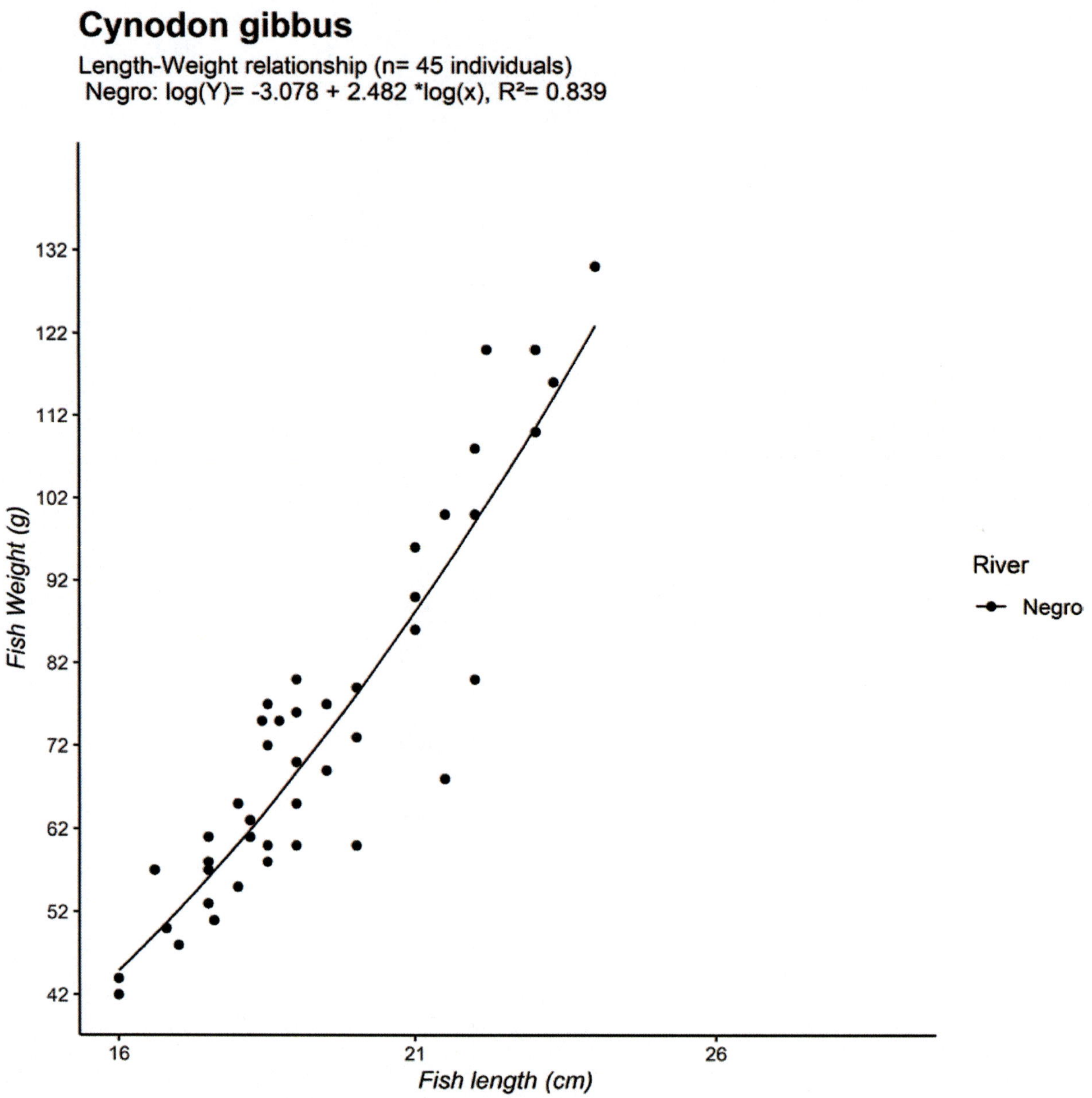

Fig. 4.106 Length-weight relationship for *Cynodon gibbus* sampled in Negro River

Fish species: *Hydrolycus wallacei* Toledo-Piza Menezes & Santos 1999

Common name(s): peixe-cachorro, pirandirá

Size of fish sampled (standard length, both rivers): 16 cm

Fisheries: peixe cachorro (*Acestrorhynchus* spp. and larger species of Cynodontidae) is not important to fisheries and has few or no commercial value, being only occasionally caught in the Tapajós and Negro rivers. Biomass caught is in ◘ Table 3.3 (▶ Chap. 3).

◘ **Fig. 4.107** *Hydrolycus wallacei* sampled in the Negro River. (Photo: Renato Silvano)

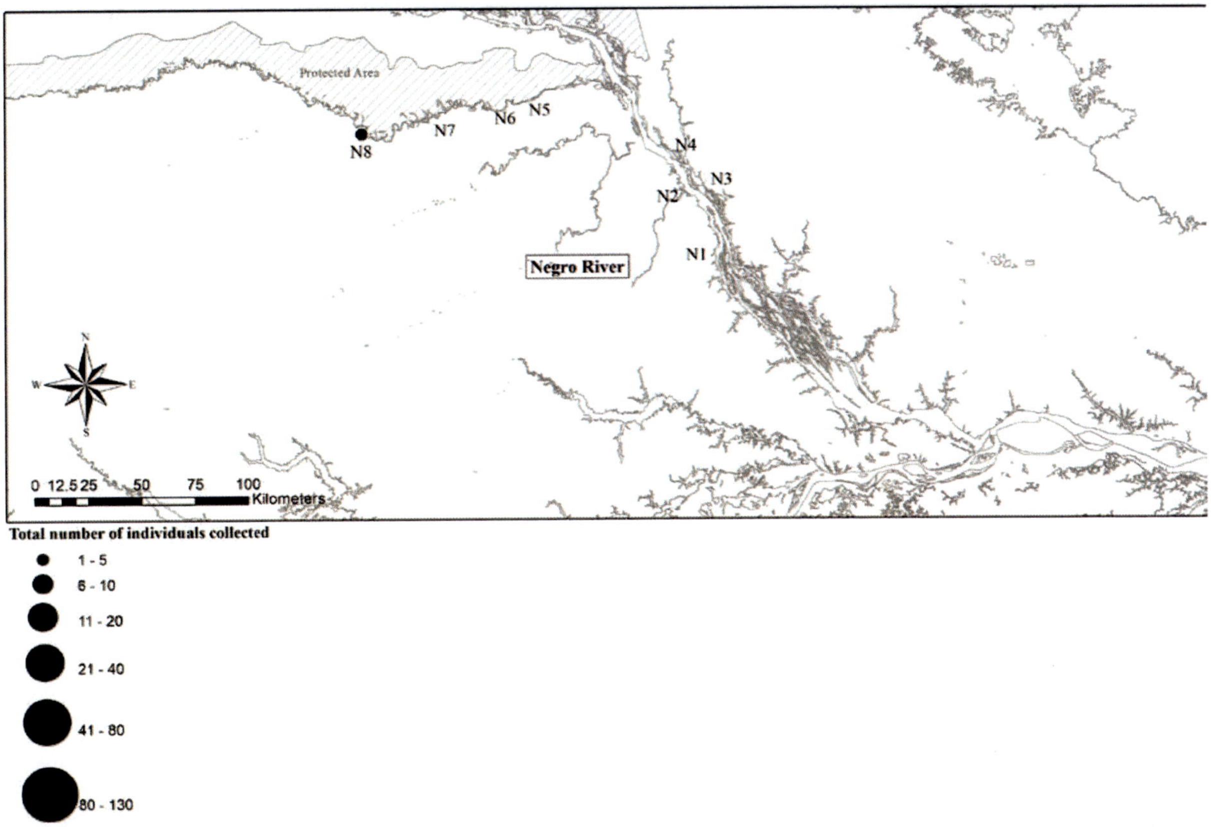

◘ **Fig. 4.108** Distribution of individuals of *Hydrolycus wallacei* sampled in the Negro River

Fish species: *Rhaphiodon vulpinus* Spix & Agassiz 1829

Common name(s): peixe-cachorro, peixe-facão

Size of fish sampled (standard length, both rivers): min. = 28 cm, max. = 46.2 cm, mean = 35.7 cm, median = 36.5 cm

Fisheries: peixe cachorro (*Acestrorhynchus* spp. and larger species of Cynodontidae) is not important to fisheries and has few or no commercial value, being only occasionally caught in the Tapajós and Negro rivers. Biomass caught is in ◘ Table 3.3 (► Chap. 3).

◘ **Fig. 4.109** *Rhaphiodon vulpinus* sampled in the Tapajós River. (Photo: Renato Silvano)

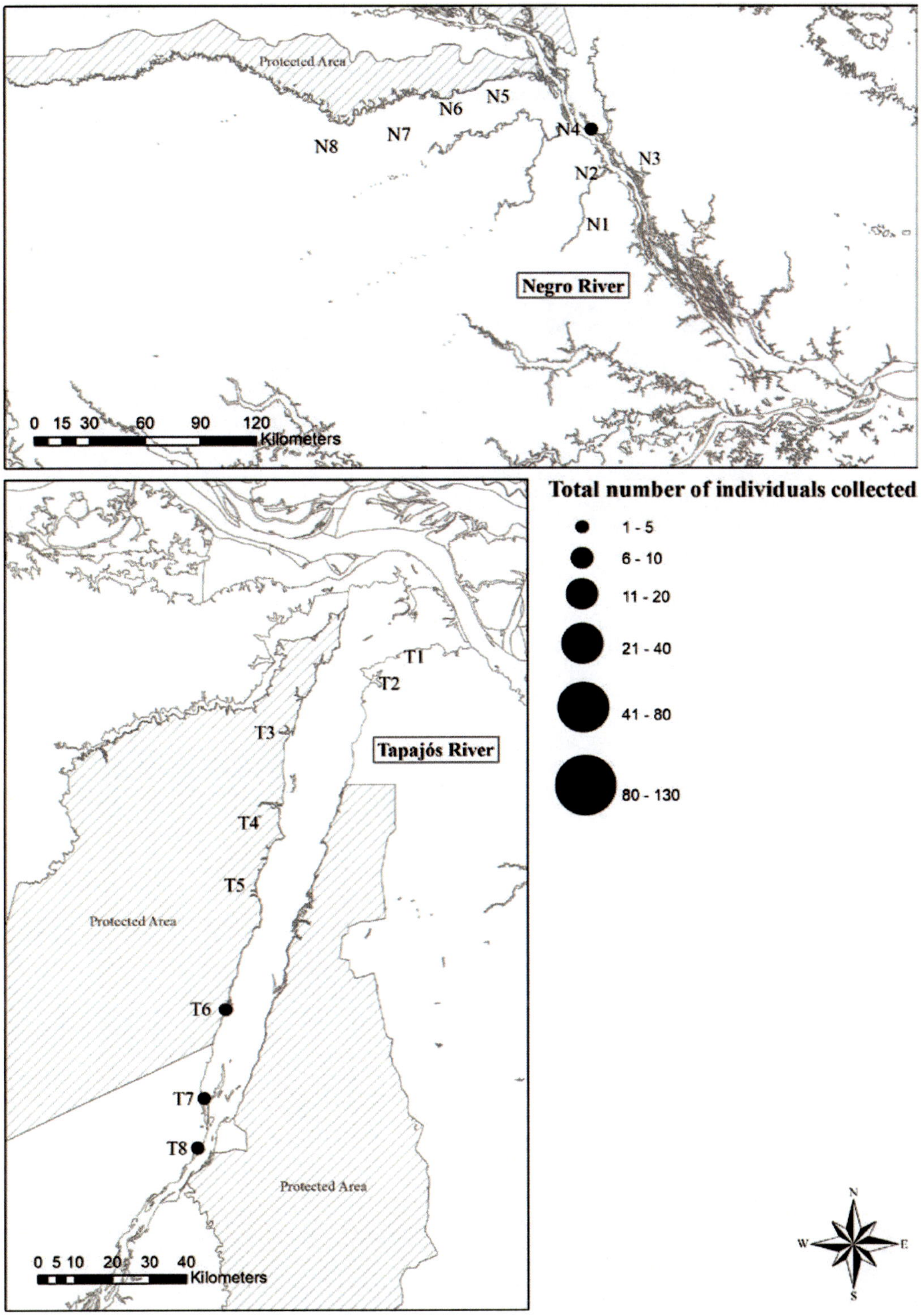

Fig. 4.110 Distribution of individuals of *Rhaphiodon vulpinus* sampled in the Negro and Tapajós rivers

Fish species: ***Roestes molossus*** **(Kner 1858)**

Common name(s): peixe-cachorro, pirandirá

Size of fish sampled (standard length, both rivers): mean = 13.6 cm, median = 14 cm

Fisheries: not relevant to fisheries, due to its small size.

4

Fig. 4.111 *Roestes molossus* sampled in the Negro River. This species now belongs to the family Acestrorhynchidae. (Photo: Renato Silvano)

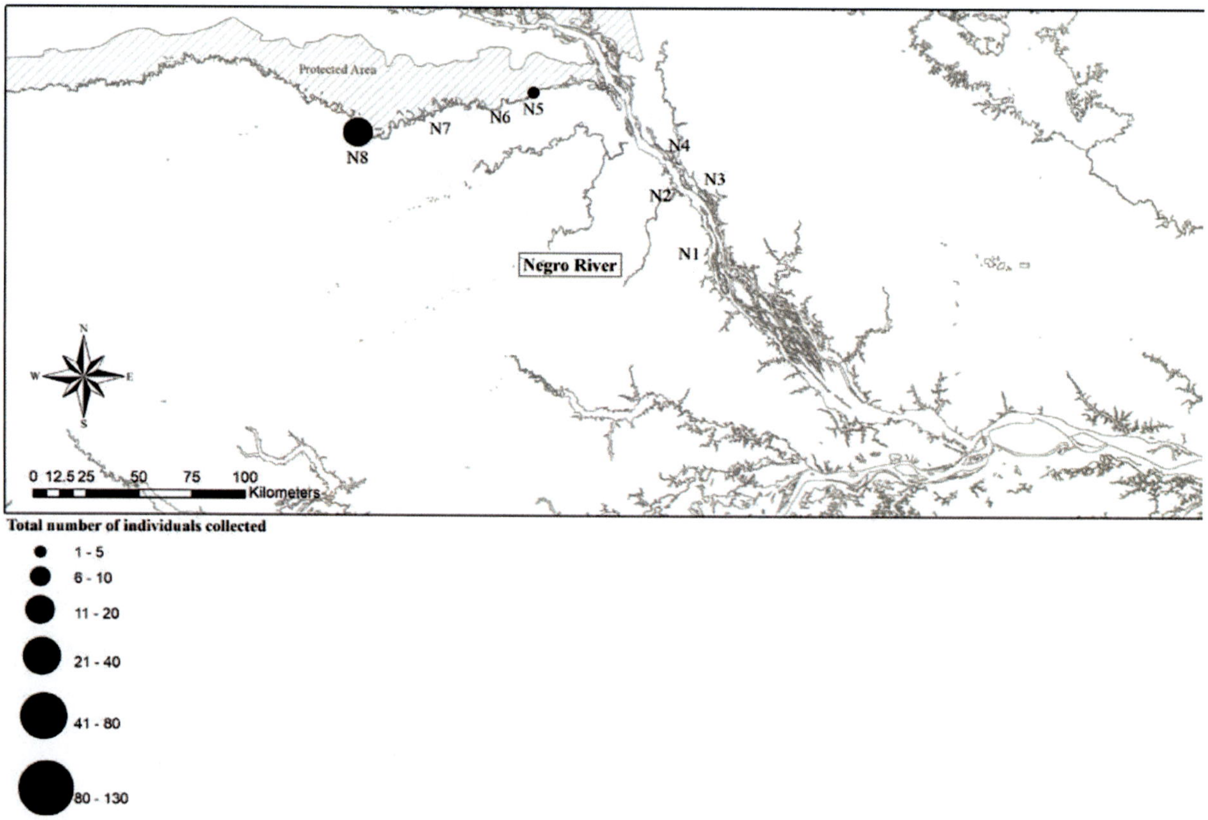

Fig. 4.112 Distribution of individuals of *Roestes molossus* sampled in the Negro River

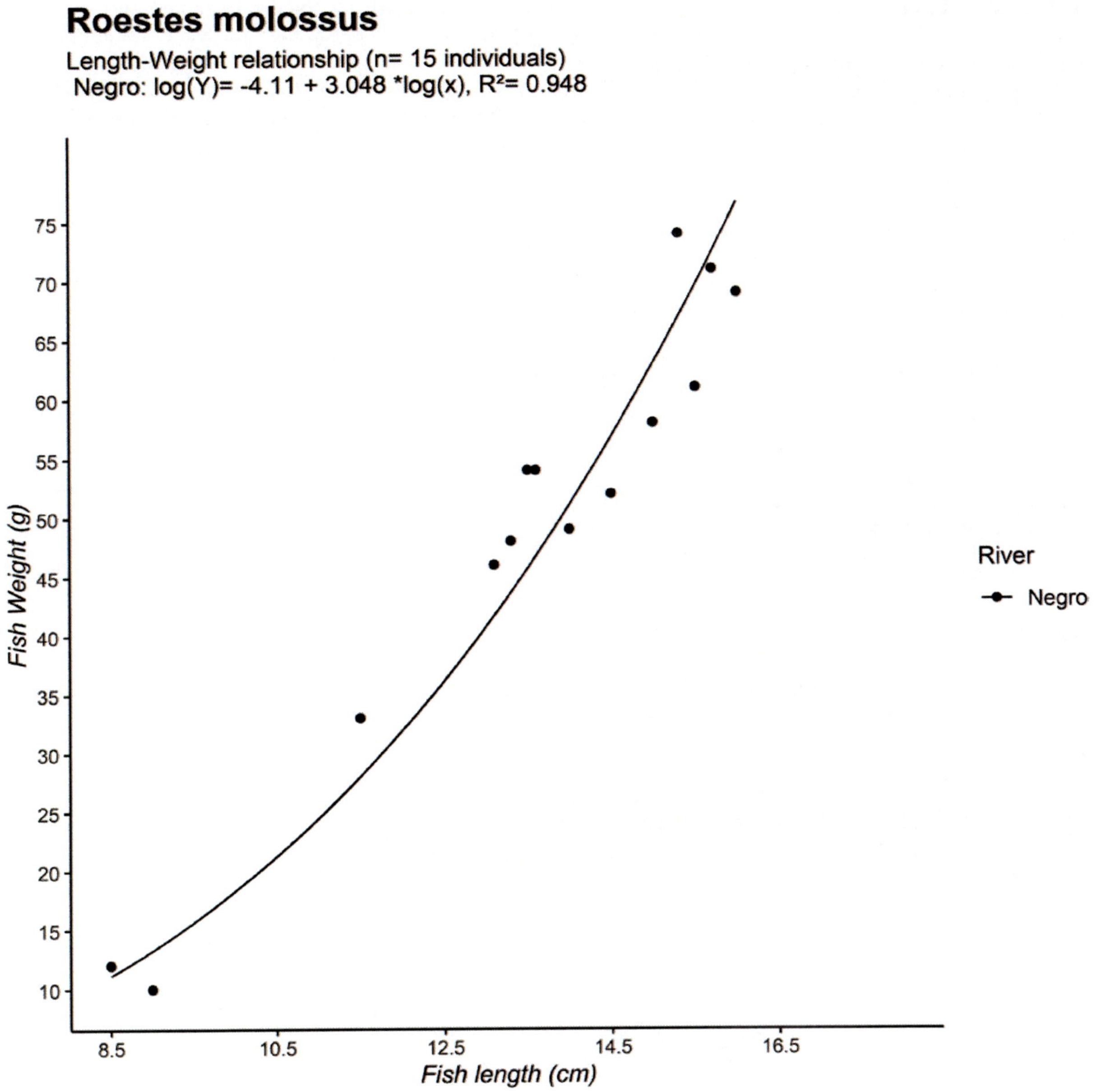

Fig. 4.113 Length-weight relationship for *Roestes molossus* sampled in Negro river

4.8 Family Erythrinidae

4.8.1 Summary of Biology and Ecology (Literature)

4

Diversity: this family has three genera and 19 species. *Hoplerythrinus* and *Erythrinus* are restricted in distribution to the Amazon basins, Orinoco, and Guyanese region; on the other hand, the genus *Hoplias* has a wide distribution found in all the lowlands of the Neotropics. Some authors consider the species of this family as primitive within the order Characiformes.

Morphology: they are medium-sized fish with elongated and cylindrical body, large head, and conical teeth. They are characterized by having rounded caudal fin and lack of fat fin. Larger fishes of this family may reach 49 cm of standard length (SL). They have variable color, often reddish-brown lighter toward the belly. Their head has three elongated spots that extend from the eye to the operculum. All fins have several irregular black or dark brown bands. Their mouth has two to three canines.

Habitat: probably it is a complex of species, widely distributed, from Costa Rica to Argentina in most of the lowlands. Fishes of this family may occur in pipes, lagoons, puddles, flood forests, estuaries, beaches, and backwaters.

Ecology and feeding: fishes from the genus *Hoplias* are carnivorous that feeds on other fish, besides aquatic and terrestrial insects that fall into the water. They are predators that generally stay hidden to ambush its prey. Those of the genus *Hoplerythrinus* are gregarious and move in shoals, whereas those of the genus *Hoplias* are territorial and solitary. Fishes of this family are able to breathe atmospheric oxygen through their vascularized bladder, being able to survive anoxic conditions. All fishes from this family are predators, solitary, and territorial.

Reproduction: low survival of juveniles and adults, high fecundity, long life, total spawning one per season, and large fluctuation in population density are reported. These fishes can reach sexual maturity at 12 months. The reproduction begins at the beginning of the rainy season, but it can be extended in the high-water season.

Economic importance: fishes from the genus *Hoplias* are important to artisanal fisheries in Amazonian rivers, mostly for consumption and also for commerce. Fishes from this family may be also used as ornamental fish in Colombia and Peru.

References: Géry (1977), McConell (1987), Taphorn (1992), Machado-Allison (1993), Castellanos (2002), Lasso (2004), Galvis et al. (2006, 2007a,b), Mojica et al. (2012).

Fish species: *Hoplias malabaricus* **(Bloch 1794)**

Common name(s): traíra

Size of fish sampled (standard length, both rivers): mean = 24 cm, median = 23.8 cm

Fig. 4.114 *Hoplias malabaricus* sampled in the Tapajós River. (Photo: Renato Silvano)

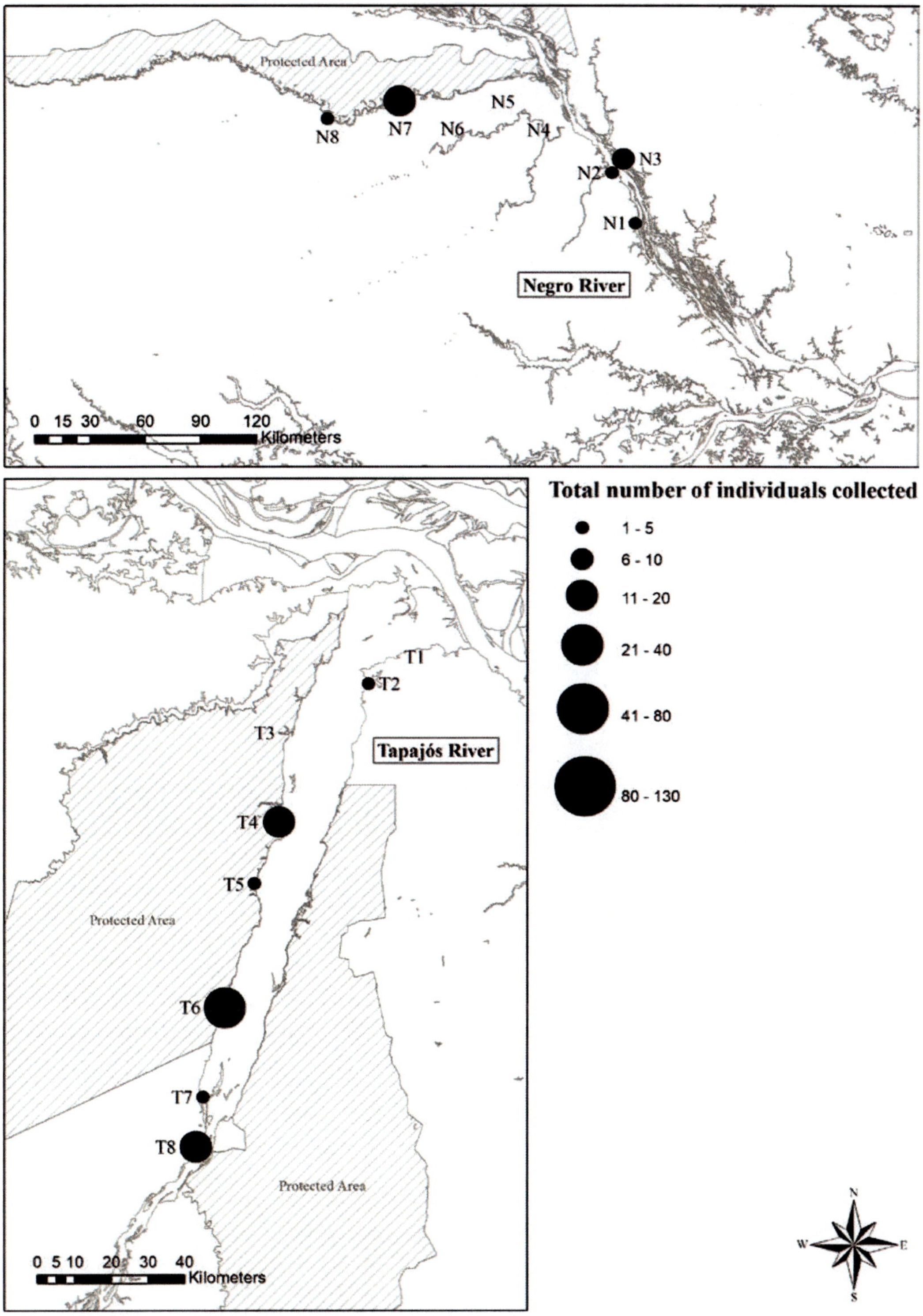

Fig. 4.115 Distribution of individuals of *Hoplias malabaricus* sampled in the Negro and Tapajós rivers

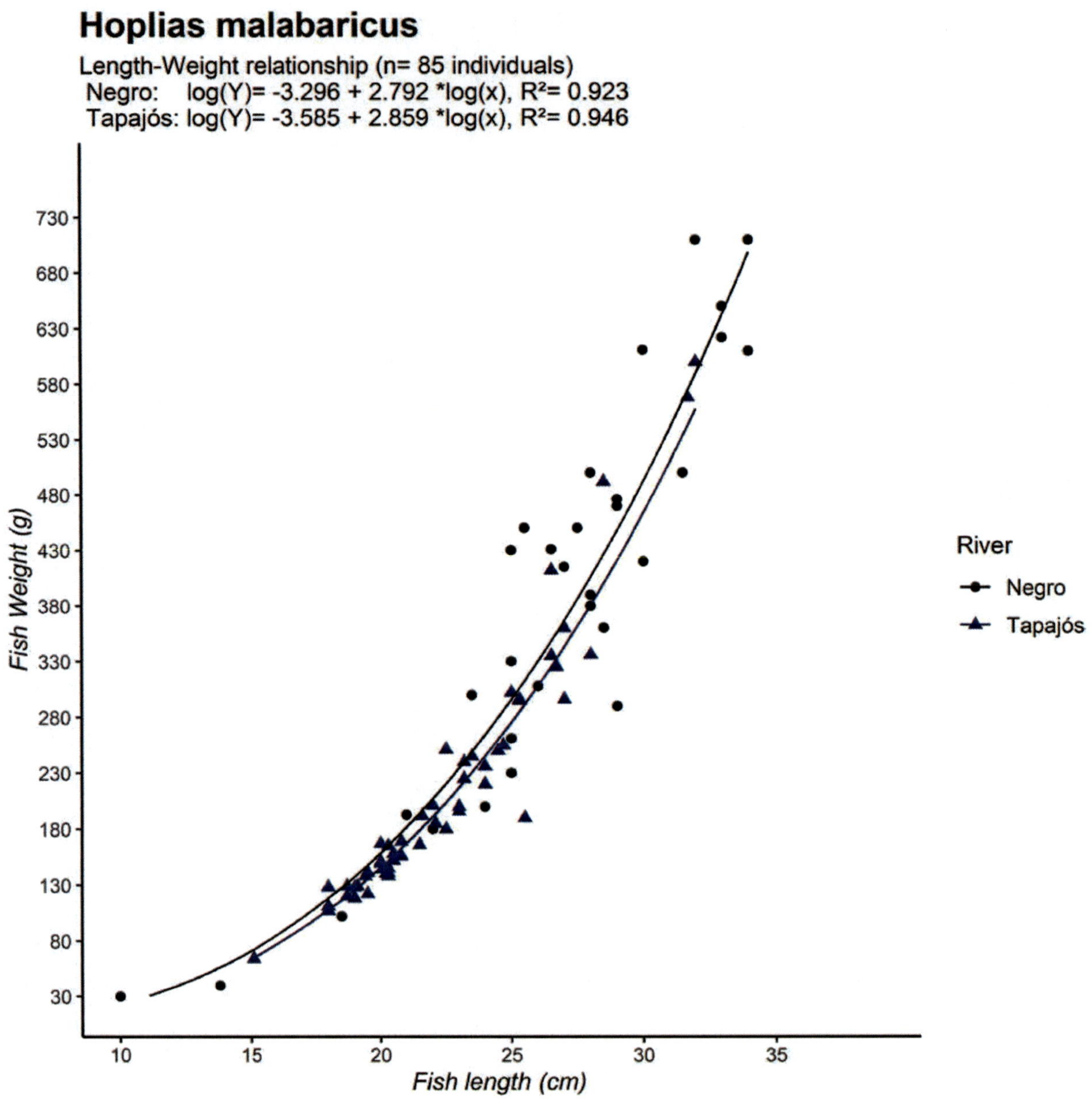

Fig. 4.116 Length-weight relationship for *Hoplias malabaricus* sampled in Tapajós and Negro rivers

Fisheries: The traíra (*Hoplias malabaricus*) is important to fisheries for both consumption and commercial purposes, especially in Negro River. Biomass caught is in Table 3.3 (▶ Chap. 3).

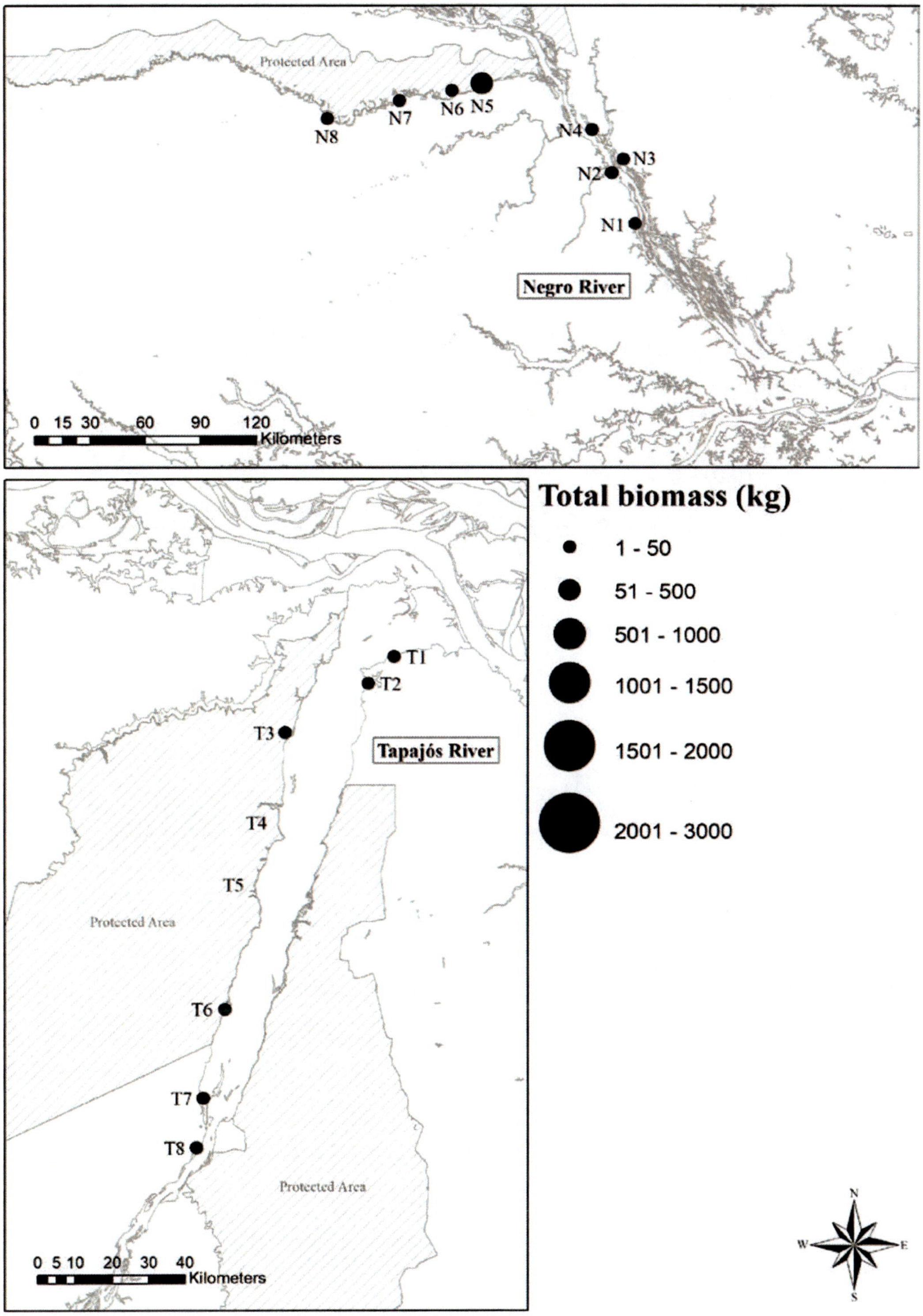

Fig. 4.117 Distribution of biomass of traíra (*H. malabaricus*) caught in small-scale fisheries in the Negro and Tapajós rivers

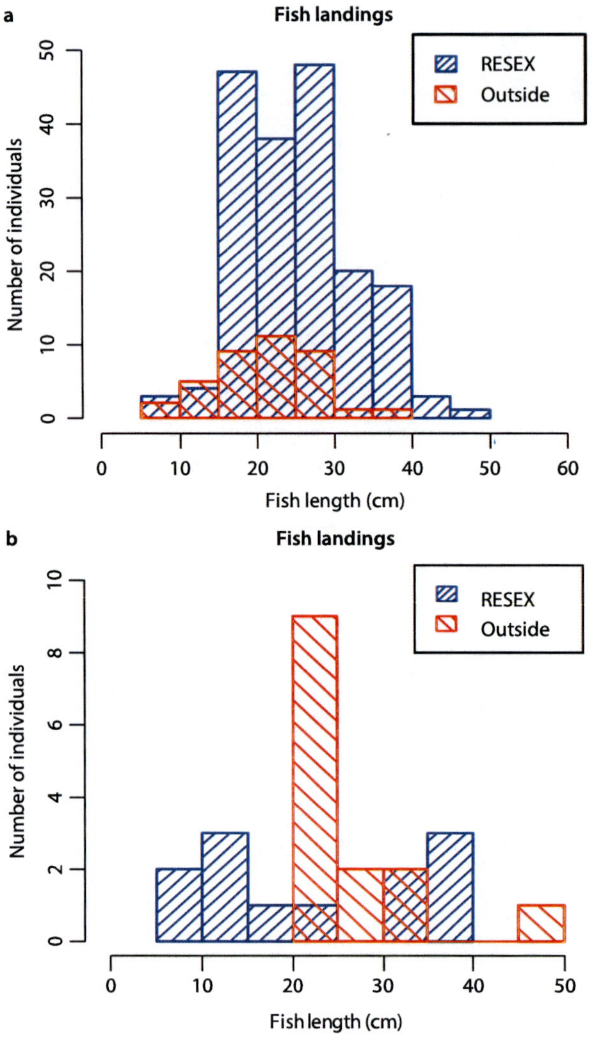

Fig. 4.118 Size distribution of traíra (*H. malabaricus*) caught in small-scale fisheries in 16 communities inside and outside the Extractive Reserves (RESEX) in the rivers **a** Negro, N Resex = 182, N outside = 38 and **b** Tapajós, N Resex = 12, N outside = 7

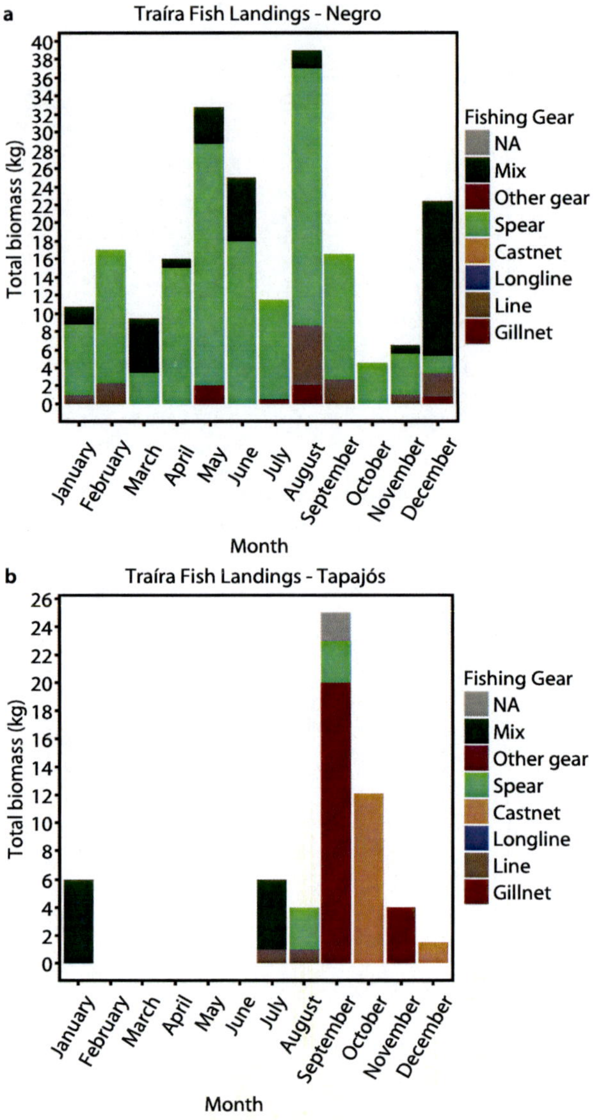

Fig. 4.119 Seasonal variation of biomass of traíra (*H. malabaricus*) caught per fishing gear in small-scale fisheries in16 communities in the rivers **a** Negro and **b** Tapajós. NA: fishing gear not informed; Mix: more than one gear was used

4.9 Family Hemiodontidae

4.9.1 Summary of Biology and Ecology (Literature)

Diversity: this family can be divided into two subfamilies: Anodontinae and Hemiodontinae. Currently, there are 33 valid species and five genera.

Habitat: these fishes occur in Guyana, French Guiana, Suriname, and the Amazon, Paraná-Paraguay, Orinoco, and Tocantins basins. They are also found in other independent drainages, including the Amapá, Arguari, Itapecuru, Mearim, and Parnaíba. They are usually found in pelagic habitats of lakes, small streams (igarapés), and rivers.

Ecology and feeding: most species feed on larvae and nymphs of insects, detritus, filamentous algae, superior plants, periphyton, or plankton. Species of the genus *Anodus* are planktivorous that capture small cladocerans, copepods, and rotifers with their long and numerous gill rakers. Species of the genus *Hemiodus* are iliophages/detritivores or herbivores. Species of this family form large schools and many undergo migrations.

Morphology: fishes of this family have small- to medium-sized bodies with fusiform shape, a characteristic of good swimmers. They are easily distinguished from other Characiformes due to a series of synapomorphies: eyes almost completely covered (excepting a small oval vertical slit or small orifice over or on the pupil) with an adipose eyelid; operculum deeply concave dorsally; supra-pectoral groove, restricted dorsally by the cleithrum and a series of three or more elongated scales; and pelvic fins with 9 to 11 rays.

Reproduction: hemiodontids form schools and undergo reproductive migrations. Most species are total spawners that spawn during raising water periods.

Economic importance: species of this family usually do to not have commercial importance to fisheries, but during some periods of the year, when schools are migrating, their density can become high in some locations. In these periods, species of larger size, such as *Anodus, Argonectes*, and *Hemiodus*, can be exploited by local riverine communities, especially in the Tapajós River. Some bars in touristic regions (such as the beaches in the Tapajós River) offer fried hemiodontids to their customers. Some small species of this family, especially those of the genus *Hemiodus*, are also appreciated by aquarists and can thus be relevant to the aquarium trade.

References: Roberts (1971), Santos et al. (2006), Silva et al. (2008), Langeani (2013), Eschmeyer (1998), Ribeiro et al. (2016), Queiroz et al. (2013), Hallwass et al. (2020).

Fish species: *Anodus elongatus* **Agassiz 1829**
Common name(s): charuto, cubiu, orana, peixe-cavalo

Fig. 4.120 *Anodus elongatus* sampled in the Negro River. (Photo: Renato Silvano)

Fish species: *Anodus orinocensis* (Steindachner 1887)

Common name(s): charuto, cubiu

Fig. 4.121 *Anodus orinocensis* sampled in the Negro River. (Photo: Renato Silvano)

Observation: abundance and size data were grouped for these species *Anodus orinocensis* was collected only in Negro River and *A. elongatus* was collected in both rivers; see Table 1.1 (▶ Chap. 1).

Size of fish sampled (standard length, both rivers): min. = 13 cm, max. = 18.2 cm, mean = 16 cm, median = 16.5 cm

Fisheries: charuto (several species of Hemiodontidae) is exploited by fisheries in both rivers, mostly for food, but this fish has commercial value in some communities in the Tapajós River; see *Argonectes longiceps* for charuto fisheries data. Biomass caught is in Table 3.3 (▶ Chap. 3).

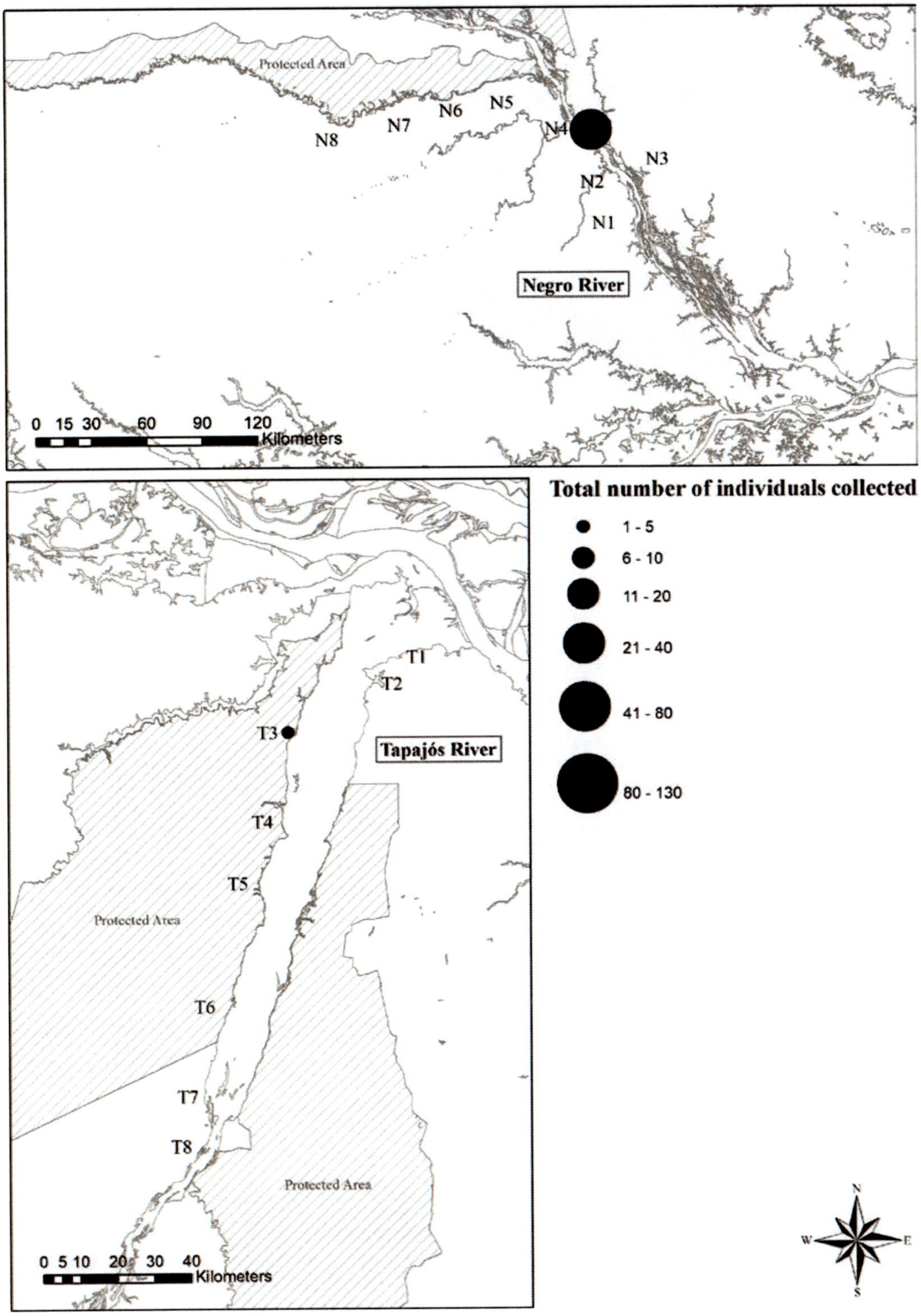

Fig. 4.122 Distribution of individuals of *Anodus* spp. sampled in the Negro and Tapajós rivers

Fish species: *Argonectes longiceps* (Kner 1858)

Common name(s): charuto, cubiu, peixe-cavalo

Size of fish sampled (standard length, both rivers): mean = 13.6 cm, median = 12.5 cm

Fig. 4.123 *Argonectes longiceps* sampled in the Negro River. (Photo: Renato Silvano)

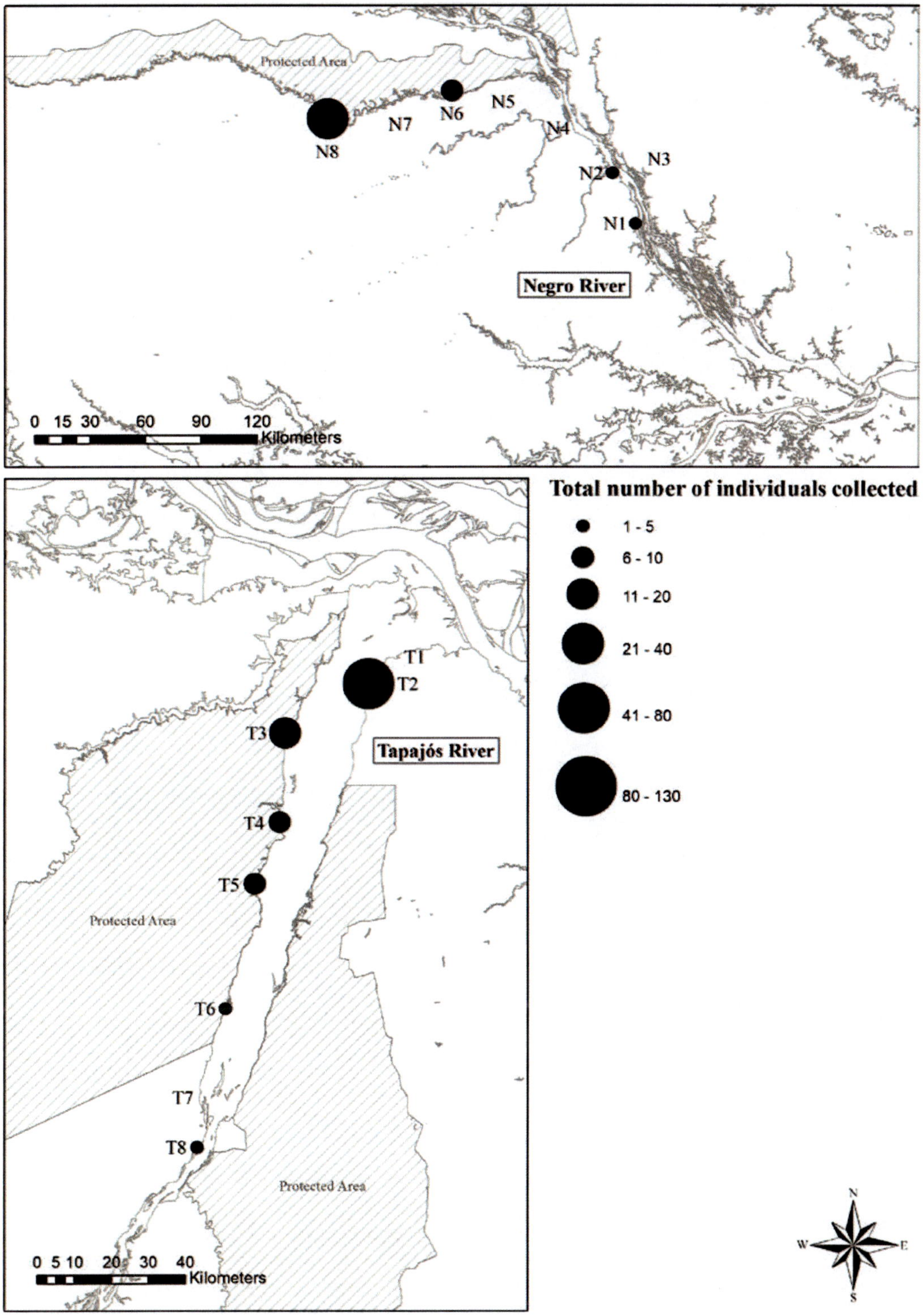

Fig. 4.124 Distribution of individuals of *Argonectes longiceps* sampled in the Negro and Tapajós rivers

4

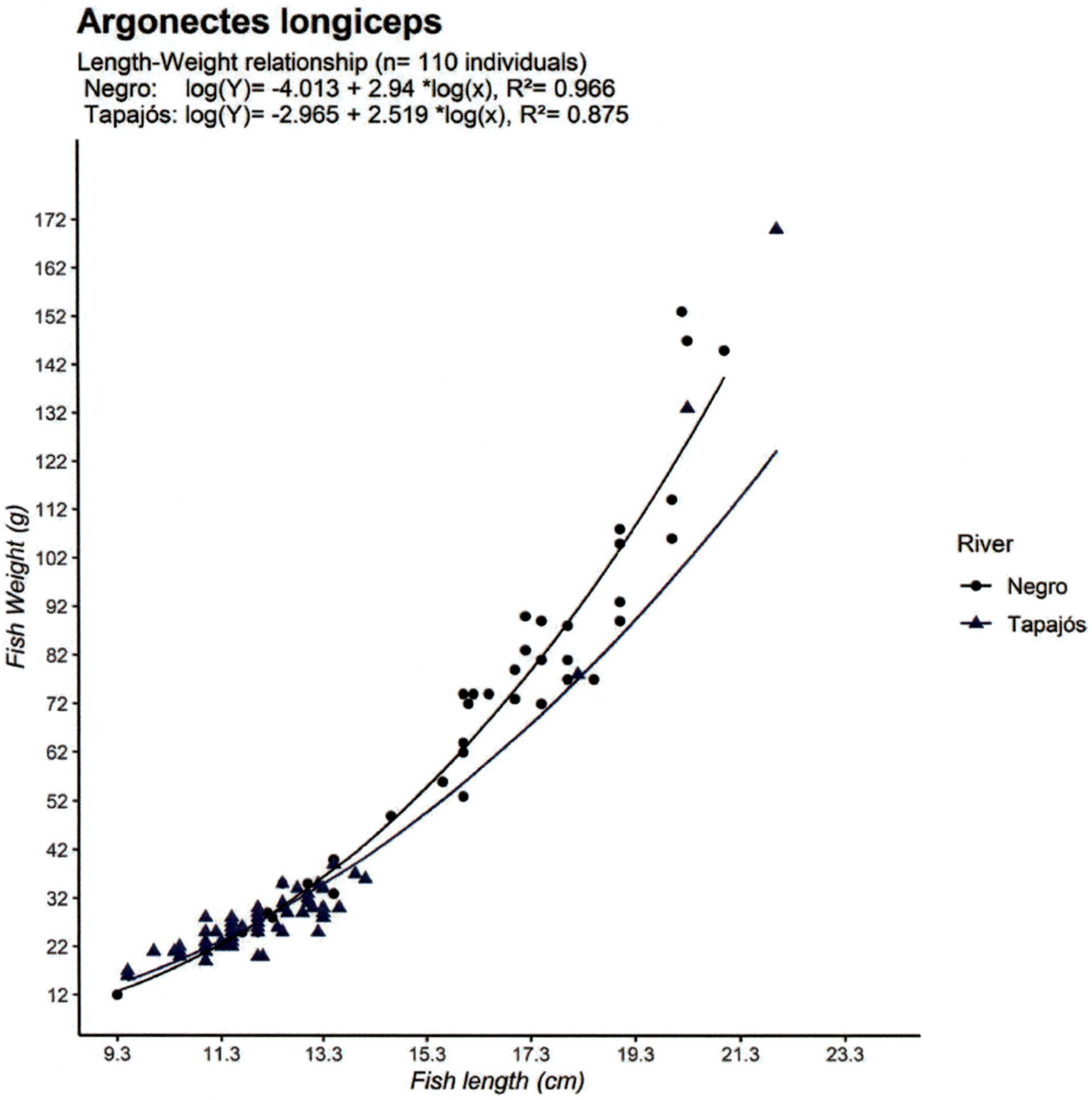

Fig. 4.125 Length-weight relationship for *Argonectes longiceps* sampled in Negro and Tapajós rivers

Fisheries: charuto (several species of Hemiodontidae) is exploited by fisheries in both rivers, mostly for food, but this fish has commercial value in some communities in the Tapajós River. Biomass caught is in Table 3.3 (► Chap. 3).

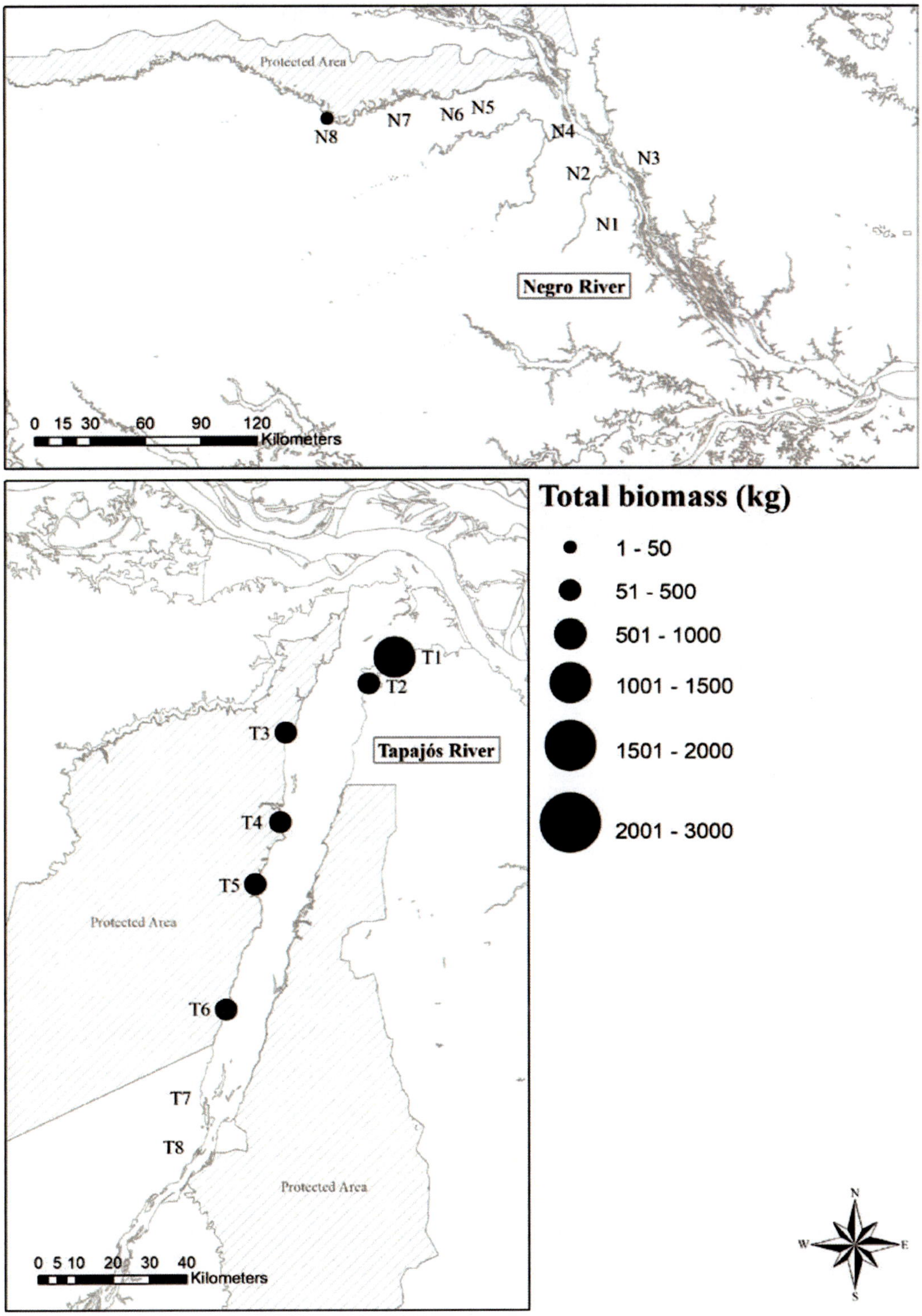

Fig. 4.126 Distribution of biomass of charuto (Hemiodontidae) caught in small-scale fisheries in the Negro and Tapajós rivers

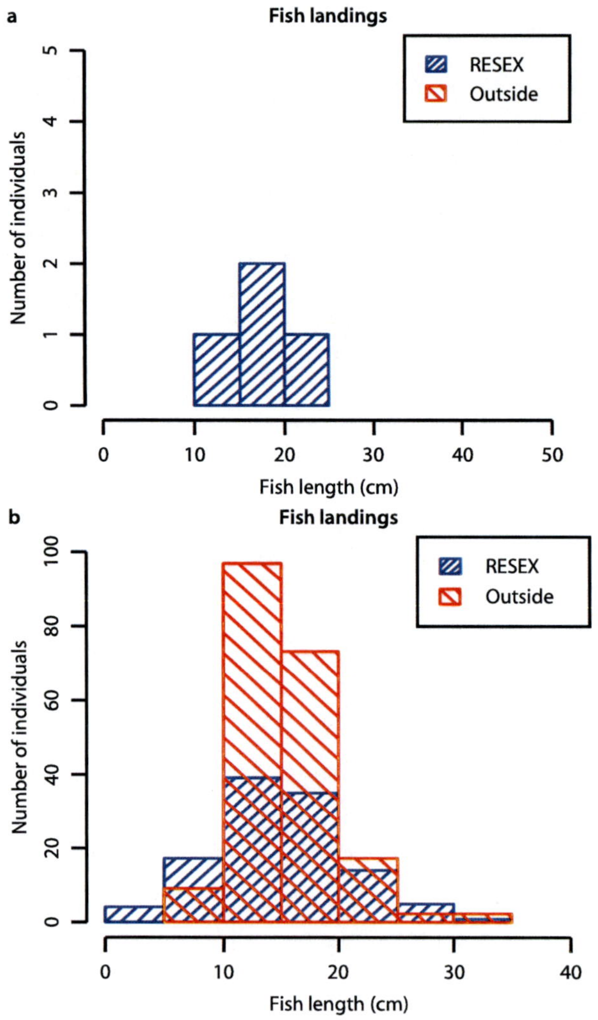

Fig. 4.127 Size distribution of charuto (Hemiodontidae) caught in small-scale fisheries in 16 communities inside and outside the Extractive Reserves (RESEX) in the rivers **a** Negro, N Resex = 4, N outside = 0 and **b** Tapajós, N Resex = 115, N outside = 200

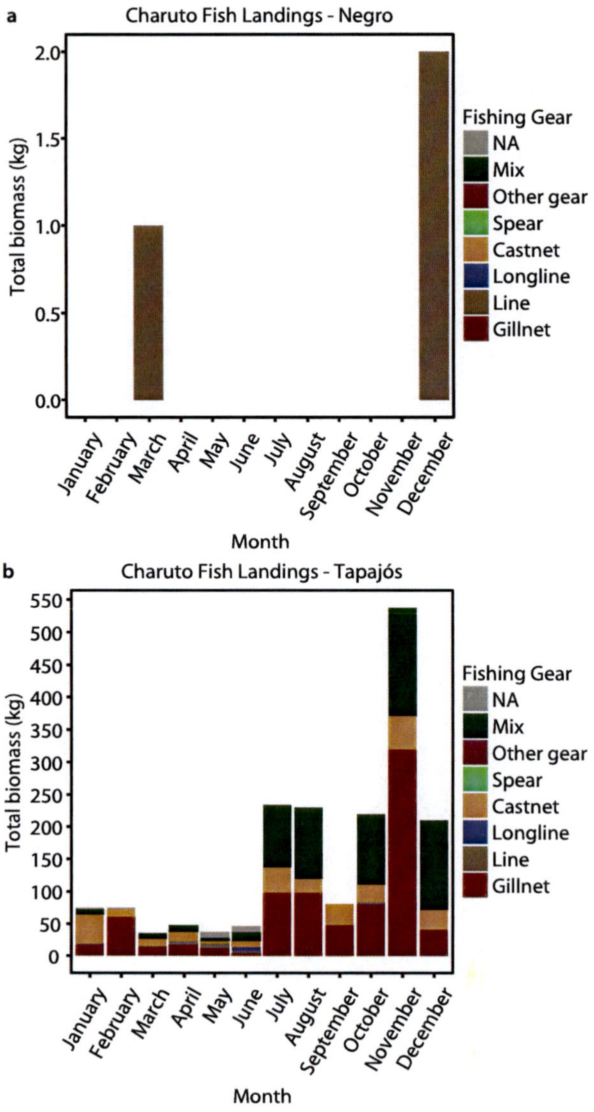

Fig. 4.128 Seasonal variation of biomass of charuto (Hemiodontidae) caught per fishing gear in small-scale fisheries in 16 communities in the rivers **a** Negro and **b** Tapajós. NA: fishing gear not informed; Mix: more than one gear was used

Fish species: *Bivibranchia fowleri* (Steindachner 1908)

Common name(s): charuto

Size of fish sampled (standard length, both rivers): mean = 12.2 cm, median = 11.8 cm

Fisheries: charuto (several species of Hemiodontidae) is exploited by fisheries in both rivers, mostly for food, but this fish has commercial value in some communities in the Tapajós River; see *Argonectes longiceps* for charuto fisheries data. Biomass caught is in ◘ Table 3.3 (▶ Chap. 3).

◘ **Fig. 4.129** *Bivibranchia fowleri* sampled in the Tapajós River. (Photo: Renato Silvano)

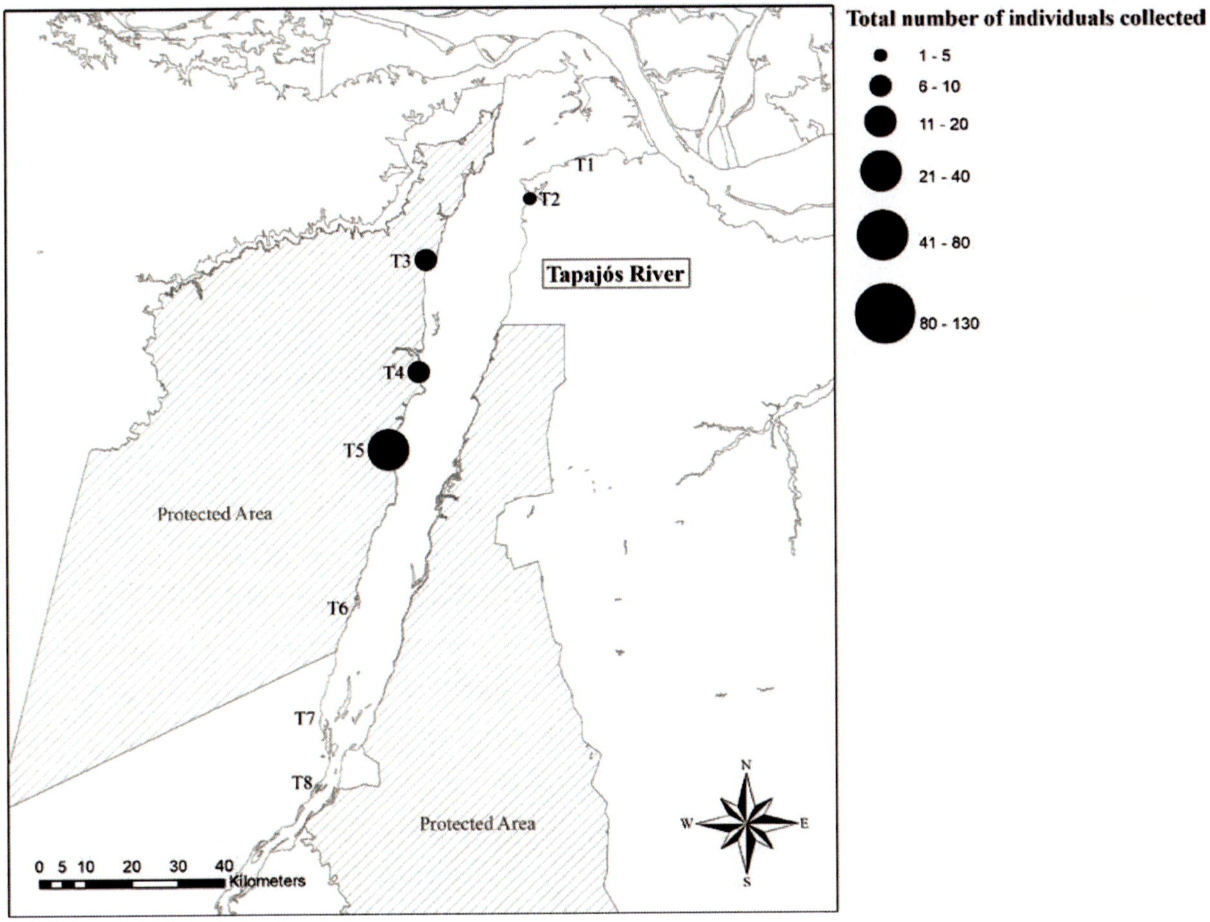

◘ **Fig. 4.130** Distribution of individuals of *Bivibranchia fowleri* sampled in the Tapajós river

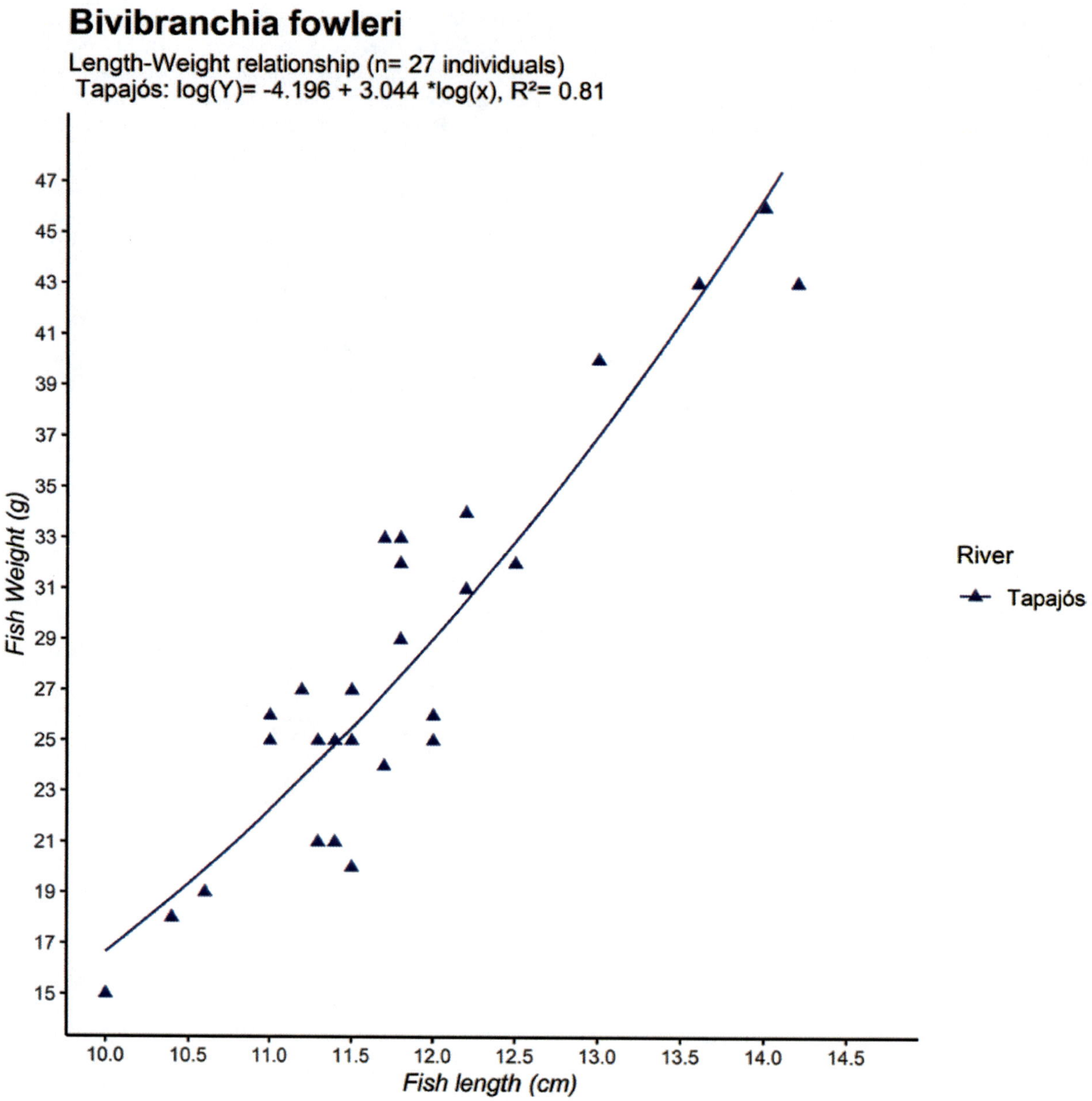

Fig. 4.131 Length-weight relationship for *Bivibranchia fowleri* sampled in the Tapajós river

Fish species: *Hemiodus argenteus* Pellegrin 1909

Common name(s): charuto, charuto-escama-fina, piraruíra

Fig. 4.132 *Hemiodus argenteus* sampled in the Tapajós River. (Photo: Renato Silvano)

Fish species: *Hemiodus atranalis* (Fowler 1940)

Common name(s): charuto

Fig. 4.133 *Hemiodus atranalis* sampled in the Negro River. (Photo: Renato Silvano)

Fish species: *Hemiodus gracilis* Günther 1864

Common name(s): charuto, charuto-rabo-vermelho, cubiu

Fig. 4.134 *Hemiodus gracilis* sampled in the Tapajós River. (Photo: Renato Silvano)

Fish species: *Hemiodus immaculatus* Kner 1858

Common name(s): charuto, charuto-do-rabo-preto, piraruira-do-rabo-preto, cubiu, orana

Fig. 4.135 *Hemiodus immaculatus* sampled in the Negro River. (Photo: Renato Silvano)

Fish species: *Hemiodus unimaculatus* (Bloch 1794)

Common name(s): charuto, cubiu, orana, piraruira

Fig. 4.136 *Hemiodus unimaculatus* sampled in the Negro River. (Photo: Renato Silvano)

Fish species: *Hemiodus* sp.

Common name(s): charuto, cubiu

■ **Fig. 4.137** *Hemiodus* sp. sampled in the Tapajós River. (Photo: Renato Silvano)

Observation: abundance and size data were grouped for all these species, due to the difficulty in distinguishing them during sampling: *Hemiodus argenteus* and *H. tocantinensis* (no photo) were collected only in Tapajós River; *H. goeldii* (no photo) was collected only in Negro River; *H. gracilis, H. microlepis* (no photo), *H. immaculatus, H. unimaculatus, H. atranalis*, and *Hemiodus* sp. were collected in both rivers; see ■ Table 1.1 (▶ Chap. 1).

Size of fish sampled (standard length, both rivers): min. = 8.6 cm, max. = 23.5 cm, mean = 14.5 cm, median = 15 cm

Fisheries: charuto (several species of Hemiodontidae) is exploited by fisheries in both rivers, mostly for food, but this fish has commercial value in some communities in the Tapajós River; see *Argonectes longiceps* for charuto fisheries data. Biomass caught is in ■ Table 3.3 (▶ Chap. 3).

4

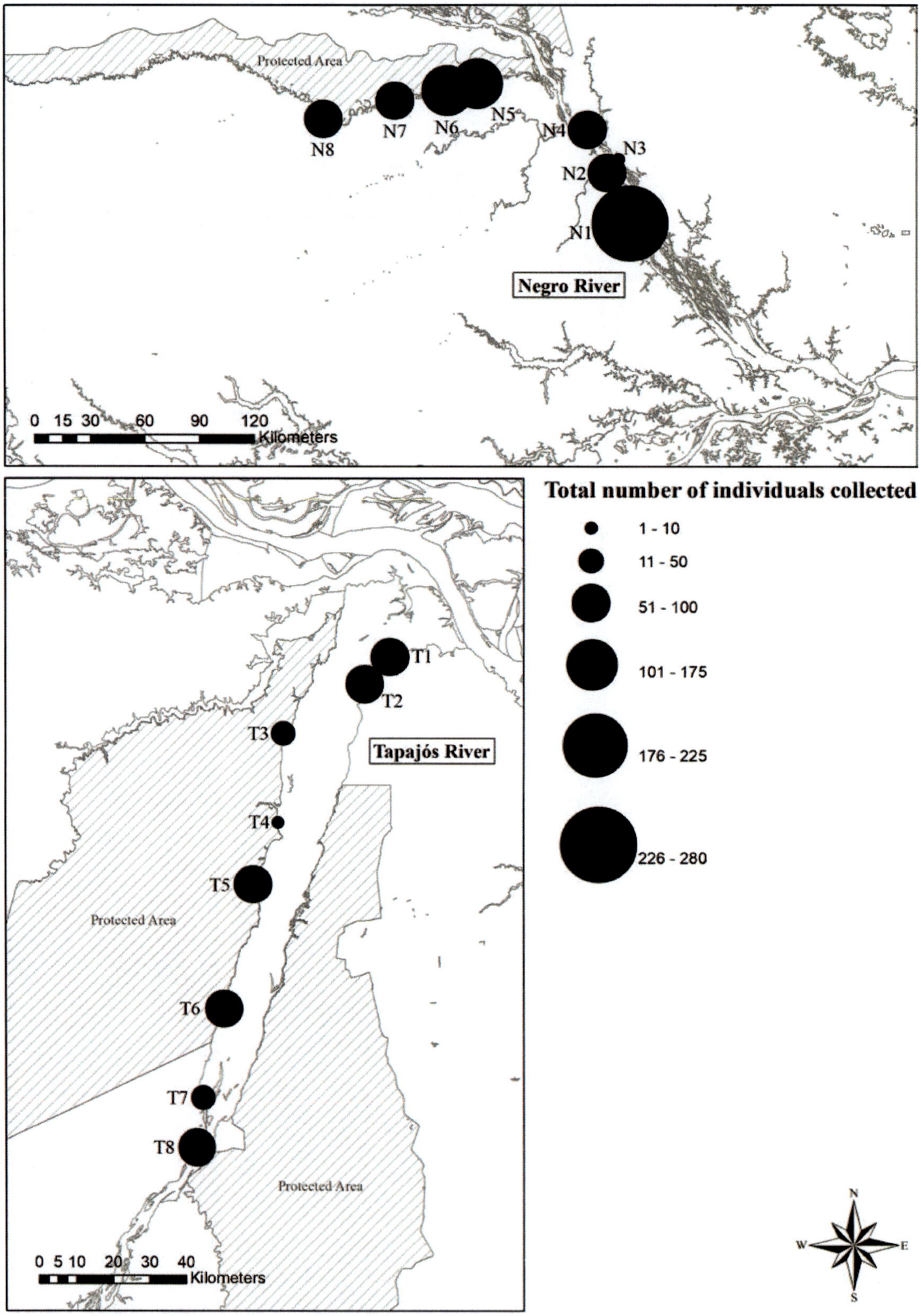

Fig. 4.138 Distribution of individuals of *Hemiodus* spp. sampled in the Negro and Tapajós rivers

Fish species: *Micromischodus sugillatus* Roberts 1971

Common name(s): charuto, cubiu, apirá

Size of fish sampled (standard length, both rivers): mean = 13.3 cm, median = 13.1 cm

Fisheries: charuto (several species of Hemiodontidae) is exploited by fisheries in both rivers, mostly for food, but this fish has commercial value in some communities in the Tapajós River; see *Argonectes longiceps* for charuto fisheries data. Biomass caught is in ◘ Table 3.3 (► Chap. 3).

◘ **Fig. 4.139** *Micromischodus sugillatus* sampled in the Negro River. (Photo: Renato Silvano)

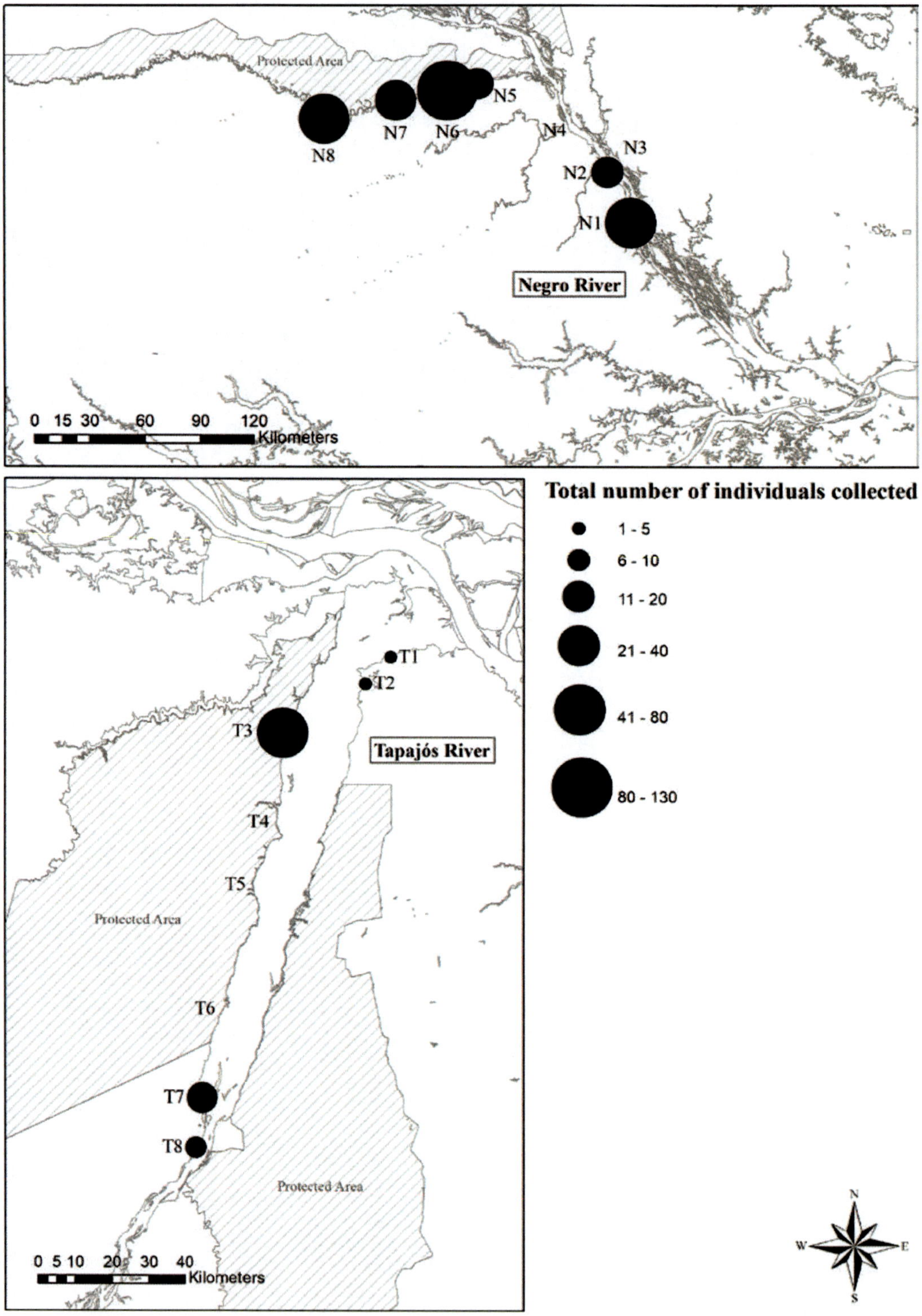

Fig. 4.140 Distribution of individuals of *Micromischodus sugillatus* sampled in the Negro and Tapajós rivers

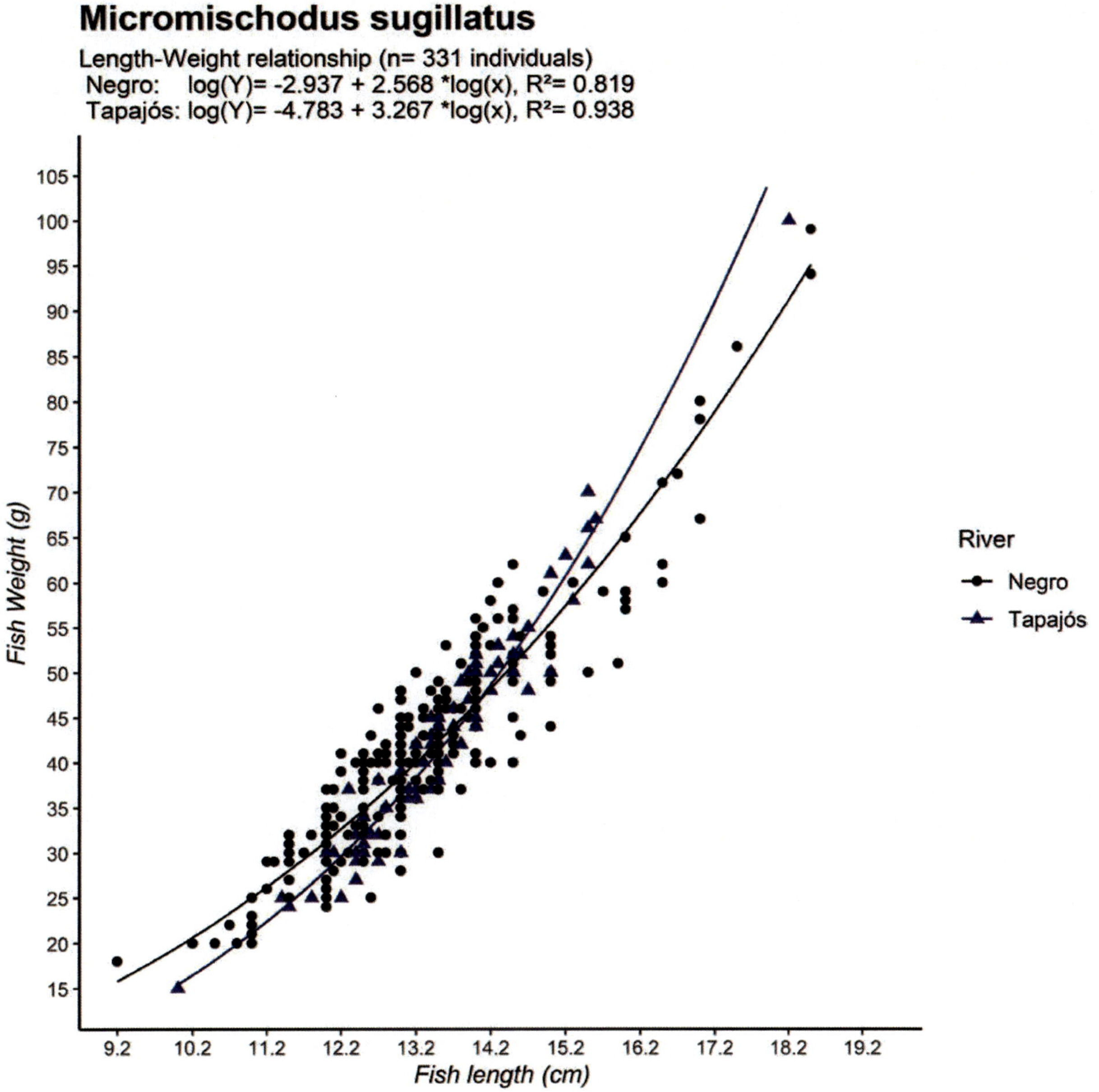

Fig. 4.141 Length-weight relationship for *Micromischodus sugillatus* sampled in Negro and Tapajós rivers

4.10 Family Prochilodontidae

4.10.1 Summary of Biology and Ecology (Literature)

Diversity: the flannel-mouthed characins are considered a monophyletic family composed of 21 valid species distributed in the main drainage systems of South America, including the rivers Magdalena, Orinoco, Amazonas, São Francisco, and Paraná-Paraguay. The Prochilodontidae family is separated into three genera: *Ichthyoelephas* (two spp.), *Prochilodus* (13 spp.) and *Semaprochilodus* (six spp.), but only *Prochilodus* and *Semaprochilodus* occur in the medium-lower Amazon River.

Habitat: fishes of this family can be found in a variety of habitats, including the channel of major rivers, floodplain lakes, and even rapid streams. They can also be found in waters with different physical-chemical properties, such as clear, white, and black water rivers.

Morphology: species of this family present a medium to large body size (max 80 cm in *Prochilodus lineatus*) with a fusiform shape. Their mouths are protactile with numerous tiny teeth and fleshy lips that function as a sucker disk. Scales are usually large. Intestines are long and can reach more than three times fish body length. A spine that is either bifurcate or pointed anteriorly is located in the base of the dorsal fin of *Semaprochilodus* and *Prochilodus* and *Ichthyoelephas,* respectively. Species of the *Semaprochilodus* genus possess distinct stripes across the caudal and anal fins.

Ecology and feeding: these fishes are essentially detritivores and/or herbivores, consuming seeds, fruits, periphyton, algae, plant remains, and fine detritus located on the benthic zone and on the surface of submerged objects in the pelagic zone (e.g., logs, macrophytes). Studies suggest that some species conduct trophic migrations linking rivers with different levels of productivity. For instance, *Semaprochilodus* is known to move seasonally from white water rivers (such as the Amazon River), which are richer in nutrients, to poor black and clear water rivers (such as the Negro and Tapajós rivers, respectively). It is hypothesized that sedentary predators that live in poor water rivers, such as the tucunarés (*Cichla*), may rely on *Semaprochilodus* species as a seasonal supply of food. In this sense, prochilodontids may have a key role in the nutrient cycling and transport of energy and mass in the trophic webs of South America freshwaters. Given their characteristic of having long migratory routes, prochilodontids may be sensitive to the construction of hydroelectric dams and reservoirs that reduce fluvial connection. A study based on fishers' knowledge indicates that a species of *Semaprochilodus* had its abundance drastically reduced to a point of local extinction in the lower regions of the Tocantins river (Amazon basin) downstream from a large reservoir.

Reproduction: fishes of this family are seasonal reproducers that possess high fecundity. Some species can produce more than 600 thousand eggs per reproductive event. They are usually known to undertake long migrations to spawn, overcoming geographic barriers, such as rapids and waterfalls.

Economic Importance: prochilodontids are very important and are intensely exploited by both commercial and small-scale fisheries in South America and in the Amazon. The genus *Prochilodus* is among the most important commercial fish in several Amazonian regions, and the genus *Semaprochilodus* is one of the most important food fish for Amazonian people.

References: Ribeiro and Petrere Jr (1990), Berra (2001), Benedito and Araujo (2002), Castro and Vari (2004), Hoeinghaus et al. (2006), Nelson et al. (2006), Taylor et al. (2006), Mcintyre et al. (2007), Hallwass et al. (2013a), Hallwass and Silvano (2016), Piedade et al. (2016), Keppeler et al. (2017), Froese and Pauly (2018), Fricke et al. (2019).

Fish species: *Semaprochilodus insignis* (Jardine 1841)

Common name(s): jaraqui-escama-grossa

Size of fish sampled (standard length, both rivers): mean = 15.1 cm, median = 15.2 cm

Fisheries: The jaraquis (*Semaprochilodus* spp.) are among the most important fishes to fishers in both rivers, for consumption and commerce. Biomass caught is in ◘ Table 3.3 (▸ Chap. 3).

◘ **Fig. 4.142** *Semaprochilodus insignis* sampled in the Tapajós River. (Photo: Renato Silvano)

4

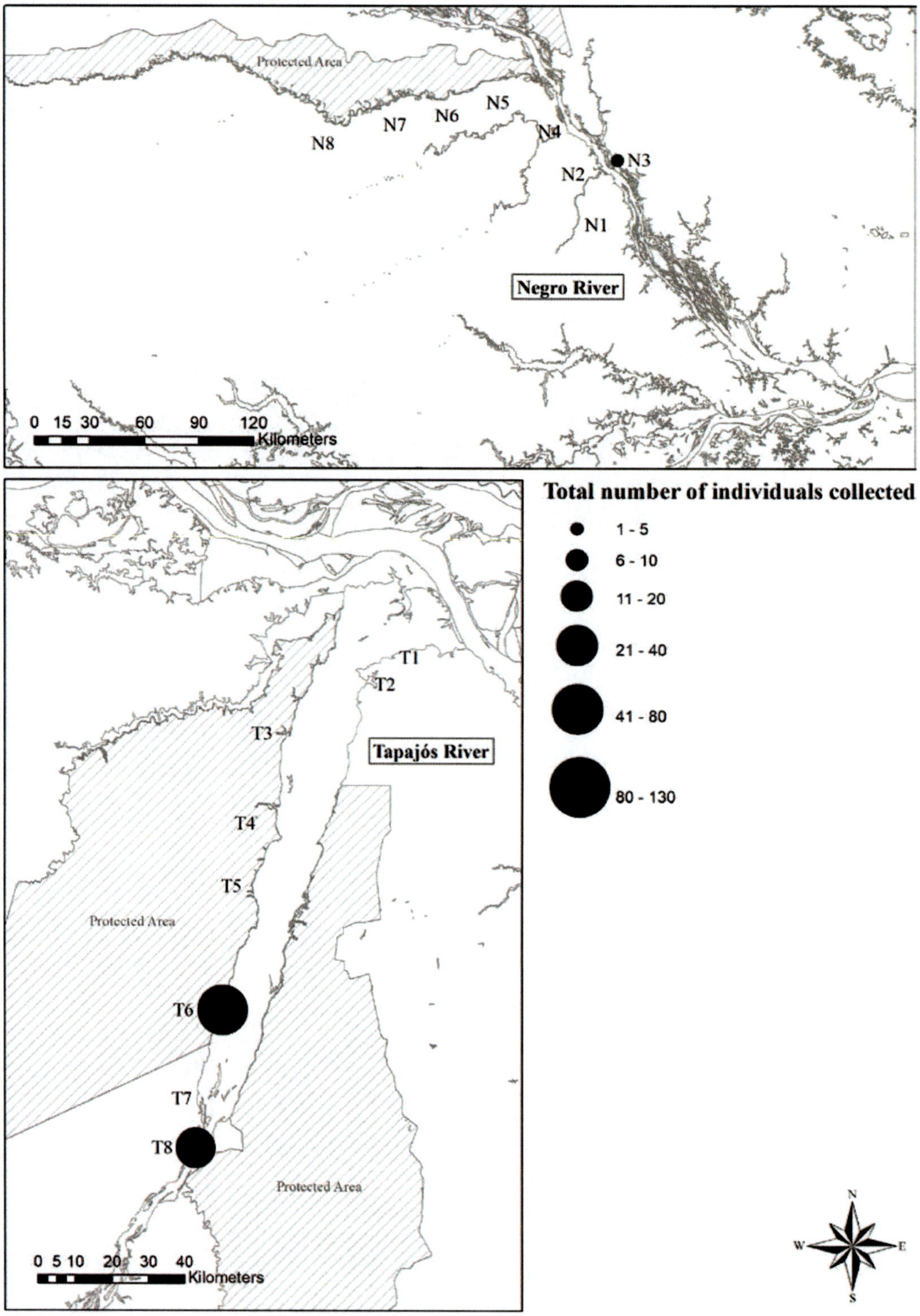

Fig. 4.143 Distribution of individuals of *Semaprochilodus insignis* sampled in the Negro and Tapajós rivers

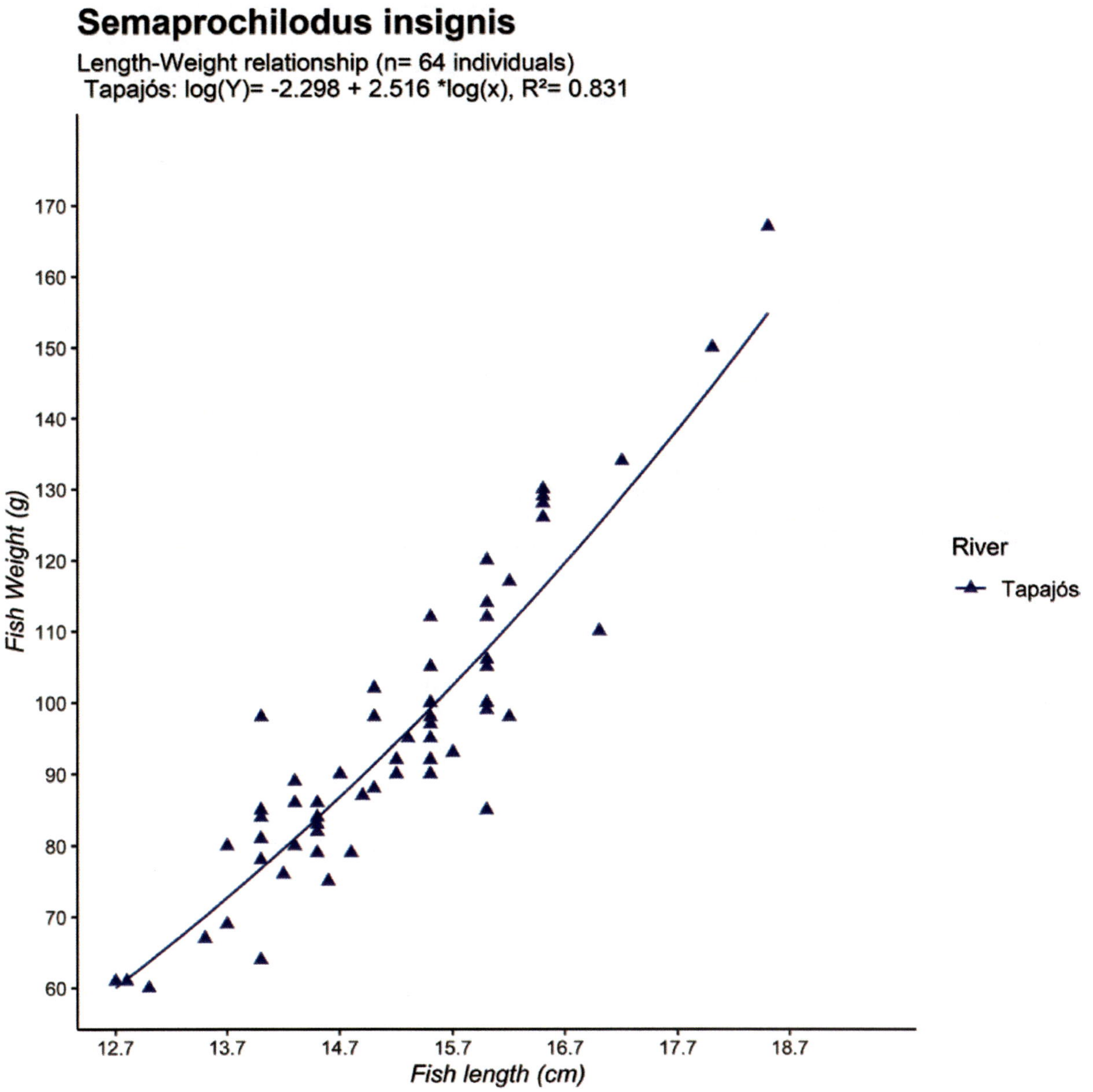

Fig. 4.144 Length-weight relationship for *Semaprochilodus insignis* sampled in the Tapajós River

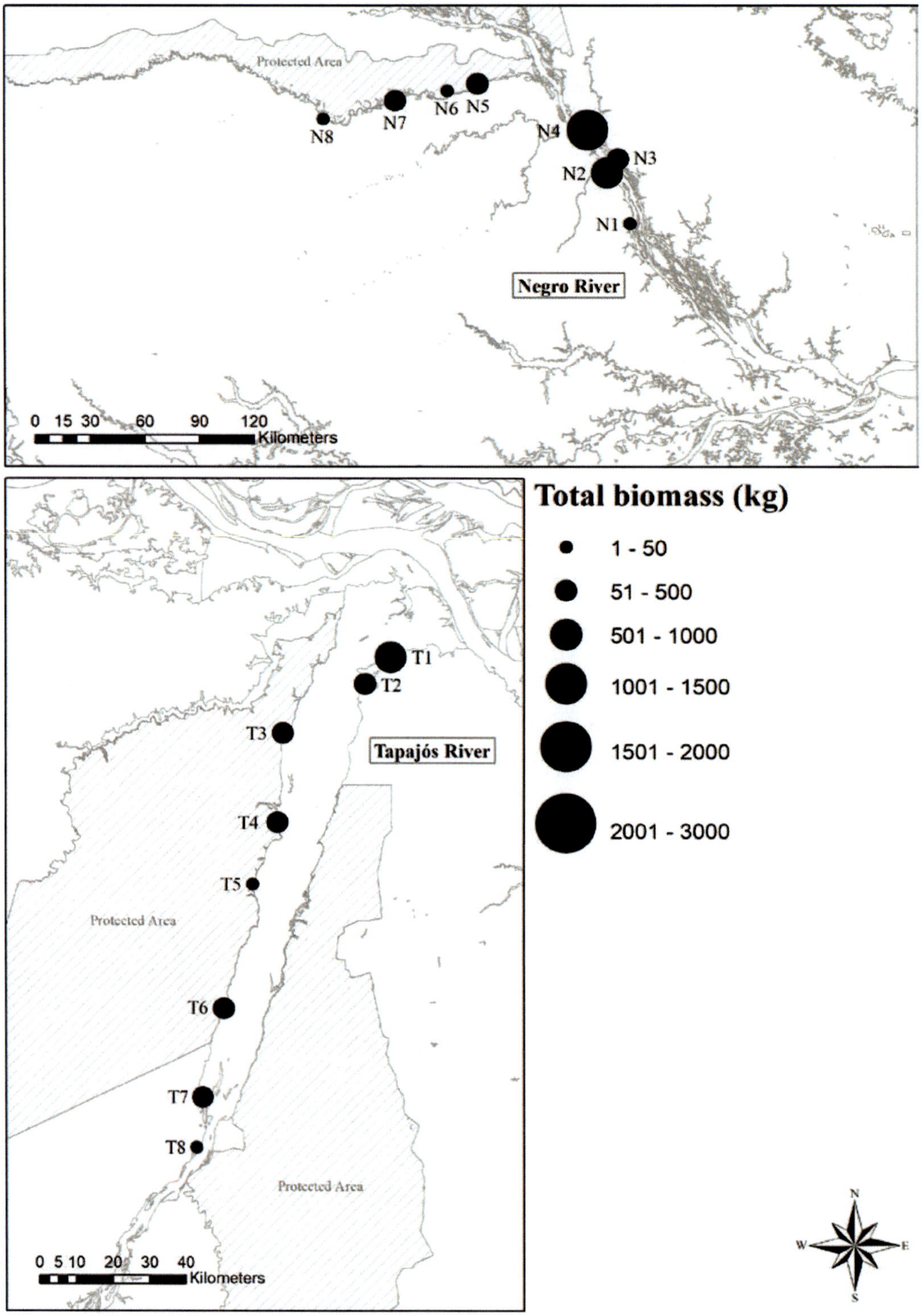

Fig. 4.145 Distribution of biomass of jaraquis (*Semaprochilodus* spp.) caught in small-scale fisheries in the Negro and Tapajós rivers

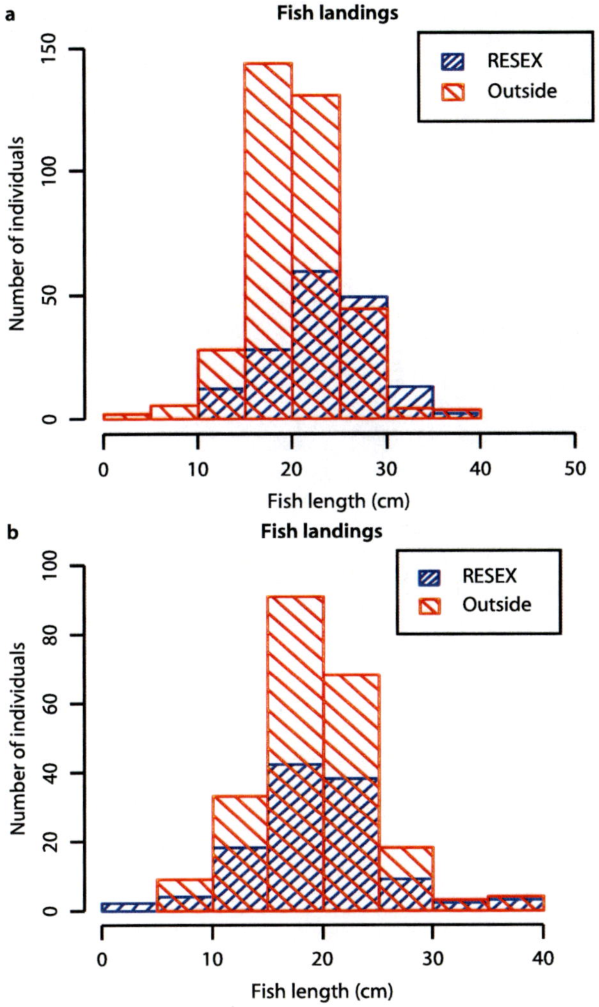

Fig. 4.146 Size distribution of jaraquis (*Semaprochilodus* spp.) caught in small-scale fisheries in 16 communities inside and outside the Extractive Reserves (RESEX) in the rivers **a** Negro, N Resex = 163, N outside = 361 and **b** Tapajós, N Resex = 118, N outside = 226

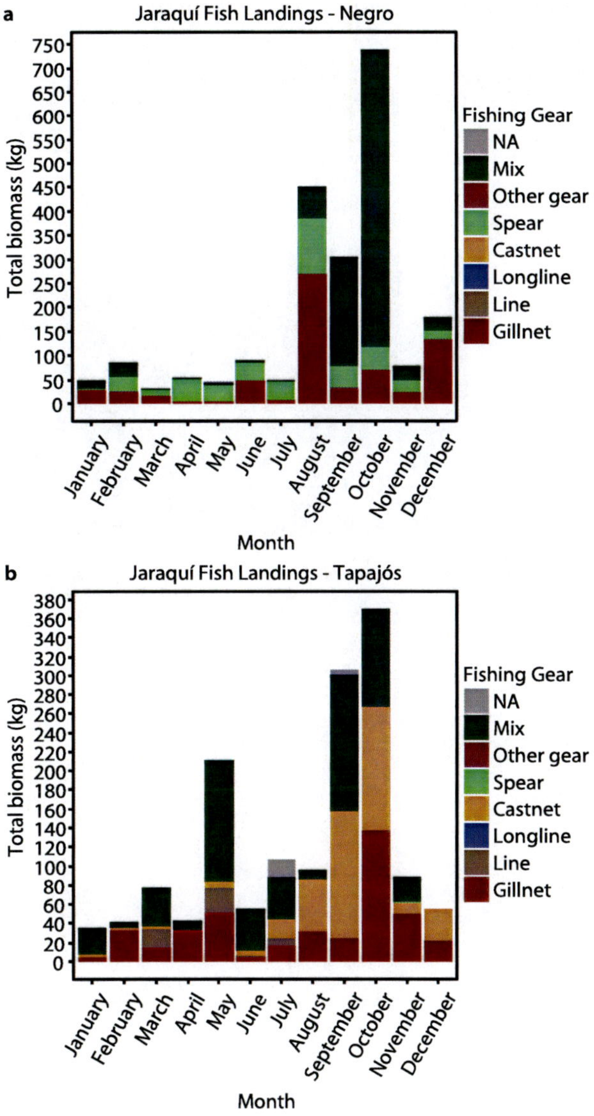

Fig. 4.147 Seasonal variation of biomass of jaraquis (*Semaprochilodus* spp.) caught per fishing gear in small-scale fisheries in16 communities in the rivers **a** Negro and **b** Tapajós. NA: fishing gear not informed; Mix: more than one gear was used

Fish species: *Semaprochilodus taeniurus* (Valenciennes 1821)

Common name(s): jaraqui-escama-fina

Size of fish sampled (standard length, both rivers): mean = 19.1 cm, median = 20 cm

Fisheries: The jaraquis (*Semaprochilodus* spp.) are among the most important fishes to fishers in both rivers, for consumption and commerce; see *Semaprochilodus insignis* for jaraqui fisheries data. Biomass caught is in ◘ Table 3.3 (► Chap. 3).

◘ **Fig. 4.148** *Semaprochilodus taeniurus* sampled in the Negro River. (Photo: Renato Silvano)

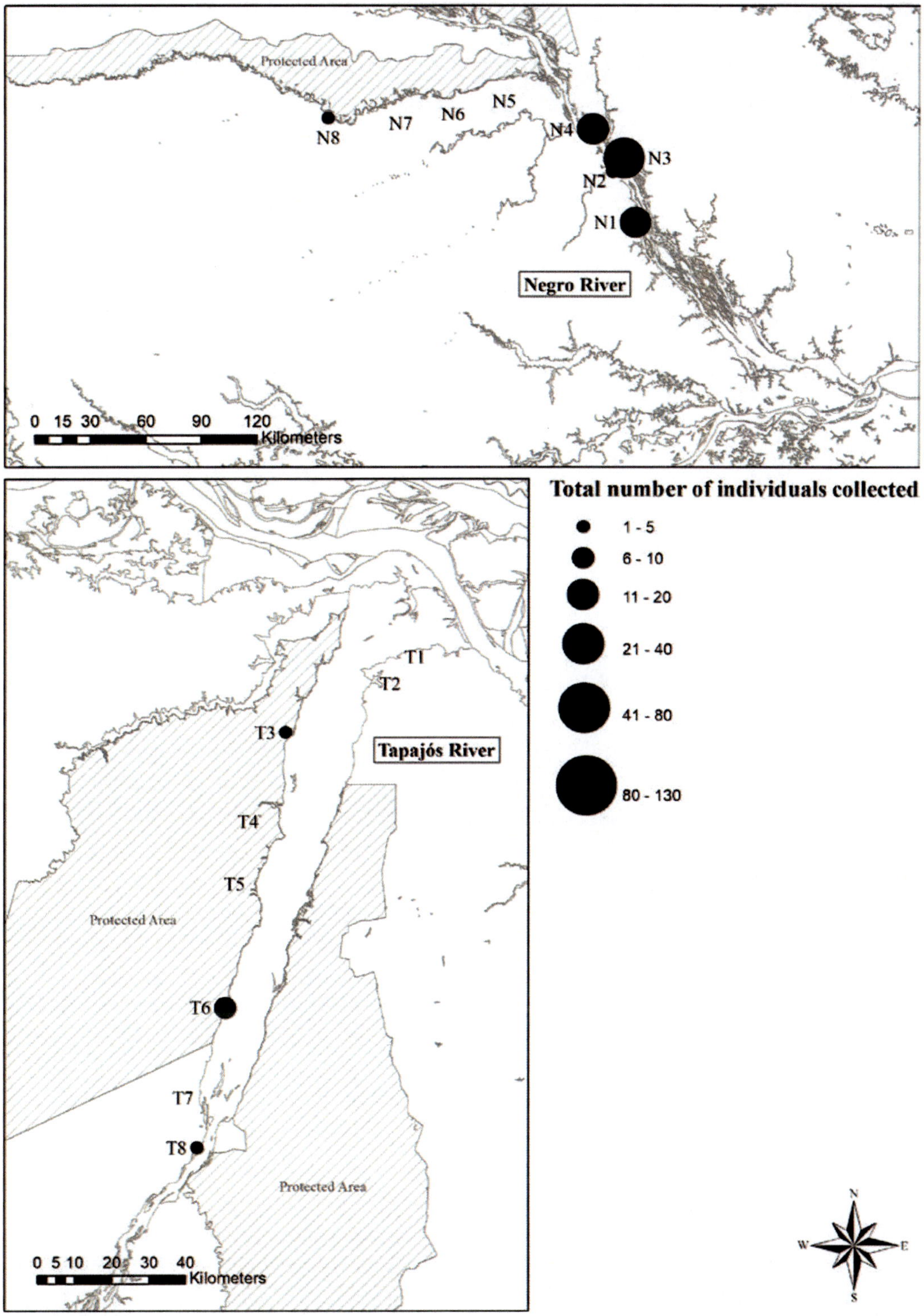

Fig. 4.149 Distribution of individuals of *Semaprochilodus taeniurus* sampled in the Negro and Tapajós rivers

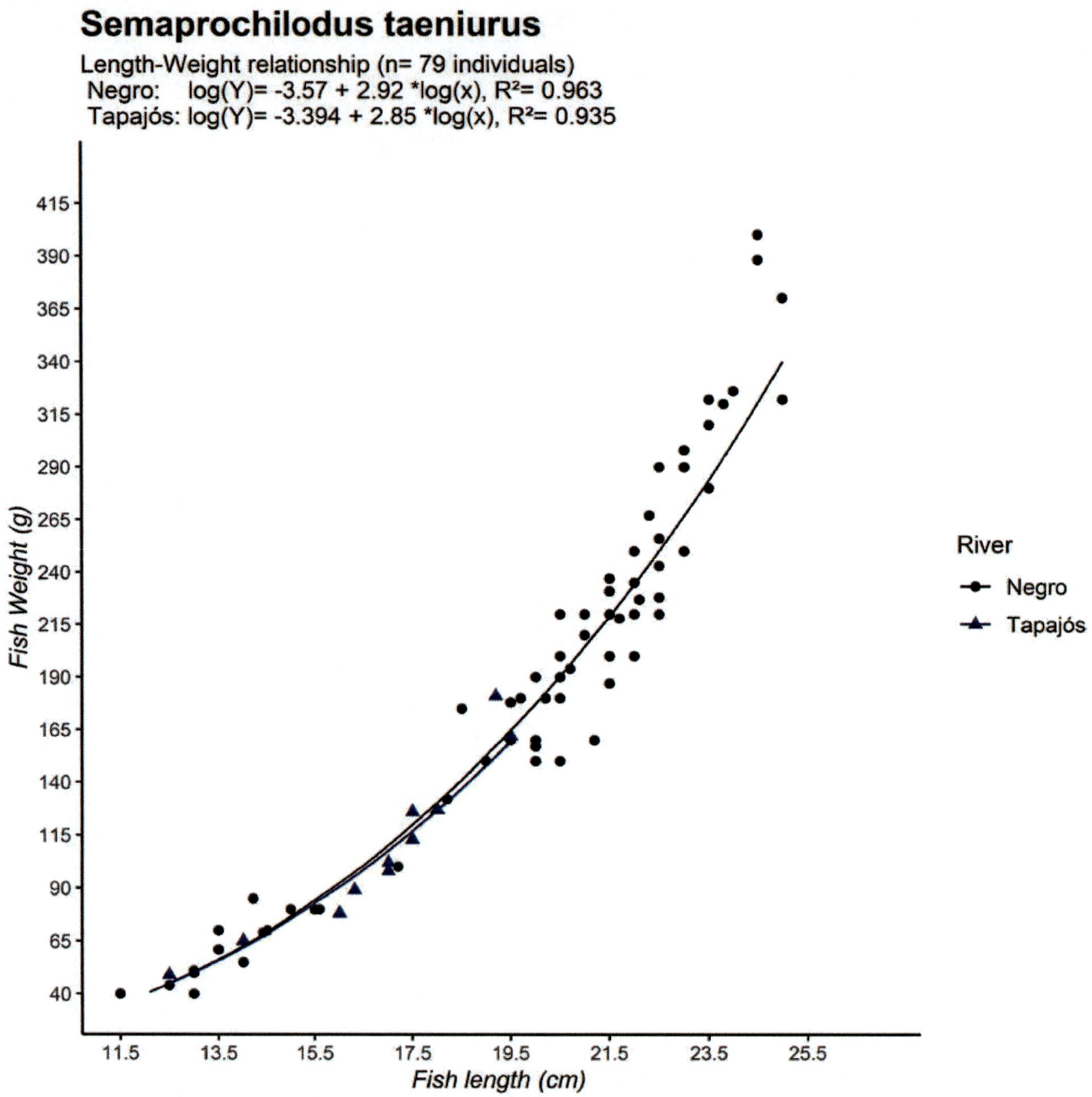

Fig. 4.150 Length-weight relationship for *Semaprochilodus taeniurus* sampled in Tapajós and Negro rivers

4.11 Family Serrasalmidae

4.11.1 Summary of Biology and Ecology (Literature)

Diversity: this family of South American fishes is composed of 101 valid species and 16 genera. Their relation with other Characiformes is still a matter of debate, since morphological and molecular studies have generated conflicting results. For a long time, the Serrasalmidae family was considered a subfamily (Serrasalminae) within the family Characidae. The current consensus is that the family Serrasalmidae is an independent monophyletic group. The family can be divided into three main clades: the pacus (*Colossoma, Mylossoma* and *Piaractus*), the true piranhas (*Pygopristis, Pygocentrus, Pristobrycon, Catoprion*, and *Serrasalmus*), and the *Myleus*-like pacus (*Myleus*). The pacus are the less diverse clade and diverged from the rest of the family around the Late Cretaceous, more than 68 million years ago. The famous piranhas of the genus *Serrasalmus* and *Pygocentrus* are suggested to be a younger clade originated in the Miocene, around 15–20 million years ago. More taxonomic work is required to understand the relationships between the species of this family, especially the ones of the genera *Myloplus, Mylesinus, Tometes* and *Pristobrycon*, which do not seem to constitute monophyletic groups.

Habitat: fishes of this family are usually found forming schools in lakes and rivers of slow flow. Studies have shown that the population of piranhas tend to increase in abundance after the construction of dams in large rivers, as dams and reservoirs transform a lotic system into a lentic one. Some herbivorous species may also inhabit rapids, such as members of the genus *Myleus, Ossubtus*, and *Tometes*.

Morphology: pacus and piranhas possess a serrated keel located in the anterior ventral portion of the body and typical compressed body shape (short and deep). The scales are numerous and small. Most species present a single small spine located anteriorly to the dorsal fin.

Ecology and feeding: members of this family vary in terms of diet. Some species, such as the pacus, are mainly herbivorous, feeding on fruits, leaves, and seeds. Many piranhas are generalists and may feed on a variety of prey, including fish, other vertebrates, and arthropods. Species of the genus *Catoprion* are known to feed primarily on fish scales. Some species undertake trophic migrations. Herbivores species that are highly mobile and that include seeds and fruits in their diet may also function as seed dispersers, allowing gene flow among plant populations and facilitating plant establishment. The large seed eaters of the genus *Colossoma* and *Piaractus* are important seed dispersers, but their ecological role of dispersing seeds of many trees over large distances has been compromised by intense fishing of larger individuals of these fishes.

Reproduction: there are many knowledge gaps related to the reproduction of serrasalmids. Some species are known to migrate laterally (between floodplain lakes and the river) and/or vertically (upstream-downstream) to reproduce. Most species of this family seem to present a periodic (seasonal) or equilibrium life history strategy.

Economic importance: many species of the family Serrasalmidae, especially the pacus, are fished by riverine communities for subsistence and also commercially. The tambaqui (*Colossoma macropomum*) is among the most commercially valuable fish in the Brazilian Amazon. This species, which has large body size, slow growth, and low fecundity, has showed population declines due to overfishing in the last decades. Programs to recover the tambaqui populations include local co-management initiatives of protected lakes. There have been also efforts to raise this fish in aquaculture to supply the demand for its tasteful meat.

References: Rodrigues et al. (1978), Goulding (1980), Goulding and Carvalho (1982), Berra (2001), Nelson et al. (2006), Agostinho et al. (2008), Silvano et al. (2008, 2009), Anderson et al. (2011), Correa and Winemiller (2014), Costa-Pereira and Galetti (2015), Keppeler et al. (2017, 2018), Andrade et al. (2018), Arantes et al. (2018), Fricke et al. (2019), Hallwass et al. (2013b), Thompson et al. (2014), Tregidgo et al. (2017).

Fish species: *Catoprion absconditus* Mateussi, Melo & Oliveira 2020

Common name(s): piranha-queixuda, piranha, piranha-pacu

Size of fish sampled (standard length, both rivers): min. = 7 cm, max. = 10.2 cm, mean = 7.8 cm, median = 8 cm

Fisheries: Not relevant to fisheries, due to its small size.

Fig. 4.151 *Catoprion absconditus* sampled in the Negro River. (Photo: Renato Silvano)

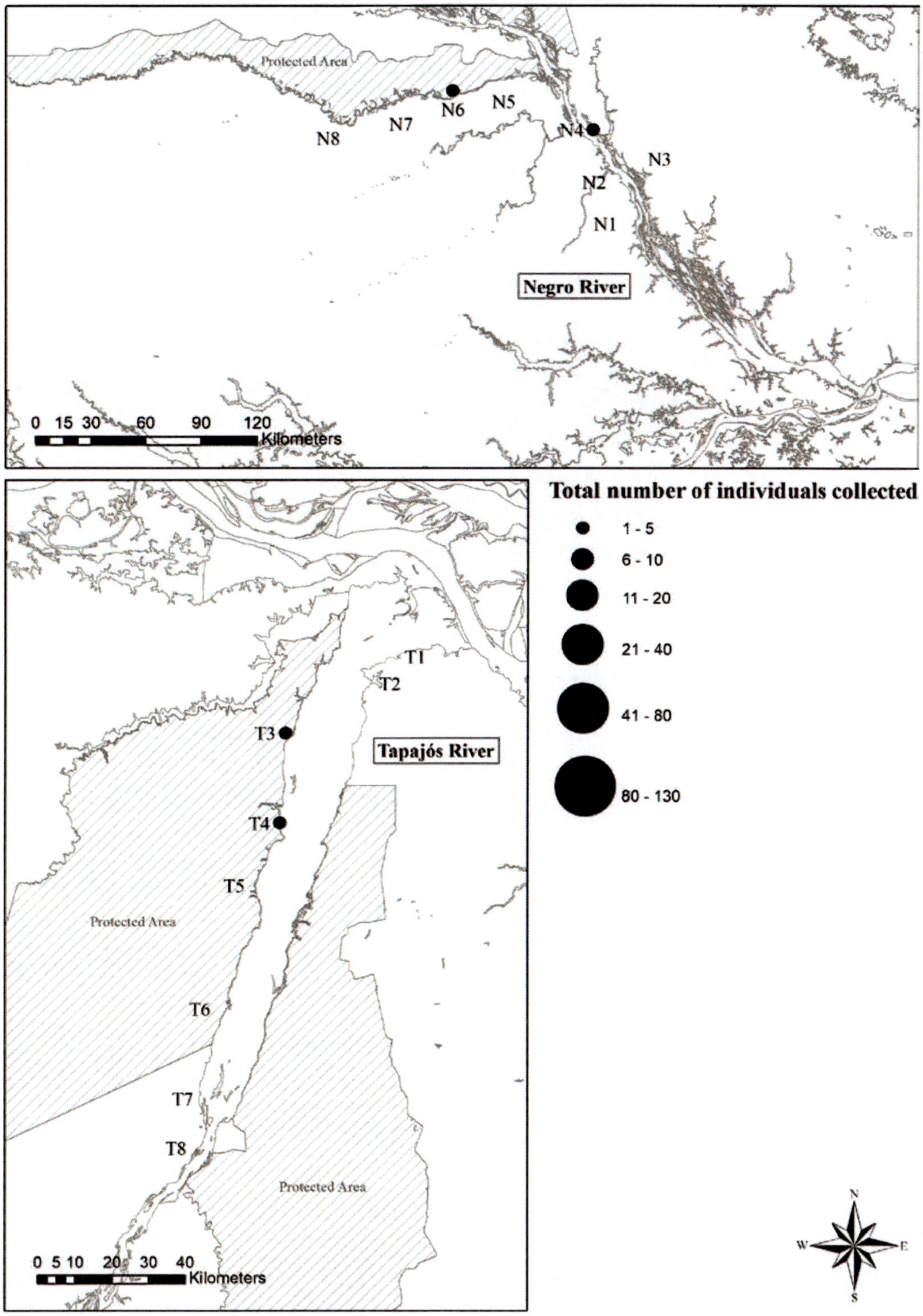

Fig. 4.152 Distribution of individuals of *Catoprion absconditus* sampled in the Negro and Tapajós rivers

Fish species: *Colossoma macropomum* (Cuvier 1816)

Common name(s): tambaqui, bocó

Size of fish sampled (standard length, both rivers): min. = 16.5 cm, max. = 25.5 cm, mean = 20.4 cm, median = 19 cm

Fisheries: The tambaqui (*Colossoma macropomum*) is a valuable commercial fish to fisheries in the Tapajós River, but this fish is not regularly caught in the Negro River fisheries. Biomass caught is in ◻ Table 3.3 (▸ Chap. 3).

◻ **Fig. 4.153** *Colossoma macropomum* sampled in the Tapajós River. (Photo: Renato Silvano)

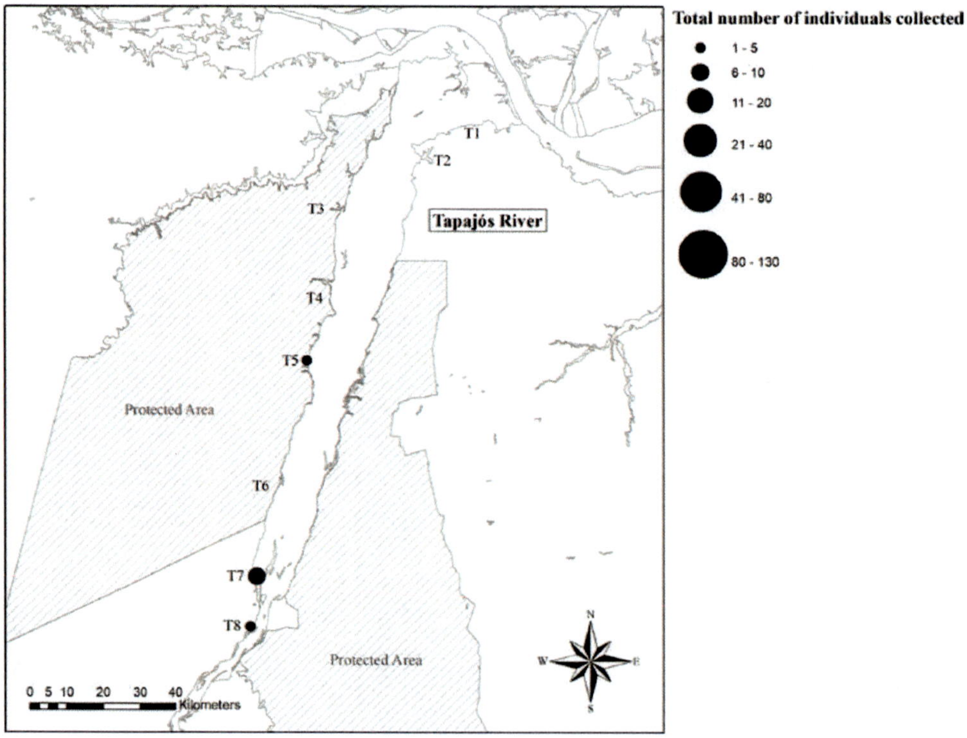

◻ **Fig. 4.154** Distribution of individuals of *Colossoma macropomum* sampled in the Tapajós River

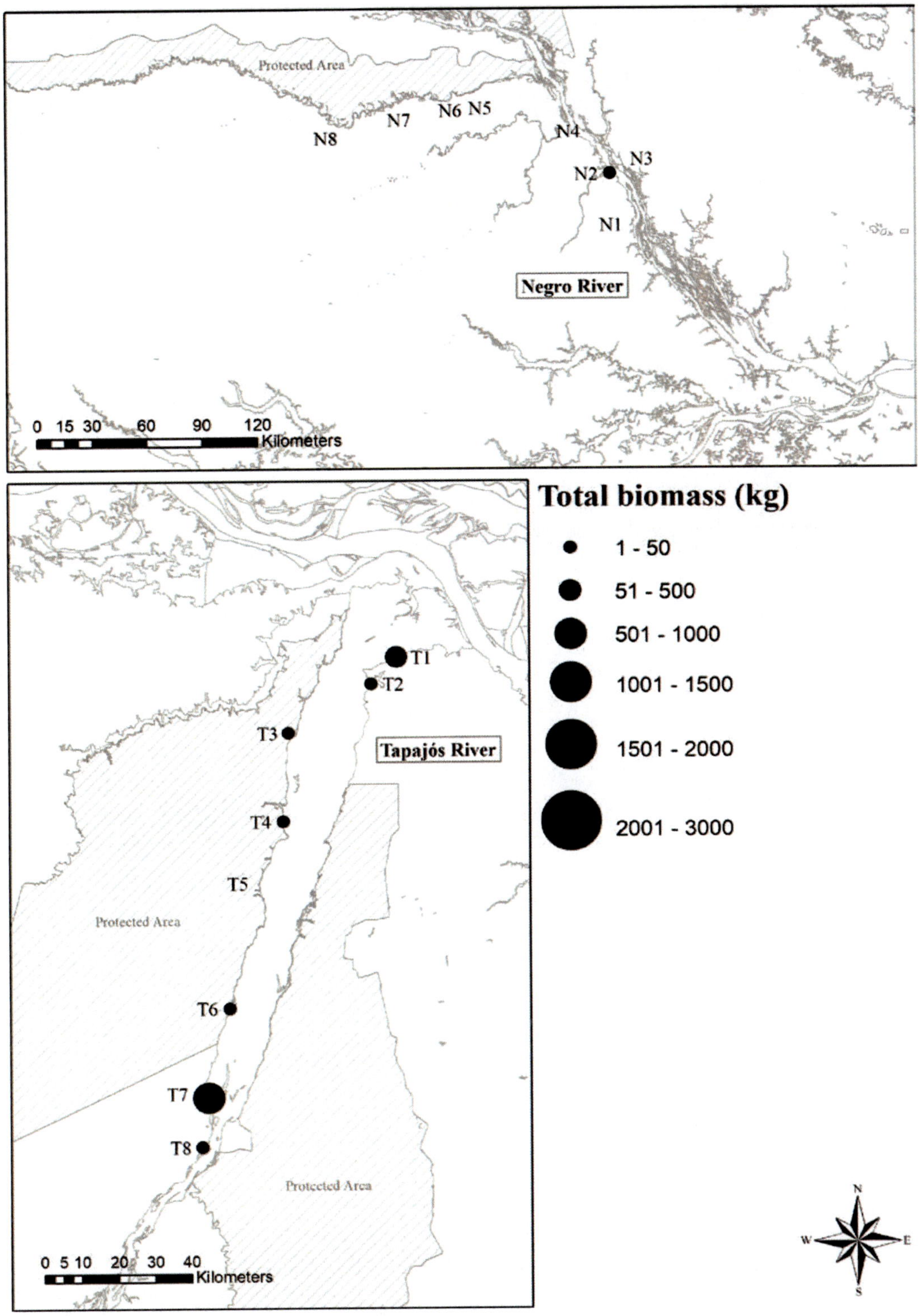

Fig. 4.155 Distribution of biomass of tambaqui (*Colossoma macropomum*) caught in small-scale fisheries in the Negro and Tapajós rivers

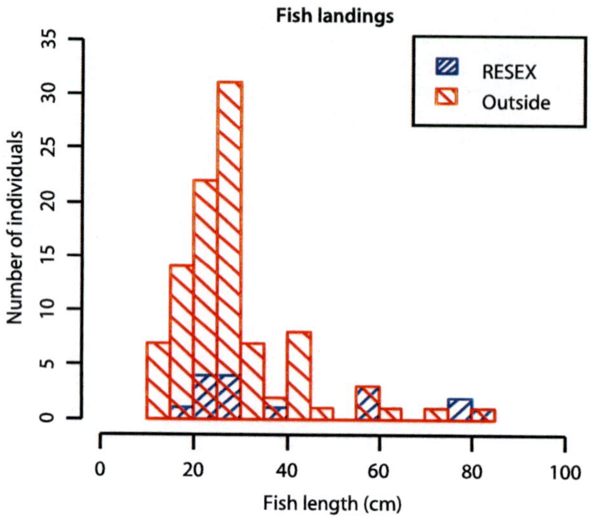

Fig. 4.156 Size distribution of tambaqui (*Colossoma macropomum*) caught in small-scale fisheries in eight communities inside and outside the Extractive Reserves (RESEX) in the Tapajós River, N Resex = 16, N outside = 98

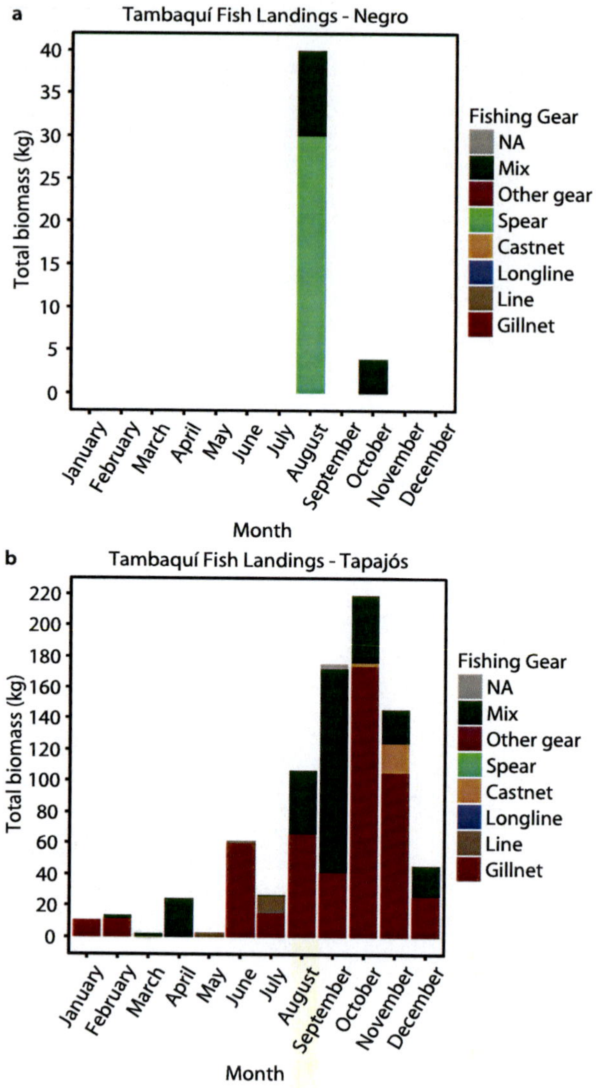

Fig. 4.157 Seasonal variation of biomass of tambaqui (*Colossoma macropomum*) caught per fishing gear in small-scale fisheries in16 communities in the rivers **a** Negro and **b** Tapajós. NA: fishing gear not informed; Mix: more than one gear was used

Fish species: *Metynnis altidorsalis* Ahl 1923

Common name(s): pacu, pacu-marreca

Size of fish sampled (standard length, both rivers): mean and median = 12.1 cm

Fisheries: pacuís (*Metynnis* spp.) are not much relevant to fisheries due to their small size, but these fishes may be occasionally caught together with the more valuable pacus (*Myloplus* spp.), mostly for food in both rivers; see *Myloplus lobatus* for pacu fisheries data. Biomass caught is in ◘ Table 3.3 (► Chap. 3).

◘ **Fig. 4.158** *Metynnis altidorsalis* sampled in the Tapajós River. (Photo: Renato Silvano)

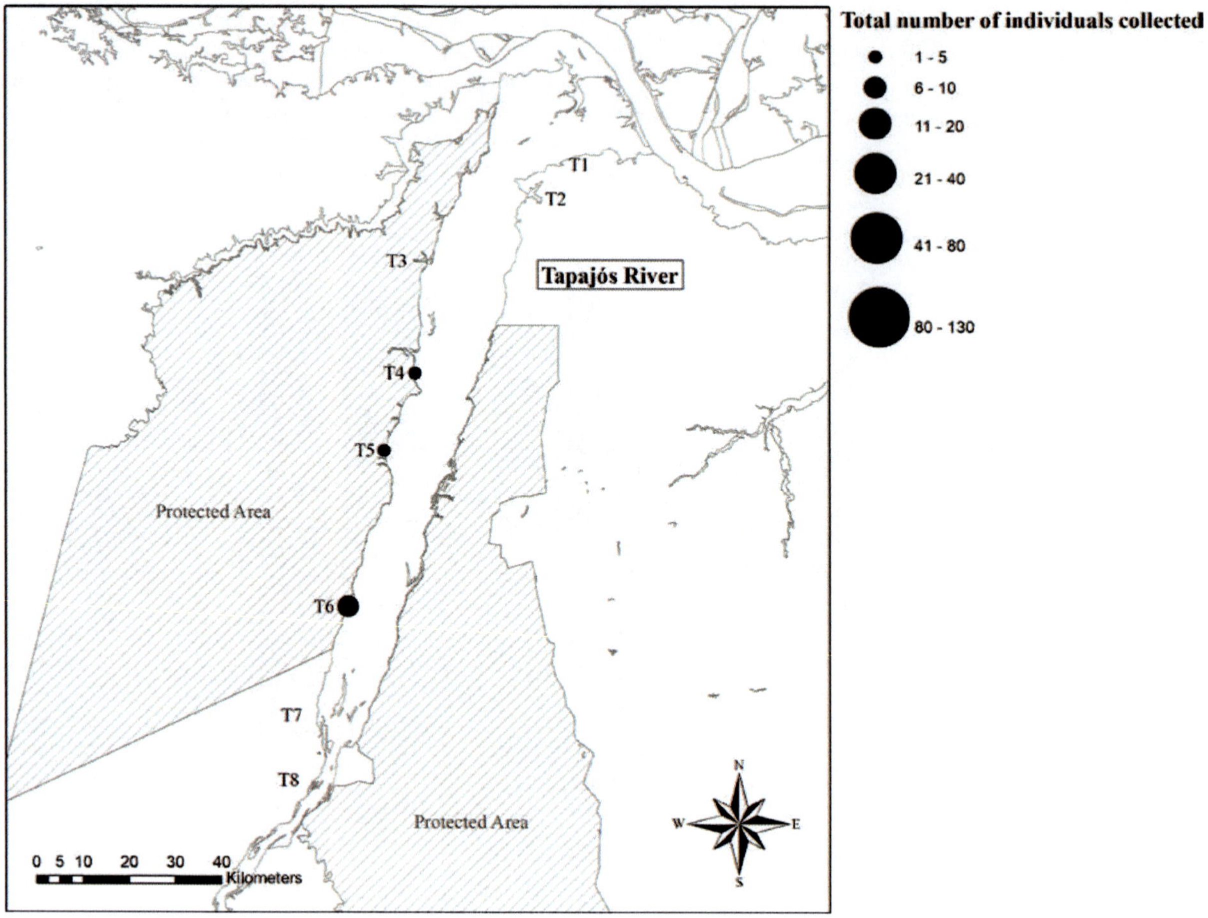

■ **Fig. 4.159** Distribution of individuals of *Metynnis altidorsalis* sampled in the Tapajós River

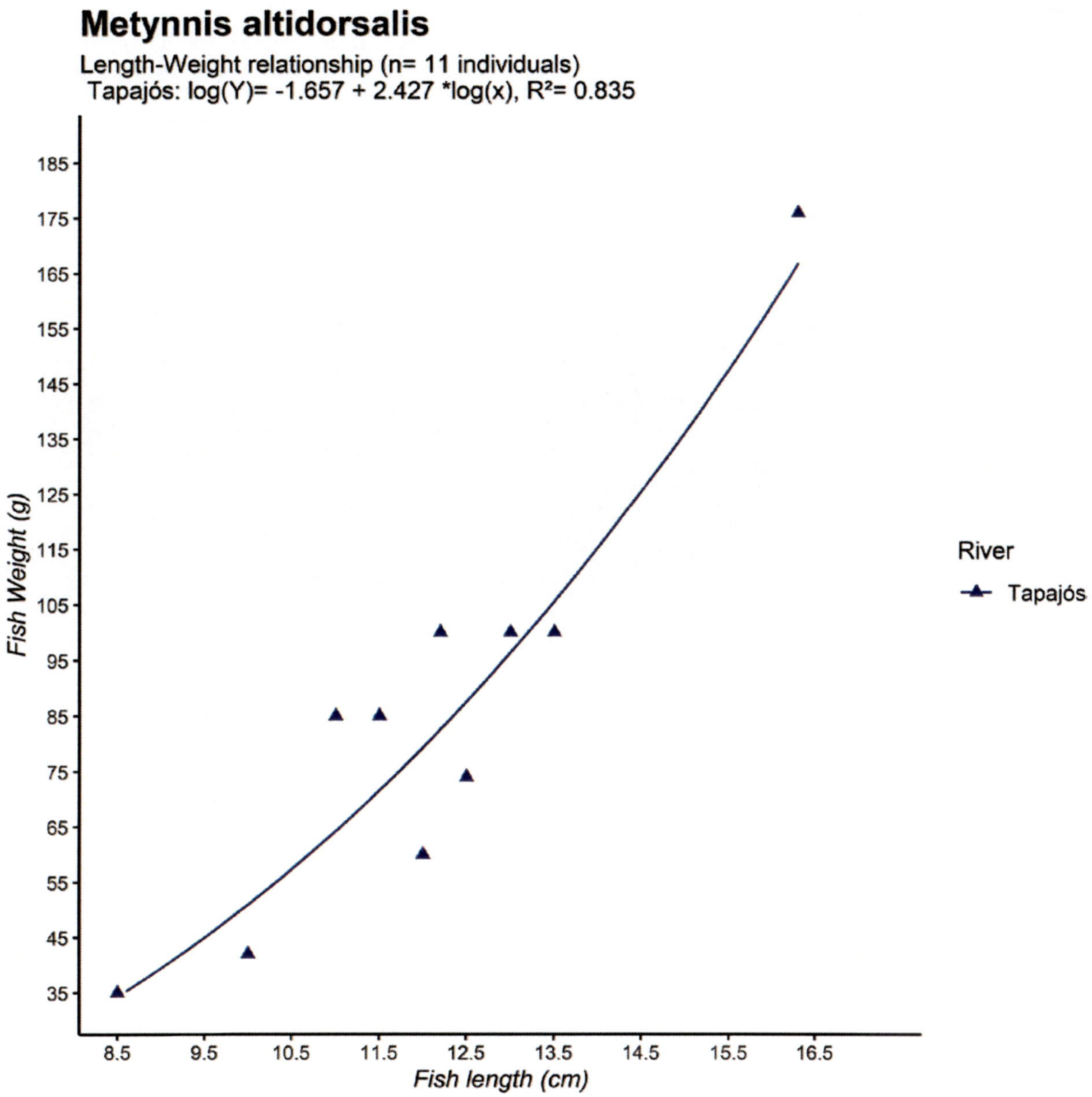

Fig. 4.160 Length-weight relationship for *Metynnis altidorsalis* sampled in the Tapajós river

4

Fish species: *Metynnis hypsauchen* (Müller & Troschel 1844)

Common name(s): pacuí, pacu, pacu-pintado

Size of fish sampled (standard length, both rivers): mean = 11.2 cm, median = 12 cm

Fisheries: pacuís (*Metynnis* spp.) are not much relevant to fisheries due to their small size, but these fishes may be occasionally caught together with the more valuable pacus (*Myloplus* spp.), mostly for food in both rivers; see *Myloplus lobatus* for pacu fisheries data. Biomass caught is in ◘ Table 3.3 (▸ Chap. 3).

◘ **Fig. 4.161** *Metynnis hypsauchen* sampled in the Tapajós River. (Photo: Renato Silvano)

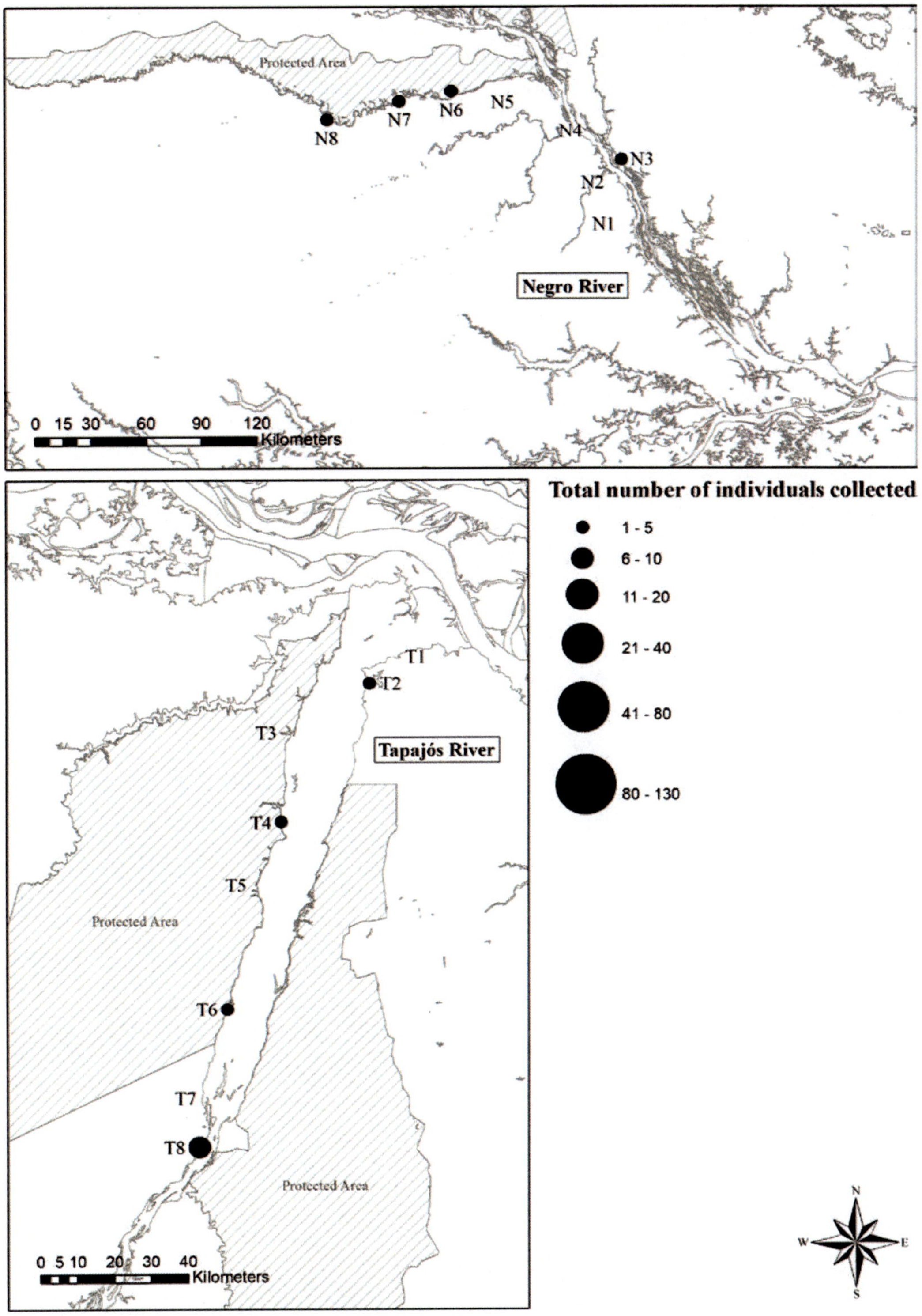

Fig. 4.162 Distribution of individuals of *Metynnis hypsauchen* sampled in the Negro and Tapajós rivers

4

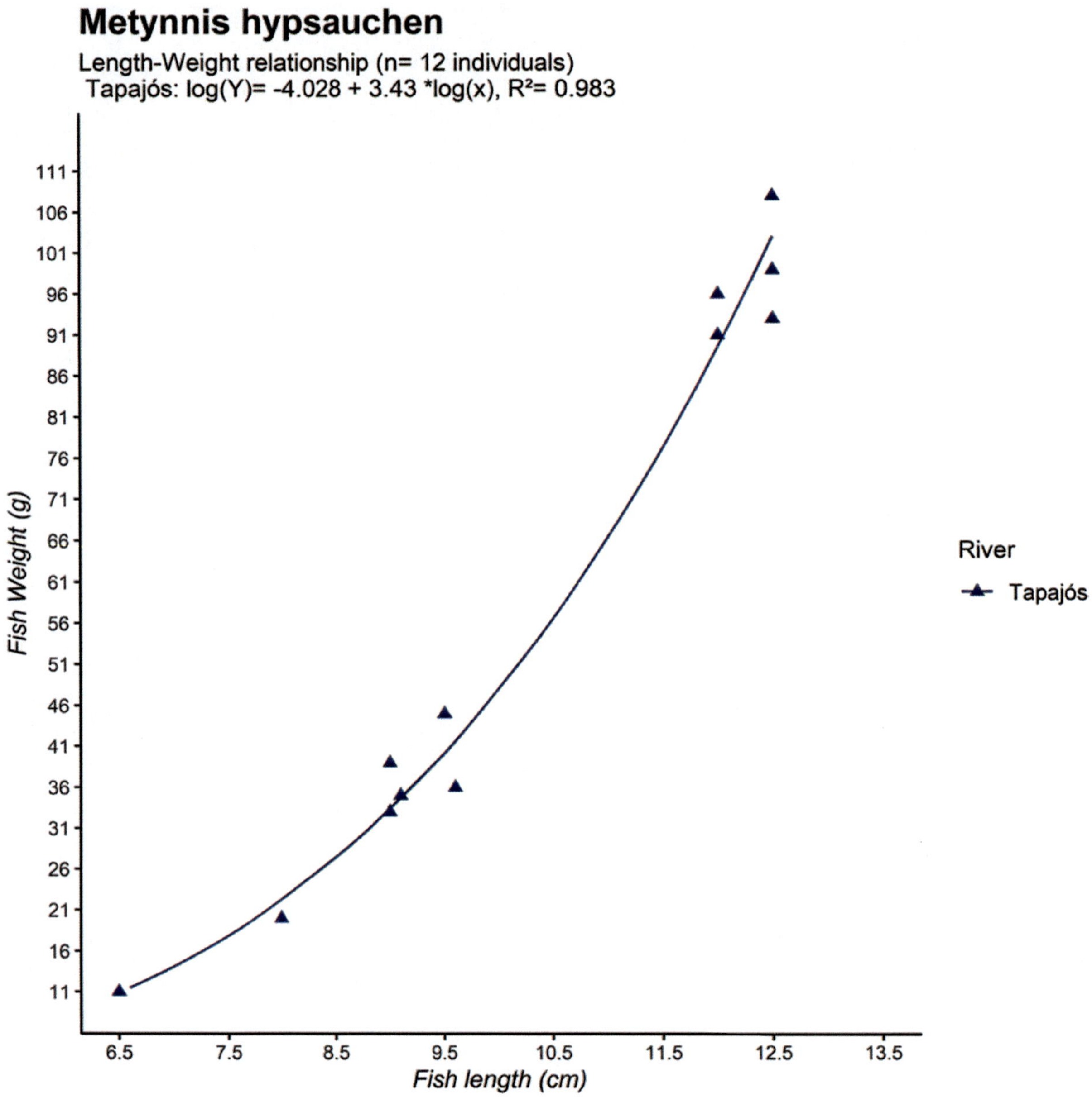

Fig. 4.163 Length-weight relationship for *Metynnis hypsauchen* sampled in the Tapajós River

Fish species: *Metynnis lippincottianus* (Cope 1870)

Common name(s): pacuí

Size of fish sampled (standard length, both rivers): mean = 9.7 cm, median = 8 cm

Fisheries: pacuís (*Metynnis* spp.) are not much relevant to fisheries due to their small size, but these fishes may be occasionally caught together with the more valuable pacus (*Myloplus* spp.), mostly for food in both rivers; see *Myloplus lobatus* for pacu fisheries data. Biomass caught is in ◘ Table 3.3 (▸ Chap. 3).

◘ **Fig. 4.164** *Metynnis lippincottianus* sampled in the Negro River. (Photo: Renato Silvano)

4

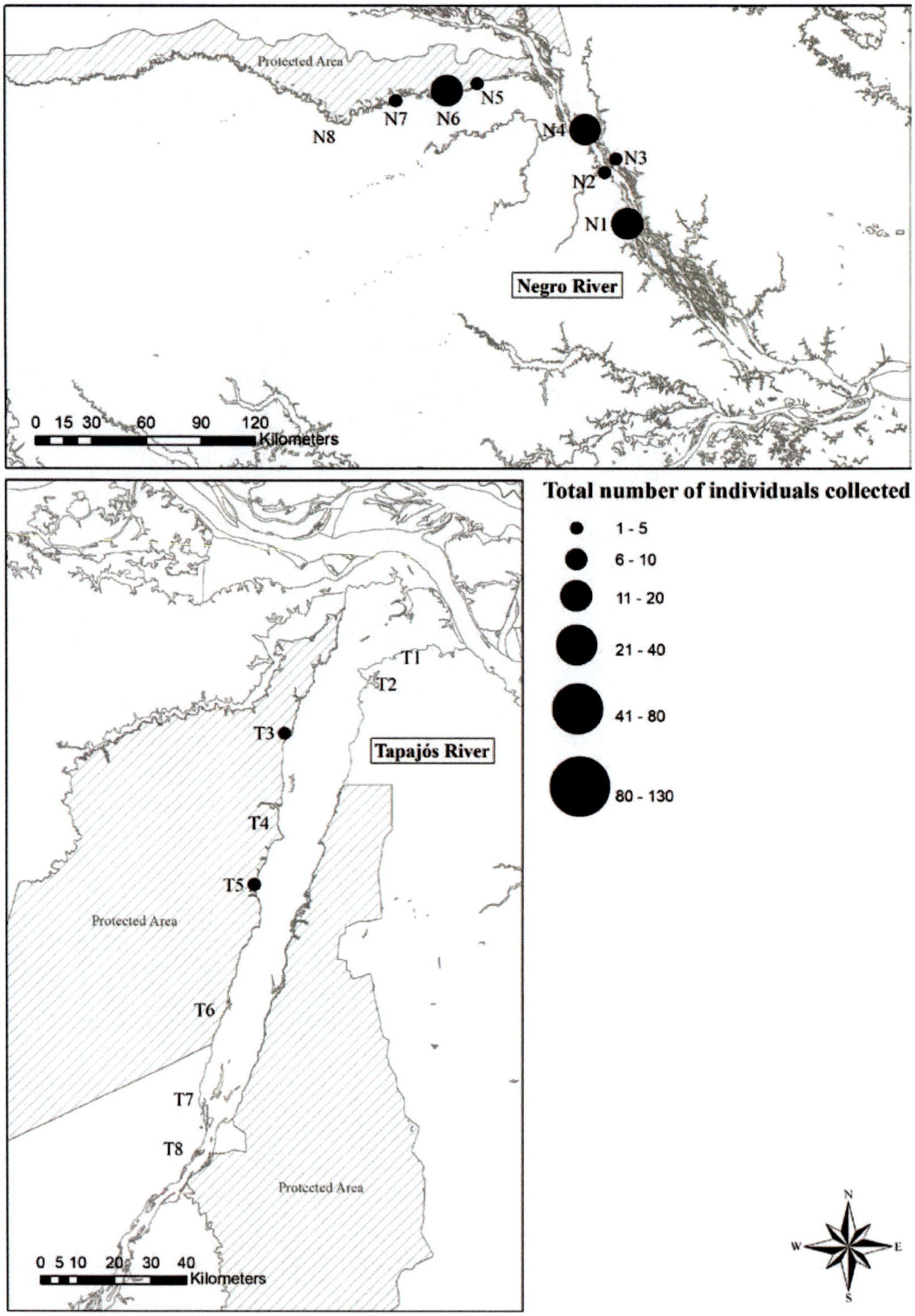

Fig. 4.165 Distribution of individuals of *Metynnis lippincottianus* sampled in the Negro and Tapajós rivers

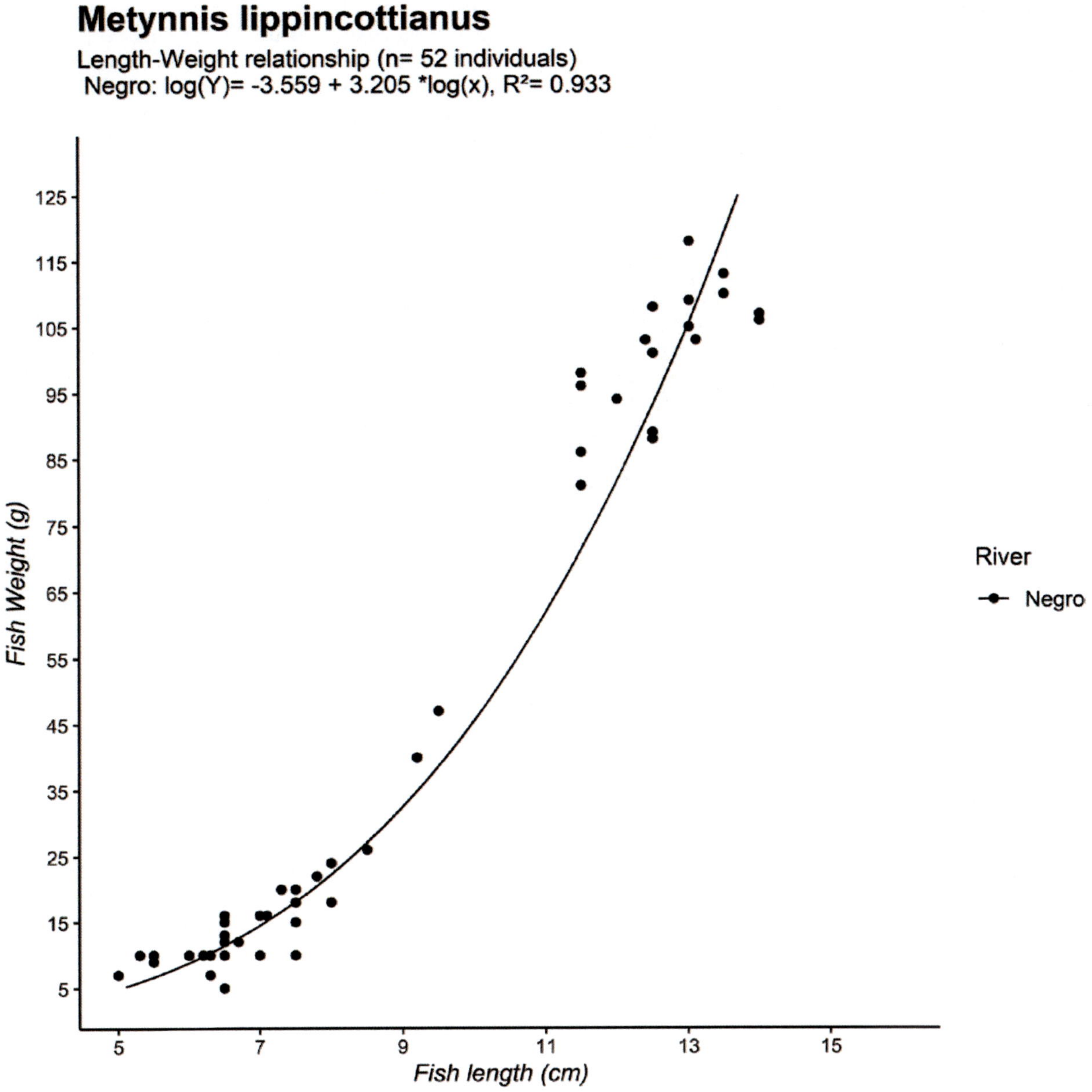

Fig. 4.166 Length-weight relationship for *Metynnis lippincottianus* sampled in Negro river

Fish species: *Metynnis melanogrammus* Ota, Rapp Py-Daniel & Jégu 2016

Common name(s): pacuí, pacu, pacu-marreca

Size of fish sampled (standard length, both rivers): min. = 9 cm, max. = 13 cm, mean and median = 11 cm

Fisheries: pacuís (*Metynnis* spp.) are not much relevant to fisheries due to their small size, but these fishes may be occasionally caught together with the more valuable pacus (*Myloplus* spp.), mostly for food in both rivers; see *Myloplus lobatus* for pacu fisheries data. Biomass caught is in ◘ Table 3.3 (► Chap. 3).

◘ **Fig. 4.167** *Metynnis melanogrammus* sampled in the Negro River. (Photo: Renato Silvano)

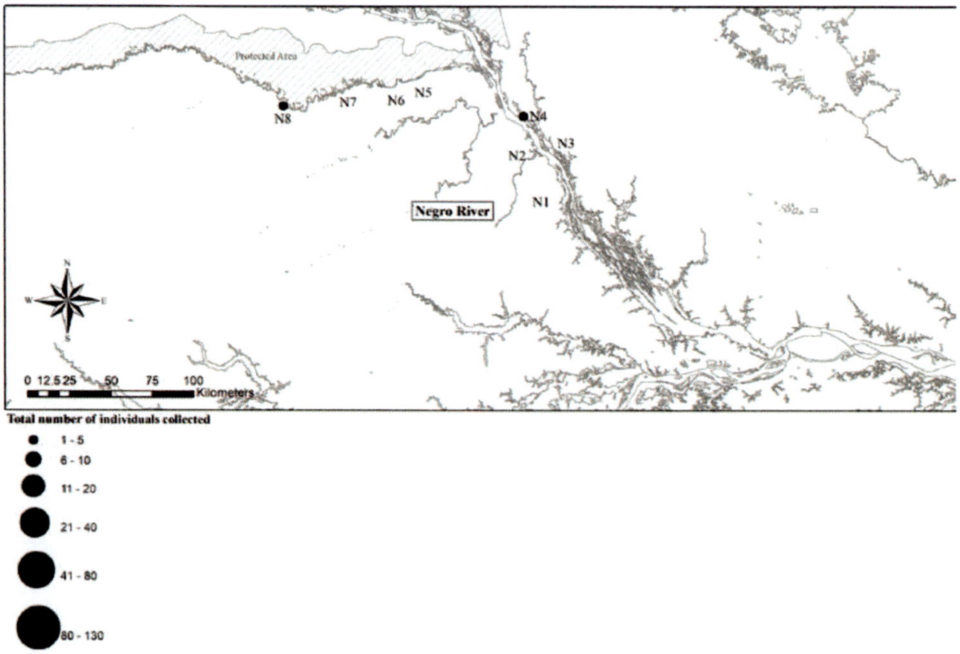

◘ **Fig. 4.168** Distribution of individuals of *Metynnis melanogrammus* sampled in the Negro River

Fish species: *Myloplus asterias* (Müller & Troschel 1844)

Common name(s): pacu, pacu-olhudo

Size of fish sampled (standard length, both rivers): mean = 14.4 cm, median = 14.5 cm

Fisheries: pacus (*Myloplus* spp.) are important fish for food and commerce to fisheries in both rivers; see *Myloplus lobatus* for pacu fisheries data. Biomass caught is in ▫ Table 3.3 (▶ Chap. 3).

▫ **Fig. 4.169** *Myloplus asterias* sampled in the Tapajós River. (Photo: Renato Silvano)

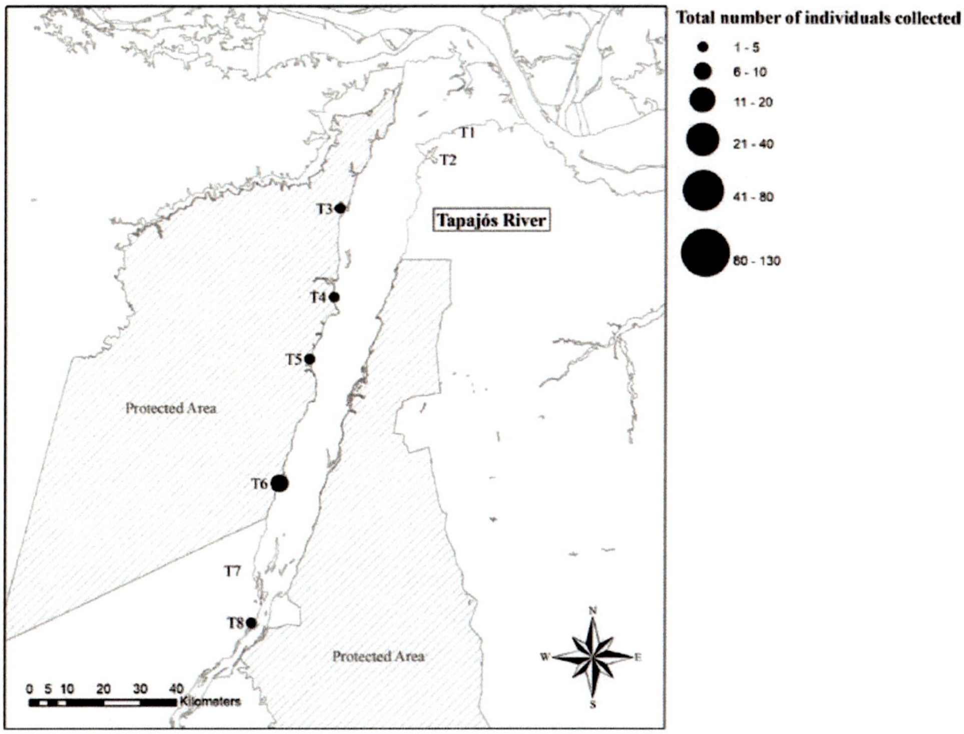

▫ **Fig. 4.170** Distribution of individuals of *Myloplus asterias* sampled in the Tapajós River

4

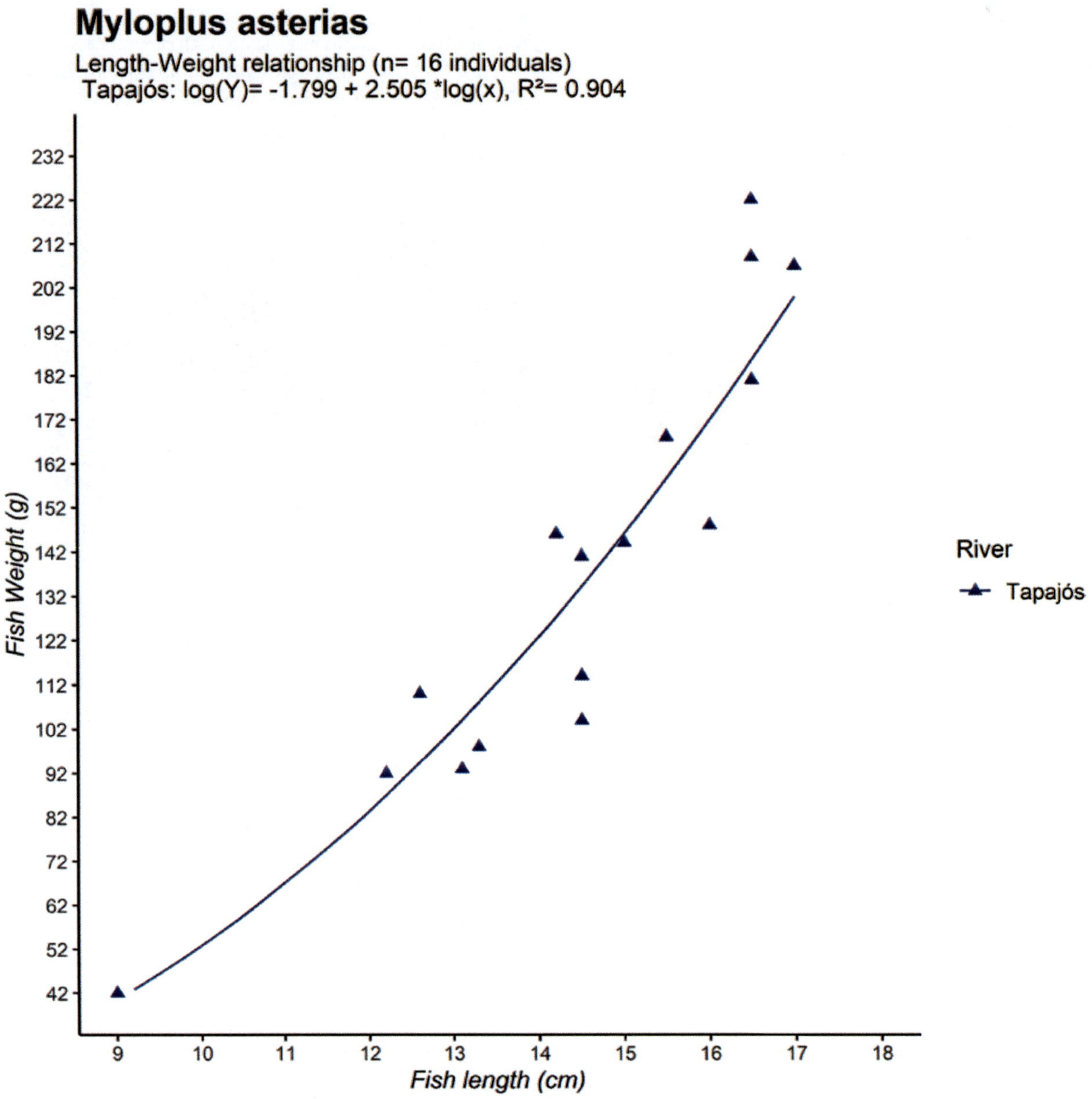

Fig. 4.171 Length-weight relationship for *Myloplus asterias* sampled in Tapajós River

Fish species: ***Myloplus lobatus*** **(Valenciennes 1850)**

Common name(s): pacu, pacu-olhudo

Size of fish sampled (standard length, both rivers): min. = 12 cm, max. = 19 cm, mean = 15.3 cm, median = 14.8 cm

Fisheries: pacus (*Myloplus* spp.) are important fish for food and commerce to fisheries in both rivers. Biomass caught is in ◘ Table 3.3 (► Chap. 3).

◘ **Fig. 4.172** *Myloplus lobatus* sampled in the Tapajós River. (Photo: Renato Silvano)

4

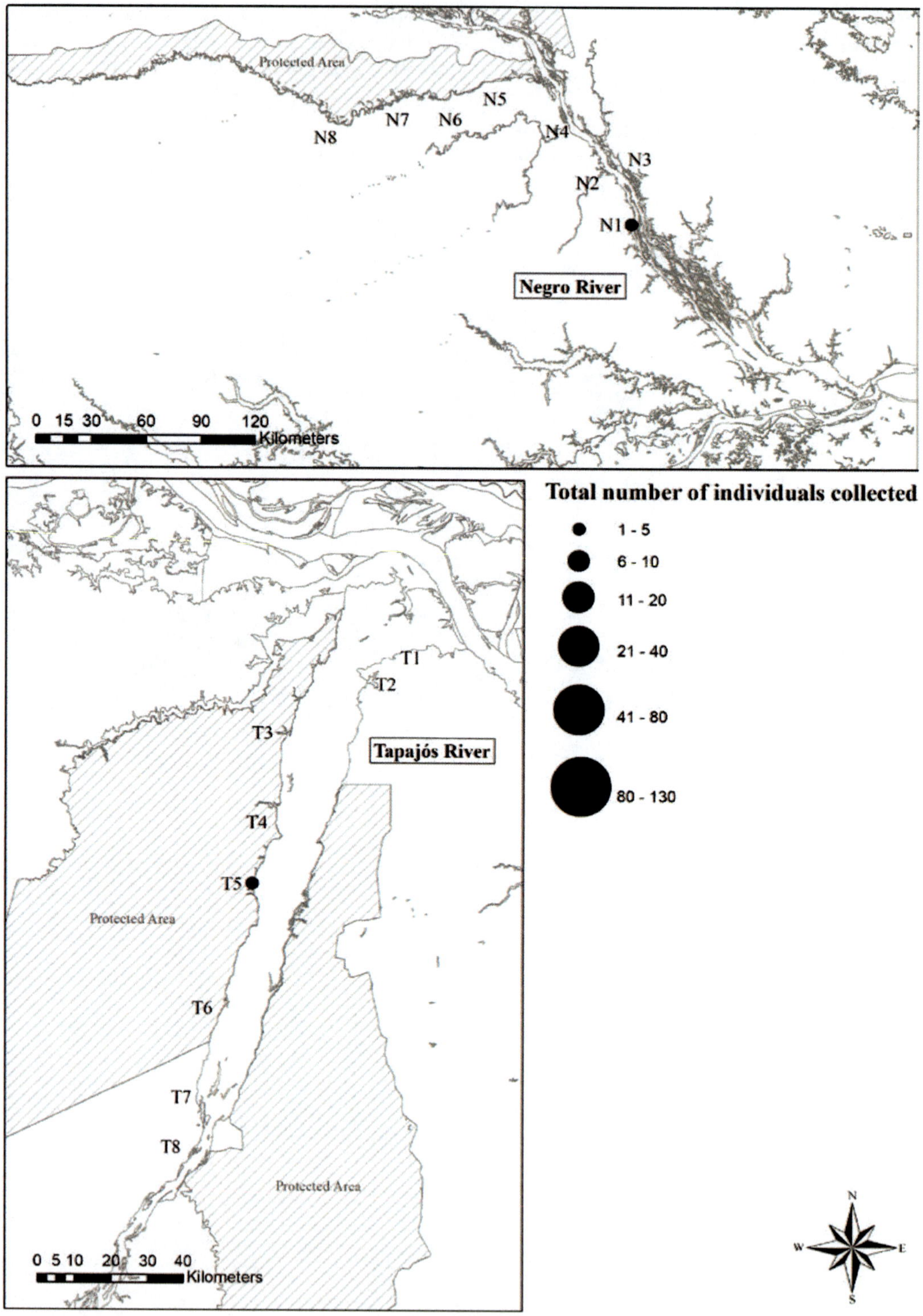

Fig. 4.173 Distribution of individuals of *Myloplus lobatus* sampled in the Negro and Tapajós rivers

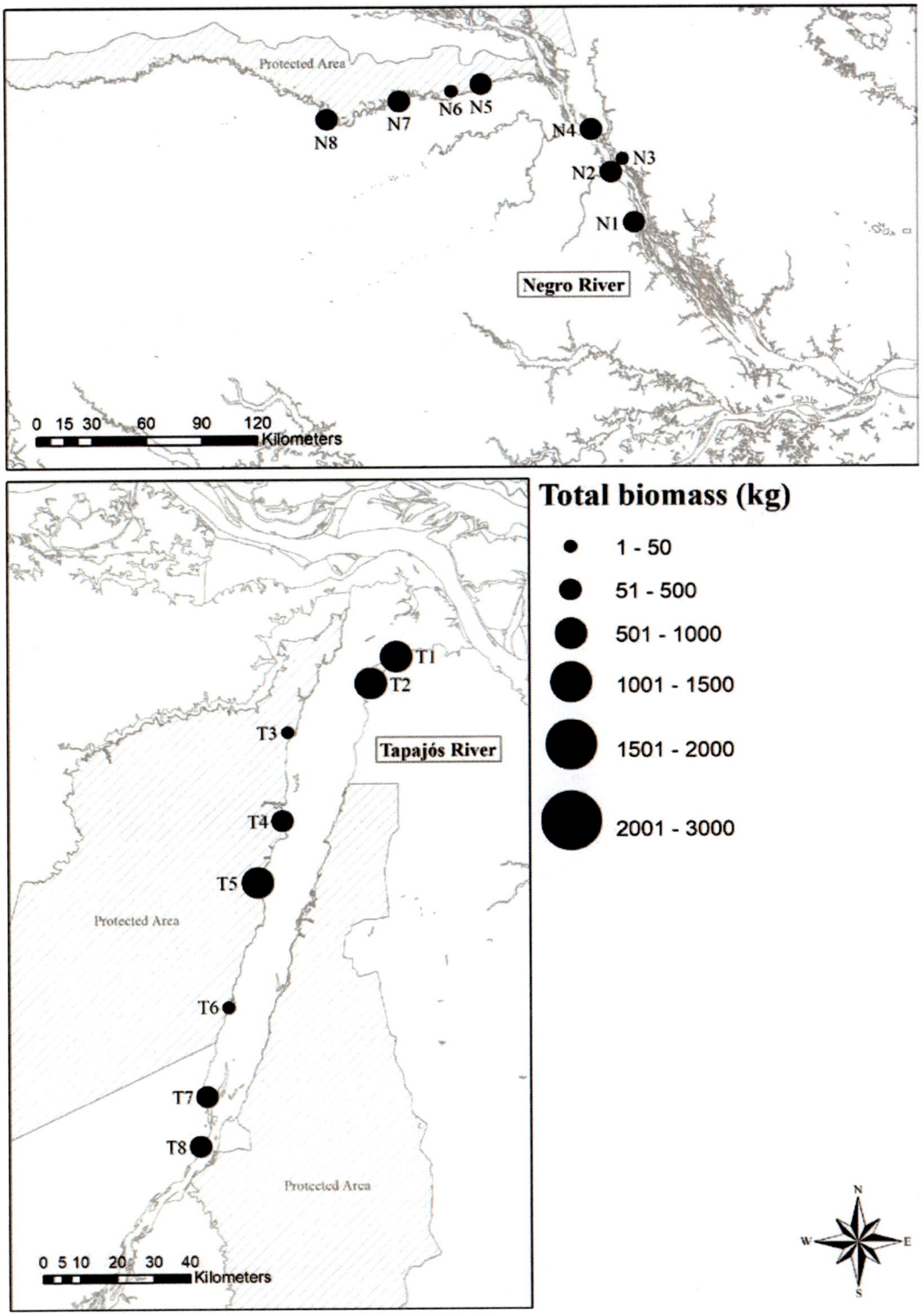

Fig. 4.174 Distribution of biomass of pacus (*Myloplus* spp.) caught in small-scale fisheries in the Negro and Tapajós rivers

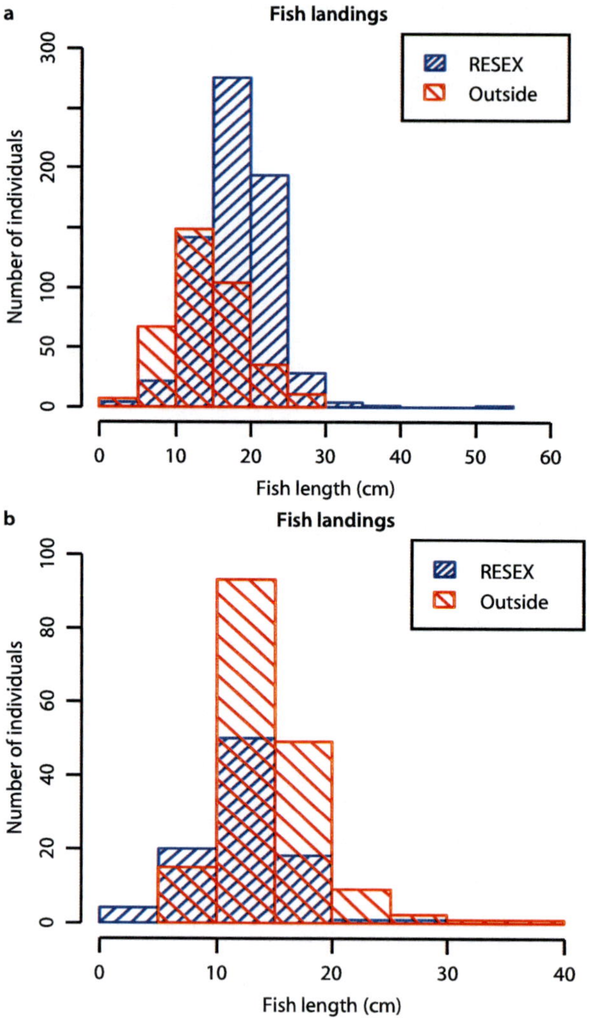

Fig. 4.175 Size distribution of pacus (*Myloplus* spp.) caught in small-scale fisheries in 16 communities inside and outside the Extractive Reserves (RESEX) in the rivers **a** Negro, N Resex = 671, N outside = 372 and **b** Tapajós, N Resex = 94, N outside = 170

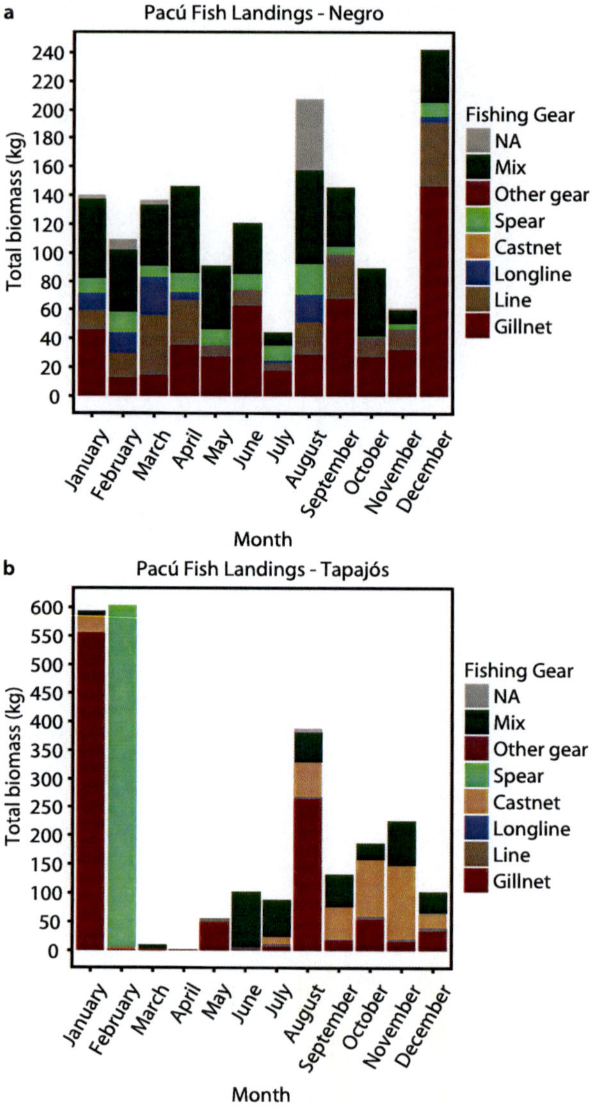

Fig. 4.176 Seasonal variation of biomass of pacus (*Myloplus* spp.) caught per fishing gear in small-scale fisheries in 16 communities in the rivers **a** Negro and **b** Tapajós. NA: fishing gear not informed; Mix: more than one gear was used

Fish species: *Myloplus lucienae* Andrade, Ota, Bastos & Jégu 2016

Common name(s): pacu, pacu-tiuí, pacu-tuí

Size of fish sampled (standard length, both rivers): min. = 7.5 cm, max. = 9.4 cm, mean = 8.3 cm, median = 8.1 cm

Fisheries: pacus (*Myloplus* spp.) are important fish for food and commerce to fisheries in both rivers; see *Myloplus lobatus* for pacu fisheries data. Biomass caught is in Table 3.3 (▶ Chap. 3).

Fig. 4.177 *Myloplus lucienae* sampled in the Negro River. (Photo: Renato Silvano)

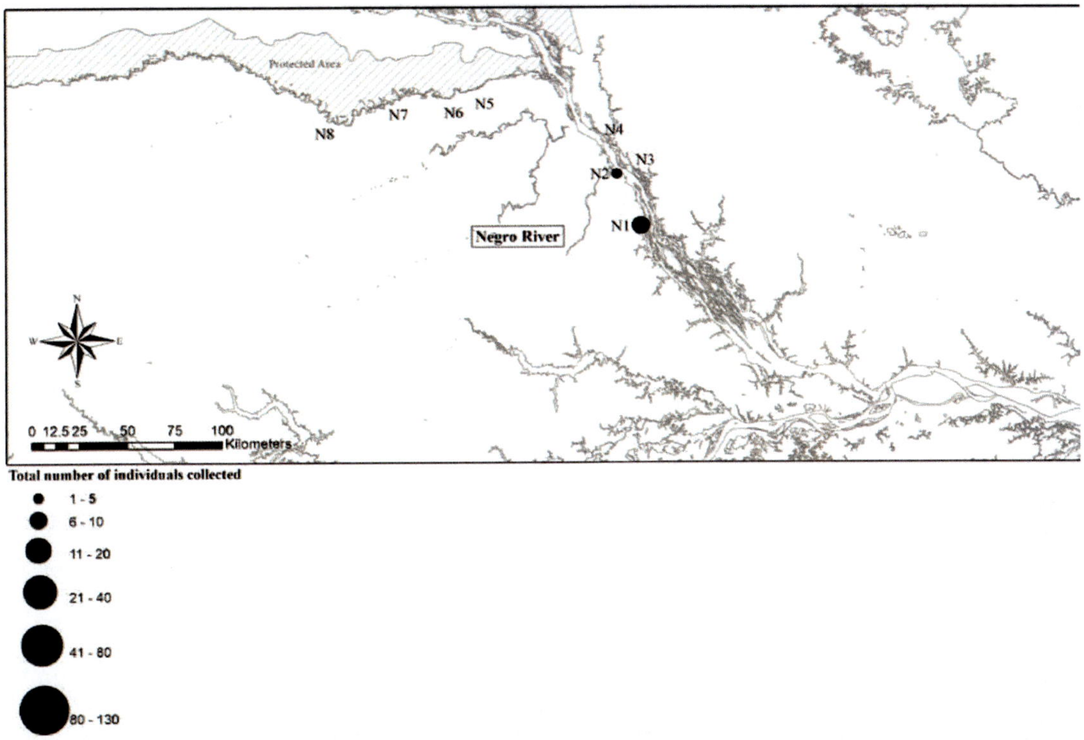

Fig. 4.178 Distribution of individuals of *Myloplus lucienae* sampled in the Negro River

Fish species: ***Myloplus schomburgkii*** **(Jardine 1841)**

Common name(s): pacu, pacu-listrado

Size of fish sampled (standard length, both rivers): min. = 6.5 cm, max. = 16 cm, mean = 11.3 cm, median = 11.3 cm

Fisheries: pacus (*Myloplus* spp.) are important fish for food and commerce to fisheries in both rivers; see *Myloplus lobatus* for pacu fisheries data. Biomass caught is in ◘ Table 3.3 (► Chap. 3).

◘ **Fig. 4.179** *Myloplus schomburgkii* sampled in the Negro River. (Photo: Renato Silvano)

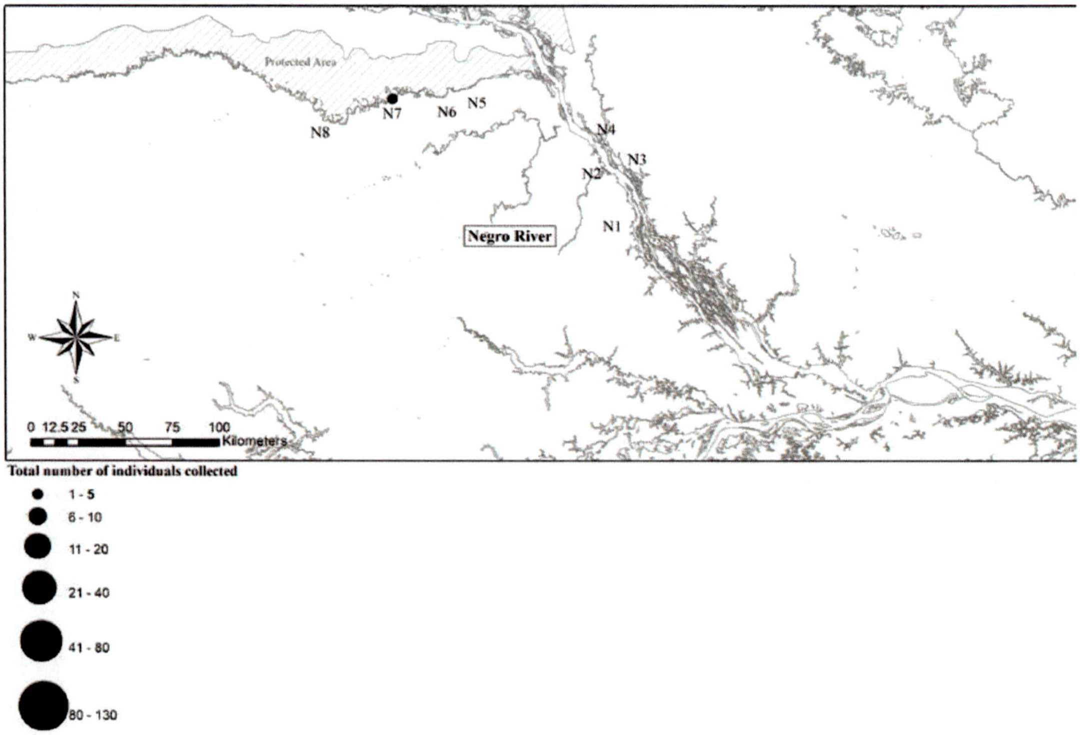

◘ **Fig. 4.180** Distribution of individuals of *Myloplus schomburgkii* sampled in the Negro River

Fish species: *Myloplus nigrolineatus* Ota, Machado, Andrade, Collins, Farias & Hrbek 2020

Common name(s): pacu, pacu-galo, pacu-vermelho

Size of fish sampled (standard length, both rivers): mean = 16.9 cm, median = 17.3 cm

Fisheries: pacus (*Myloplus* spp.) are important fish for food and commerce to fisheries in both rivers; see *Myloplus lobatus* for pacu fisheries data. Biomass caught is in ◘ Table 3.3 (► Chap. 3).

◘ **Fig. 4.181** *Myloplus nigrolineatus* male sampled in the Negro River. (Photo: Renato Silvano)

◘ **Fig. 4.182** *Myloplus nigrolineatus* female sampled in the Negro River. (Photo: Renato Silvano)

4

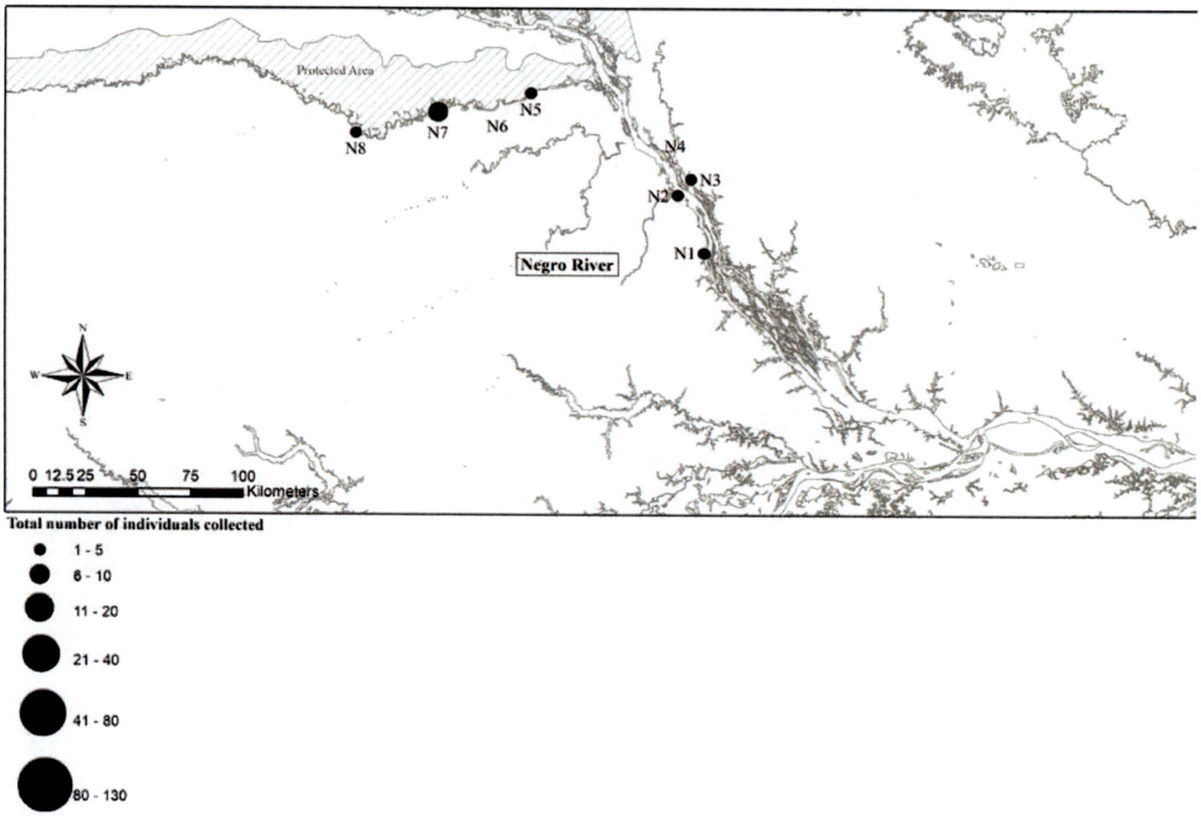

Fig. 4.183 Distribution of individuals of *Myloplus nigrolineatus* sampled in the Negro River

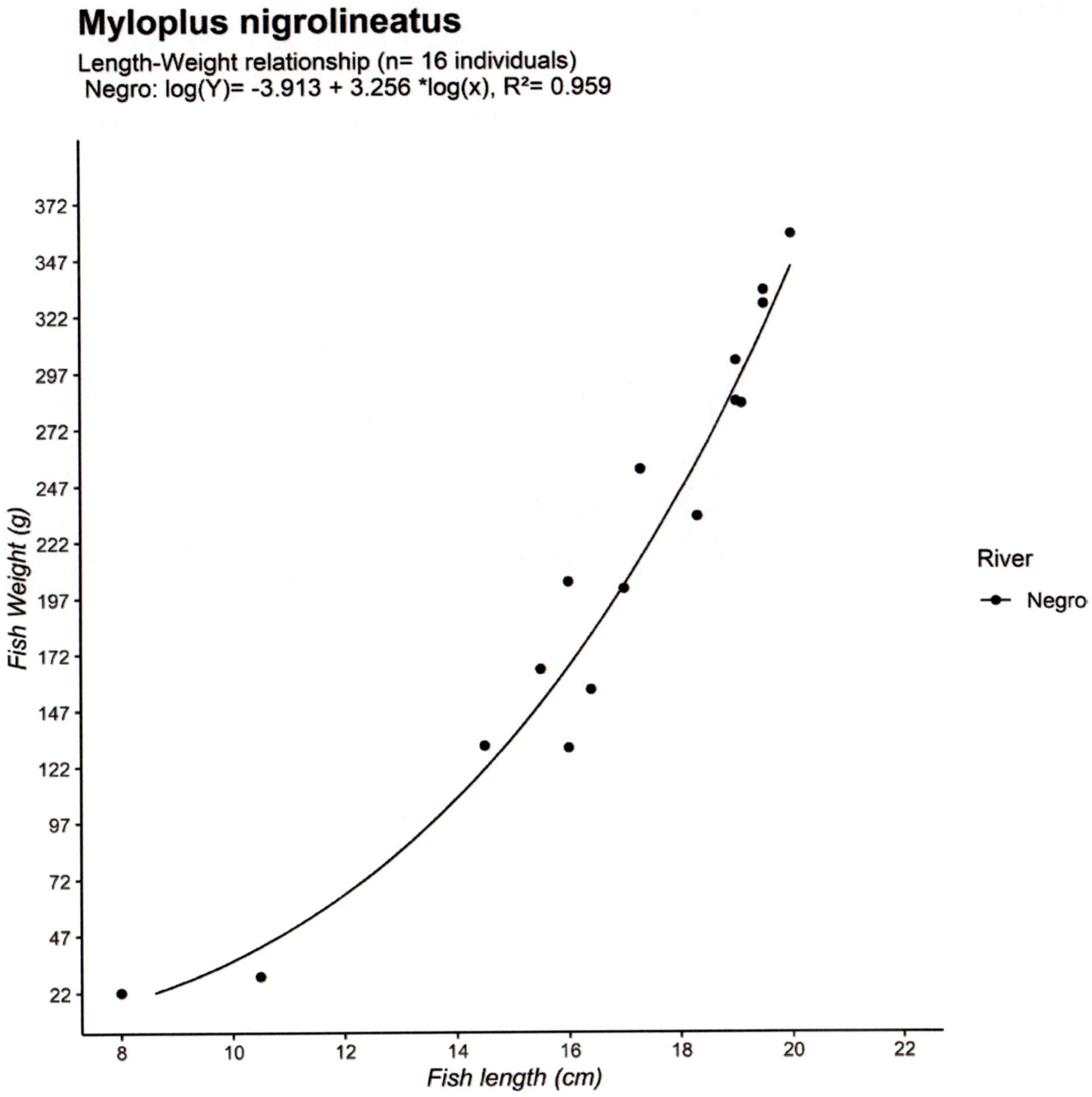

Fig. 4.184 Length-weight relationship for *Myloplus nigrolineatus* sampled in the Negro river

Fish species: *Pristobrycon maculipinnis* Fink & Machado-Allison 1992

Common name(s): piranha-chidaua, chidaua

Size of fish sampled (standard length, both rivers): min. = 12 cm, max. = 15.2 cm, mean and median = 13.6 cm

Fisheries: piranhas (several species of Serrasalmidae, especially *Serrasalmus* spp. and *Pygocentrus nattereri*) are occasionally caught by fishers in both rivers, mostly for consumption; see *Serrasalmus* spp. for piranha fisheries data. Biomass caught is in Table 3.3 (▶ Chap. 3).

Fig. 4.185 *Pristobrycon maculipinnis* sampled in the Negro River. (Photo: Renato Silvano)

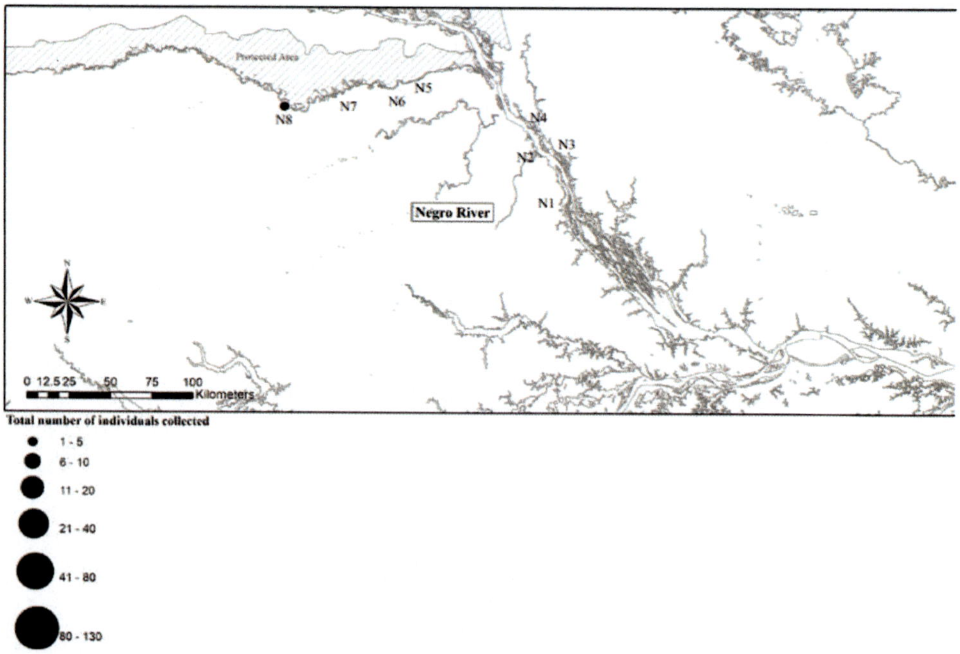

Fig. 4.186 Distribution of individuals of *Pristobrycon maculipinnis* sampled in the Negro River

Fish species: *Pristobrycon* cf. *scapularis*

Common name(s): piranha, piranha-branca, piranha-mapurá

Size of fish sampled (standard length, both rivers): min. = 8 cm, max. = 10 cm, mean and median = 9 cm

Fisheries: piranhas (several species of Serrasalmidae, especially *Serrasalmus* spp. and *Pygocentrus nattereri*) are occasionally caught by fishers in both rivers, mostly for consumption; see *Serrasalmus* spp. for piranha fisheries data. Biomass caught is in ◘ Table 3.3 (► Chap. 3).

◘ **Fig. 4.187** *Pristobrycon* cf. *scapularis* sampled in the Tapajós River. (Photo: Renato Silvano)

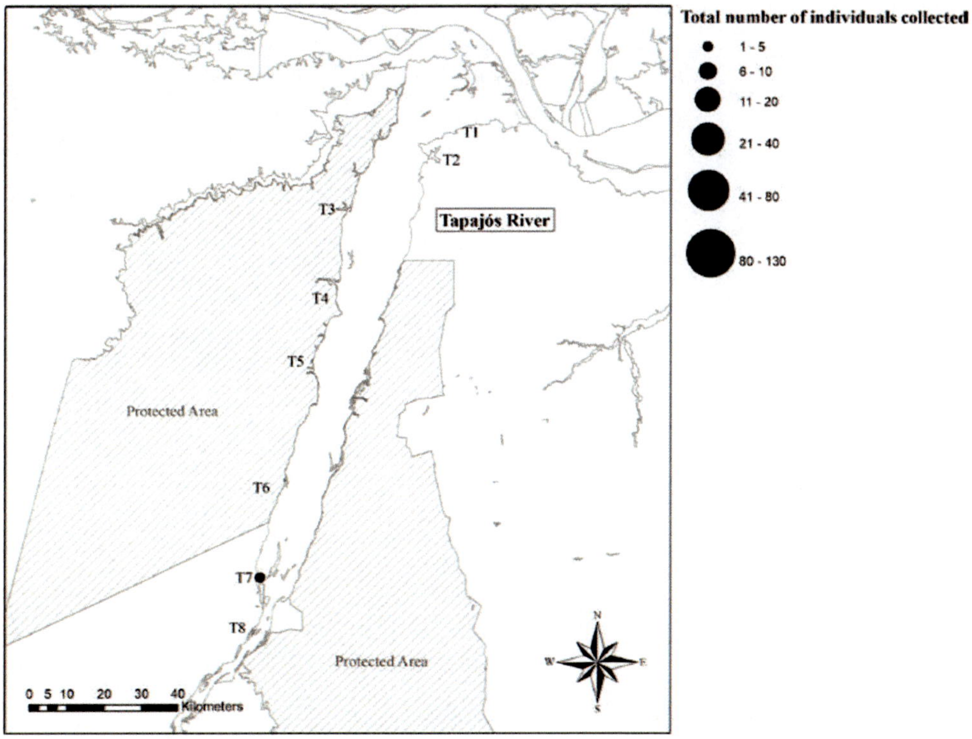

◘ **Fig. 4.188** Distribution of individuals of *Pristobrycon* cf. *scapularis* sampled in the Tapajós River

Fish species: *Pygocentrus nattereri* Kner 1858

Common name(s): piranha-caju

Size of fish sampled (standard length, both rivers): mean = 13.8 cm, median = 13.4 cm

Fisheries: piranhas (several species of Serrasalmidae, especially *Serrasalmus* spp. and *Pygocentrus nattereri*) are occasionally caught by fishers in both rivers, mostly for consumption; see *Serrasalmus* spp. for piranha fisheries data. Biomass caught is in ◘ Table 3.3 (▶ Chap. 3).

◘ **Fig. 4.189** *Pygocentrus nattereri* (red coloration) sampled in the Tapajós River. (Photo: Renato Silvano)

◘ **Fig. 4.190** *Pygocentrus nattereri* (dark coloration) sampled in the Tapajós River. (Photo: Renato Silvano)

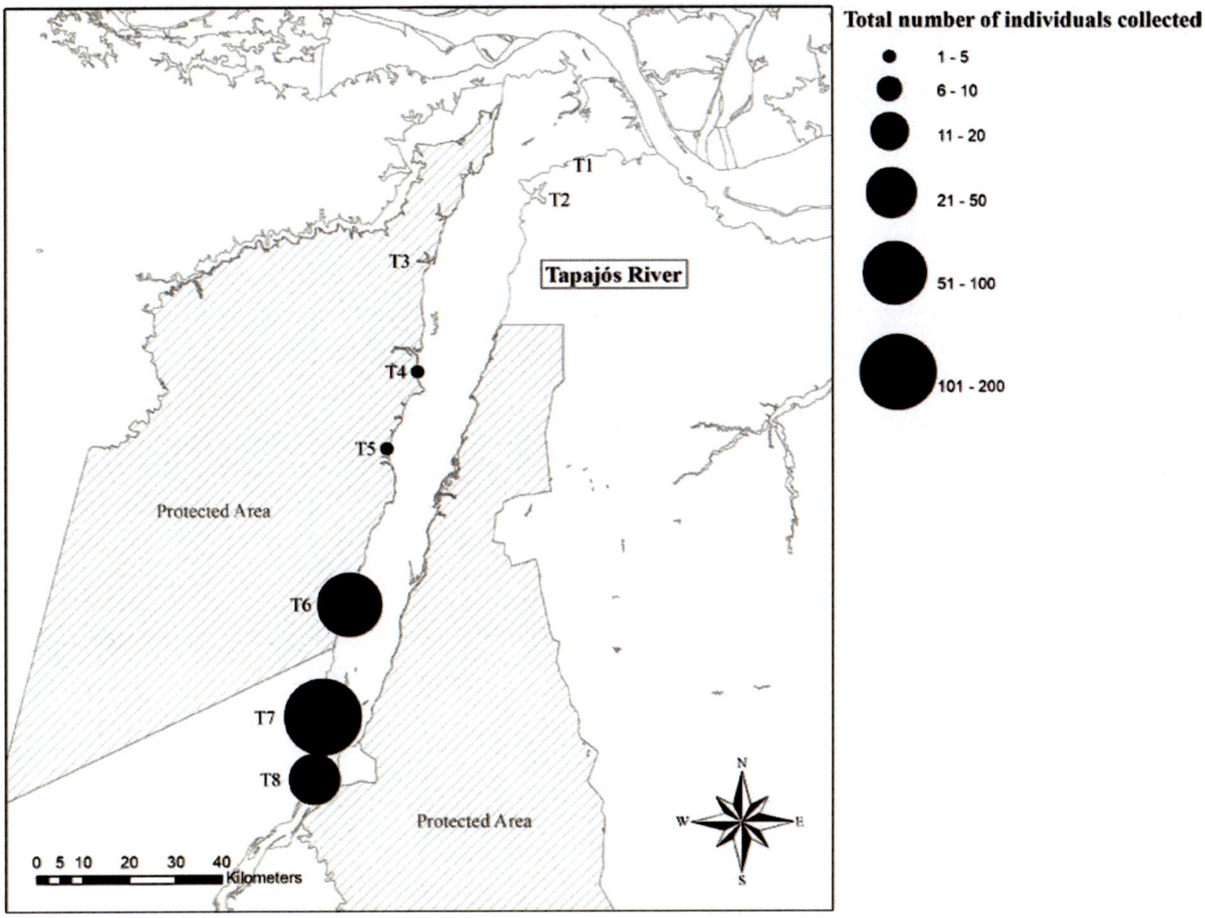

■ **Fig. 4.191** Distribution of individuals of *Pygocentrus nattereri* sampled in the Tapajós River

4

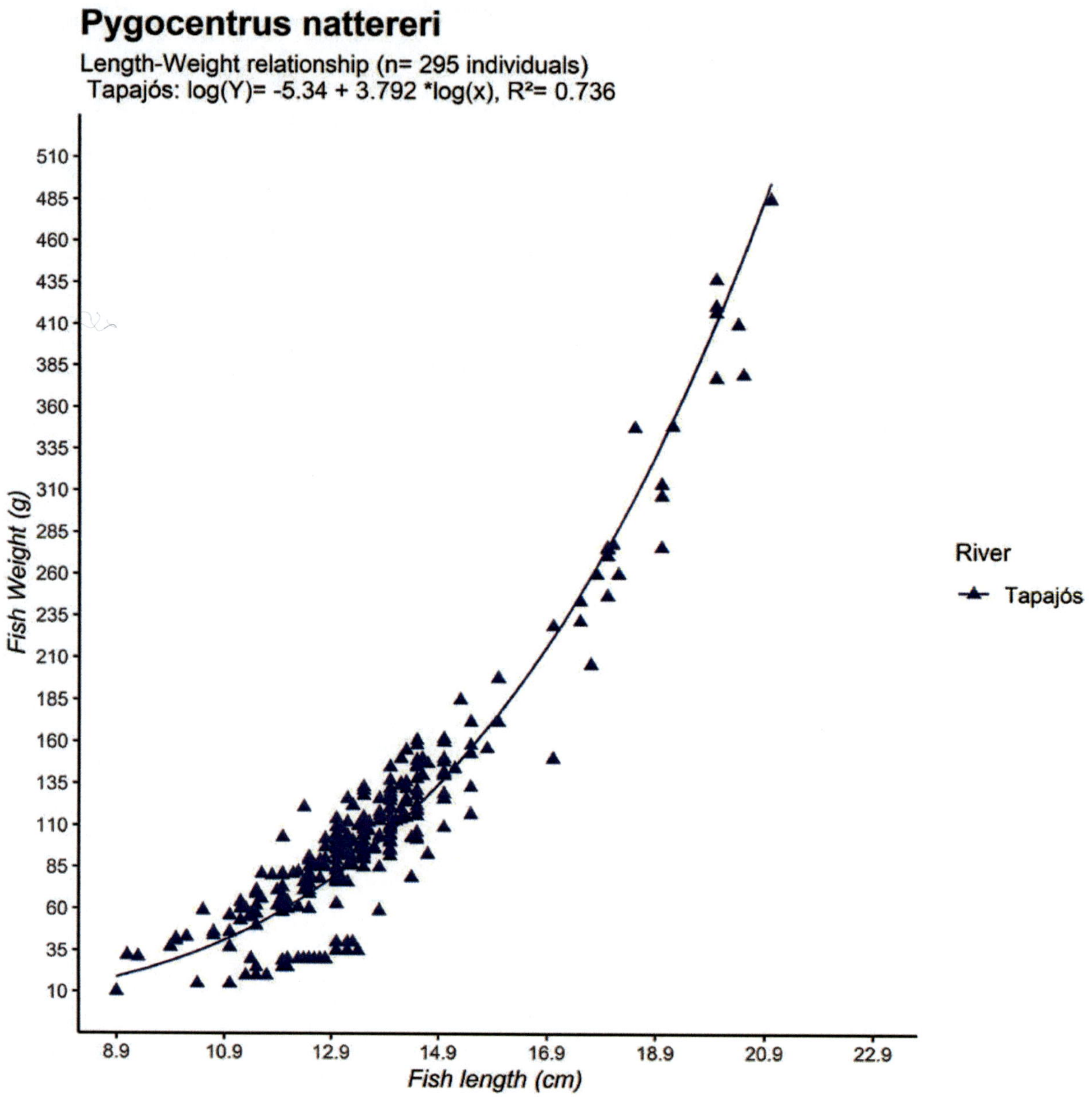

Fig. 4.192 Length-weight relationship for *Pygocentrus nattereri* sampled in the Tapajós River

Fish species: *Pygopristis denticulata* (Cuvier 1819)

Common name(s): piranha, piranha-chidaua, piranha-mafurá, mapurá

Size of fish sampled (standard length, both rivers): min. = 8 cm, max. = 10.7 cm, mean = 9.6 cm, median = 9.8 cm

Fisheries: piranhas (several species of Serrasalmidae, especially *Serrasalmus* spp. and *Pygocentrus nattereri*) are occasionally caught by fishers in both rivers, mostly for consumption; see *Serrasalmus* spp. for piranha fisheries data. Biomass caught is in ◘ Table 3.3 (▶ Chap. 3).

◘ **Fig. 4.193** *Pygopristis denticulata* sampled in the Tapajós River. (Photo: Renato Silvano)

4

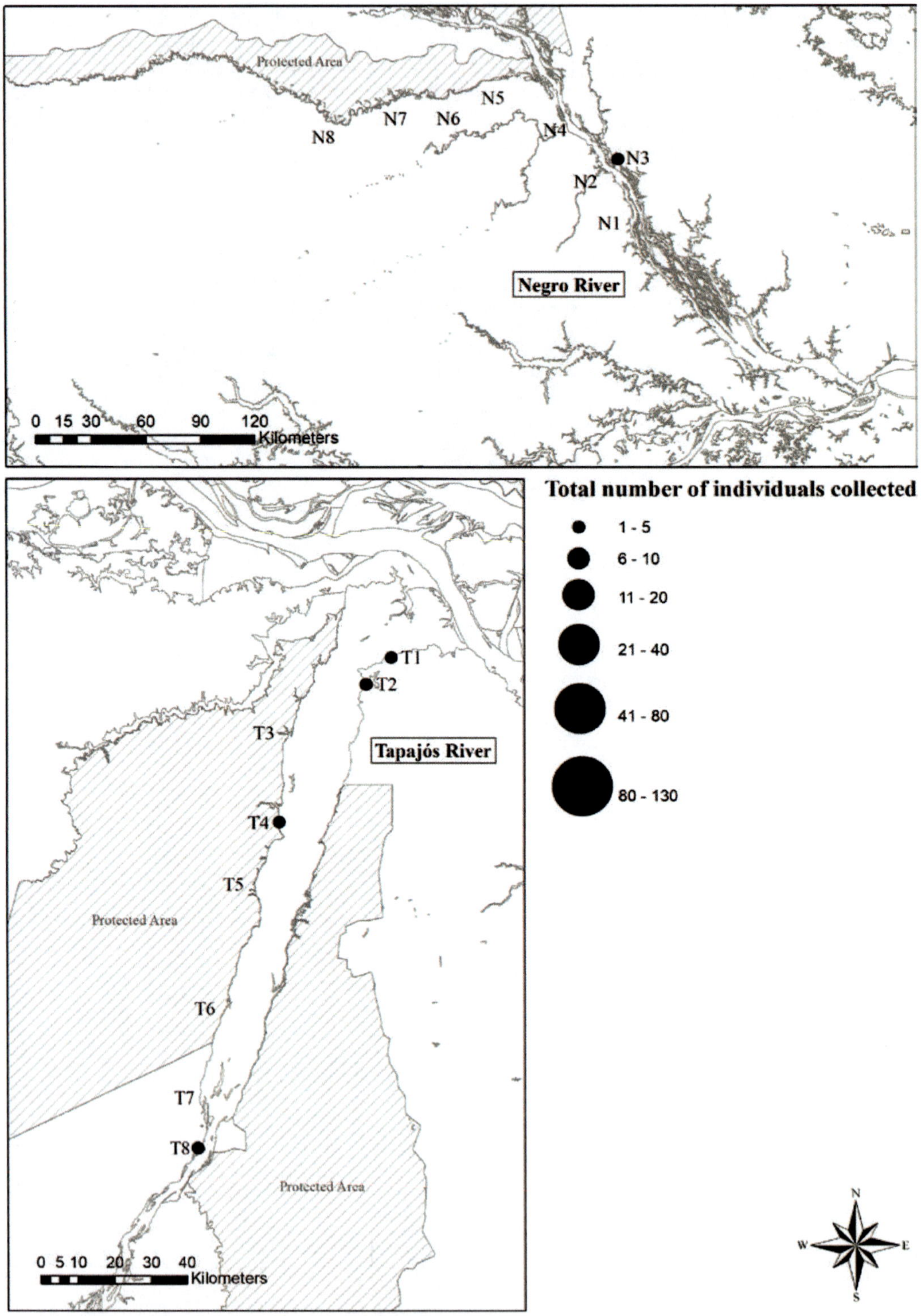

Fig. 4.194 Distribution of individuals of *Pygopristis denticulata* sampled in the Negro and Tapajós rivers

Fish species: *Serrasalmus elongatus* **Kner 1858**

Common name(s): piranha-mucura

Fig. 4.195 *Serrasalmus elongatus* sampled in the Negro River. (Photo: Renato Silvano)

Fish species: *Serrasalmus gouldingi* **Fink & Machado-Allison 1992**

Common name(s): piranha, piranha-branca

Fig. 4.196 *Serrasalmus gouldingi* sampled in the Negro River. (Photo: Renato Silvano)

Fish species: *Serrasalmus hollandi* Eigenmann 1915

Common name(s): piranha, piranha-branca

Fig. 4.197 *Serrasalmus hollandi* sampled in the Tapajós River. (Photo: Renato Silvano)

Fish species: *Serrasalmus maculatus* Kner 1858

Common name(s): piranha, piranha-amarela, mapurá

Fig. 4.198 *Serrasalmus maculatus* sampled in the Tapajós River. (Photo: Renato Silvano)

Fish species: *Serrasalmus manueli* (Fernández-Yépez & Ramírez 1967)

Common name(s): piranha, piranha-branca

Fig. 4.199 *Serrasalmus manueli* sampled in the Negro River. (Photo: Renato Silvano)

Fish species: *Serrasalmus medinai* Ramírez 1965

Common name(s): piranha

Fig. 4.200 *Serrasalmus medinai* sampled in the Negro River. (Photo: Renato Silvano)

Fish species: *Serrasalmus odyssei* Hubert & Renno 2010

Common name(s): piranha

Fig. 4.201 *Serrasalmus odyssei* sampled in the Negro River. (Photo: Renato Silvano)

Fish species: *Serrasalmus rhombeus* (Linnaeus 1766)

Common name(s): piranha-preta, piranha-fula

Fig. 4.202 *Serrasalmus rhombeus* sampled in the Tapajós River. (Photo: Renato Silvano)

Fish species: *Serrasalmus serrulatus* **(Valenciennes 1850)**

Common name(s): piranha-pacu, piranha-redonda

■ **Fig. 4.203** *Serrasalmus serrulatus* sampled in the Negro River. (Photo: Renato Silvano)

Fish species: *Serrasalmus eigenmanni* **Norman 1929**

Common name(s): piranha, piranha-branca

■ **Fig. 4.204** *Serrasalmus eigenmanni* sampled in the Negro River. (Photo: Renato Silvano)

Observation: abundance and size data were grouped for all these species, due to the difficulty in distinguishing them during sampling: *Serrasalmus elongatus, S. altuvei, S. gouldingi, S. medinai, S. odyssei*, and *S. serrulatus* were collected only in Negro River; *S. hollandi*, and *S. maculatus* were collected only in Tapajós River, while *S. manueli, S. eigenmanni* and *S. rhombeus* were collected in both rivers; see ■ Table 1.1 (▶ Chap. 1).

Size of fish sampled (standard length, both rivers): min. = 4.5 cm, max. = 29.5 cm, mean = 11.3 cm, median = 10.5 cm

Fisheries: piranhas (several species of Serrasalmidae, especially *Serrasalmus* spp. and *Pygocentrus nattereri*) are occasionally caught by fishers in both rivers, mostly

4

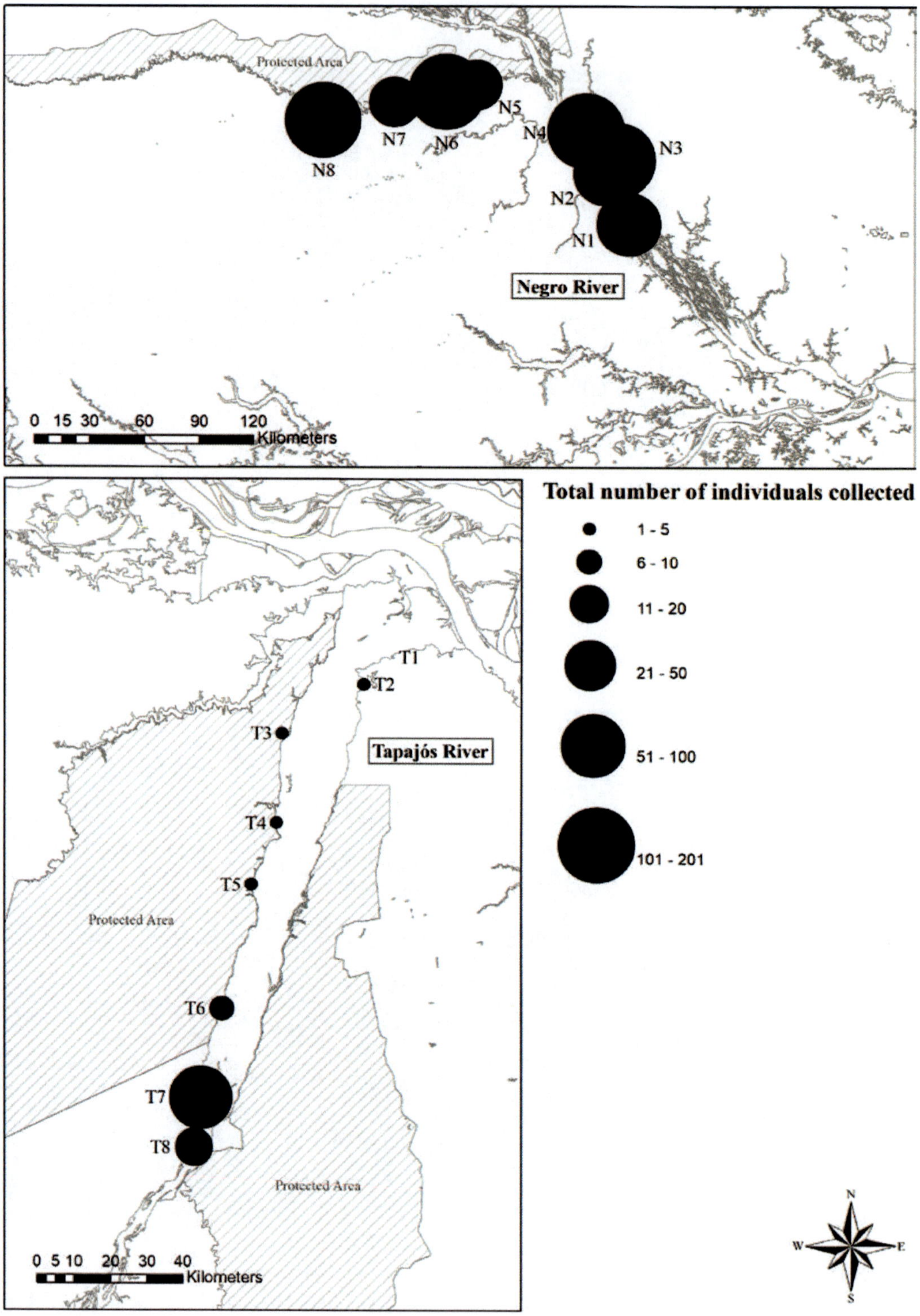

Fig. 4.205 Distribution of individuals of *Serrasalmus* spp. sampled in the Negro and Tapajós rivers

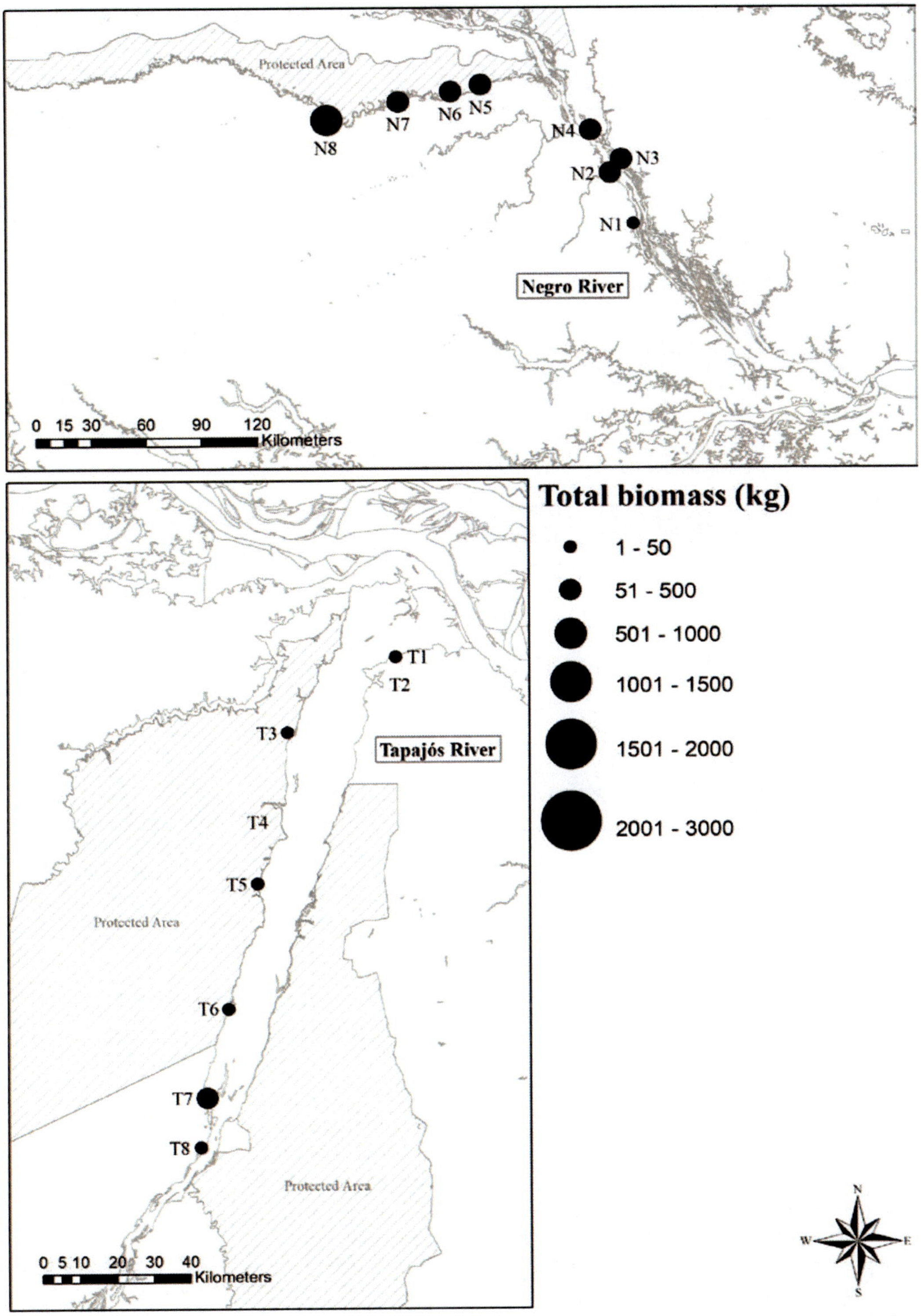

Fig. 4.206 Distribution of biomass of piranhas (Serrasalmidae) caught in small-scale fisheries in the Negro and Tapajós rivers

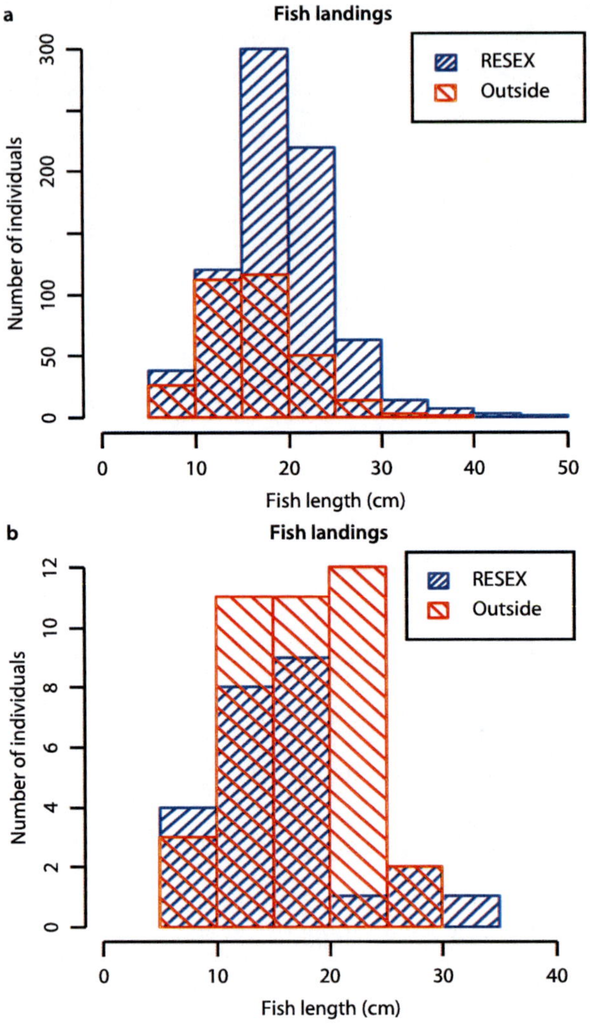

Fig. 4.207 Size distribution of piranhas (Serrasalmidae) caught in small-scale fisheries in 16 communities inside and outside the Extractive Reserves (RESEX) in the rivers **a** Negro, N Resex = 764, N outside = 320 and **b** Tapajós, N Resex = 25, N outside = 39

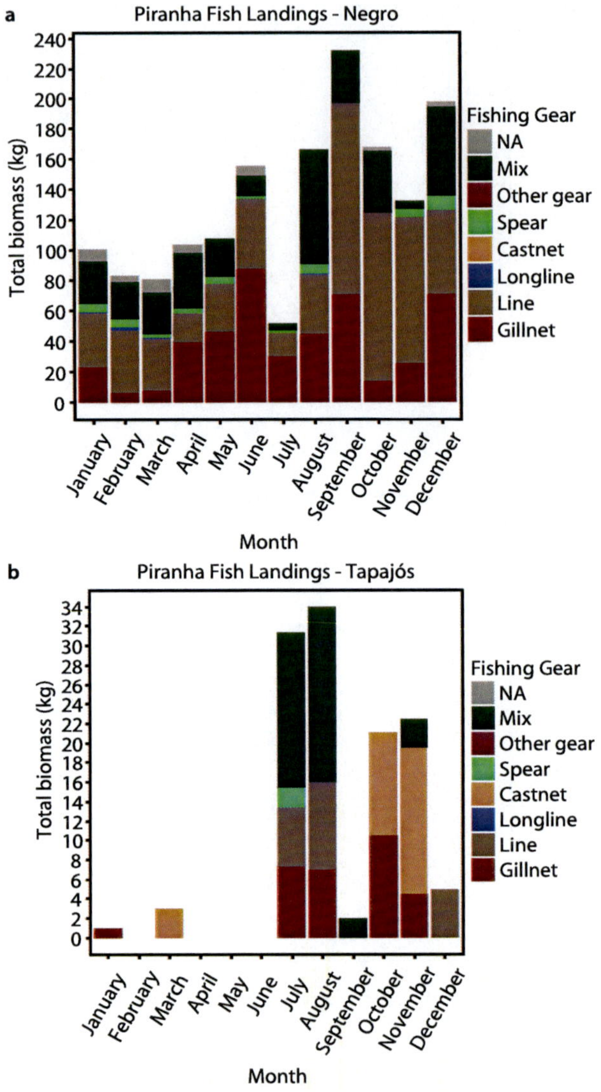

Fig. 4.208 Seasonal variation of biomass of piranhas (Serrasalmidae) caught per fishing gear in small-scale fisheries in 16 communities in the rivers **a** Negro and **b** Tapajós. NA: fishing gear not informed; Mix: more than one gear was used

for consumption. Biomass caught is in Table 3.3 (▶ Chap. 3).

References

Agostinho AA, Pelicice FM, Gomes LC (2008) Dams and the fish fauna of the Neotropical region: impacts and management related to diversity and fisheries. Brazilian Journal of Biology 68(4):1119–1132

Anderson JT, Nuttle T, Rojas JSS, Pendergast TH, Flecker AS (2011) Extremely long-distance seed dispersal by an overfished Amazonian frugivore. Proceedings of the Royal Society B 278:3329–3335

Andrade MC, Fitzgerald DB, Winemiller KO, Barbosa PS, Giarrizzo T (2018) Trophic niche segregation among herbivorous serrasalmids from rapids of the lower Xingu River, Brazilian Amazon. Hydrobiologia 829:265–280

Arantes CC, Winemiller KO, Petrere M, Castello L, Hess LL, Freitas CEC (2018) Relationships between forest cover and fish diversity in the Amazon River floodplain. Journal of Applied Ecology 55(1):386–395

Balassa GC, Fugi R, Hahn NS, Galina AB (2004) Dieta de espécies de Anostomidae (Teleostei, Characiformes) na área de influência do reservatório de Manso, Mato Grosso, Brasil. Iheringia, Série Zoologia, Porto Alegre 94:77–82

Baumgartner G, Pavanelli CS, Baumgartner D, Bifi AG, Debon T, Frana VA (2012) Peixes do baixo rio Iguaçu. Editora da Universidade Estadual de Maringá-EDUEM, Maringá, Brasil

Benedito CE, Araujo CARML (2002) Variation in the carbon isotope composition of *Semaprochilodus insignis*, a detritivorous fish associated with oligotrophic and eutrophic Amazonian rivers. Journal of Fish Biol. 60:1603–1607

Berra TM (2001) Freshwater fish distribution. Academic Press, San Diego, California

Britski HA, Sato Y, Rosa ABS (1986) Manual de identifica,co de peixes da regiao de Tres Marias (com chaves de identificacao para os peixes da bacia do Sao Francisco). Camara dos Deputados, Coordenagcio de Publicações: CODEVASF, Divisao de Piscicultura e Pesca, Brasilia, Brasil

Castellanos C (2002) Distribución especial de la comunidad de peces en una quebrada de aguas negras amazónicas, Leticia, Colombia. Departamento de Biología. Facultad de Ciencias, Universidad Nacional de Colombia, Bogotá, Colombia

Castro RMC, Vari RP (2004) Detritivores of the South American fish family prochilodontidae (Teleostei: Ostariophysi: Characiformes): A phylogenetic and revisionary study. In: Smithsonian Contributions to Zoology - Number 622. Smithsonian Contributions and Studies Series, Washington, DC

Ribeiro AC, Vilara GT, Filho JAL, Doria CRC (2016) Ecologia e biologia de peixes do rio madeira. Editora EDUFRO, Porto Velho, Brasil

Correa SB, Winemiller KO (2014) Niche partitioning among frugivorous fishes in response to fluctuating resources in the Amazonian floodplain forest. Ecology 95(1):210–224

Costa-Pereira R, Galetti M (2015) Frugivore downsizing and the collapse of seed dispersal by fish. Biological Conservation 191:809–811

Queiroz LJ, Vilara GT, Ohara WM, Pires THS, Zuanon J, Doria CR (2013) Peixes do rio madeira. Santo Antônio Energia, São Paulo, Brasil

Dos Anjos MR (2009) Ditribuição e diversidade da fauna de peixes nas sub-bacias do Maici e Ipixuna, Médio Madeira/AM, Brasil. Fundação Universidade Federal de Rondônia, p 102

Eschmeyer WN. 1998. Catalog of fishes. Special Publication, California Academy of Sciences, San Francisco. 3 Vols. 2905 p. Available at: research.calacademy.org/research/ichthyology/catalog/fishcatmain.asp. Accessed July 2019.

Frable, B. (2017). Family Curimatidae—Toothless Characins. In: van der Sleen P, Albert JS (Eds) Field guide to the fishes of the Amazon, Orinoco & Guianas. Princeton University Press, Princeton, 148–153.

Fricke R, Eschmeyer W, Fong JD. 2019. Species by family/subfamily in the catalog of fishes. California Academy of Sciences. Available at: http://researcharchive.calacademy.org/research/ichthyology/catalog/SpeciesByFamily.asp. Accessed July 2019.

Fricke, R., Eschmeyer, W. N. & Van der Laan, R. (eds) 2020. Eschmeyer's Catalog of Fishes: Genera, Species, References. (http://researcharchive.calacademy.org/research/ichthyology/catalog/fishcatmain.asp). Electronic version accessed 01 Set 2020

Froese R, Pauly D. 2018. Family Prochilodontidae - Flannel-mouth characiforms. Available at: http://www.fishbase.se/summary/FamilySummary.php?ID=598.Accessed July 2019.

Galvis G, Mojica JI, Duque SR, Castellanos C, Duarte PS, Arce M, Gutiérrez A, Jiménez LF, Santos M, Vejarano S, Arbeláez F, Prieto E, Leiva M (2006) Peces del medio Amazonas, Región de Leticia. Serie de Guías Tropicales de Campo N°5. Panamericana, Formas e Impresos, Bogotá, Colombia

Galvis G, Duarte PS, Salazar LMM, Pinto YL, Cortés ÂG, Castaño ML, Castillo CC (2007a) Peces de la Amazonía colombiana con énfasis en especies de interés ornamental. Ramon Lopez Editorial, Bogotá, Colombia

Galvis G, Mojica JI, Provenzano F, Lasso CA, Taphorn DCB, Royero R, Castillo CC, Cortés ÂG, Gutiérrez MAE, Pinto YL, Mesa LM, Duarte PS, Cipamocha CA (2007b) Peces de la Orinoquia colombiana con énfasis en especies de interés ornamental. Instituto Colombiano de Desarrollo Rural INCODER y Universidad Nacional de Colombia, Bogotá, DC

Garavello JC, Britski HA. 2003. Family Anostomidae. In: Check list of the freshwater fishes of South and Central America. By: Reis RE, Kullander SO, Ferraris CJ. EDIPUCRS, Porto Alegre, Brasil, p. 71–84.

Géry J (1961) Contributions à l'étude des poissons Characoides. 13. Structures et evolution des Anostominae. Bull. Aquat. Biol 2(19):93–112

Géry J (1977) Characoids of the world. T.F.H. Public, Neptune City, p 672

Godoy MP (1975) Peixes do Brasil – Subordem Characoidei. Bacia do rio Mogi Guassu. Editora Franciscana, Piracicaba, Brasil

Goulding M (1980) The fishes and the forest: explorations in Amazonian natural history. Univ of California Press, London, England

Goulding M, Carvalho ML (1982) Life history and management of the tambaqui (*Colossoma macropomum*, Characidae): an important Amazonian food fish. Revista Brasileira de Zoologia 1(2):107–133

Hallwass G, Silvano RAM (2016) Patterns of selectiveness in the Amazonian freshwater fisheries: Implications for management. Journal of Environmental Planning and Management 59:1537–1559

Hallwass G, Lopes PF, Juras AA, Silvano RAM (2013a) Fishers' knowledge identifies environmental changes and fish abundance trends in impounded tropical rivers. Ecological Applications 23:392–407

Hallwass G, Lopes PF, Juras AA, Silvano RAM (2013b) Behavioral and environmental influences on fishing rewards and the outcomes of alternative management scenarios for large tropical rivers. Journal of Environmental Management 128:274–282

Hallwass G, Schiavetti A, Silvano RAM (2020) Fishers' knowledge indicates temporal changes in composition and abundance of fishing resources in Amazon protected areas. Animal Conservation 23(1):36–47

Hoeinghaus DJ, Winemiller KO, Layman CA, Arrington DA, Jepsen DB (2006) Effects of seasonality and migratory prey on body condition of *Cichla* species in a tropical floodplain river. Ecology of Freshwater Fish 15:398–407

Knöeppel HA (1972) Zur ahrung tropischer Süsswasserfische aus Südamerika. Amazoniana 3:231–246

Keppeler FW, Hallwass G, Silvano RAM (2017) Influence of protected areas on fish assemblages and fisheries in a large tropical river. Oryx 51(2):268–279

Keppeler FW, Souza AC, Hallwass G, Begossi A, Almeida MC, Isaac VJ, Silvano RAM (2018) Ecological influences of human population size and distance to urban centres on fish communities in tropical lakes. Aquatic Conservation: Marine and Freshwater Ecosystems 28(5):1030–1043

Langeani F (2013) Hemiodontidae. In: Queiroz LJ, Vilara GT, Ohara WM, THS P, Zuanon J, Doria CR (eds) Peixes do Rio Madeira, vol 1. Dialeto, São Paulo, Brasil, pp 192–205

Lasso C (2004) Los peces de la estación biológica El Frío y Caño Guaritico (Estado Apure), Llanos del Orinoco, Venezuela. Publicaciones del Comité Español del Programa Hombre y Biosfera-Red IberoMab ed. UNESCO, Sevilla, Spain

Lima FCT, Pires THS, Ohara WM, Jerep FC, Carvalho FR, Marinho MMF, Zuanon J. 2013. Characidae. In: Peixes do rio madeira. By: Queiroz LJ, Torrent-vilara G, Ohara WM, Pires THS, Zuanon J, Doria CRC. Vol. 1. Editora: Dialeto Latin American Documentary, São Paulo, Brazil. Cap. 19, p. 212.

McConell RL (1987) Ecological studies in tropical fish communities. Cambridge University Press, Cambridge, Reino Unido

Machado-Allison A (1993) Los peces de los llanos de Venezuela: un ensayo sobre su historia natural. Universidad Central de Venezuela, Caracas, Venezuela

4

Mago-Leccia F (1970) Estudios preliminares sobre la ecologia de los peces de los llanos de Venezuela. Acta Biol. Venezuelica 7(1):71–102

Mcintyre PB, Jones LE, Flecker AS, Vanni MJ (2007) Fish extinctions alter nutrient recycling in tropical freshwaters. PNAS 104:4461–4466

Melo, B. & Sidlauskas, B. (2017) Family Chilodontidae—Headstanders. In: van der Sleen P, Albert JS (Eds) Field guide to the fishes of the Amazon, Orinoco & Guianas. Princeton University Press, Princeton, 141–142.

Menezes NA (1992) Redefinição taxonômica das espécies de *Acestrorhynchus* do grupo lacustris com a descrição de uma espécie (Osteichthyes, Characiformes, Characidae). Comun. Mus. Ciênc. PUCRS, Sér. Zool., Porto Alegre, pp 39–54

Mojica JI, Usma J, Leon RA, Lasso C (2012) Libro Rojo de peces dulceacuícolas de Colombia. Instituto de Investigación de Recursos Biológicos Alexander von Humboldt, Instituto de Ciencias Naturales de la Universidad Nacional de Colombia, WWF Colombia, Universidad de Manizales, Bogotá

Nelson JS. 1994 . Fishes of the world. Editora Wiely, New York.

Nelson JS, Grande TC, Wilson MVH (2006) Fishes of the world. John Wiley & Sons, Hoboken, USA

Ohara WM, Lima FCT, Salvador GN, Andrade MC (2017) Peixes do Rio Teles Pires: diversidade e guia de identificação. Museu de Zoologia da Universidade de São. Paulo/Universidade Federal do Pará/Museu de Zoologia da Universidade Estadual de Campinas, São Paulo, Campinas

Piedade MTS, Parolin P, Junk WJ (2016) Phenology, fruit production and seed dispersal of *Astrocaryum jauari* (Arecaceae) in Amazonian black water floodplains. Revista de Biología Tropical. 54:1171–1178

Ribeiro MCLB, Petrere M Jr (1990) Fisheries ecology and management of the Jaraqui (*Semaprochilodus taeniurus, S. insignis*) in Central Amazonia. Regulated Rivers Research and Management 5:195–215

Roberts TR (1971) *Micromischodus sugillatus*, a new hemiodontid characin fish from Brazil, and its relationship to the Chilodontidae. Breviora 67:1–25

Rodrigues JD, Mota A, Moraes MN, Ferreira AE (1978) Curvas de maturação gonadal e crescimento de fêmeas de pirambeba, *Serrasalmus spilopleura* Kner, 1859 (Pisces, Characiformes). Boletim do Instituto de Pesca 5(2):51–63

Saccol-Pereira AS, Milani PCC, Fialho CB (2006) First record of *Acestrorhynchus pantaneiro* Menezes, 1992 (CHARACIFORMES, ACESTRORHYNCHIDAE) in the system of the laguna dos Patos, Rio Grande do Sul, Brazil. Biota Neotrop. 6(3). http://www.biotaneotropica.org.br/v6n3/pt/abstract?shortcommunication+bn00706032006. ISSN 1676-0603.

Santos GM (1980) Aspectos de sistemática e morfología de *Schizodon fasciatus* AGASSIZ, 1829; *Rhytiodus microlepis* KNER, 1859 e *R. argenteofuscus* KNER. 1859 (Osteichthyes, Characoidei, Anostomidae) do lago Janauacá - AM. Acta Amazónica 10(3):635–649

Santos GM (1981) Estudos de alimentação e hábitos alimentares de *Schizodon fasciatus*, AGASSIZ. 1829, *Rhytiodus microlepis* KNER, 1859 e *R. argenteofuscus* KNER. 1859 do lago Janauacá, AM. (Osteichthyes, Characoidei, Anostomidae). Acta Amazónica 11(2):267–228

Santos, G. M., M. Jegu, and B. Merona. (1984). Catálogo de peixes comercia do Baixo Rio Tocantins–Projecto Tucurui. Electronorte/CNPQ/Instituto Nacional de Pesquisas da Amazonia (INPA), Manaus, Brazil.

Santos GM, Jegu M (1996) Inventário taxonômico dos anostomídeos (pisces, anostomidae) da bacia do rio Uatumã-AM, Brasil, com descrição de duas espécies novas. Acta Amazonica 26(3):151–184

Santos GM, Rosa PS (1998) Alimentação de *Anostomus ternetzi* e *Synaptolaemus cingulatus*, duas espécies de peixes Amazônicos com boca superior. Revista Brasileira Biologica 58:255–262

Santos G, Ferreira E, Zuanon J (2006) Peixes comerciais deManaus. Editora Pr-Vrzea, Ibama, Manaus, Brasil

Saul WG (1975) An ecological study of fishes at a site in upper Amazonian Ecuador. Proceedings of the Academy of Natural Sciences of Philadelphia 127:93–114

Silva CC, Ferreira JGE, Deus CP (2008) Dieta de cinco espécies de Hemiodontidae (Teleostei, Characiformes) na área de influência do reservatório de Balbina, rio Uatumã, Amazonas, Brasil. Iheringia Série Zoologia 98(4):465–468

Silvano RAM, Ramires M, Zuanon J (2009) Effects of fisheries management on fish communities in the floodplain lakes of a Brazilian Amazonian Reserve. Ecology of Freshwater Fish 18:156–166

Silvano RAM, Silva AL, Ceroni M, Begossi A (2008) Contributions of ethnobiology to the conservation of tropical rivers and streams. Aquatic Conservation: Marine and Freshwater Ecosystems 18(3):241–260

Taphorn DC (1992) The characiform fishes of the Apure River drainage, Venezuela. In: Biollania, edición especial, 4. Monografias Científicas del Museo de Ciencias Naturales, UNELLEZ, Guanare, Venezuela

Taylor BW, Flecker AS, Hall RO Jr (2006) Loss of a Harvested Fish Species Disrupts Carbon Flow in a Diverse Tropical River. Science 313:833–836

Thompson AW, Betancur RR, Fernández HL, Ortí G (2014) A time-calibrated, multi-locus phylogeny of piranhas and pacus (Characiformes: Serrasalmidae) and a comparison of species tree methods. Molecular Phylogenetics and Evolution 81:242–257

Toledo-Piza M (2017), Family Acestrorhynchidae. In: van der Sleen P, Albert JS (Eds) Field guide to the fishes of the Amazon, Orinoco & Guianas. Princeton University Press, Princ eton, 88 pp

Toledo-Piza M (2013) Cynodontidae. In: Peixes do Rio Madeira. By: Queiroz LJ, Vilara GT, Ohara WM, Pires THS, Zuanon J, Doria CR. VoL. 2. Editora Dialeto, São Paulo, Brail. p.62−69

Queiroz LJ, Vilara GT, Ohara WM, Pires THS, Zuanon J, Doria CR (2013b) Peixes do rio madeira. Santo Antônio Energia, Dialeto, São Paulo, Brasil

Torrente-Vilara G, Queiroz LJ, Vari RP. 2013. Chilodontidae. In: Peixes do Rio Madeira. By: Queiroz LJ, Vilara GT, Ohara WM, Pires THS, Zuanon J, Doria CR. VoL. 1. Editora Dialeto, São Paulo, Brail. p.62−69

Tregidgo DJ, Barlow J, Pompeu PS, Rocha MA, Parry L (2017) Rainforest metropolis casts 1,000-km defaunation shadow. Proceedings of the National Academy of Sciences 114:8655–8659

Zuanon J, Mendonça FP, Santo HMVE, Dias MS, Galuch AV, Akama A (2015) Guia de Peixes da Reserva Adolpho Ducke. Editora INPA, Manaus, Brasil

van der Sleen P, Albert, J. Lima, FCT, Netto-Ferreira, AL., Mattox, G. & Toledo-Piza (2017). Family Characidae—Tetras and Relatives. In: van der Sleen P, Albert JS (Eds) Field guide to the fishes of the Amazon, Orinoco & Guianas. Princeton University Press, Princeton, 92–93.

Vari RP, Castro RMC, Raredon SJ. 1995. The Neotropical fish family Chilodontidae (Teleostei: Characiformes): a phylogenetic study and a revision of *Caenotropus* Günther. Smithsonian Contributions to Zoology, 577: 1–32

Vari RP, Ropke CP. 2013. Curimatidae. In: Peixes do Rio Madeira. By: Queiroz LJ, Vilara GT, Ohara WM, Pires THS, Zuanon J, Doria CR. VoL. 2. Editora Dialeto, São Paulo, Brail. p.112–136

Winterbottom R (1974) A new species of anostomid characoid fish, *Anostomus spiloclistron*, from the Nickerie river system of western Surinam (Pisces, Cypriniformes, Anostomidae). Beaufortia 21:153–163

Atlas of Fish of Tapajós and Negro Rivers II: Gymnotiformes and Siluriformes

Renato A. M. Silvano, Pedro P. Nitschke, Kaluan C. Vieira, Paula Nagl, Astrid T. R. Martínez, Márcia C. F. Dutra, Friedrich W. Keppeler, Junior A. Chuctaya, Henrique N. Oliveira, and Marcelo C. Andrade

Contents

Renato A. M. Silvano (ed.), *Fish and Fisheries in the Brazilian Amazon*, https://doi.org/10.1007/978-3-030-49146-8_5

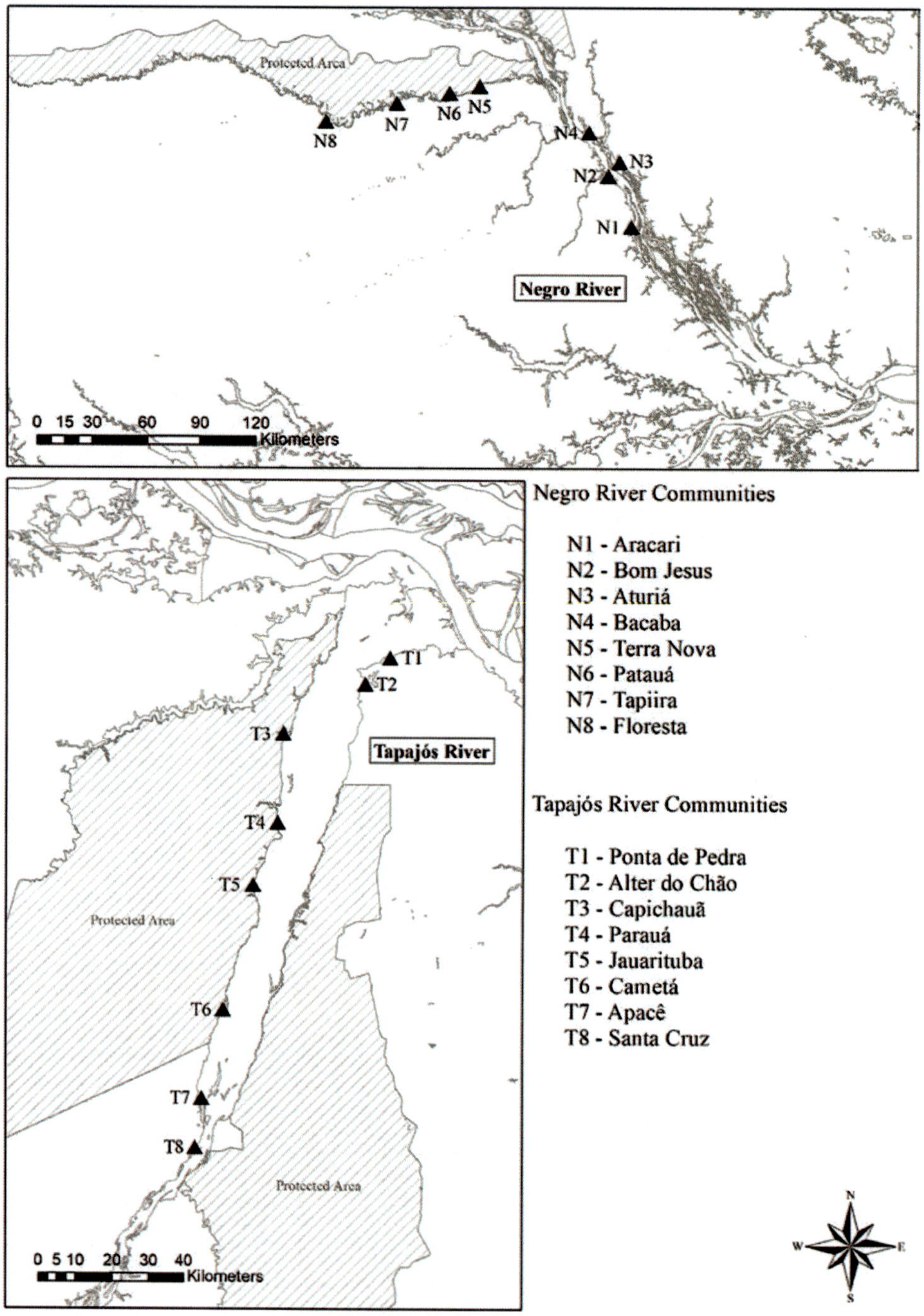

Fig. 5.1 Map of the studied region, showing the 16 studied communities with codes for community localization in maps showing the occurrence of fish species

5.1 Order Gymnotiformes

5.1.1 Family Rhamphichthyidae

5.1.1.1 Summary of Biology and Ecology (Literature)

Diversity: family composed of 28 species and five genera, including *Gymnorhamphichthys* (six species), *Hypopygus* (nine), *Iracema* (one), *Rhamphichthys* (nine), and *Steatogenys* (three).

Habitat: fishes of this family are restricted to the South America continent. They usually forage at night and hide during the day. Many species tend to dig themselves in the sandy bottom of rivers to hide from visual predators.

Feeding: Although different genera have some morphological differences in the snout (*Hypopygus* and *Steatogenys* have a short snout, while the other genera possess a long, tubular snout), their diet consists mainly on insect larvae.

Morphology: Members of the family Rhamphichthyidae have a long, lateral compressed body, possess 12–26 precaudal vertebrae, have long anal fin extending from the pectoral fins to almost the posterior tip of the body, and do not have teeth on the lower jaw.

Reproduction: no information.

Economic importance: these fishes are not exploited in Amazonian fisheries and do not have commercial relevance.

References: Marrero & Winemiller (1993), Nelson (1994), Winemiller & Adite (1997), Crampton et al. (2004), Carvalho et al. (2011), Ferraris et al. (2017).

Fish species: ***Rhamphichthys pantherinus* Castelnau 1855**

Common name(s): sarapó

Fig. 5.2 *Rhamphichthys pantherinus* sampled in the Tapajós River. (Photo: Renato Silvano)

Fish species: ***Rhamphichthys rostratus* (Linnaeus 1766)**

Fig. 5.3 *Rhamphichthys rostratus* sampled in the Tapajós River. (Photo: Renato Silvano)

Common name(s): sarapó, tui, itui

Observation: abundance and size data were grouped for these two species (*Rhamphichthys* spp.) due to the difficulty in distinguishing them during sampling. Both species occurred in both rivers; see Table 1.1 (► Chap. 1).

Size of fish sampled (standard length, both rivers): min. = 21.5 cm, max. = 78.5 cm, mean = 53.6 cm, median = 55.8 cm.

Fisheries: Not relevant to fisheries in both rivers.

5

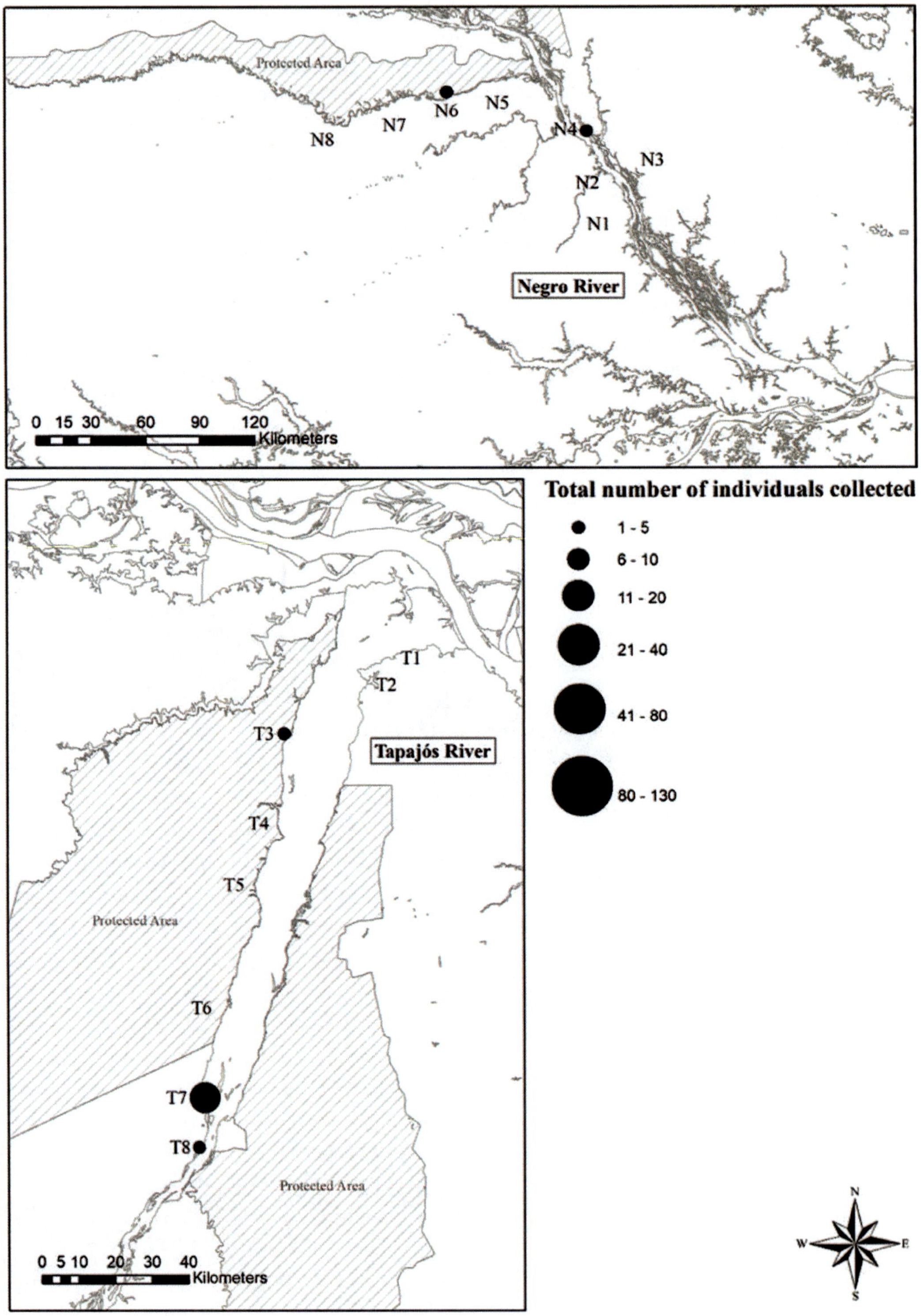

Fig. 5.4 Distribution of individuals of *Rhamphichthys* spp. sampled in the Negro and Tapajós rivers

5.1.2 Family Sternopygidae

5.1.2.1 Summary of Biology and Ecology (Literature)

Diversity: family composed of 44 species, distributed in seven genera: *Archolaemus* (six species), *Distocyclus* (two), *Eigenmannia* (16), *Humboldtichthys* (one), *Japigny* (one), *Rhabdolichops* (ten), and *Sternopygus* (eight).

Habitat: these fishes can be found in both lotic and lentic environments throughout South America, except Chile.

Feeding: members of the family Sternopygidae are omnivores and feed on invertebrates. Microcrustaceans, aquatic insect larvae, and nymphs are common items in their diet.

Morphology: glass knifefishes, as commonly known, possess multiple rows of tiny villiform teeth in both jaws, absence of urogenital papilla, large eyes with a diameter equal or larger than the distance between the nostrils, and dorsal fin originating in the isthmus. Many species are compressed laterally and some have translucid bodies. Total length varies according to the species, from 120 mm in *Sternopygus vicentespelaea* to up to 1.4 m in *Sternopygus macrurus*.

Reproduction: no information.

Economic importance: these fishes are not exploited in Amazonian fisheries and do not have commercial relevance.

References: Lundberg & Stager (1985), Albert (2001, 2003), Giora et al. (2005), Ferraris et al. (2017).

Fish species: *Eigenmannia macrops* (Boulenger 1897)

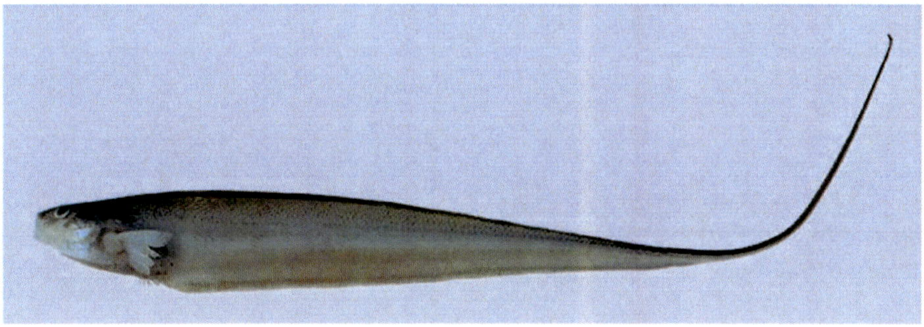

Fig. 5.5 *Eigenmannia macrops* sampled in the Tapajós River. (Photo: Renato Silvano)

Common name(s): sarapó

Size of fish sampled (standard length, both rivers): min. = 17.4 cm, max. = 38 cm, mean = 29.3 cm, median = 29.4 cm

Fisheries: Not relevant to fisheries in both rivers.

5

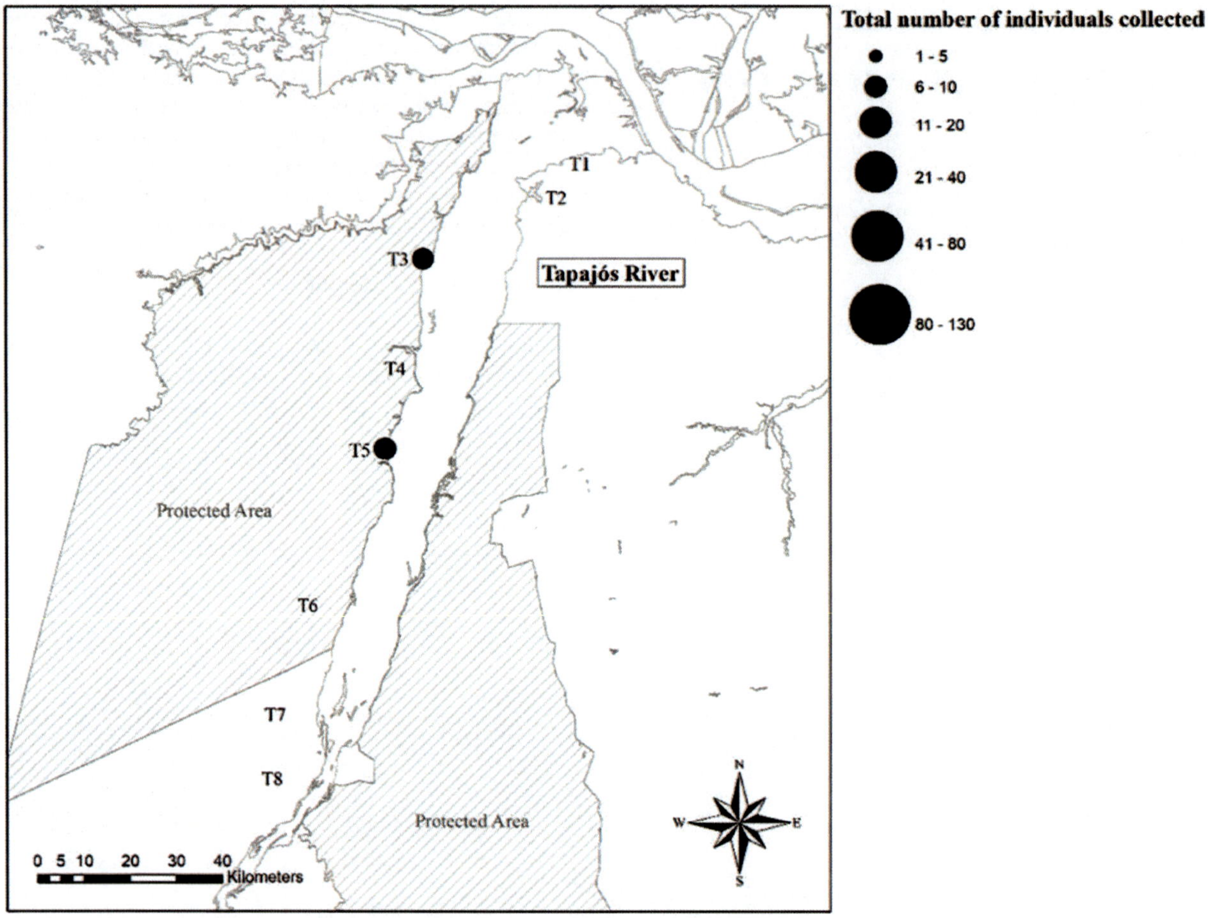

Fig. 5.6 Distribution of individuals of *Eigenmannia macrops* sampled in the Tapajós River

Fish species: ***Rhabdolichops troscheli* (Kaup 1856)**

Common name(s): sarapó

Size of fish sampled (standard length, both rivers): min. = 27 cm, max. = 35.4 cm, mean = 31.3 cm, median = 31.5 cm

Fisheries: Not relevant to fisheries in both rivers.

Fig. 5.7 *Rhabdolichops troscheli* sampled in the Tapajós River. (Photo: Renato Silvano)

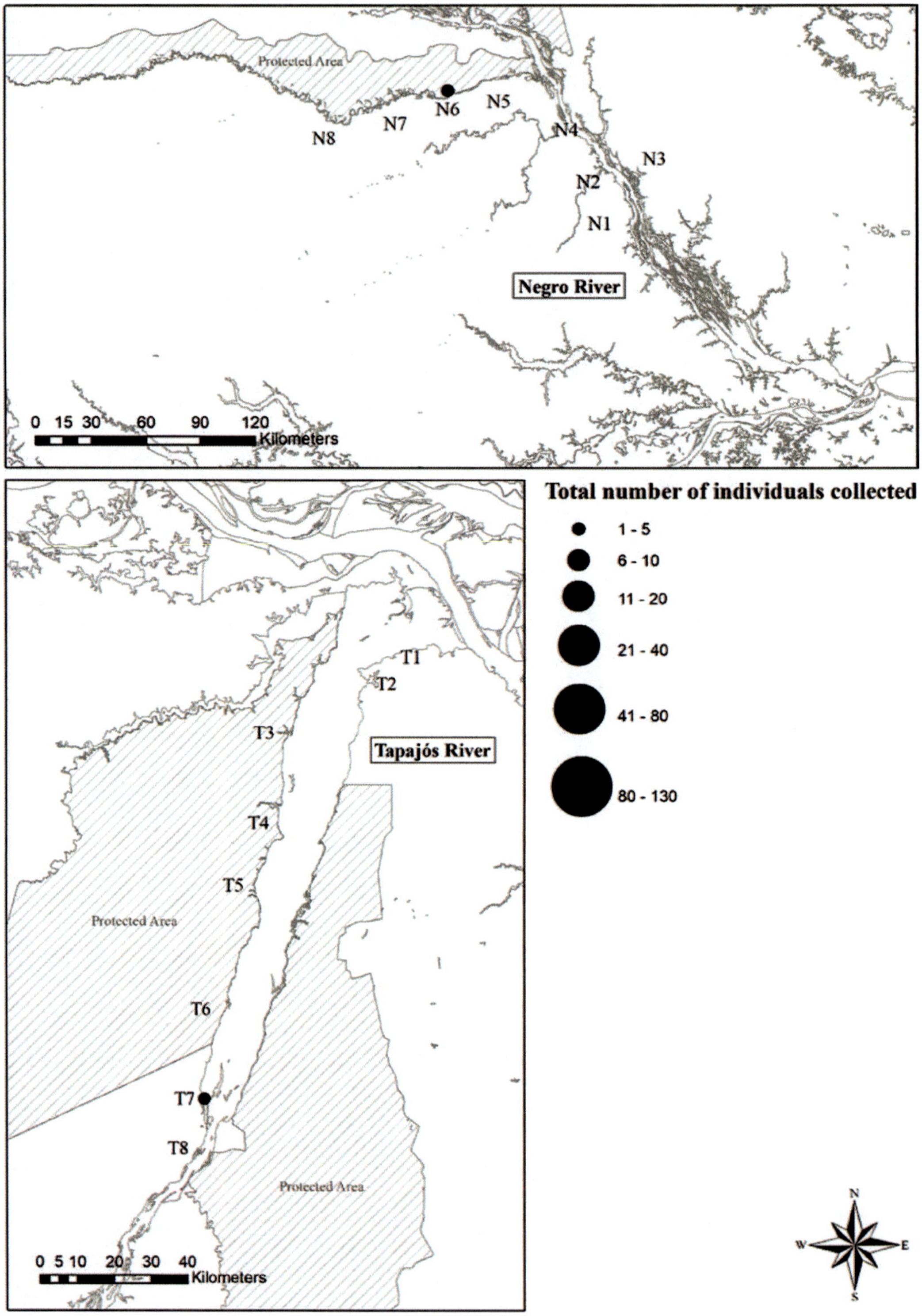

Fig. 5.8 Distribution of individuals of *Rhabdolichops troscheli* sampled in the Negro and Tapajós rivers

5.2 Order Siluriformes

5.2.1 Family Auchenipteridae

5.2.1.1 Summary of Biology and Ecology (Literature)

Diversity: family composed of 124 valid species and 20 genera restricted to the Neotropical region (South America and Panamá).

Habitat: species of this family are found mainly in freshwaters, including floodplain rivers and small streams.

Ecology and feeding: fishes of this family are generalist predators that have a crepuscular or nocturnal habit, feeding on zooplankton, aquatic invertebrates, and other fishes. During the day, these fishes usually hide in cracks, burrows, and behind logs and sticks. Fishes of this family have sharp spines in the pectoral and dorsal fins; these spines can be locked in erected position to difficult predation. This strategy of defense has caused accidents with people being stick by the spines of these fish when bathing with buckets of water in Amazonian rivers during the afternoon, when these fish are swimming near the water surface, searching for food.

Morphology: these are small- to medium-size catfishes that lack scales covering their body. The dorsal region of the body, between the head and the dorsal fin origin, is covered by saturated bony plates; the adipose fin is small; the eyes are covered by adipose tissue, without free orbit; and a dorsal fin located in the anterior portion of the body, right behind the head, is found. Their head has two to three pairs of barbells attached, one in the maxilla and one/two in the mentalis; all of these barbells are short with a length not passing the middle of the body. A channel cavity can be found in the infra-orbital region, where the maxilla barbells are attached.

Reproduction: fishes of this family show marked sexual dimorphism. Males have the first anal fin ray modified, which is probably related to internal fertilization. Internal fertilization requires diverse internal and external adaptations, such as spermatozoa with the presence of spermatophores, which is an adaptation that consists of the union of dozens of spermatozoan in packages or bundles. It is believed that females are capable of retaining packages of spermatozoan in their ovaries for a long period and trigger the fertilization and spawning after finding a suitable place for the offspring.

Economic importance: little or no commercial importance, due to the small size of most species and difficulty in handling because of the spines.

References: Ferraris (2003), Daniel and Fernandes (2005), Sazima et al. (2005), Akama and Ribeiro (2013), Cella-Ribeiro et al. (2016), Queiroz et al. (2013a, b), Ohara et al. (2017).

Fish species: *Ageneiosus akamai* **Ribeiro, Rapp Py-Daniel & Walsh 2017**

Common name(s): mandubé, maparai

Fig. 5.9 *Ageneiosus akamai* sampled in the Negro River. (Photo: Renato Silvano)

Fish species: *Ageneiosus apiaka* Ribeiro, Rapp Py-Daniel & Walsh 2017

Common name(s): mandubé, maparai

Fig. 5.10 *Ageneiosus apiaka* sampled in the Tapajós River. (Photo: Renato Silvano)

Fish species: *Ageneiosus dentatus* Kner 1858

Common name(s): mandubé

Fig. 5.11 *Ageneiosus dentatus* sampled in the Tapajós River. (Photo: Renato Silvano)

Fish species: *Ageneiosus inermis* (Linnaeus 1766)

Common name(s): mandubé

Fig. 5.12 *Ageneiosus inermis* sampled in the Negro River. (Photo: Renato Silvano)

Fish species: *Ageneiosus intrusus* Ribeiro, Rapp Py-Daniel & Walsh 2017

Common name(s): mandubé

Fig. 5.13 *Ageneiosus intrusus* sampled in the Negro River. (Photo: Renato Silvano)

Fish species: *Ageneiosus lineatus* Ribeiro, Rapp Py-Daniel & Walsh 2017

Common name(s): mandubé

Fig. 5.14 *Ageneiosus lineatus* sampled in the Negro River. (Photo: Renato Silvano)

Fish species: *Ageneiosus polystictus* Steindachner 1915

Common name(s): mandubé

Fig. 5.15 *Ageneiosus polystictus* sampled in the Negro River. (Photo: Renato Silvano)

Fish species: *Ageneiosus ucayalensis* Castelnau 1855

Common name(s): mandubé

Fig. 5.16 *Ageneiosus ucayalensis* sampled in the Negro River. (Photo: Renato Silvano)

Observation: abundance and size data were grouped for all *Ageneiosus* species, due to the difficulty in distinguishing them during sampling: *A. apiaka* and *A. dentatus* were collected only in Tapajós River; *A. inermis, A. intrusus, A. lineatus, A. polystictus, A. ucayalensis*, and *A. vittatus* were collected only in Negro River, while *A. akamai* was collected in both rivers; see Table 1.1 (▶ Chap. 1).

Size of fish sampled (standard length, both rivers): min. = 5.9 cm, max. = 38 cm, mean and median = 18 cm

Fisheries: the mandubés (*Ageneiosus* spp.) have low commercial value and are seldom exploited by fishers in the studied Amazonian rivers.

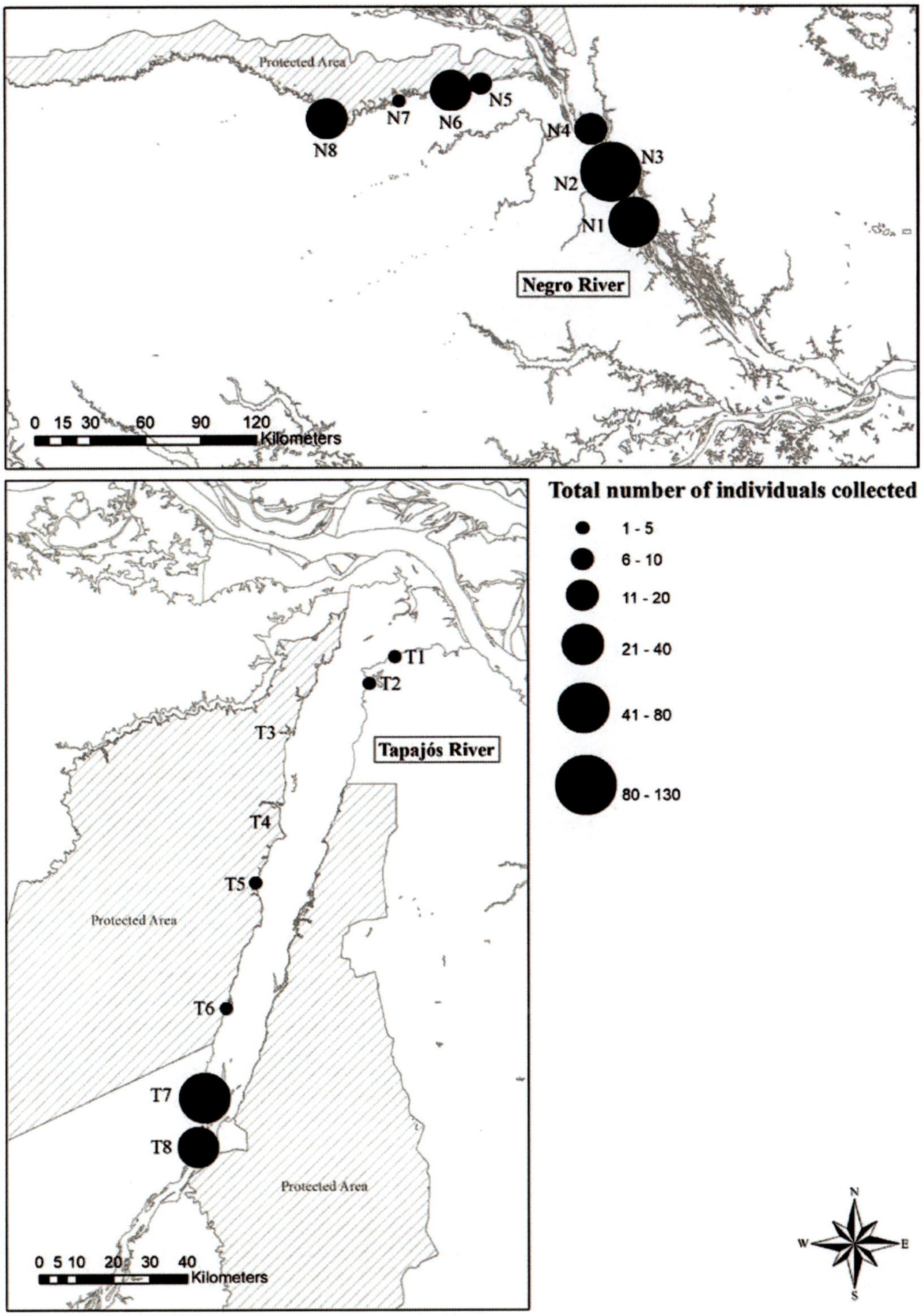

Fig. 5.17 Distribution of individuals of *Ageneiosus* spp. sampled in the Negro and Tapajós rivers

Fish species: *Auchenipterichthys coracoideus* (Eigenmann & Allen 1942)

Common name(s): orora

Size of fish sampled (standard length, both rivers): min. = 9.3 cm, max. = 22 cm, mean = 11.9 cm, median = 10.9 cm

Fisheries: not relevant to fisheries.

Fig. 5.18 *Auchenipterichthys coracoideus* sampled in the Tapajós River. (Photo: Renato Silvano)

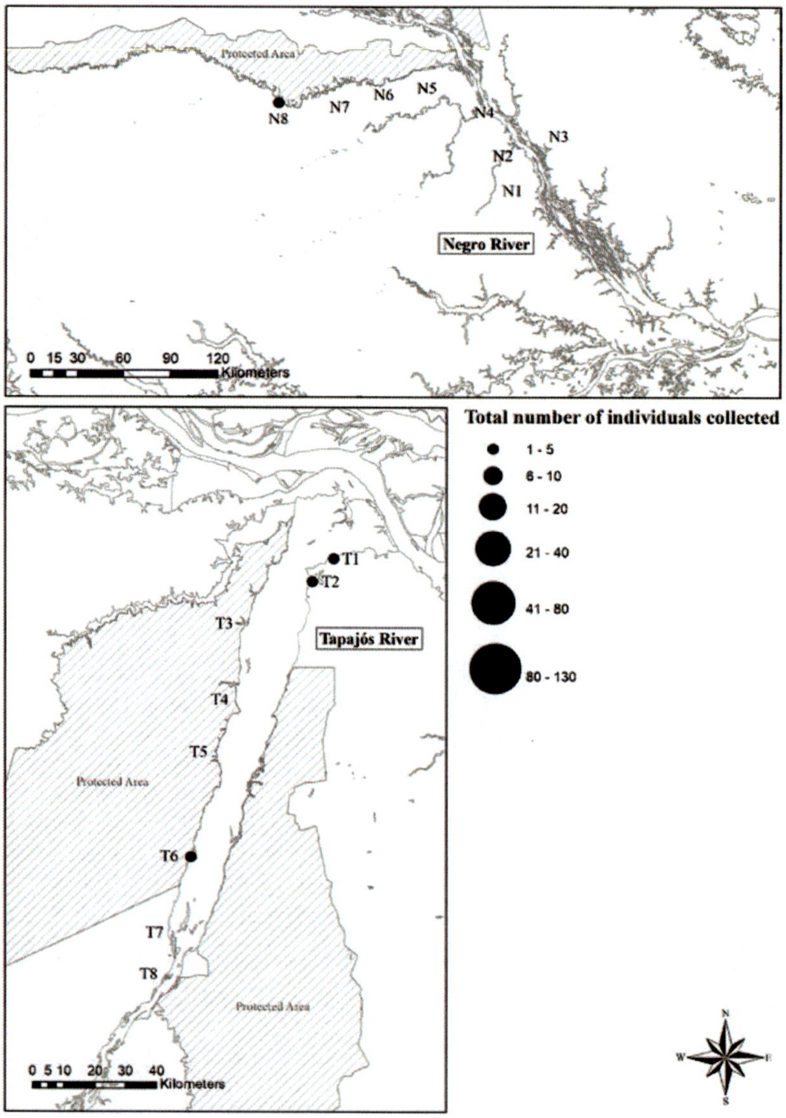

Fig. 5.19 Distribution of individuals of *Auchenipterichthys coracoideus* sampled in the Negro and Tapajós rivers

Fish species: *Auchenipterichthys longimanus* (Günther 1864)

Common name(s): cabeça-de-ferro

Size of fish sampled (standard length, both rivers): mean = 9.5 cm, median = 9 cm

Fisheries: not relevant to fisheries.

Fig. 5.20 *Auchenipterichthys longimanus* sampled in the Negro River. (Photo: Renato Silvano)

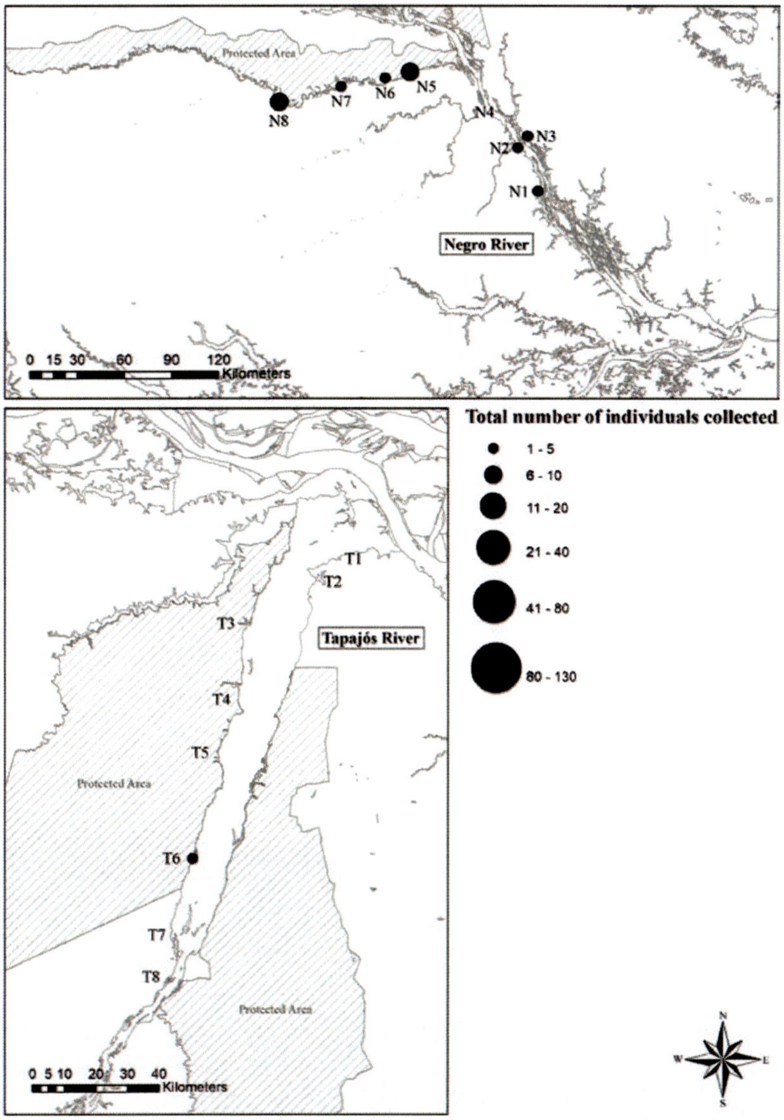

Fig. 5.21 Distribution of individuals of *Auchenipterichthys longimanus* sampled in the Negro and Tapajós rivers

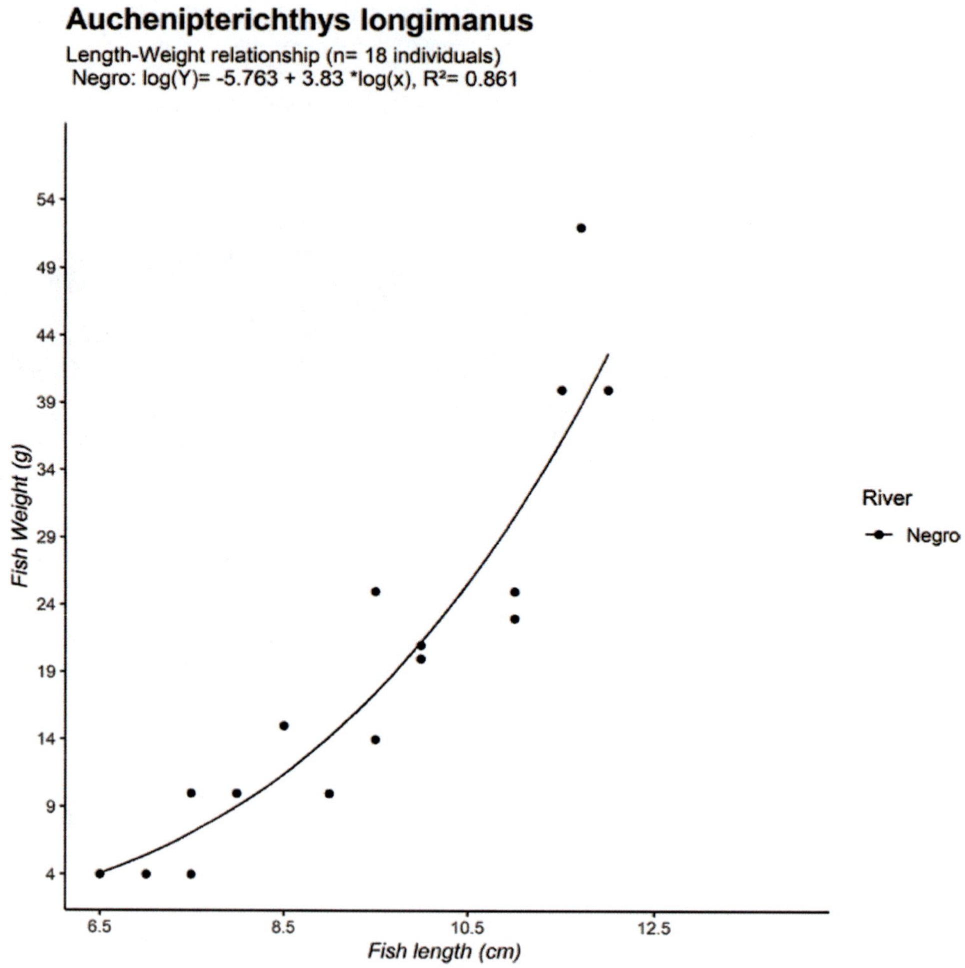

Fig. 5.22 Length-weight relationship for *Auchenipterichthys longimanus* sampled in Negro River

5

Fish species: *Auchenipterus nuchalis* (Spix & Agassiz 1829)

Common name(s): mandubé, orora

Size of fish sampled (standard length, both rivers): mean = 12.8 cm, median = 12.5 cm

Fisheries: not relevant to fisheries.

Fig. 5.23 *Auchenipterus nuchalis* sampled in the Tapajós River. (Photo: Renato Silvano)

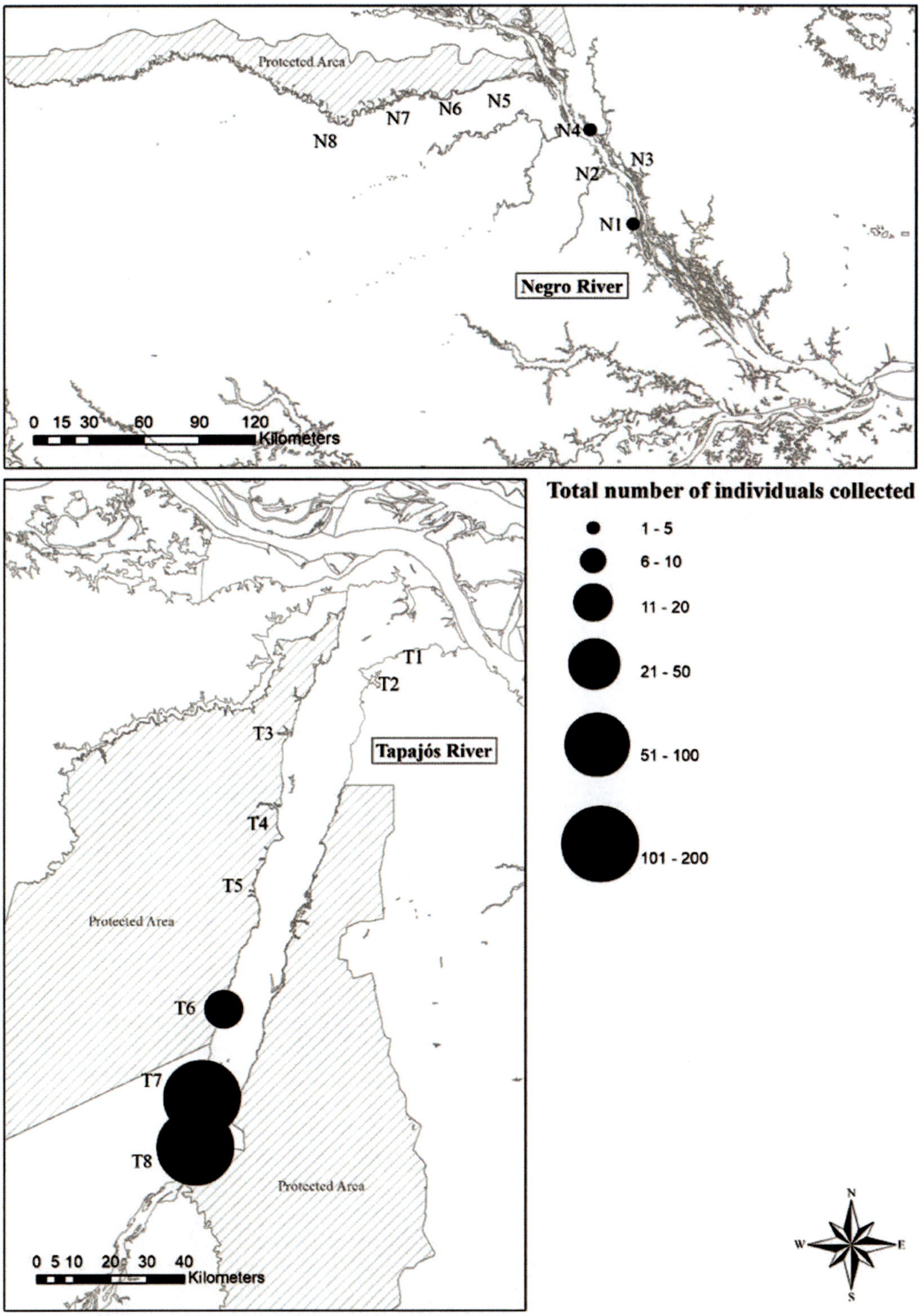

Fig. 5.24 Distribution of individuals of *Auchenipterus nuchalis* sampled in the Negro and Tapajós rivers

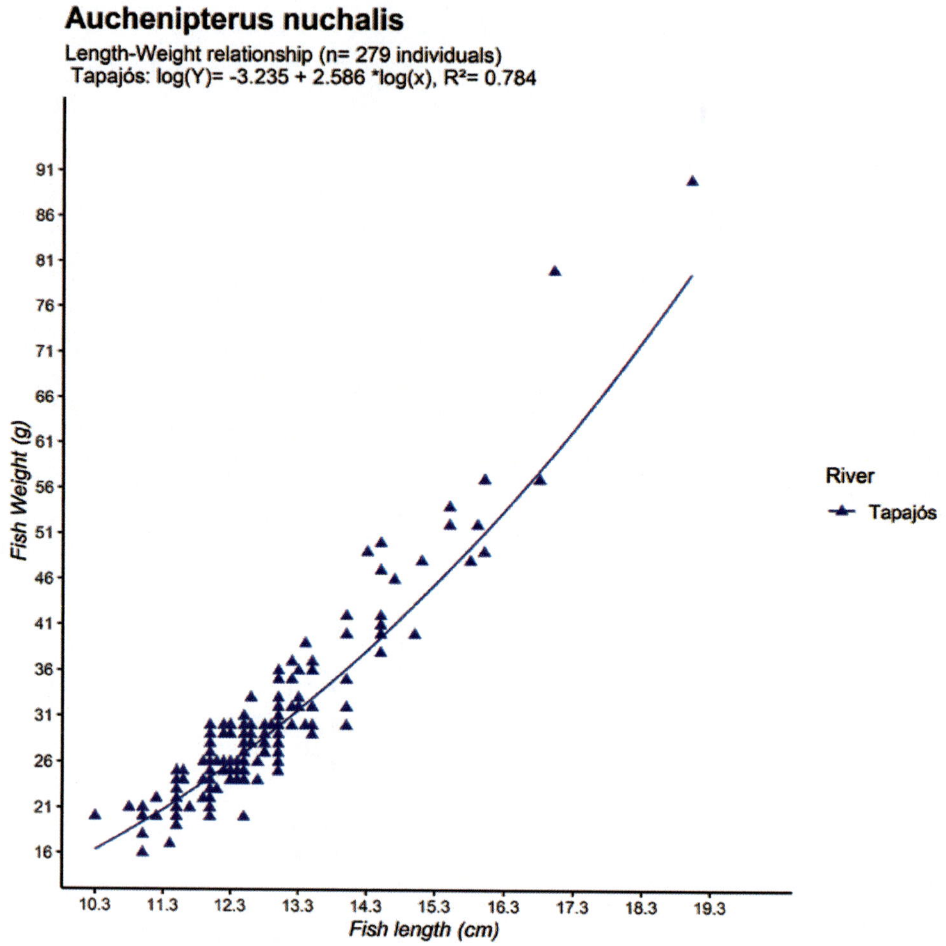

Fig. 5.25 Length-weight relationship for *Auchenipterus nuchalis* sampled in Tapajós River

---

Fish species: *Centromochlus heckelii* **(De Filippi 1853)**

Common name(s): mandi, carataí

Size of fish sampled (standard length, both rivers): mean = 7.7 cm, median = 6.9 cm

Fisheries: not relevant to fisheries.

Fig. 5.26 *Centromochlus heckelii* sampled in the Tapajós River. (Photo: Renato Silvano)

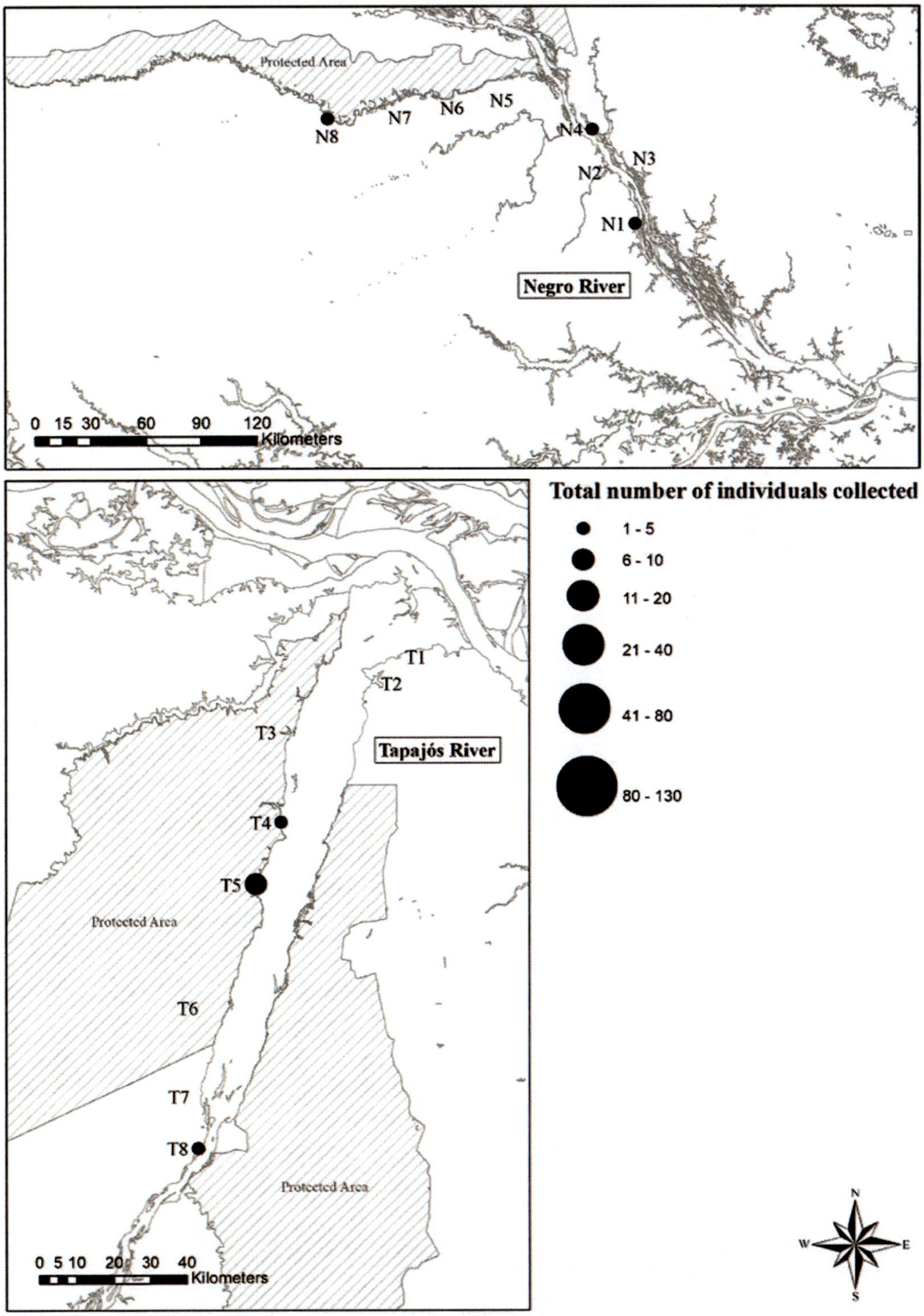

Fig. 5.27 Distribution of individuals of *Centromochlus heckelii* sampled in the Negro and Tapajós rivers

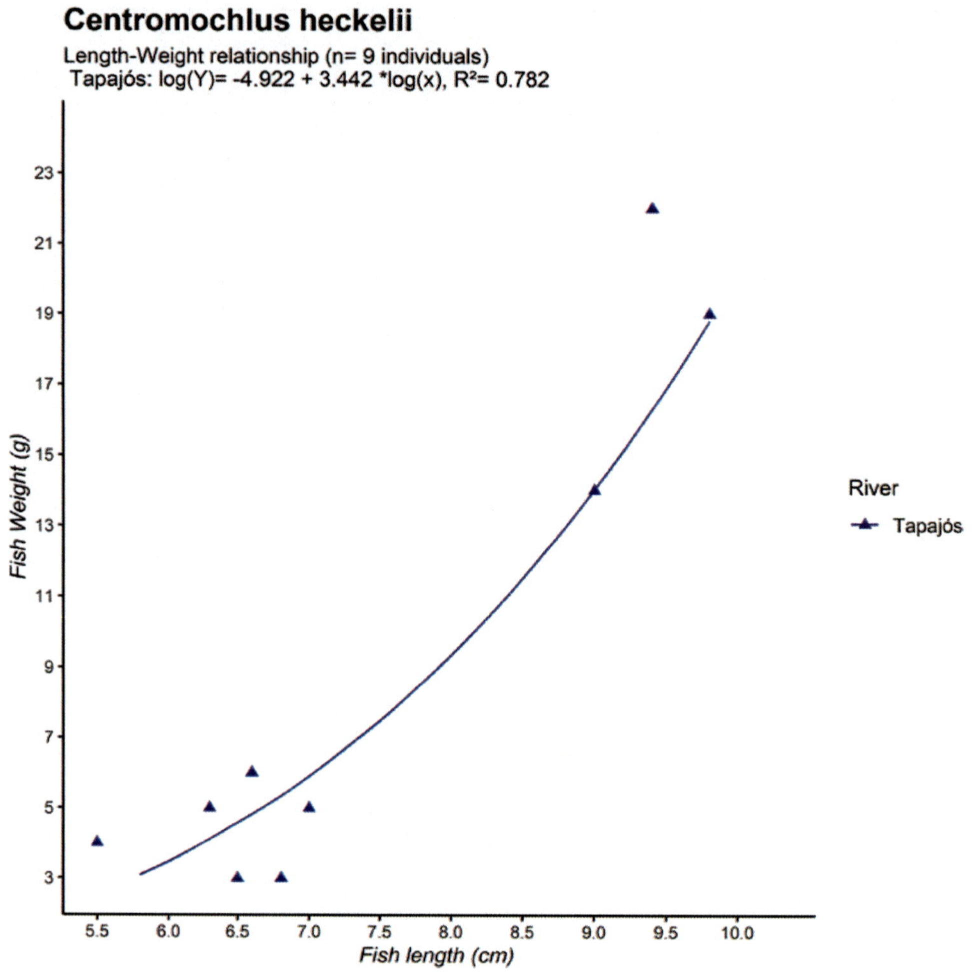

Fig. 5.28 Length-weight relationship for *Centromochlus heckelii* sampled in Tapajós River

Fish species: *Tatia* sp.

Common name(s): not available

Observation: abundance and size data were grouped for all species (*Tatia* spp.), due to the difficulty in distinguishing them during sampling: *Tatia* cf. *aulopygia*, *Tatia* cf. *Intermedia*, and *Tatia intermedia* were collected only in Negro River, while *Tatia* sp. was collected only in Tapajós River; see Table 1.1 (▶ Chap. 1).

Size of fish sampled (standard length, both rivers): min. = 4 cm, max. = 9.5 cm, mean = 6.8 cm, median = 7 cm

Fisheries: not relevant to fisheries.

Fig. 5.29 *Tatia* sp. sampled in the Tapajós River. (Photo: Renato Silvano)

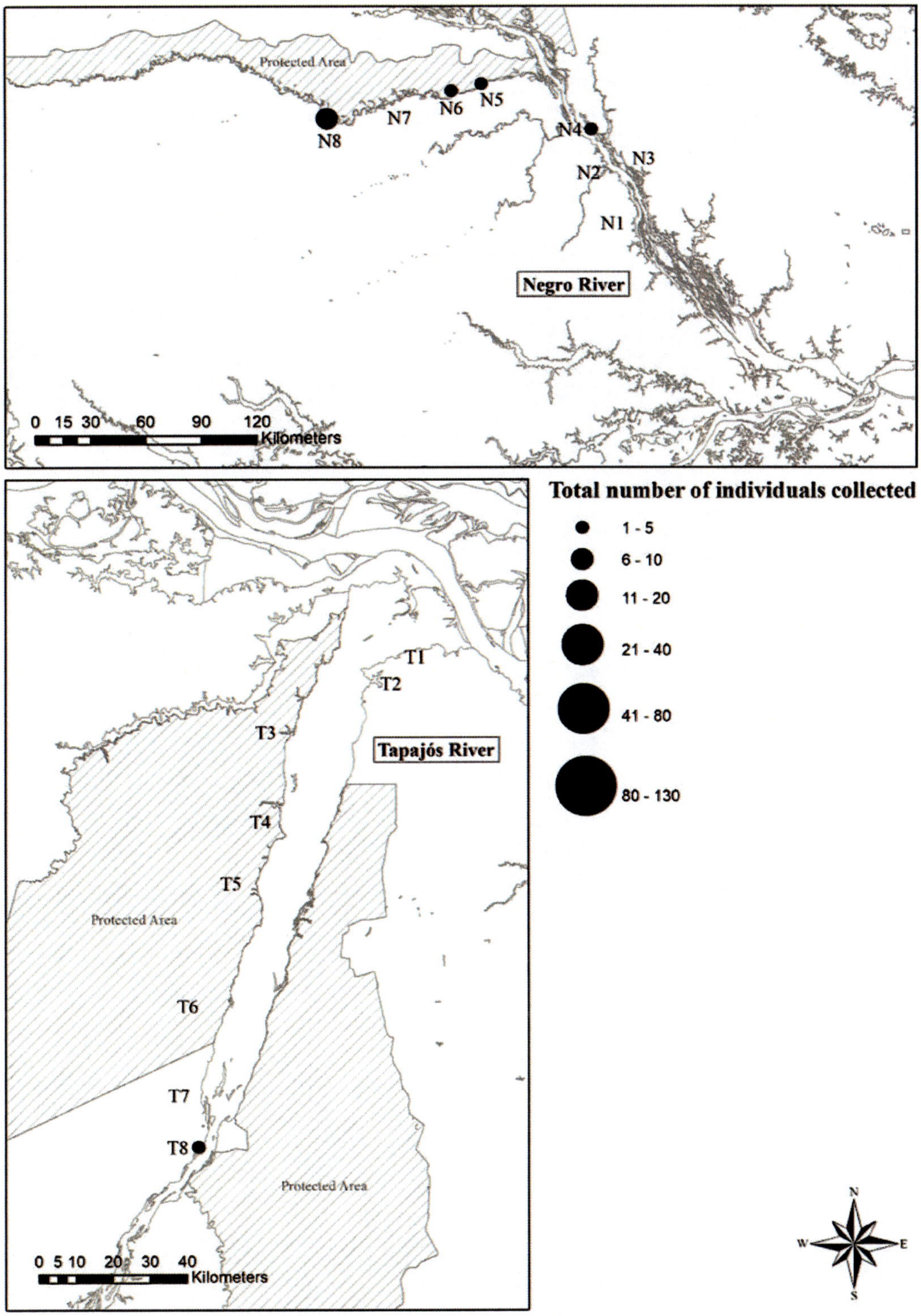

Fig. 5.30 Distribution of individuals of *Tatia* spp. sampled in the Negro and Tapajós rivers

5

Fish species: *Trachelyopterus galeatus* (Linnaeus 1766)

Common name(s): orora, anujá, cabeça-de-ferro

Observation: abundance and size data were grouped for the two species (*Trachelyopterus galeatus* and *T. coriaceus*), due to the difficulty in distinguishing them during sampling: *T. galeatus* was collected in both rivers and *T. coriaceus* was collected only in Tapajós River; see ◘ Table 1.1 (▶ Chap. 1).

Size of fish sampled (standard length, both rivers): min. = 6.5 cm, max. = 15 cm, mean = 10.8 cm, median = 10.6 cm

Fisheries not relevant to fisheries.

◘ **Fig. 5.31** *Trachelyopterus galeatus* sampled in the Tapajós River. (Photo: Renato Silvano)

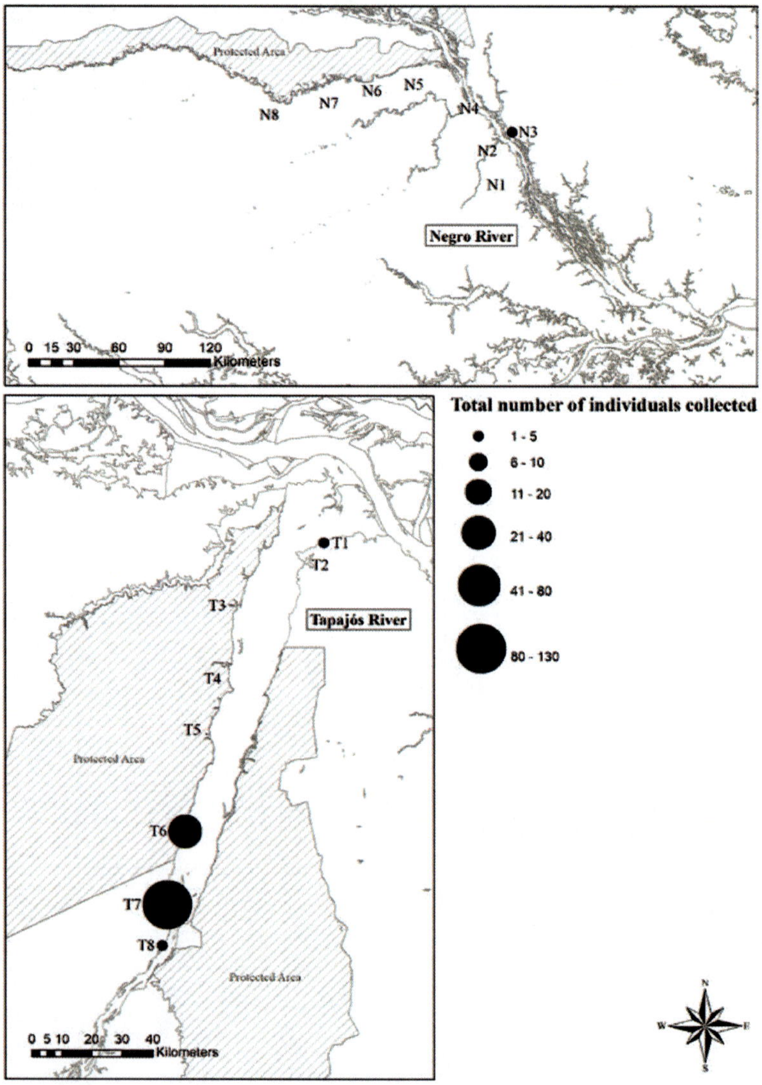

◘ **Fig. 5.32** Distribution of individuals of *Trachelyopterus* spp. sampled in the Negro and Tapajós rivers

Fish species: *Trachycorystes trachycorystes* (Valenciennes 1840)

Common name(s): anujá, cabeça-de-ferro

Size of fish sampled (standard length, both rivers): mean = 18.6 cm, median = 21.4 cm

Fisheries: not relevant to fisheries.

Fig. 5.33 *Trachycorystes trachycorystes* sampled in the Negro River. (Photo: Renato Silvano)

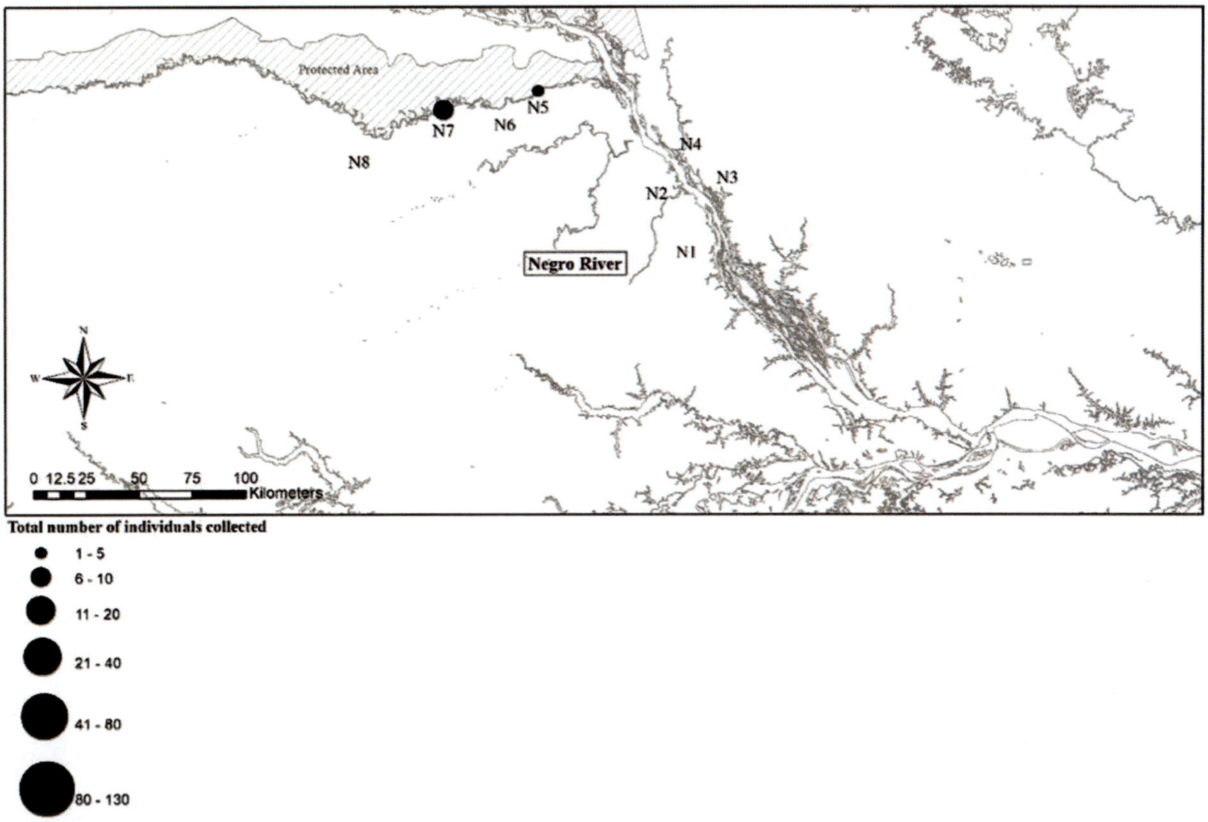

Fig. 5.34 Distribution of individuals of *Trachycorystes trachycorystes* sampled in the Negro River

5

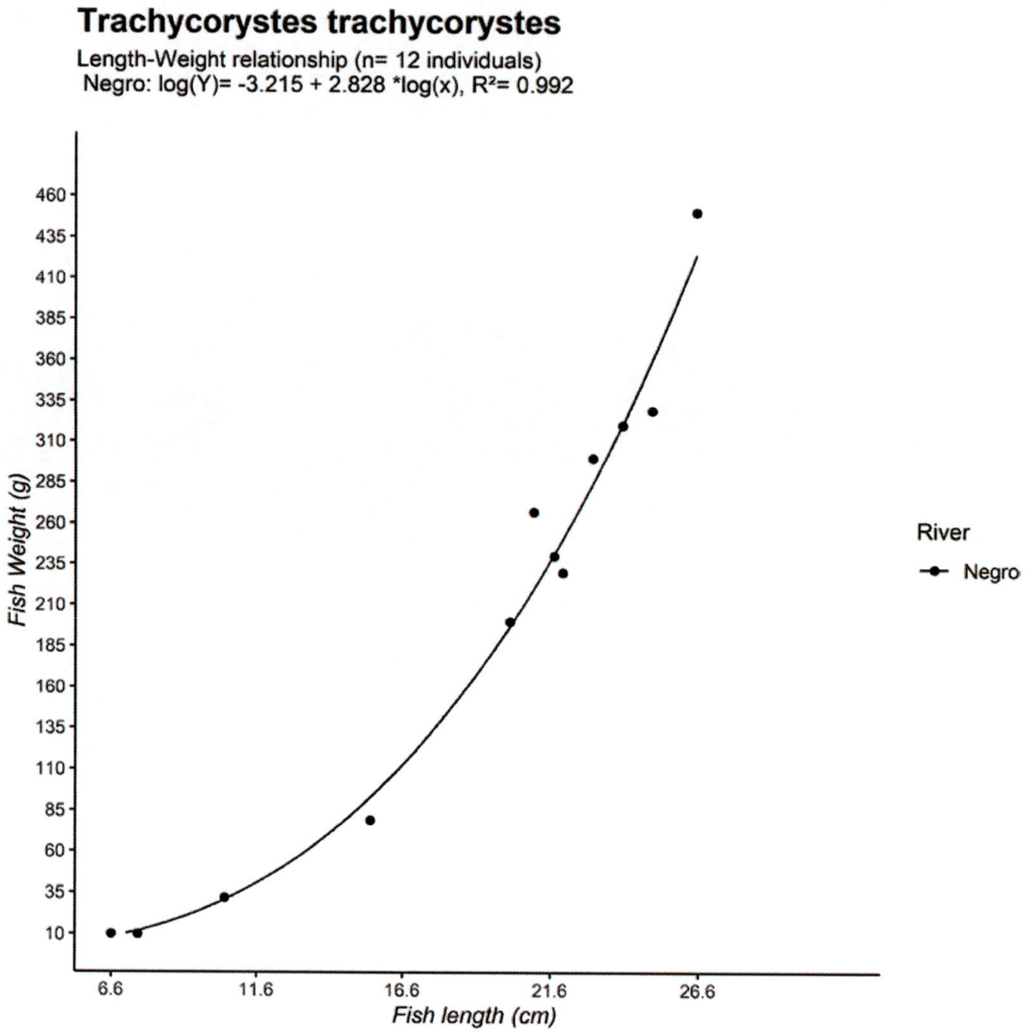

Fig. 5.35 Length-weight relationship for *Trachycorystes trachycorystes* sampled in Negro River

---

Fish species: *Trachelyichthys decaradiatus* Mees 1974

Common name(s): xaroque, orora

Size of fish sampled (standard length, both rivers): min. = 7 cm, max. = 9.5 cm, mean and median = 8.3 cm

Fisheries: not relevant to fisheries.

Fig. 5.36 *Trachelyichthys decaradiatus* sampled in the Tapajós River. (Photo: Renato Silvano)

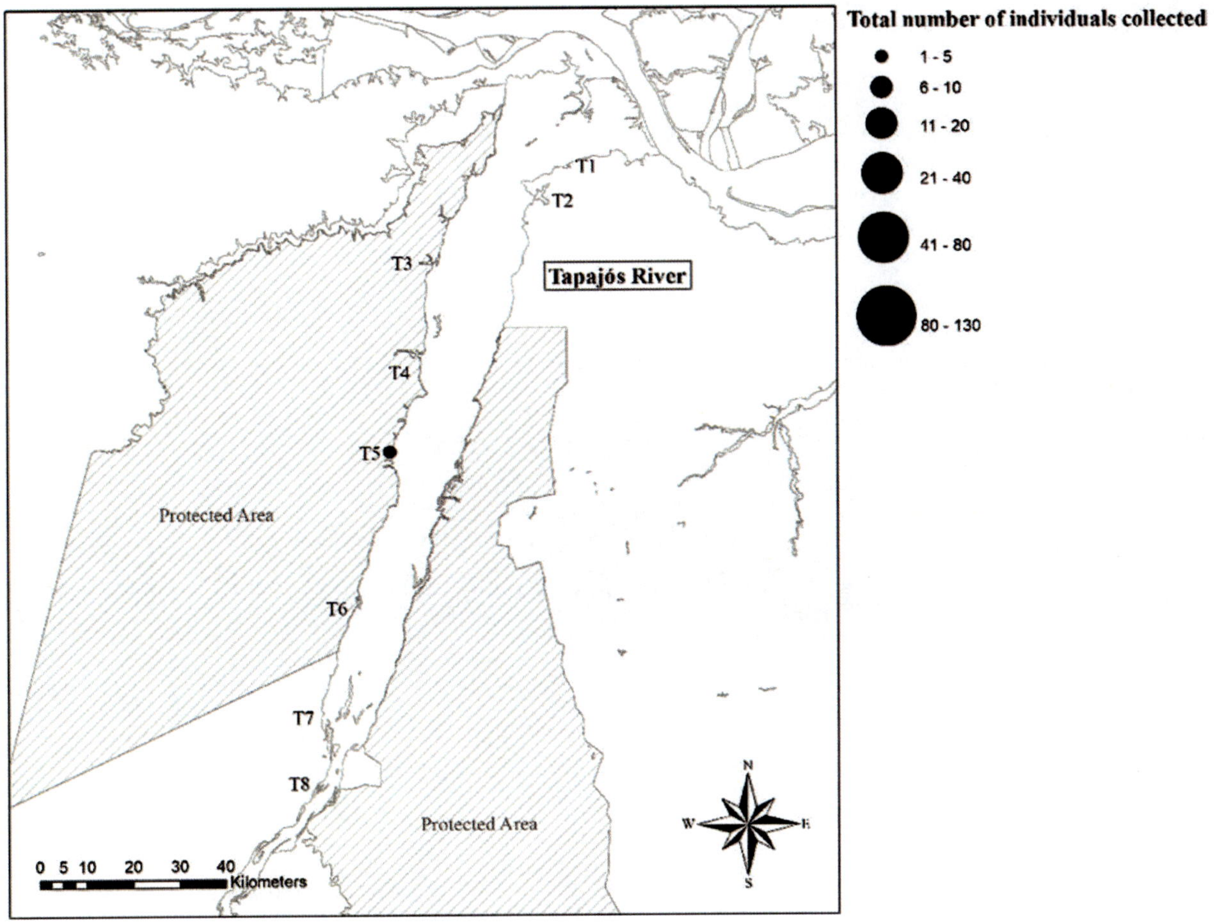

Fig. 5.37 Distribution of individuals of *Trachelyichthys decaradiatus* sampled in the Tapajós River

5.2.2 Family Callichthyidae

5.2.2.1 Summary of Biology and Ecology (Literature)

Diversity: this family is composed of 220 valid species distributed in two subfamilies: Callichthyinae and Corydoradinae.

Habitat: fishes of this family can be found in a variety of environments, including rivers with a fast flow, small streams with steady waters and low levels of oxygen, and even temporary ponds with acid water located within forests. Their distribution includes the countries of South America and Panamá.

Ecology and feeding: callichthyids are capable of swallow atmospheric air and, consequently, maintain their hydrostatic balance. Besides that, they are capable of breathing atmospheric air via their intestines, which allow them to survive in environments with very low levels of oxygen. Their bone plates protect them from dehydration, and it is not uncommon to see these animals crossing land to reach other nearby water bodies. The species from the genus *Hoplosternum* are omnivores, feeding on aquatic and terrestrial invertebrates and detritus.

Morphology: species with small to medium body size (25 to 250 mm SL); they possess a series of two lateral bony plates. The subfamily Callichthyinae encompasses species with larger body size, depressed body, and long maxillary barbels, which can reach the origin of their pelvic fins. The subfamily Corydoradinae includes species of smaller body size, deep body, and short maxillary barbels, not passing the origin of the pectoral fins.

Reproduction: some species build floating nests with bubbles, sticks, and leaves and then spawn on this structure. Males change their behavior completely and become very aggressive during the reproductive period. Females of other species stand perpendicularly to the males and, with the mouth close to the male's urogenital opening, ingest their sperm, which pass through the females' digestive trait and then are excreted on the oocytes located near the pelvic fins. During the reproductive period, the pelvic fins are turned over, forming a pouch to carry the eggs.

Economic importance: many species of this family are popular among aquarists and are thus important in the aquarium trade. In some regions of South America, they are consumed by riverine people. However, species of this family do not have commercial relevance in the Amazonian fisheries.

References: Mayer (1950), Gee and Graham (1978), Burgess (1989), Kohda et al. (1995), Hostache & Mol (1998), Reis (1997), Reis (1998), Santos et al. (2004), Queiroz et al. (2013a, b).

Fish species: *Hoplosternum littorale* (Hancock 1828)

Common name(s): tamoata

Size of fish sampled (standard length, both rivers): mean = 12.9 cm, median = 12.4 cm

Fig. 5.38 *Hoplosternum littorale* collected in the Lower Amazon (no photos with natural coloration were available for the sampled rivers). (Photo: Renato Silvano)

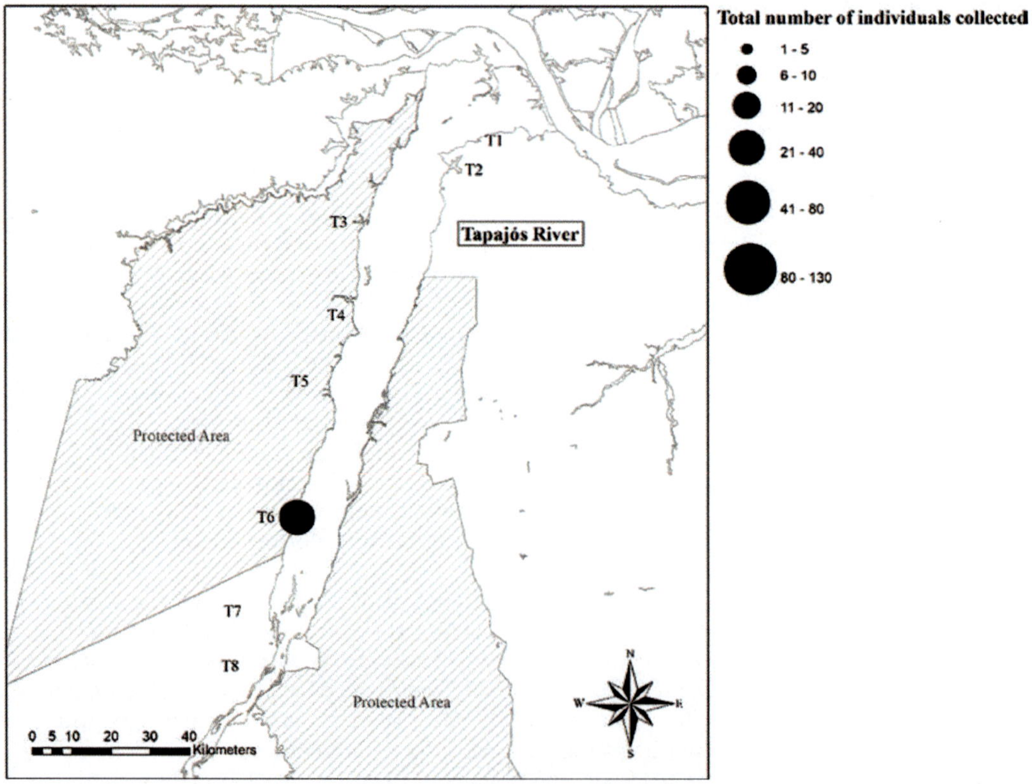

Fig. 5.39 Distribution of individuals of *Hoplosternum littorale* sampled in the Tapajós River

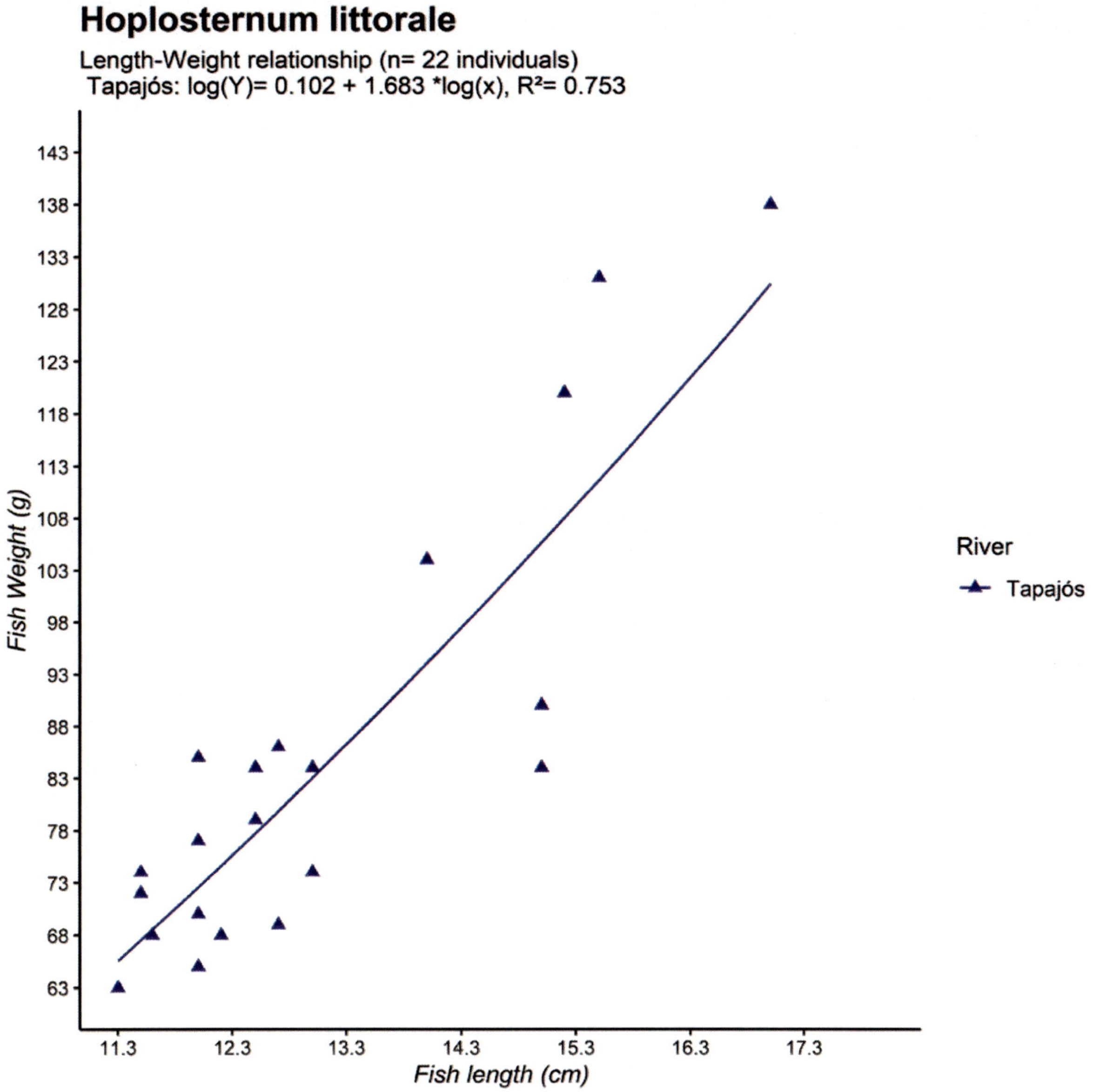

Fig. 5.40 Length-weight relationship for *Hoplosternum littorale* sampled in Tapajós River

Fisheries: The tamoata (*Hoplosternum littorale*) is occasionally caught in fisheries in the Tapajós River, mostly for consumption. Biomass caught is in ◘ Table 3.3 (▶ Chap. 3).

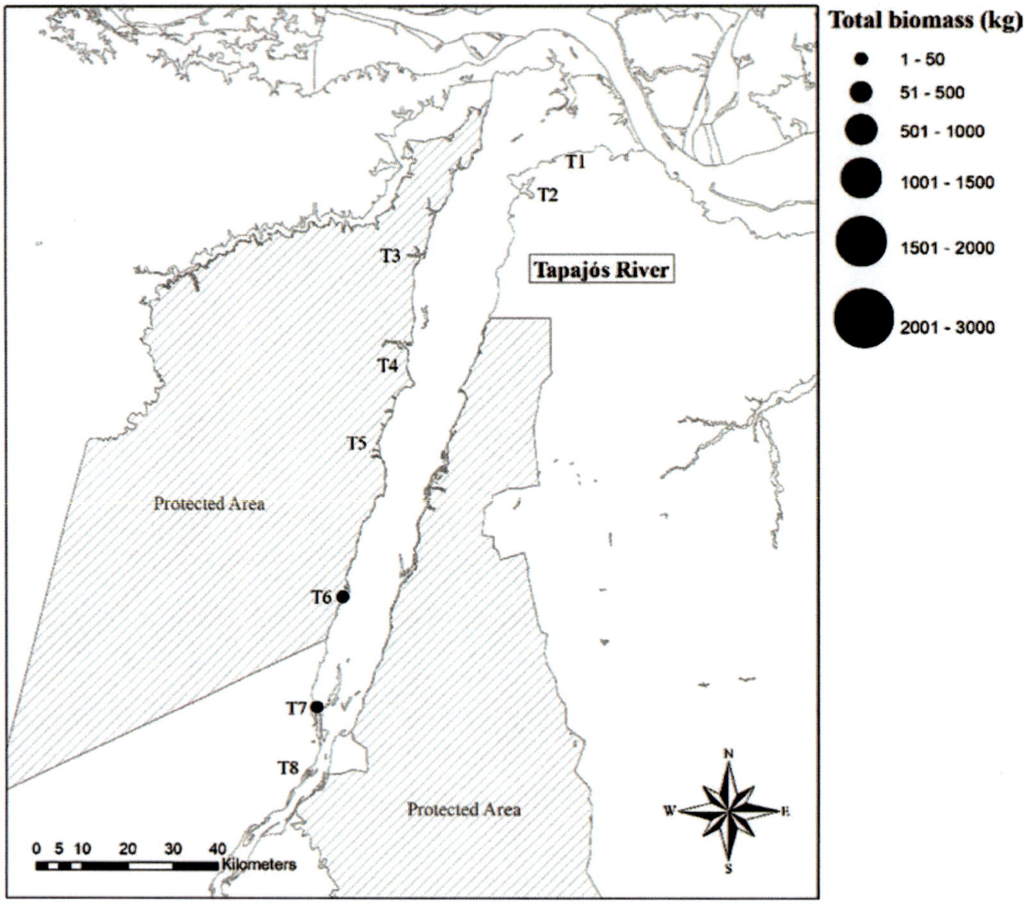

◘ **Fig. 5.41** Distribution of biomass of tamoata (*Hoplosternum littorale*) caught in small-scale fisheries in the Tapajós River

◘ **Fig. 5.42** Seasonal variation of biomass of tamoata (*Hoplosternum littorale*) caught per fishing gear in small-scale fisheries in eight communities in the Tapajós River. NA: fishing gear not informed; Mix: more than one gear was used

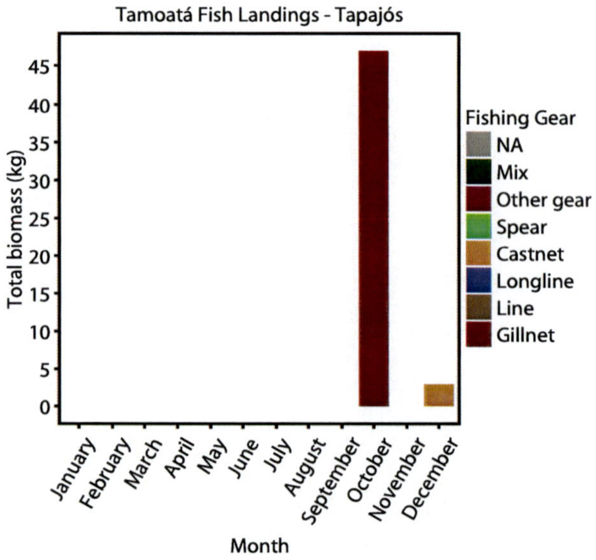

5.2.3 Family Cetopsidae

5.2.3.1 Summary of Biology and Ecology (Literature)

Diversity: this family contains 43 valid species separated in two subfamilies: Cetopsinae and Helogeneinae.

Habitat: these catfishes can be found from the north to the central regions of South America. Most of the family diversity are found in the Amazon basin.

Ecology and feeding: most species feed on allochthonous insects at the water surface, especially at night when they are more active. Some species, including those from the genus *Cetopsis*, are adapted to feed on chunks of flesh from vulnerable, injured, or even dead animals, thus acting as necrophages.

Morphology: fishes of this family can hardly be identified by external morphology, but their general body shape is typical and hard to mistaken. These fishes have a naked body, without bone plates, spines, or integumentary teeth (odontodes). They also lack adipose fin and nasal barbels. Their anal fin is long and the head has a globous aspect due to their sturdy cephalic musculature that covers part or even the whole upper part of the skull.

Reproduction: some species are sexually dimorphic, especially regarding their pectoral and dorsal fins. In these species, males usually have dorsal and pectoral fins with long filaments originated from the first fin ray. In some species, males and females have filaments, while in others the filaments are lacking in both sexes. Similarly, in some species, the convex shape of the anal fin is associated with adult males (females and immature males having a straight or concave shape), while in others this differentiation is not clear.

Economic importance: this family does not have relevance to commercial neither to subsistence fisheries.

References: Saul (1975), Goulding (1979, 1980, 1981), Baskin et al. (1980), Ferraris and Brown (1991), Lundberg and Rapp Py-Daniel (1994), Barthem and Goulding (1997), Vari and Ortega (1986), Vari et al. (2005), de Pinna et al. (2007), Queiroz et al. (2013a, b).

Fish species: *Cetopsis coecutiens* (Lichtenstein 1819)

Fig. 5.43 *Cetopsis coecutiens* sampled in the Negro River. (Photo: Renato Silvano)

Common name(s): candiru

Size of fish sampled (standard length, both rivers): min. = 11.3 cm, max. = 18.3 cm, mean = 13.6 cm, median = 12.2 cm

Fisheries: not relevant to fisheries.

5

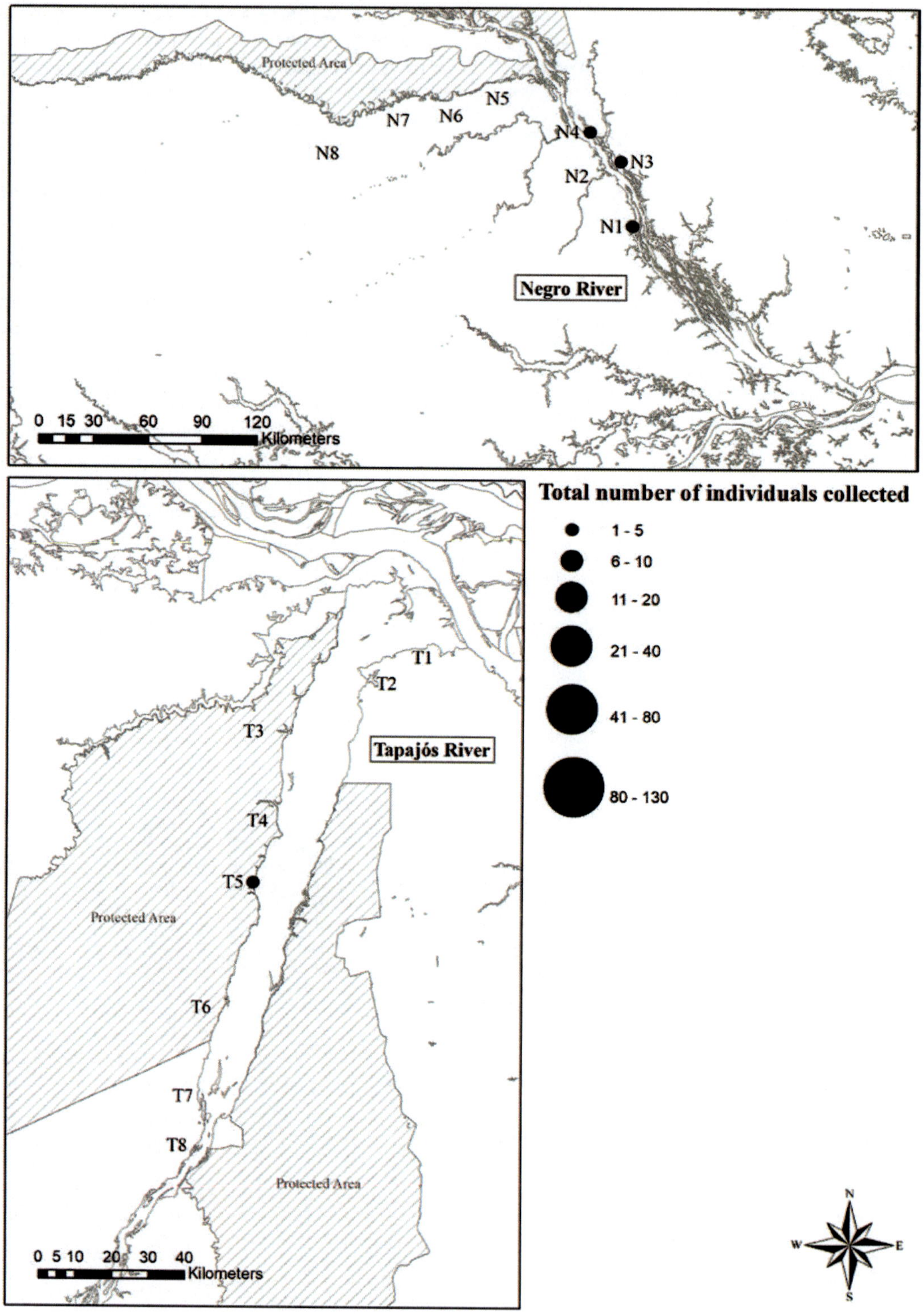

Fig. 5.44 Distribution of individuals of *Cetopsis coecutiens* sampled in the Negro and Tapajós rivers

5.2.4 Family Doradidae

5.2.4.1 Summary of Biology and Ecology (Literature)

Diversity: family composed of 96 valid species distributed in 32 genera.

Habitat: fishes of this family can be found in all South American river basins.

Ecology and feeding: most of these fishes are opportunistic and omnivorous, feeding on a high variety of food items, including fish, detritus, crustaceans, seeds, fruits, leaves, insects, and mollusks. Some species may have an important role in dispersing seeds of riparian flowering plants.

Morphology: most species have small to medium body sizes (~maximum size of 20 cm), but few can reach large sizes, such as *Oxydoras niger* and *Megalodoras uranoscopus* (~maximum size of 100 cm). The most recognizable characteristics of fishes from this family are the presence of lateral bony shields with spines and their heavily ossified head.

Reproduction: no information.

Economic importance: in the Amazon basin, some species of this family, such as *Oxydoras niger*, have been targeted by commercial and artisanal fisheries, but they usually are not among the most caught, neither among the most appreciated. Most small-sized species of this family are commercialized in the aquarium trade.

References: Goulding (1980), Ferreira et al. (1998), Birindelli (2010), Birindelli and Sousa (2013), Queiroz et al. (2013a, b).

Fish species: *Acanthodoras* cf. *cataphractus*

Common name(s): not available

Size of fish sampled (standard length, both rivers): min. = 8 cm, max. = 12 cm, mean = 11 cm, median = 11.5 cm

Fisheries: not relevant to fisheries.

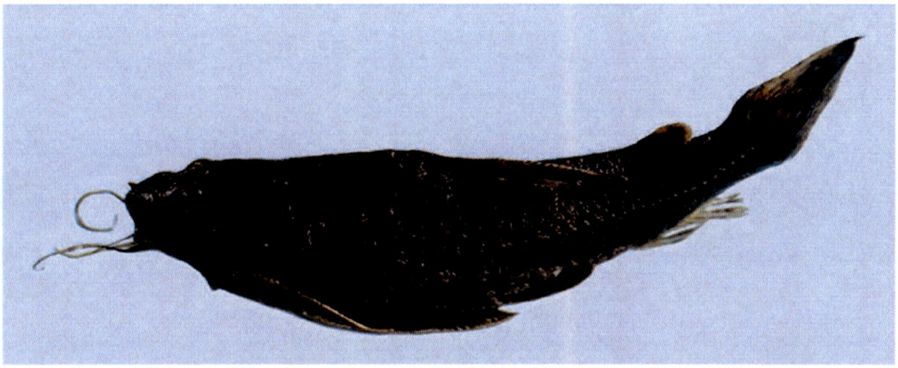

Fig. 5.45 *Acanthodoras* cf. *cataphractus* sampled in the Negro River. (Photo: Renato Silvano)

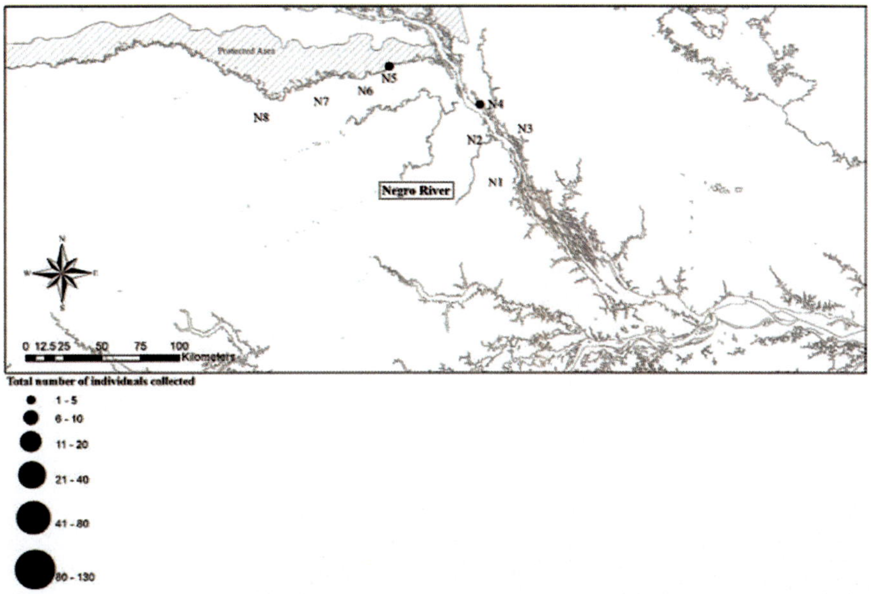

Fig. 5.46 Distribution of individuals of *Acanthodoras* cf. *cataphractus* sampled in the Negro River

Fish species: ***Amblydoras affinis*** **(Kner 1855)**

Common name(s): xaroque

Fisheries: not relevant to fisheries.

Size of fish sampled (standard length, both rivers): 7.6 cm

Fig. 5.47 *Amblydoras affinis* sampled in the Tapajós River. (Photo: Renato Silvano)

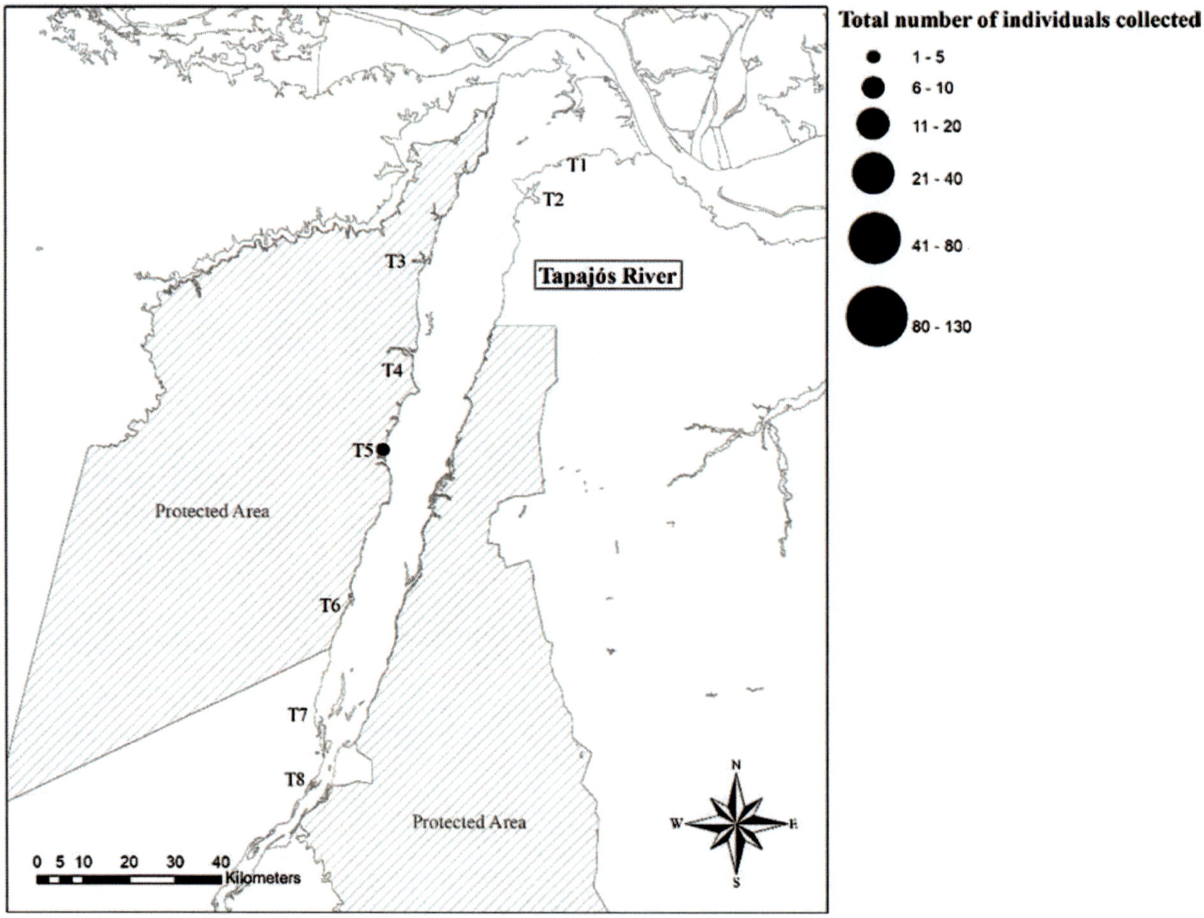

Fig. 5.48 Distribution of individuals of *Amblydoras affinis* sampled in the Tapajós River

Fish species: *Anadoras weddellii* **(Castelnau 1855)**

Fisheries: not relevant to fisheries.

Common name(s): reco-reco, xaroque, guerreiro

Size of fish sampled (standard length, both rivers): mean = 9.1 cm, median = 9 cm

Fig. 5.49 *Anadoras weddellii* sampled in the Tapajós River. (Photo: Renato Silvano)

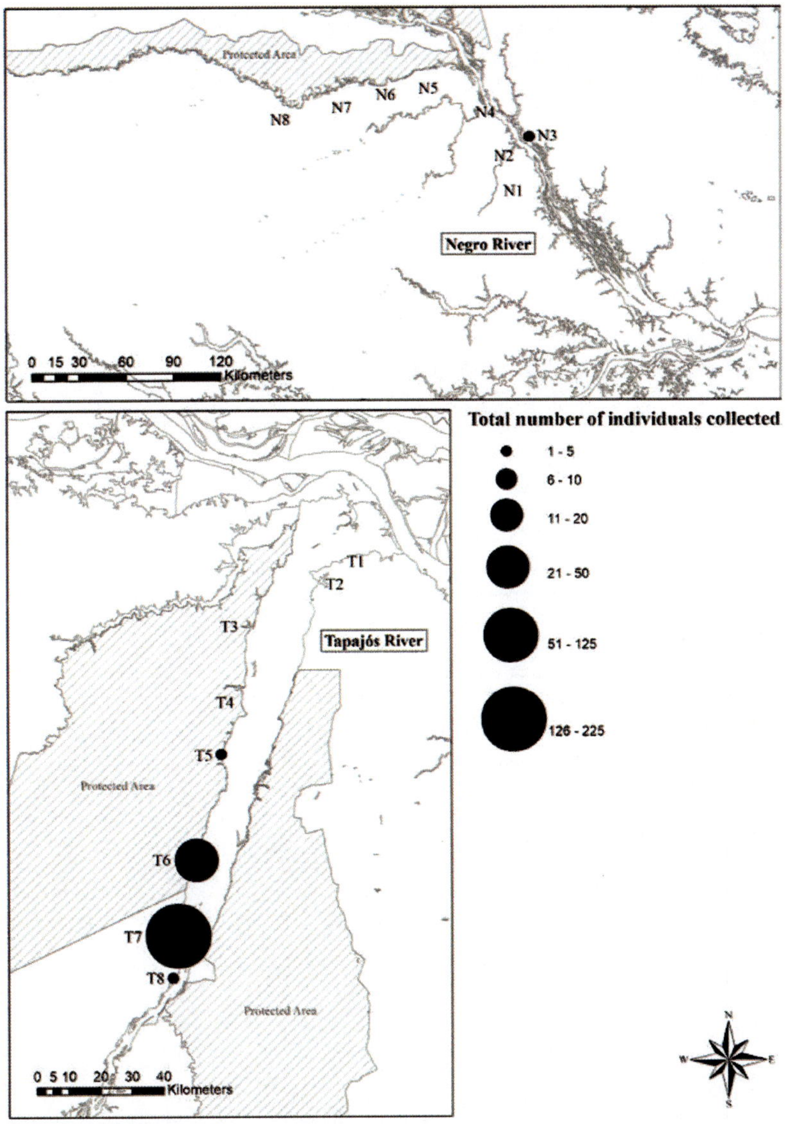

Fig. 5.50 Distribution of individuals of *Anadoras weddellii* sampled in the Negro and Tapajós rivers

5

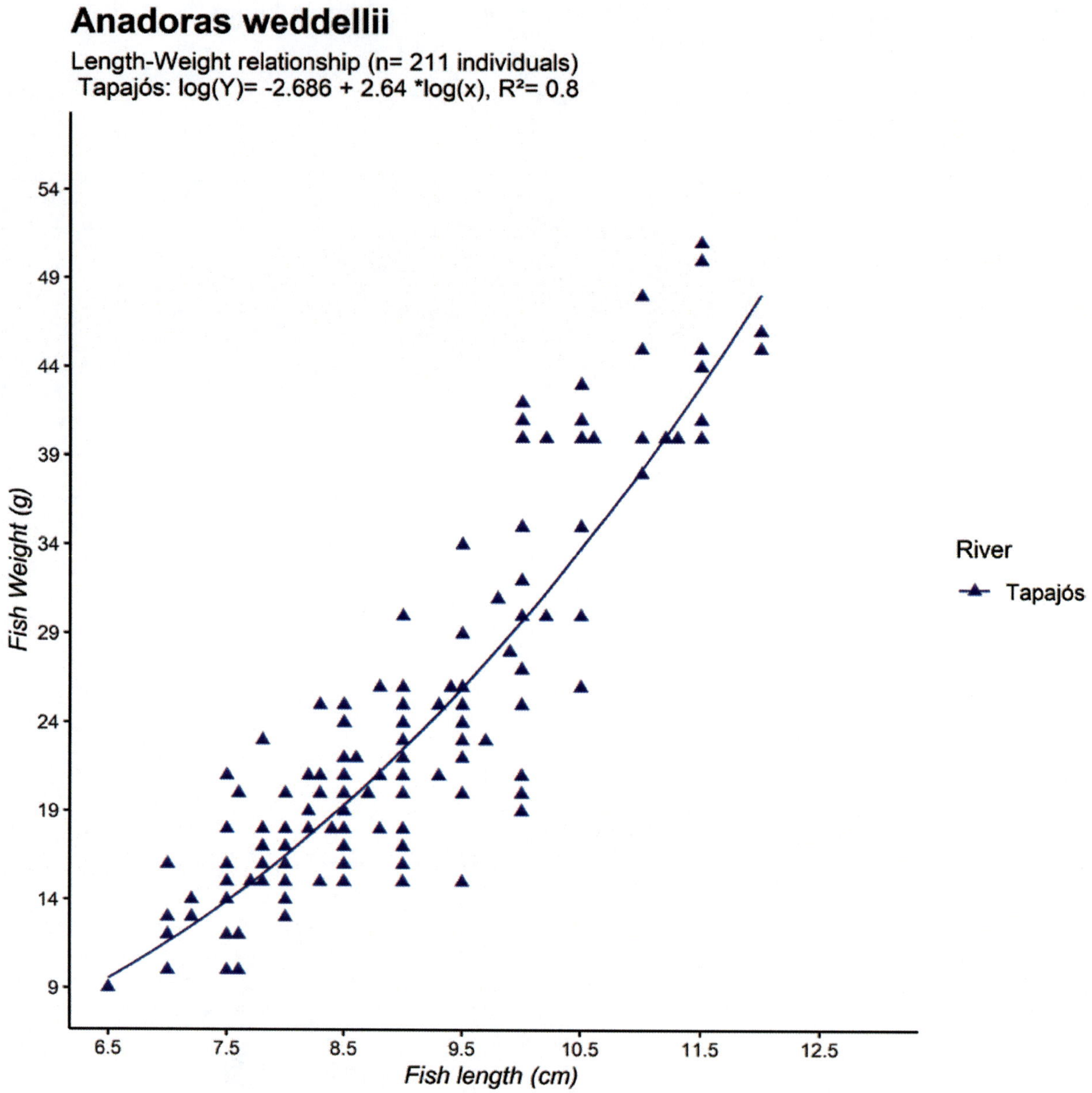

Fig. 5.51 Length-weight relationship for *Anadoras weddellii* sampled in Tapajós River

Fish species: *Anduzedoras oxyrhynchus* **(Valenciennes 1821)**

Common name(s): cuiú

Fisheries: not relevant to fisheries.

Size of fish sampled (standard length, both rivers): 11 cm

Fig. 5.52 *Anduzedoras oxyrhynchus* sampled in the Negro River. (Photo: Renato Silvano)

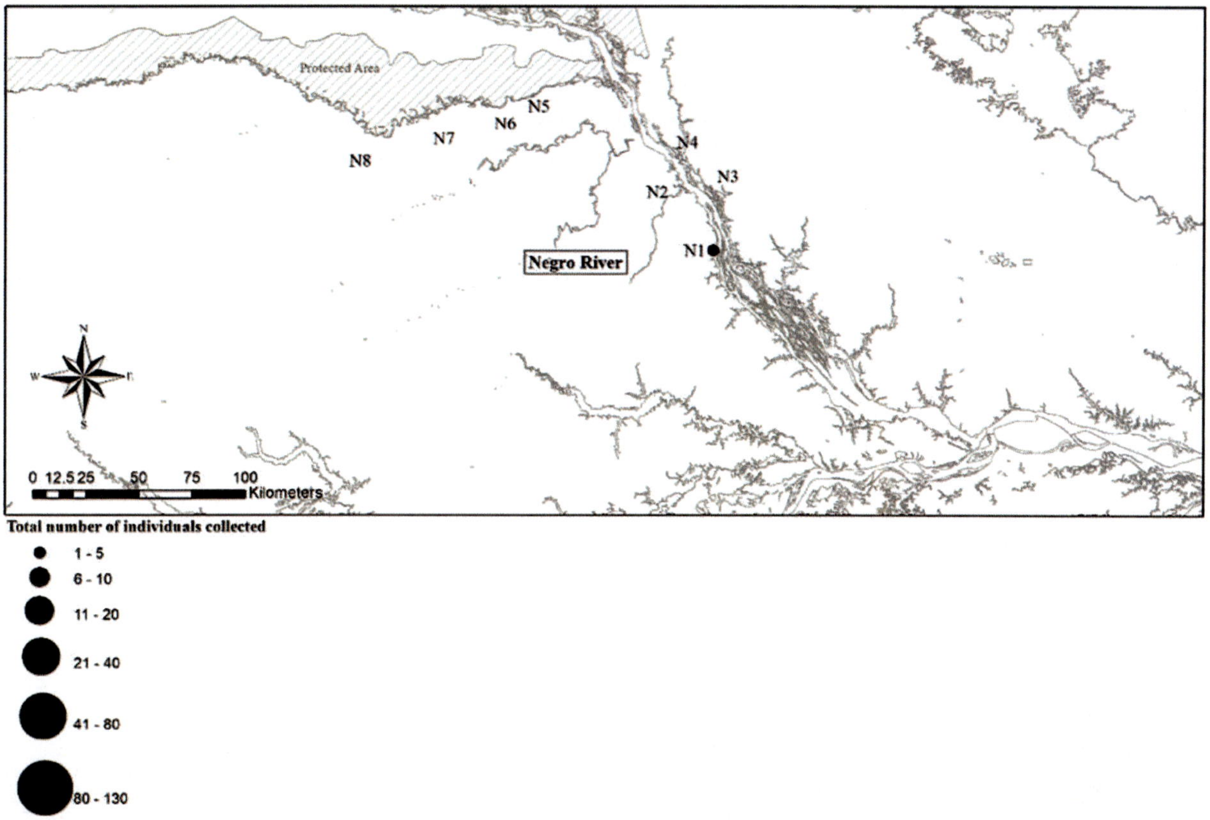

Fig. 5.53 Distribution of individuals of *Anduzedoras oxyrhynchus* sampled in the Negro River

Fish species: ***Astrodoras asterifrons*** **(Kner 1853)**

Common name(s): reco-reco, rebeca

Fisheries: not relevant to fisheries.

Size of fish sampled (standard length, both rivers): min. = 4 cm, max. = 6 cm, mean = 5 cm, median = 5 cm

Fig. 5.54 *Astrodoras asterifrons* sampled in the Negro River. (Photo: Renato Silvano)

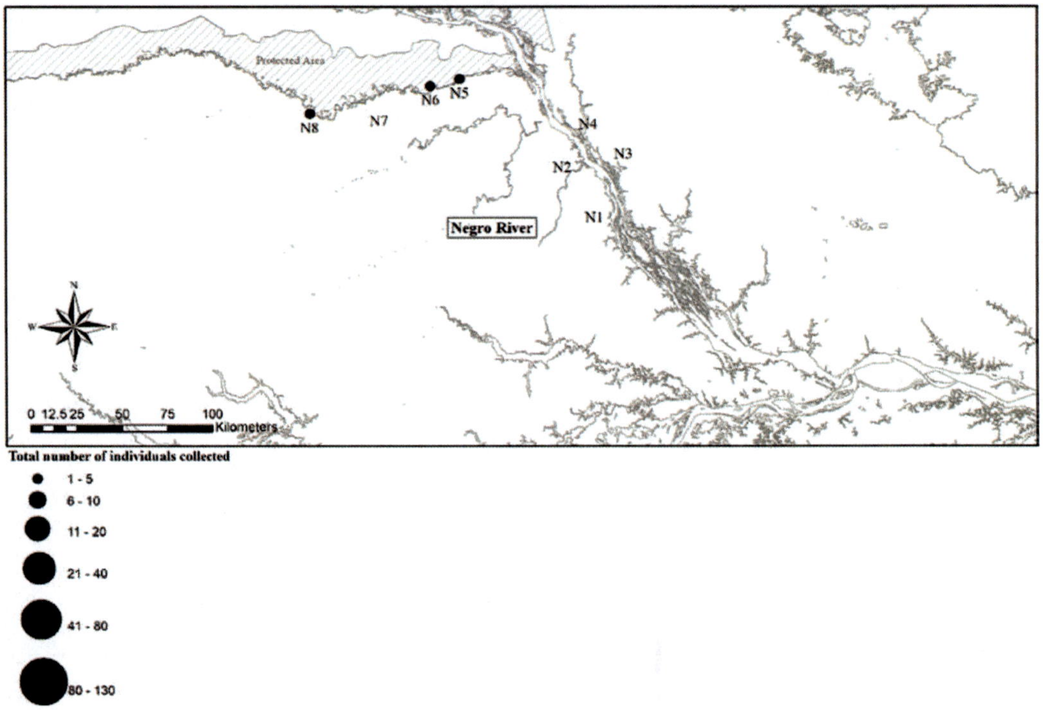

Fig. 5.55 Distribution of individuals of *Astrodoras asterifrons* sampled in the Negro River

Fish species: *Hassar orestis* **(Steindachner 1875)**

Common name(s): botinho

Fisheries: not relevant to fisheries.

Size of fish sampled (standard length, both rivers): mean = 10.2 cm, median = 9.5 cm

Fig. 5.56 *Hassar orestis* sampled in the Tapajós River. (Photo: Renato Silvano)

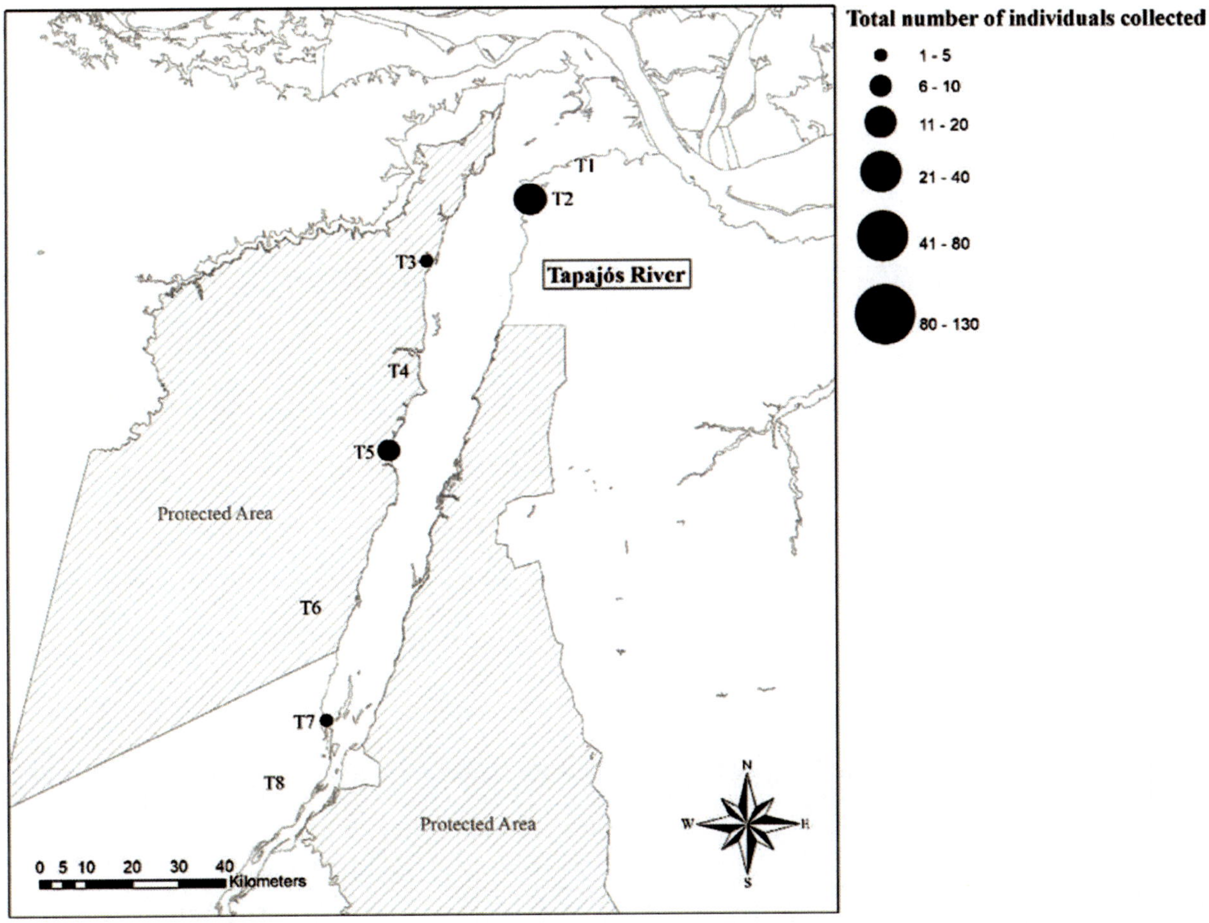

Fig. 5.57 Distribution of individuals of *Hassar orestis* sampled in the Tapajós River

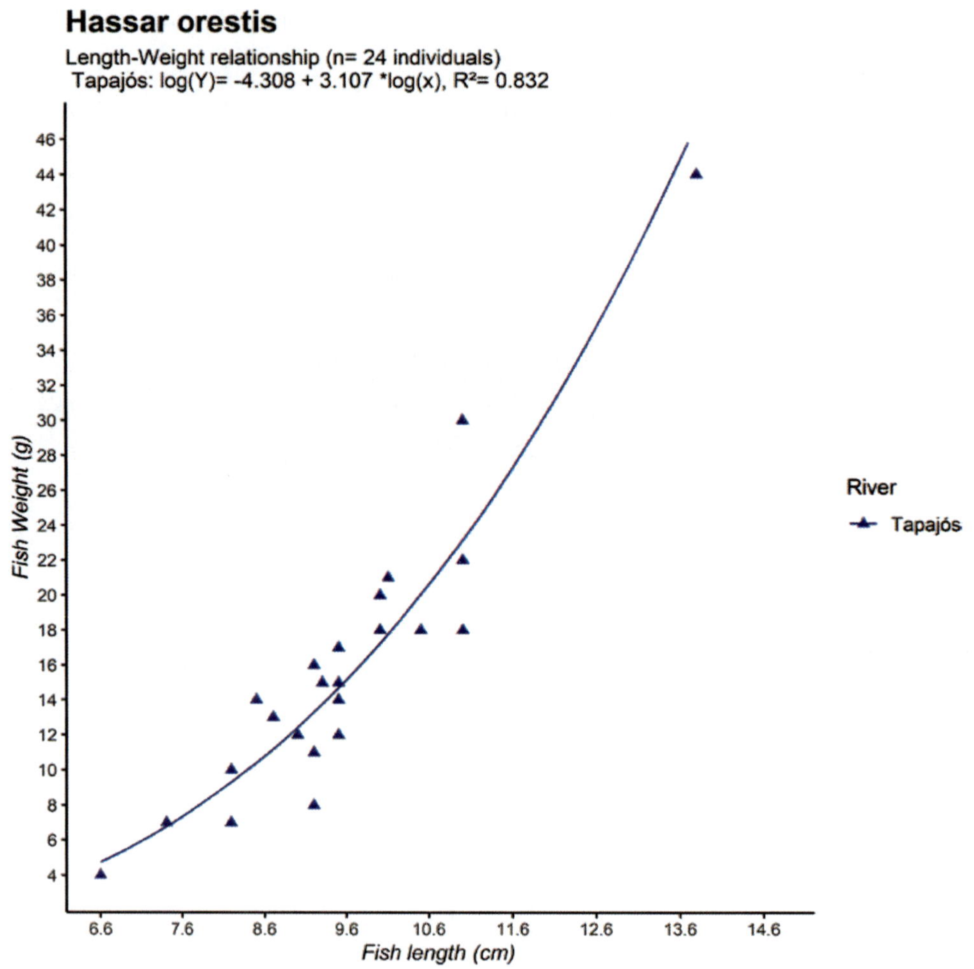

Fig. 5.58 Length-weight relationship for *Hassar orestis* sampled in Tapajós River

Fish species: *Megalodoras uranoscopus* (Eigenmann & Eigenmann 1888)

Common name(s): bacu, bacu-pedra

Fig. 5.59 *Megalodoras uranoscopus* sampled in the Negro River. (Photo: Renato Silvano)

Size of fish sampled (standard length, both rivers): min. = 27.5 cm, max. = 53 cm, mean and median = 40.3 cm

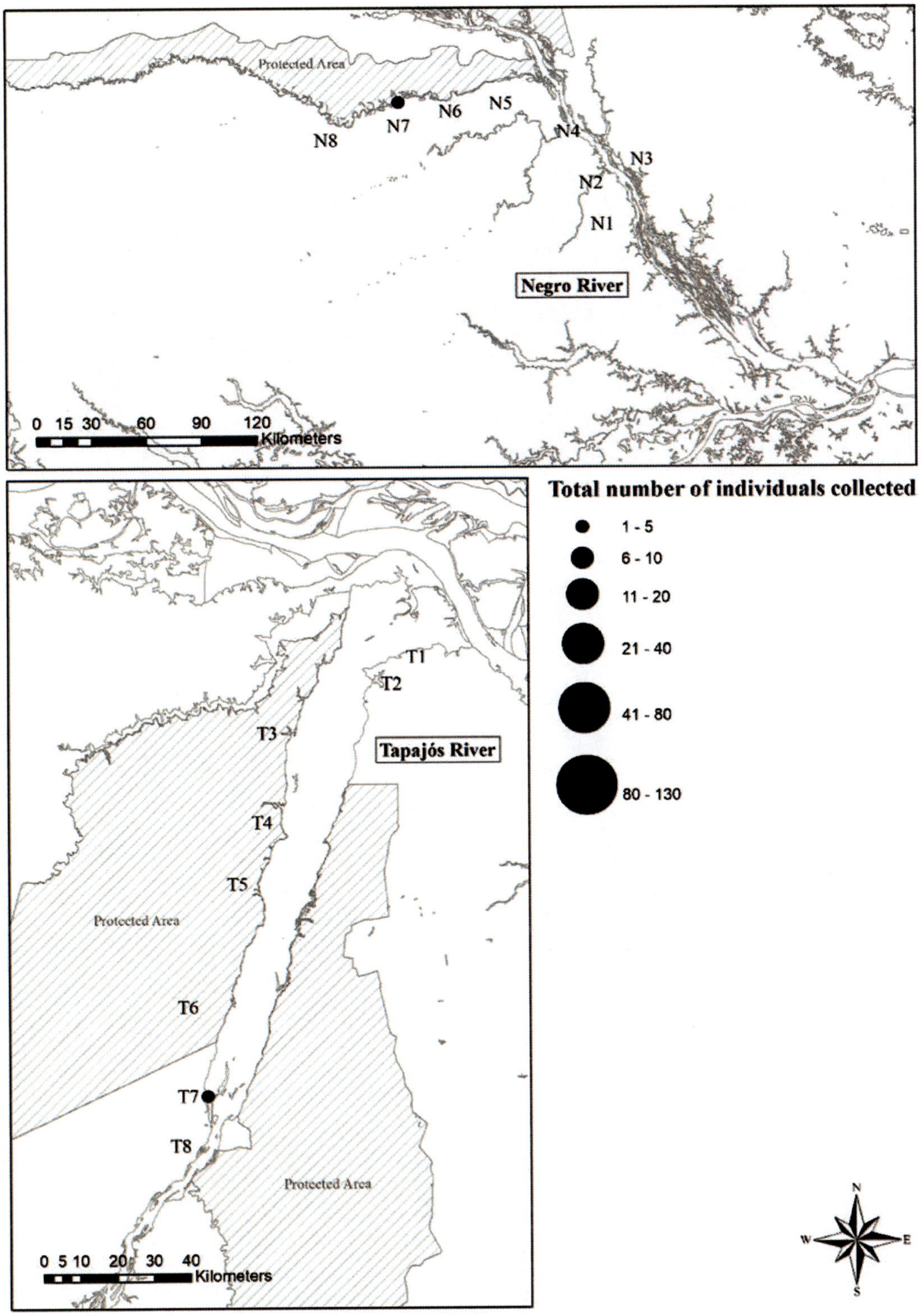

■ **Fig. 5.60** Distribution of individuals of *Megalodoras uranoscopus* sampled in the Negro and Tapajós rivers

Fisheries: The bacus (*Megalodoras uranoscopus* and other species from the family Doradidae) are only occasionally caught in fisheries. Biomass caught is in ◘ Table 3.3 (▶ Chap. 3).

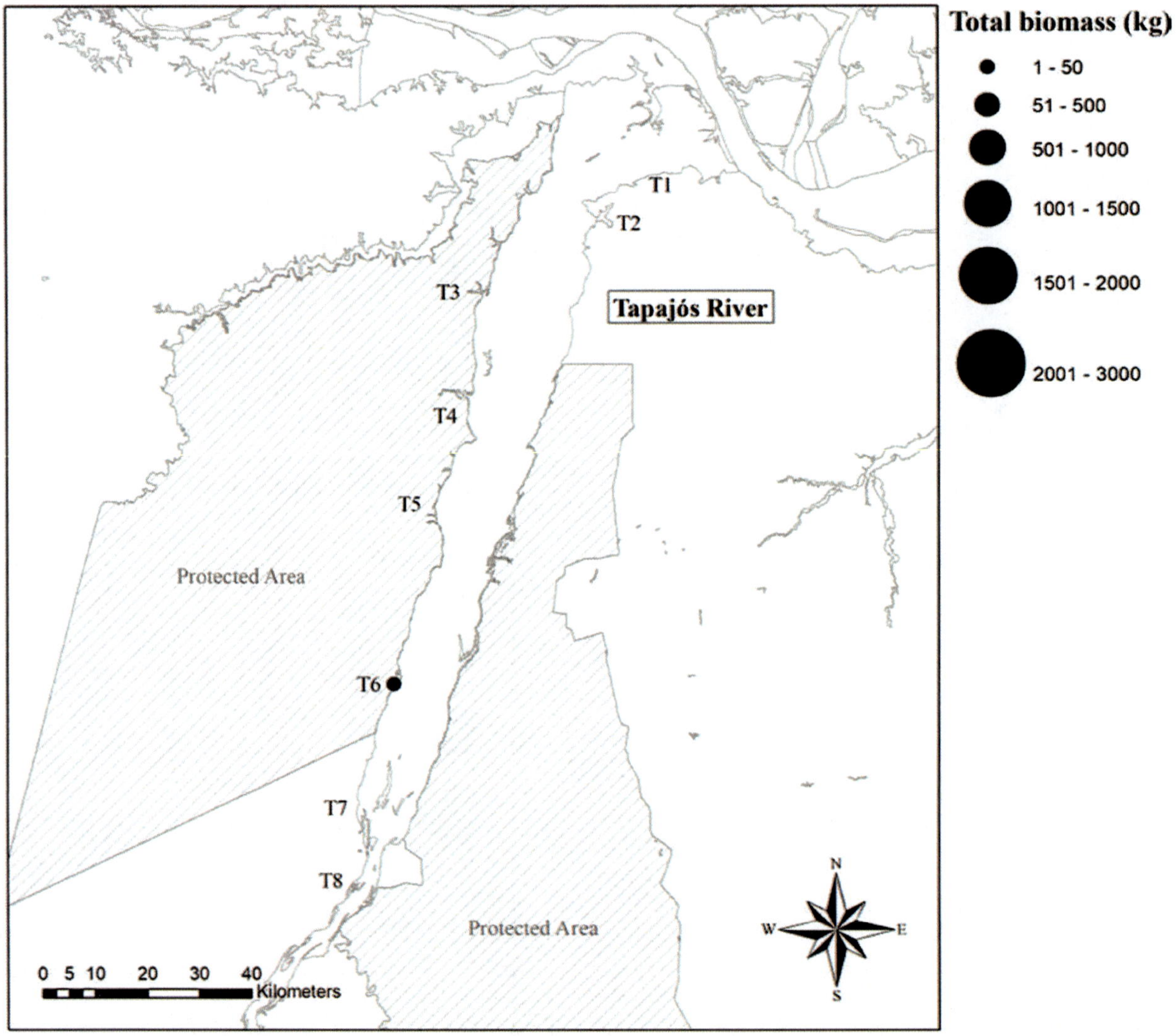

◘ **Fig. 5.61** Distribution of biomass of bacus (*Megalodoras uranoscopus* and other species from the family Doradidae) caught in small-scale fisheries in the Tapajós River

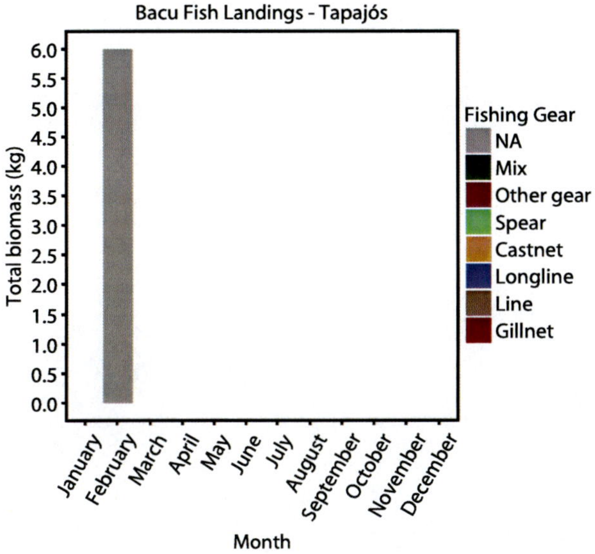

Fig. 5.62 Seasonal variation of biomass of bacus (*Megalodoras uranoscopus* and other species from the family Doradidae) caught per fishing gear in small-scale fisheries in eight communities in Tapajós River. NA: fishing gear not informed; Mix: more than one gear was used

Fish species: *Tenellus trimaculatus* (Boulenger 1898)

Common name(s): botinho, carataí

Fig. 5.63 *Tenellus trimaculatus* sampled in the Tapajós River. (Photo: Renato Silvano)

Fish species: ***Tenellus ternetzi*** **(Eigenmann 1925)**
Common name(s): not available

Observation: abundance and size data were grouped for these two species (*Tenellus trimaculatus* and *Tenellus ternetzi*). The species *Tenellus ternetzi* was formerly named as *Nemadoras ternetzi.* Both species were collected in the Negro River and only *Tenellus trimaculatus* was collected in the Tapajós River.

Size of fish sampled (standard length, both rivers): min. = 6 cm, max. = 15.5 cm, mean = 9.1 cm, median = 8.8 cm

Fisheries: not relevant to fisheries.

Fig. 5.64 *Tenellus ternetzi* sampled in the Negro River. (Photo: Renato Silvano)

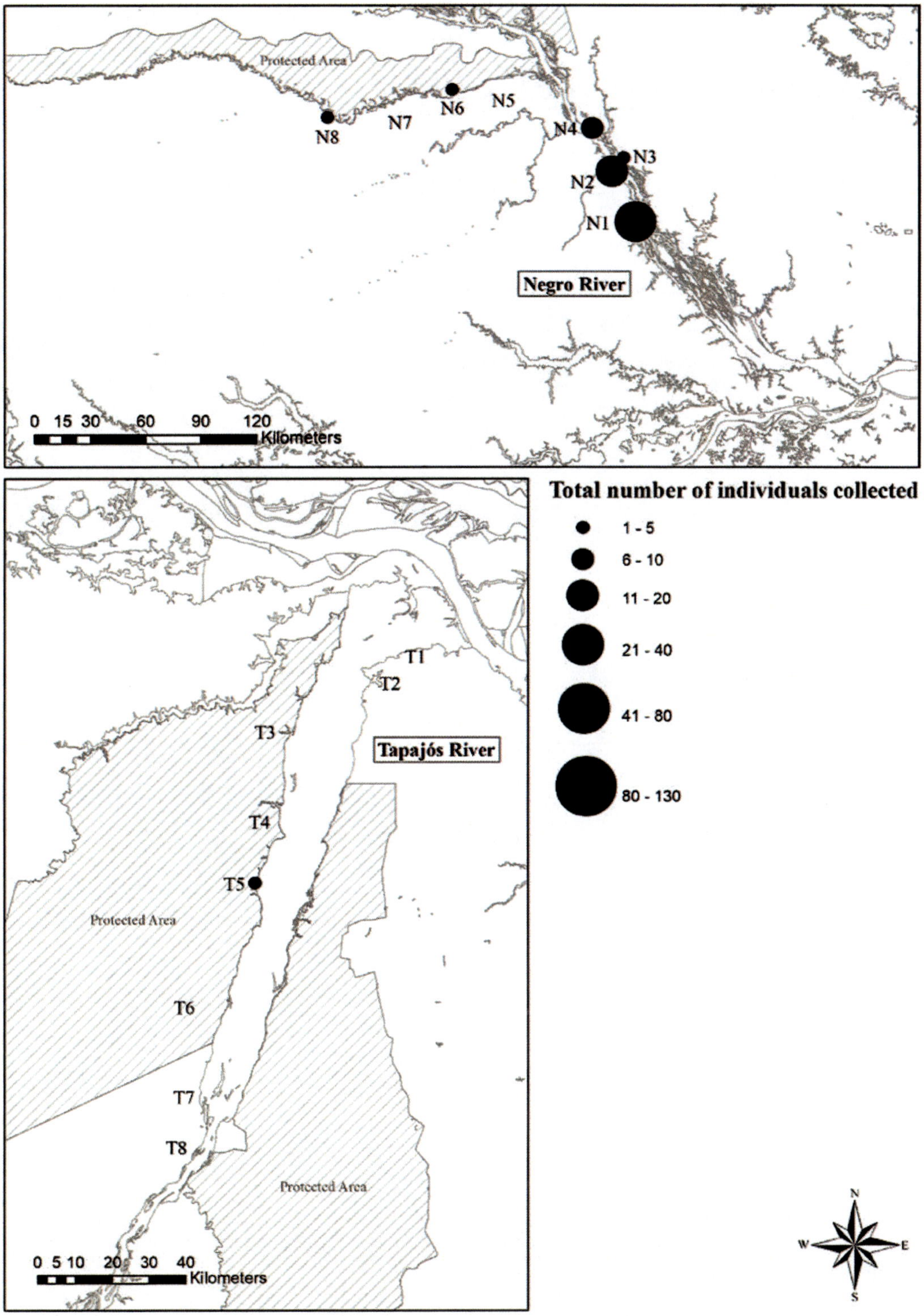

Fig. 5.65 Distribution of individuals of *Tenellus trimaculatus* and *Tenellus ternetzi* sampled in the Negro and Tapajós Rivers

Fish species: *Oxydoras niger* (Valenciennes 1821)

Common name(s): cuiu, cujuba

Size of fish sampled (standard length, both rivers): 17 cm

Fisheries: The cuiu or cujuba (*Oxydoras niger*) is occasionally caught in fisheries, and this fish has low commercial value. Biomass caught is in ◘ Table 3.3 (► Chap. 3).

◘ **Fig. 5.66** *Oxydoras niger* sampled in the Tapajós River. (Photo: Renato Silvano)

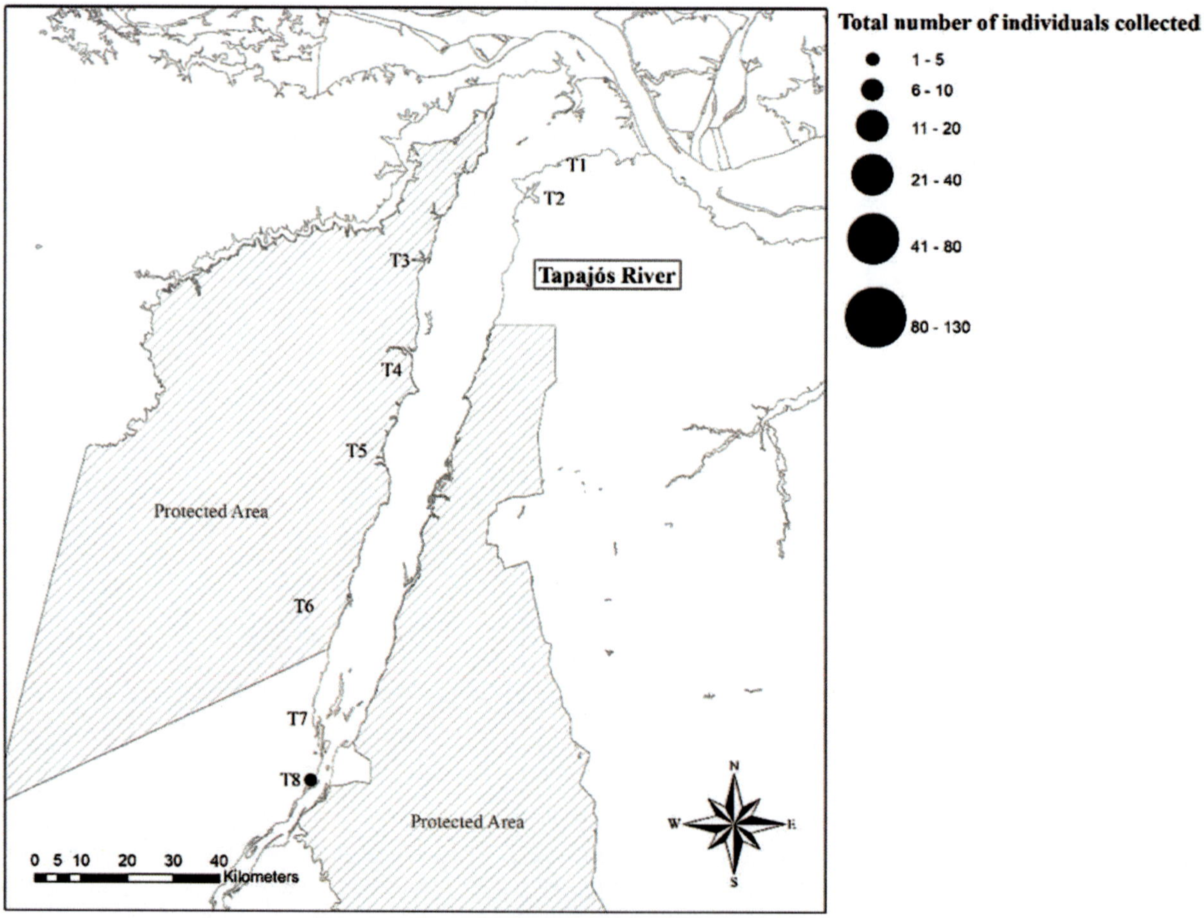

◘ **Fig. 5.67** Distribution of individuals of *Oxydoras niger* sampled in the Tapajós River

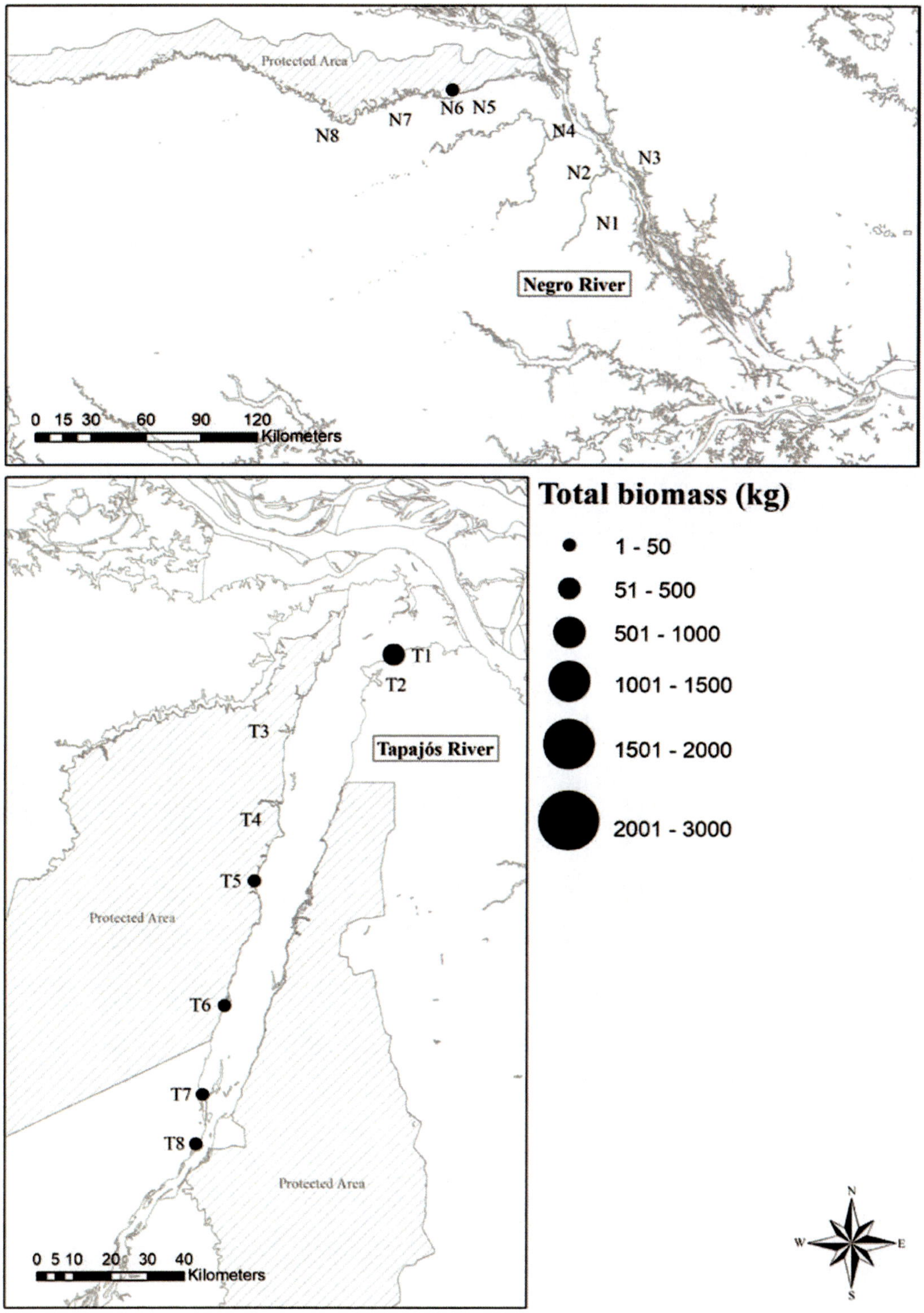

Fig. 5.68 Distribution of biomass of cuiu or cujuba (*Oxydoras niger*) caught in small-scale fisheries in the Negro and Tapajós Rivers

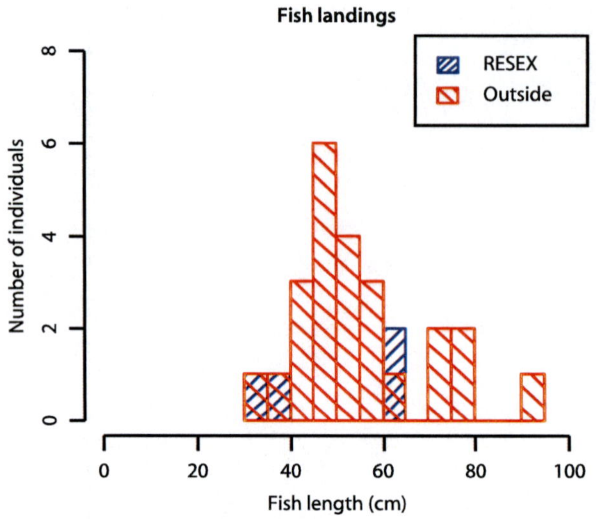

Fig. 5.69 Size distribution of cuiu or cujuba (*Oxydoras niger*) caught in small-scale fisheries in eight communities inside and outside the Extractive Reserves (RESEX) in the Tapajós River, N Resex = 4, N outside = 24

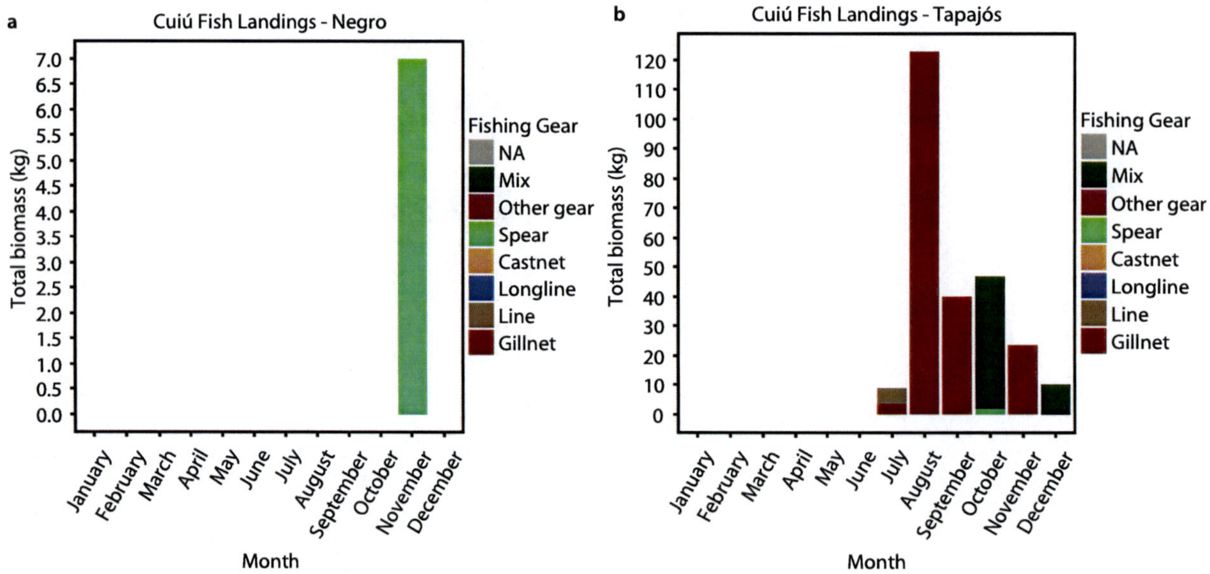

Fig. 5.70 Seasonal variation of biomass of cuiu or cujuba (*Oxydoras niger*) caught per fishing gear in small-scale fisheries in16 communities in the rivers **a** Negro and **b** Tapajós. NA: fishing gear not informed; Mix: more than one gear was used

Fish species: ***Platydoras armatulus*** **(Valenciennes 1840)**

Common name(s): bacu, xaroque

Size of fish sampled (standard length, both rivers): min. = 13.5 cm, max. = 14 cm, mean and median = 13.8 cm

Fisheries: not relevant to fisheries.

Fig. 5.71 *Platydoras armatulus* sampled in the Tapajós River. (Photo: Renato Silvano)

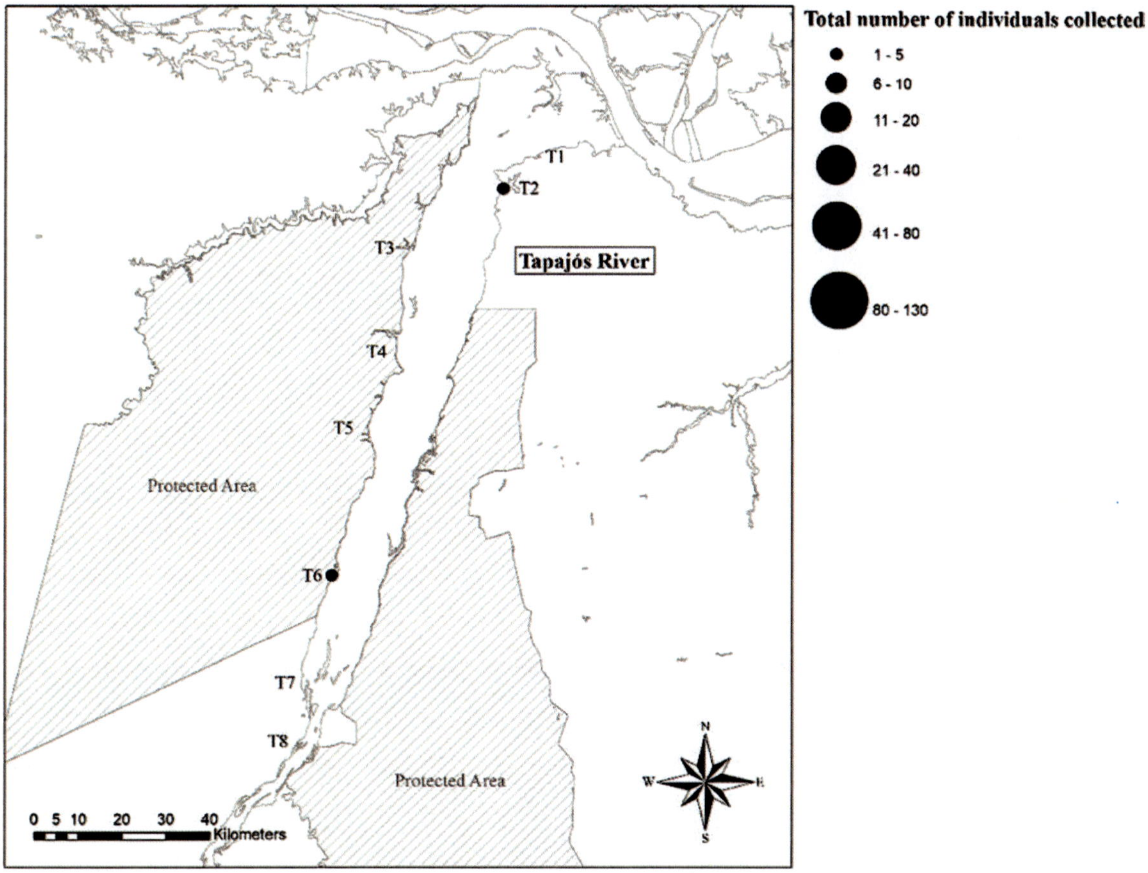

Fig. 5.72 Distribution of individuals of *Platydoras armatulus* sampled in the Tapajós River

Fish species: ***Platydoras hancockii*** **(Valenciennes 1840)**

Common name(s): reco-reco, rebeca

Size of fish sampled (standard length, both rivers): 16.5 cm

Fisheries: not relevant to fisheries.

Fig. 5.73 *Platydoras hancockii* sampled in the Negro River. (Photo: Renato Silvano)

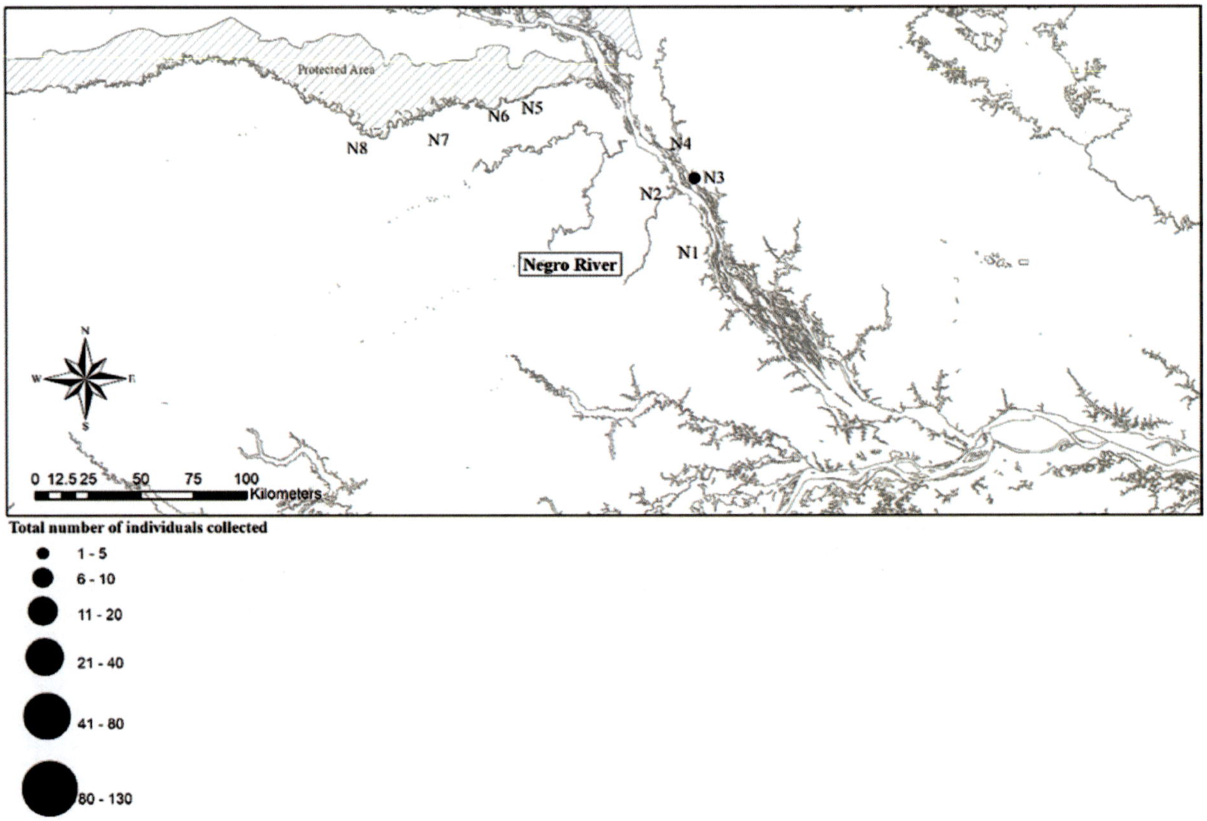

Fig. 5.74 Distribution of individuals of *Platydoras hancockii* sampled in the Negro River

Fish species: ***Trachydoras brevis*** **(Kner 1853)**

Common name(s): not available

Size of fish sampled (standard length, both rivers): 6.4 cm

Fisheries: not relevant to fisheries.

Fig. 5.75 *Trachydoras brevis* sampled in the Negro River. (Photo: Renato Silvano)

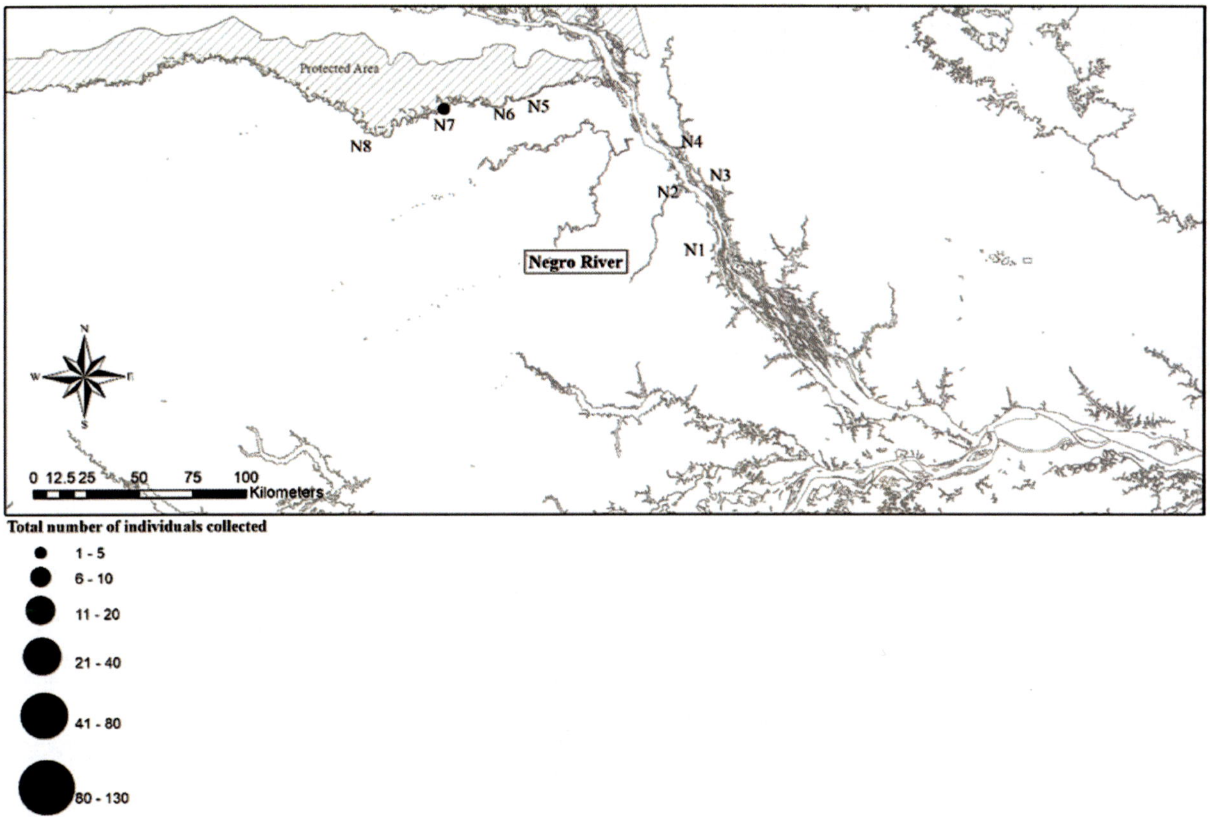

Fig. 5.76 Distribution of individuals of *Trachydoras brevis* sampled in the Negro River

5.2.5 Family Loricariidae

5.2.5.1 Summary of Biology and Ecology (Literature)

Diversity: large family of fishes with more than 988 species described, divided into 82 genera and six subfamilies.

Habitat: fishes of this family can be found throughout South America. Some species were introduced in other regions, including Central and North America.

5

Ecology and feeding: loricariids feed mainly on detritus, algae, insect, larvae, nymphs, and other microorganisms associated with the benthic zone. They inhabit the bottom of lakes, streams, and rivers, where they stay immobile and hiding under logs, rocks, and other submerge structures for most of the day, being more active during the night.

Morphology: body size varies from 2 cm to up to 90 cm in some larger species. Fishes of this family have a robust body, usually compressed dorsoventrally and covered with bone plates forming three to five series on the main body, each possessing odontodes (small bony structures that make the body harsh to the touch); abdominal region is flat; mouth is inferior with papillary lips forming a sucking disk; one or two rows of slender teeth, poorly implanted and usually with two asymmetrical cuspids or forming a scoop shape that is used to scrape the bottom, are present; a pair of short bills in the mouth corners is present; adipose fin, when present, is with a short stiff ray in the leading edge; caudal peduncle is usually long and compressed; pectoral and dorsal fin possess spines; the intestines are meandered and long to absorb nutrients from hard to digest items (plants and detritus).

Reproduction: overall, the species of this family have low fecundity and relatively large oocytes.

Economic importance: in the Amazon basin, species with larger body sizes, such as those from genera *Pterigoplichthys*, *Hypostomus*, and *Panaque*, are commonly exploited in commercial and artisanal fisheries. Many small species of this family are popular among aquarists, some being heavily exploited for aquarium trade.

References: Eschmeyer (1998), Ferreira et al. (1998), Schaefer (2003), Santos et al. (2006), Py-Daniel et al. (2007), Queiroz et al. (2013a, b), Doria and Lima (2015), Chamon and Silva (2018).

Fish species: *Ancistrus* **sp.**

Common name(s): acari-da-pedra, bodó, bodó-barbudo

Size of fish sampled (standard length, both rivers): min. = 7.2 cm, max. = 9.5 cm, mean = 8.1 cm, median = 7.9 cm

Fisheries: not relevant to fisheries, due to its small size, may be exploited as ornamental fish (aquarium trade).

Fig. 5.77 *Ancistrus* sp. sampled in the Negro River. (Photo: Renato Silvano)

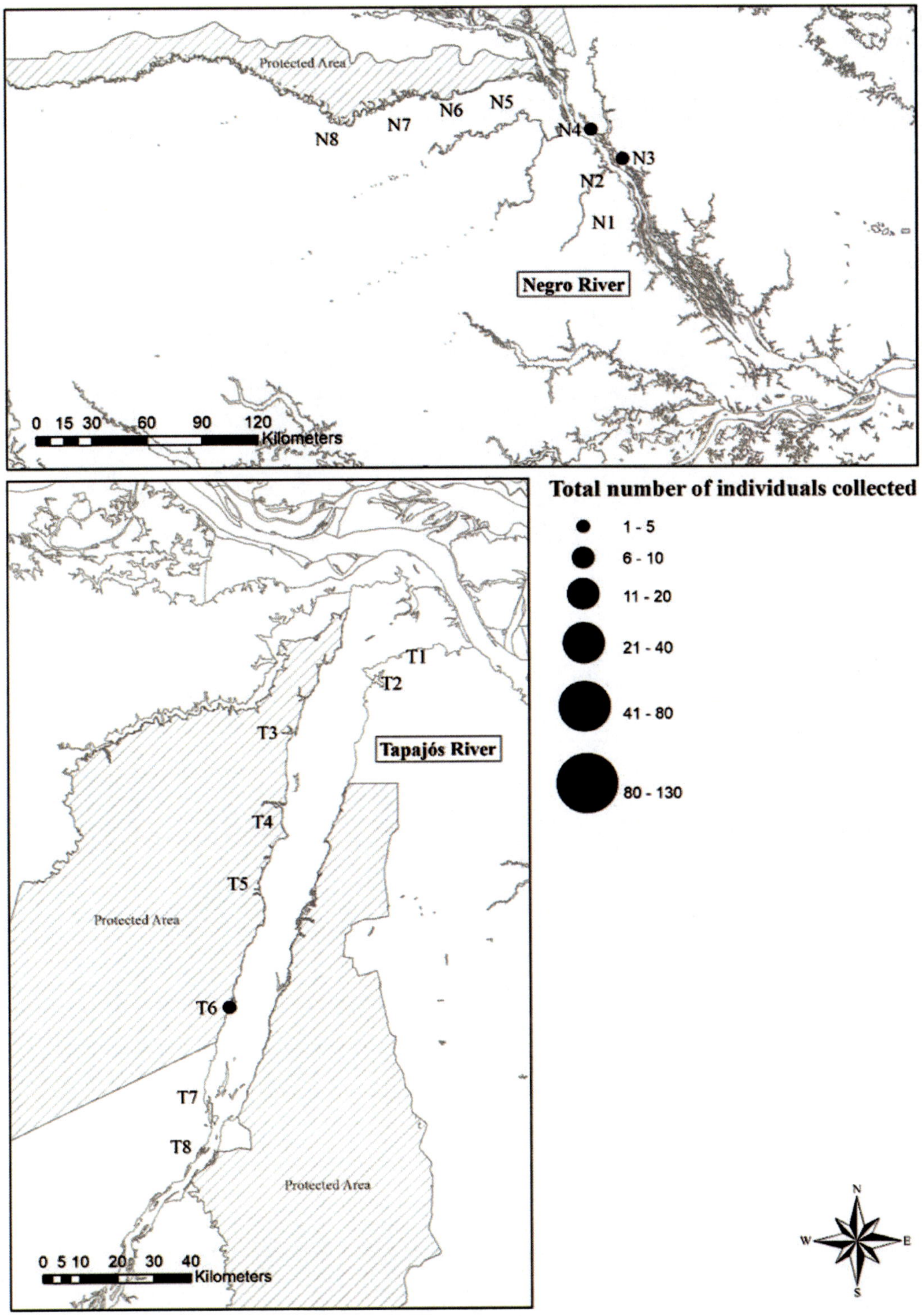

Fig. 5.78 Distribution of individuals of *Ancistrus* sp. sampled in the Negro and Tapajós rivers

Fish species: *Aphanotorulus emarginatus* (Valenciennes 1840)

Common name(s): acari, acari-pintado

Size of fish sampled (standard length, both rivers): min. = 16.4 cm, max. = 20.8 cm, mean = 18.2 cm, median = 18 cm

Fisheries: acaris (Loricariidae) are occasionally exploited by fisheries in both rivers; see data on acari fisheries in *Pterygoplichthys gibbiceps*.

Fig. 5.79 *Aphanotorulus emarginatus* sampled in the Tapajós River. (Photo: Renato Silvano)

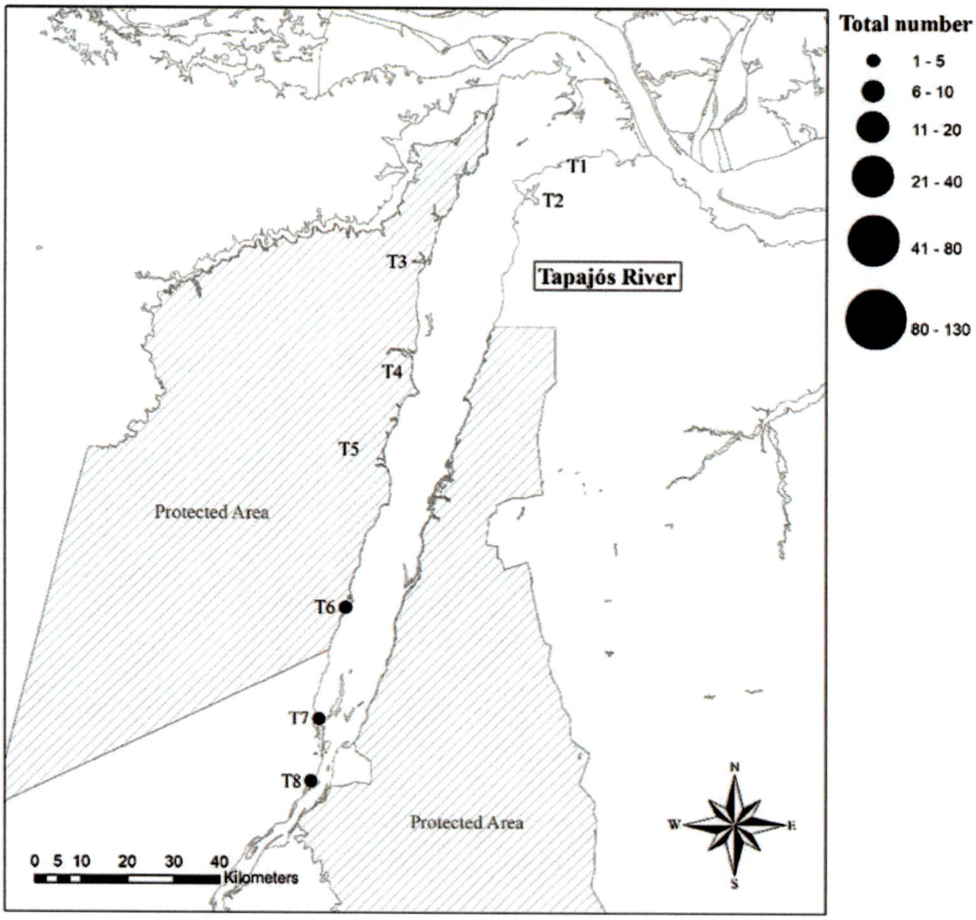

Fig. 5.80 Distribution of individuals of *Aphanotorulus emarginatus* sampled in the Tapajós River

Fish species: *Dekeyseria amazonica* Rapp Py-Daniel 1985

Common name(s): bodó

Size of fish sampled (standard length, both rivers): 15 cm

Fisheries: not relevant to fisheries, due to its small size.

Fig. 5.81 *Dekeyseria amazonica* sampled in the Negro River: **a** lateral view and **b** ventral view. (Photo: Renato Silvano)

5

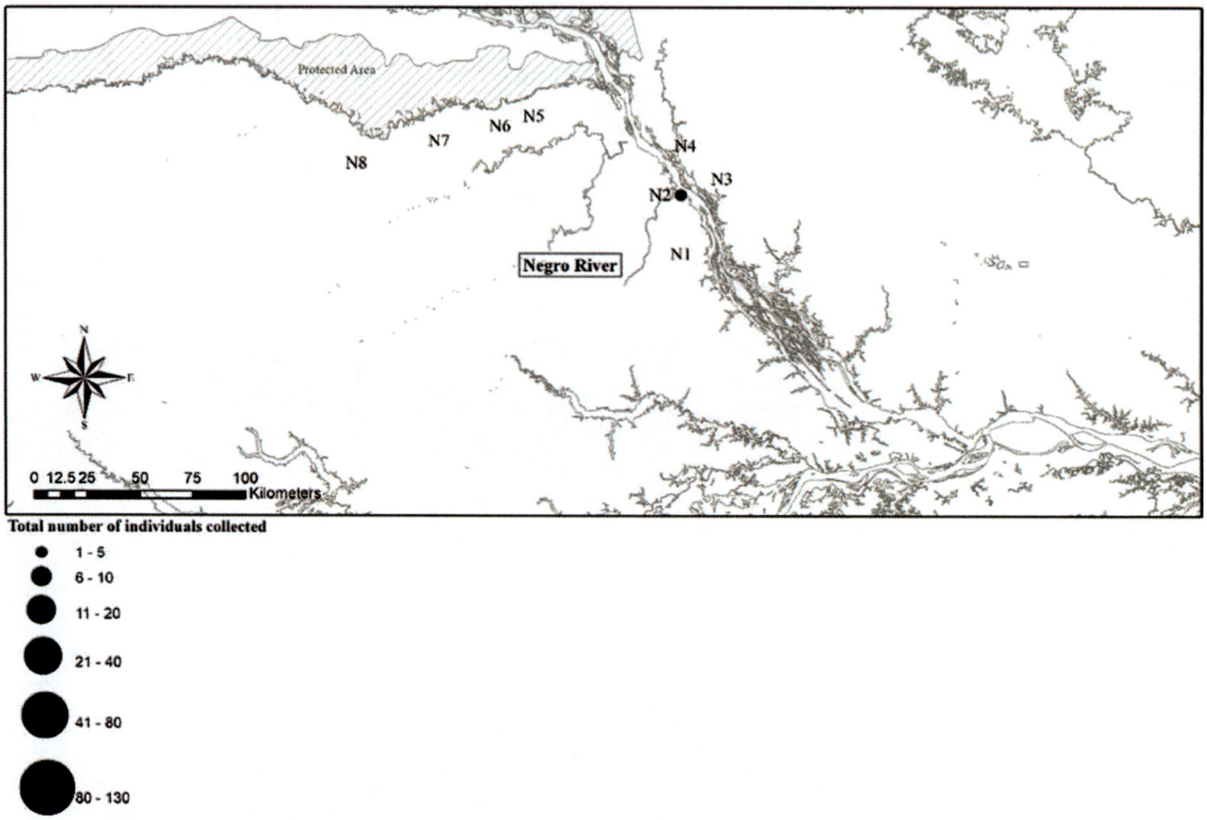

Fig. 5.82 Distribution of individuals of *Dekeyseria amazonica* sampled in the Negro River

Fish species: *Hemiodontichthys acipenserinus* **(Kner 1853)**

Common name(s): acari-cachimbo, acari-bicudo

Size of fish sampled (standard length, both rivers): min. = 7.8 cm, max. = 14.2 cm, mean = 11.7 cm, median = 11.5 cm

Fisheries: not relevant to fisheries, due to its small size.

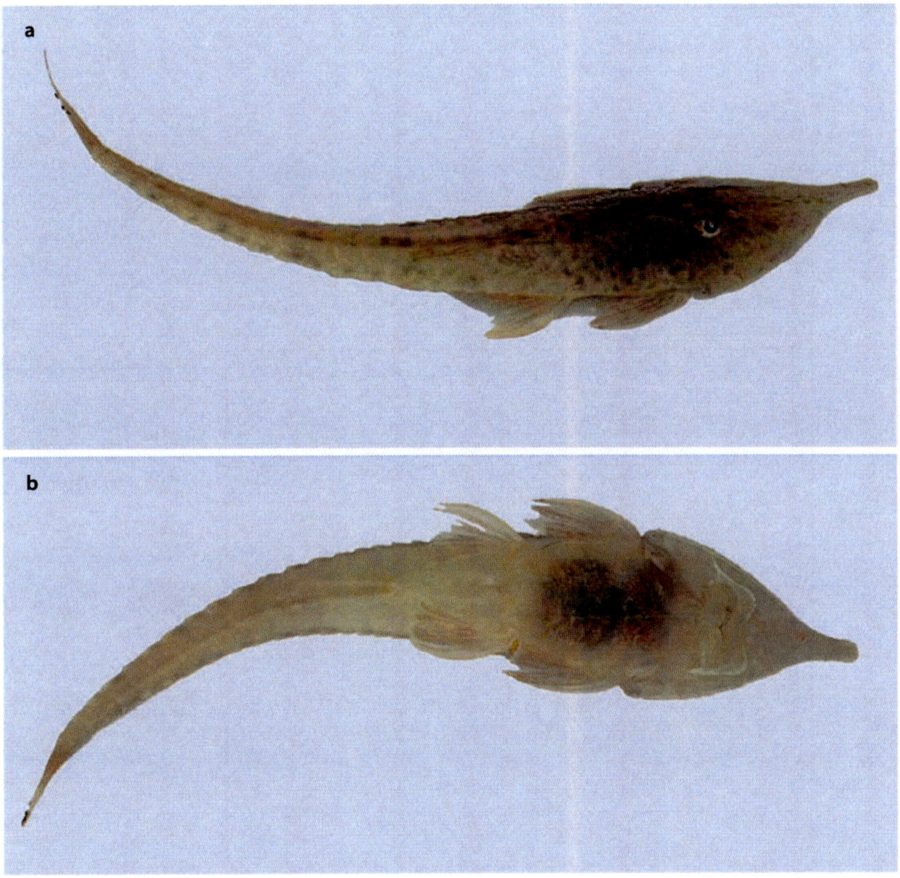

Fig. 5.83 *Hemiodontichthys acipenserinus* sampled in the Negro River: **a** lateral view and **b** ventral view. (Photos: Renato Silvano)

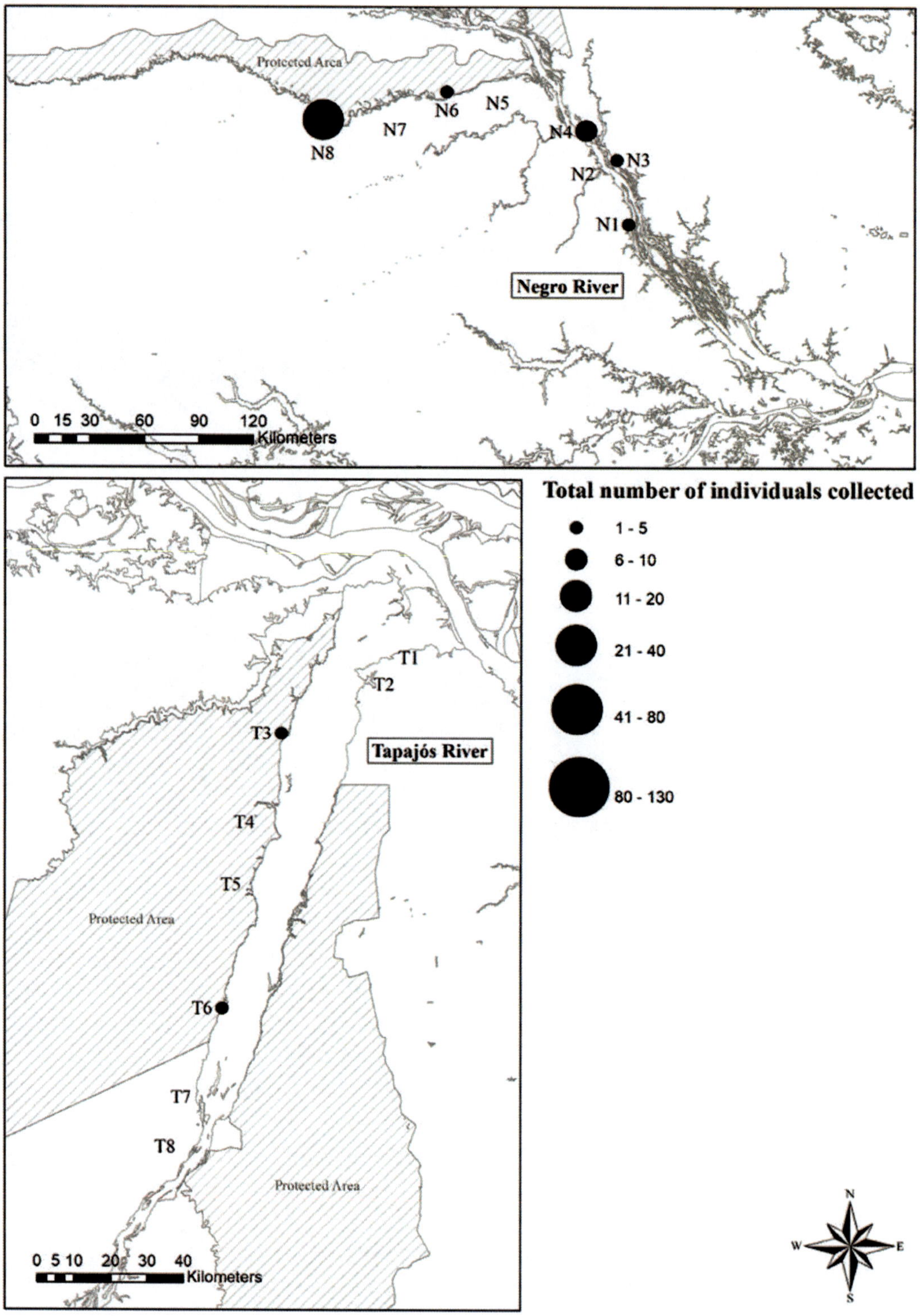

Fig. 5.84 Distribution of individuals of *Hemiodontichthys acipenserinus* sampled in the Tapajós and Negro rivers

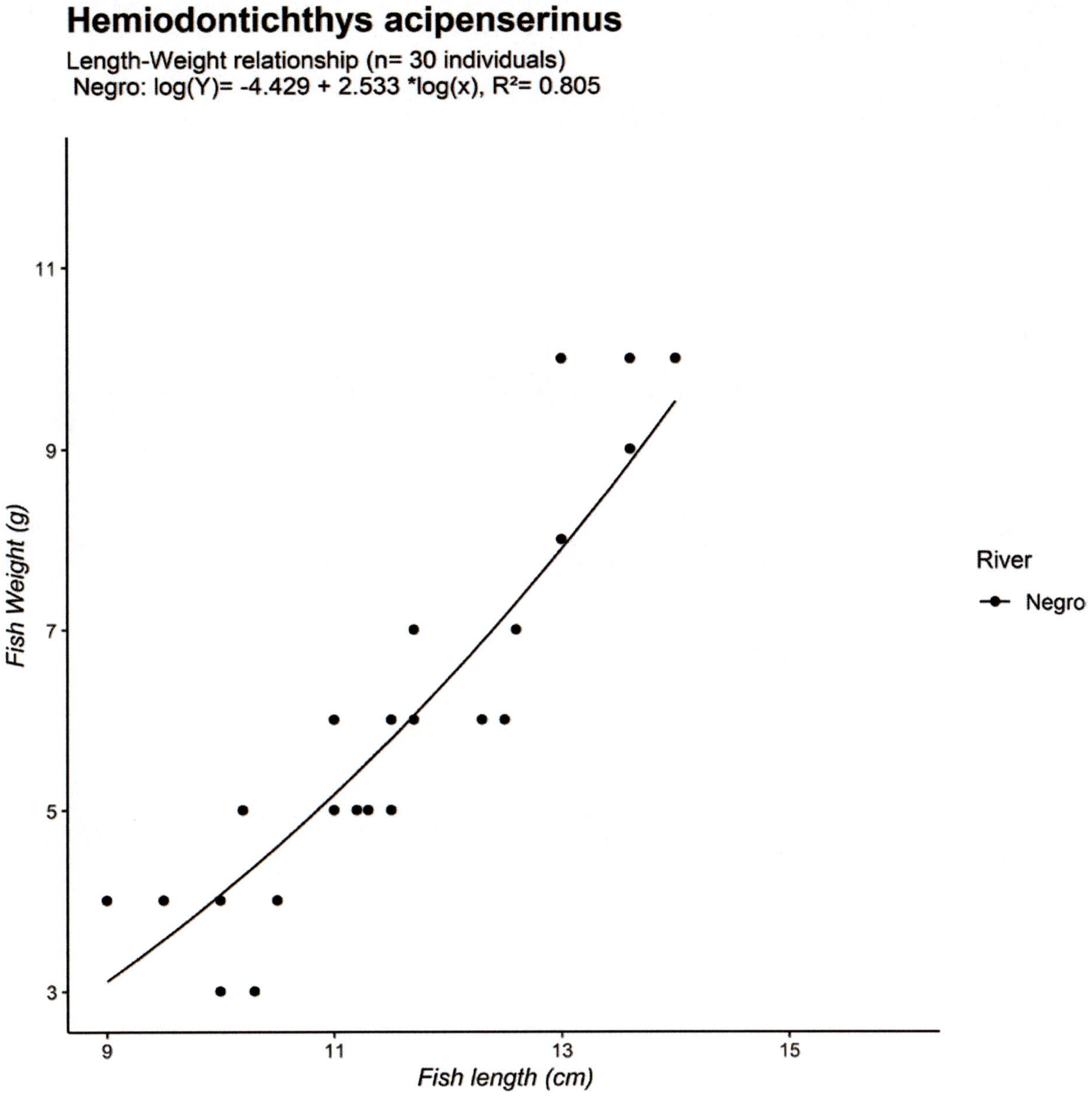

Fig. 5.85 Length-weight relationship for *Hemiodontichthys acipenserinus* sampled in the Negro River

Fish species: *Hypostomus weberi* Carvalho, Lima & Zawadzki 2010

Common name(s): bodó

Size of fish sampled (standard length, both rivers): mean and median = 17.5 cm

Fisheries: acaris (Loricariidae) are occasionally exploited by fisheries in both rivers; see data on acari fisheries in *Pterygoplichthys gibbiceps*.

Fig. 5.86 *Hypostomus weberi* sampled in the Negro River: **a** lateral view and **b** ventral view. (Photo: Renato Silvano)

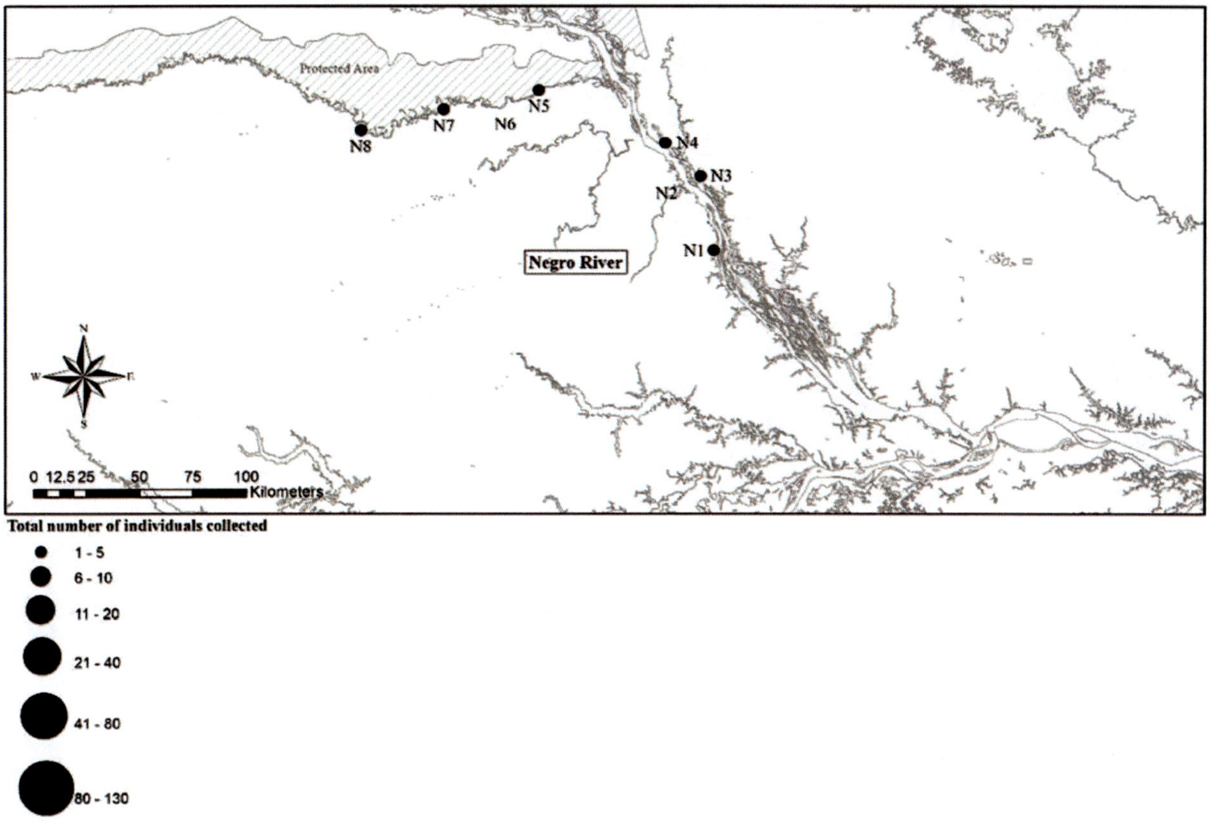

Fig. 5.87 Distribution of individuals of *Hypostomus weberi* sampled in the Negro River

5

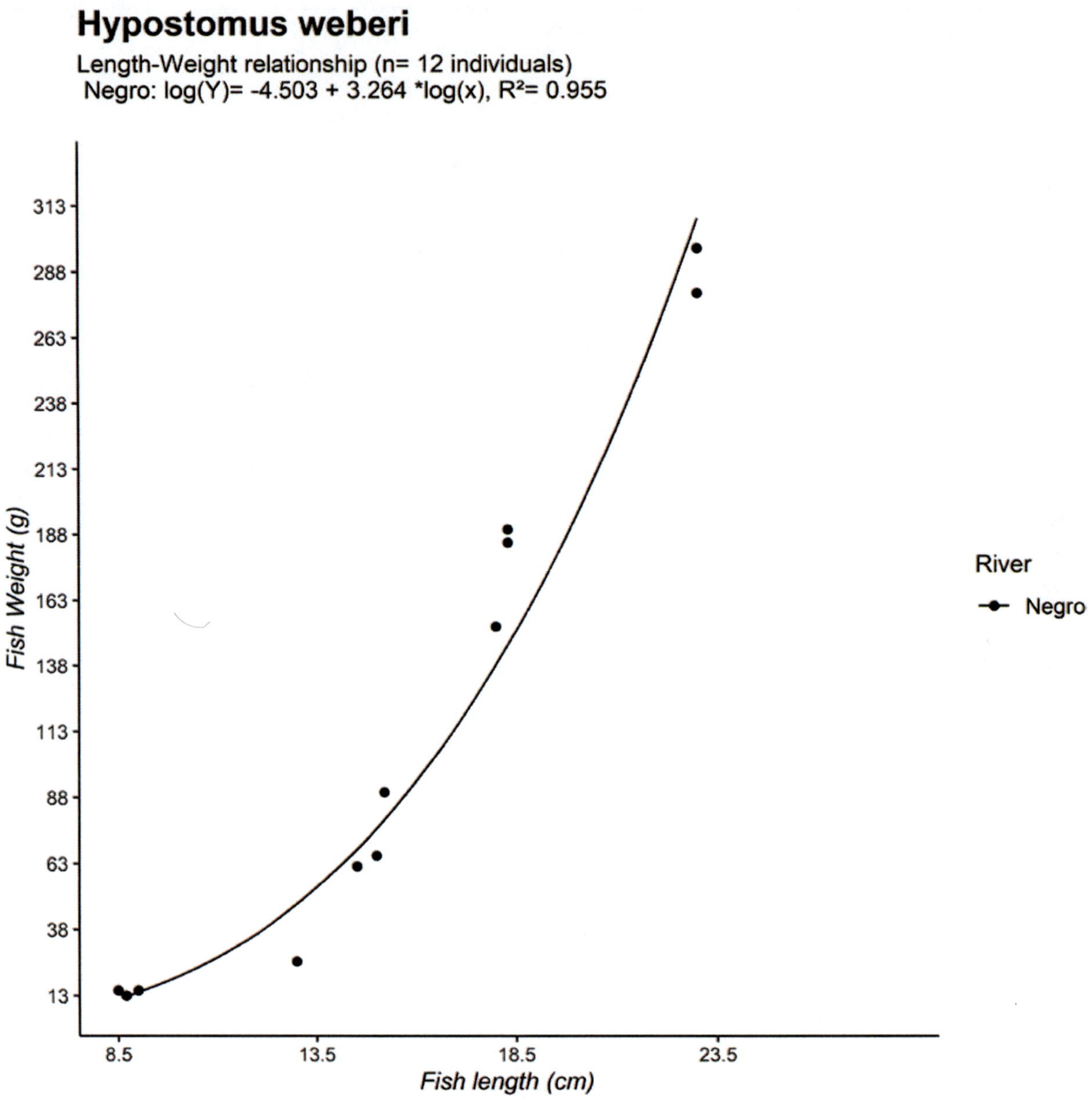

Fig. 5.88 Length-weight relationship for *Hypostomus weberi* sampled in Negro River

Fish species: *Limatulichthys griseus* (Eigenmann 1909)

Common name(s): acari-cachimbo

Size of fish sampled (standard length, both rivers): mean = 16.8 cm, median = 15.5 cm

Fisheries: not relevant to fisheries, due to its small size.

Fig. 5.89 *Limatulichthys griseus* sampled in the Negro River: **a** lateral view and **b** ventral view. (Photo: Renato Silvano)

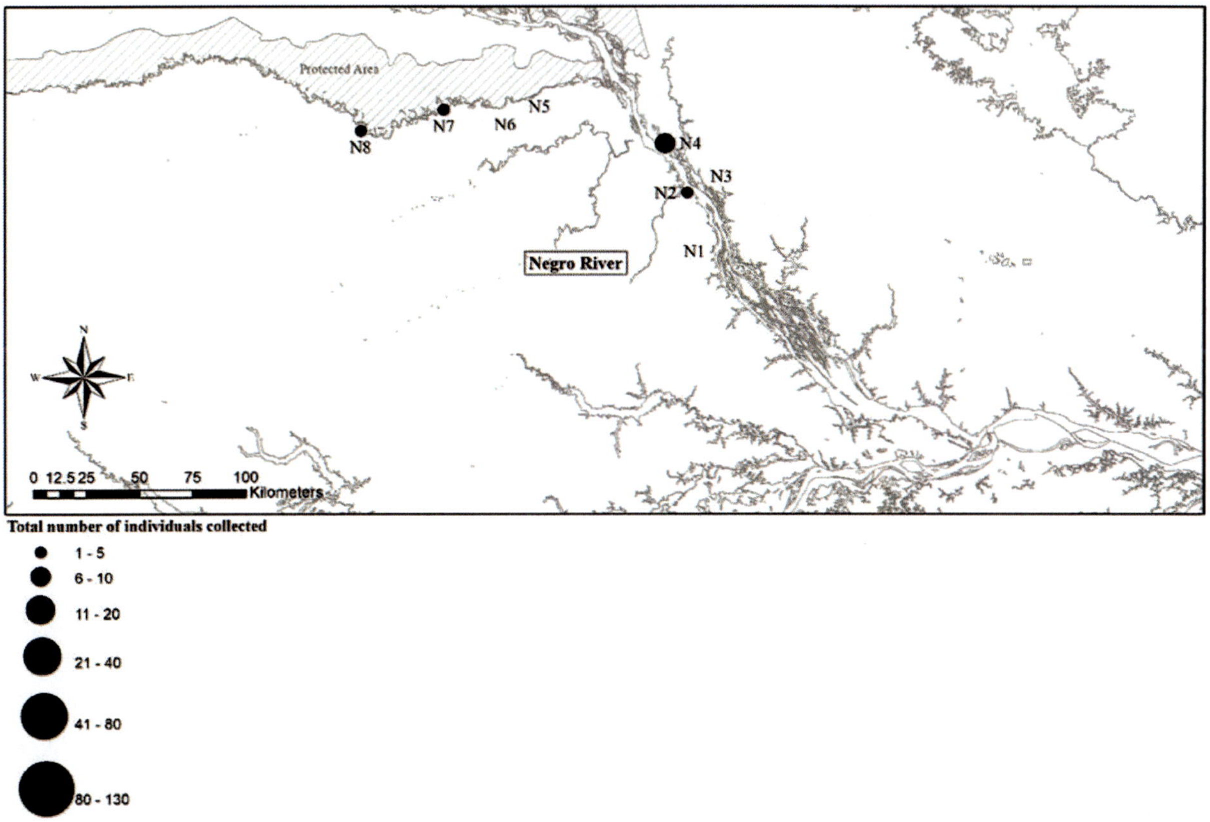

Fig. 5.90 Distribution of individuals of *Limatulichthys griseus* sampled in the Negro River

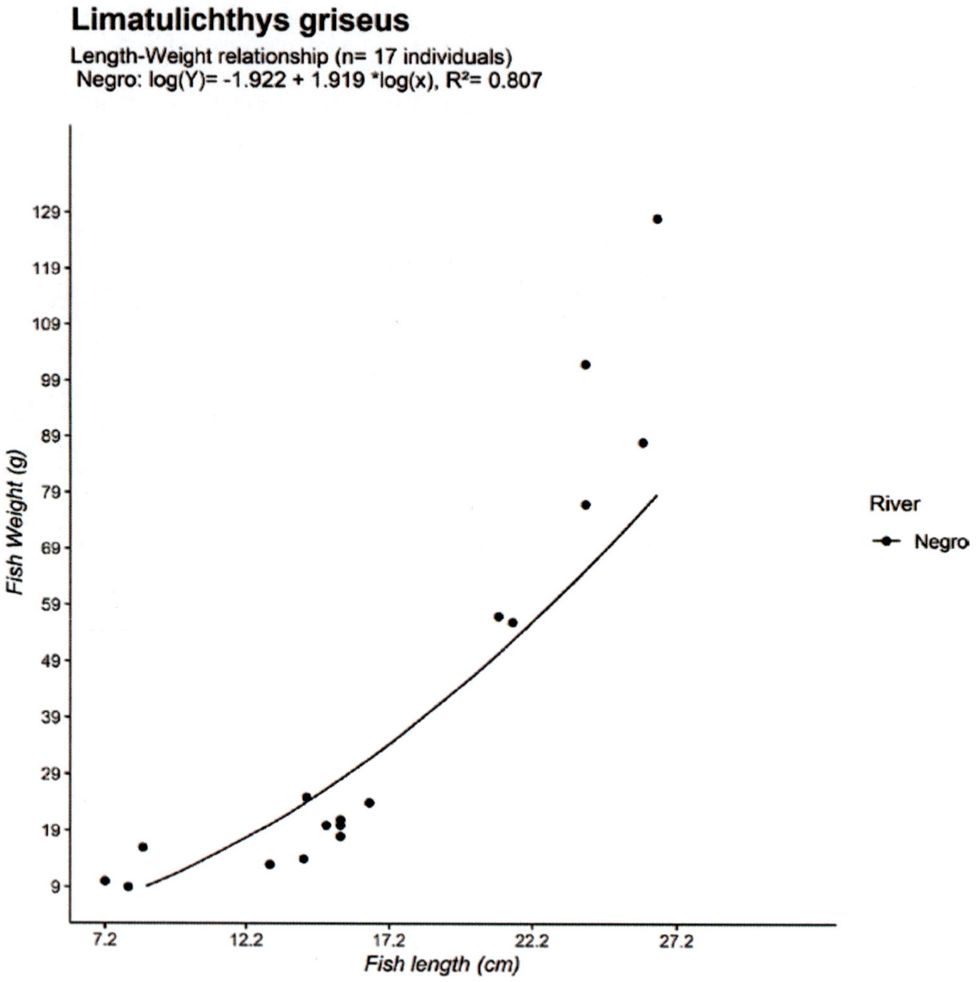

Fig. 5.91 Length-weight relationship for *Limatulichthys griseus* sampled in Negro River

Fish species: *Loricaria cataphracta* Linnaeus 1758

Common name(s): acari-cachimbo

Size of fish sampled (standard length, both rivers): mean = 17 cm, median = 17.9 cm

Fisheries: not relevant to fisheries, due to its small size.

Fig. 5.92 *Loricaria cataphracta* sampled in the Tapajós River: **a** lateral view and **b** ventral view. (Photo: Renato Silvano)

5

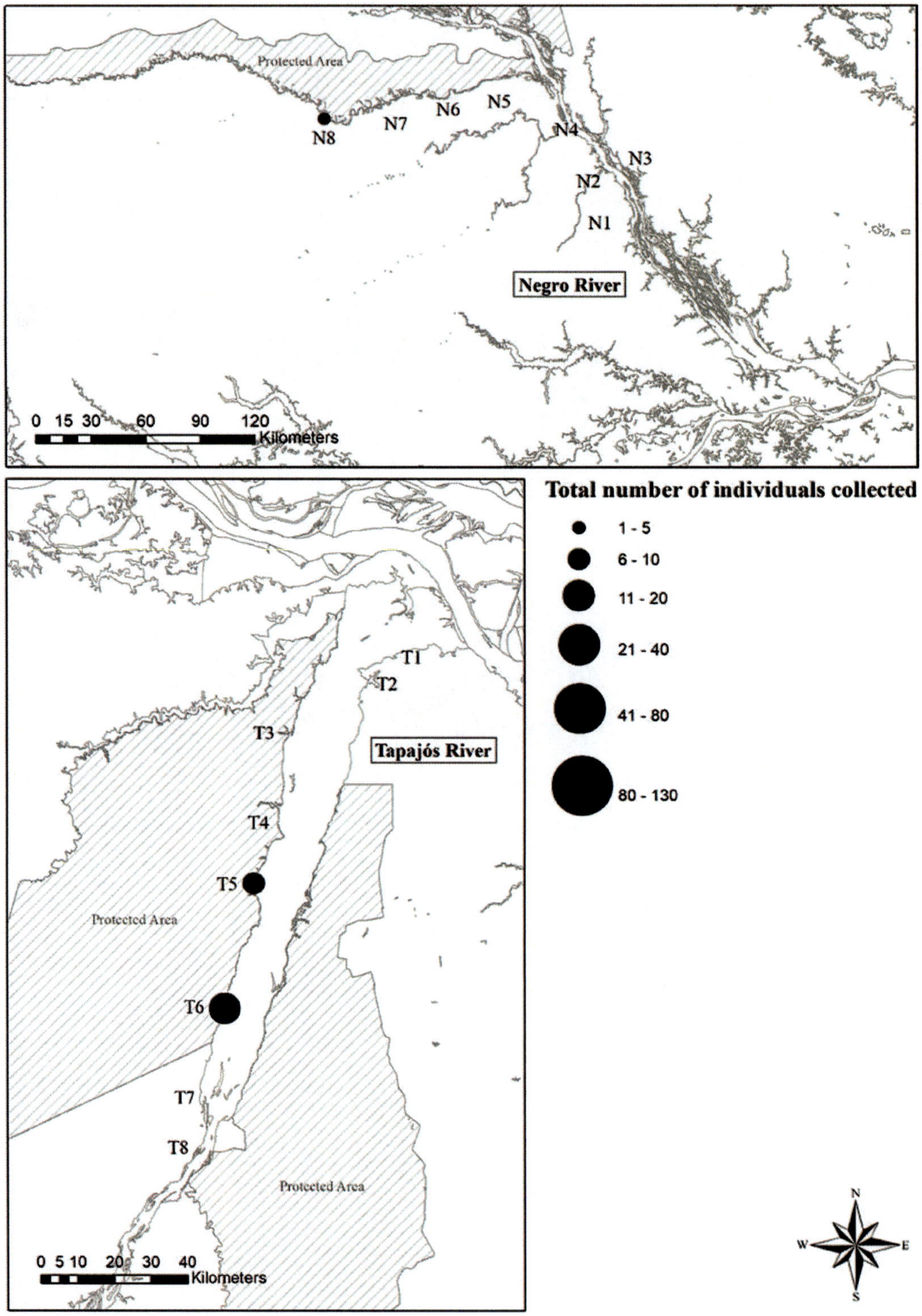

Fig. 5.93 Distribution of individuals of *Loricaria cataphracta* sampled in the Negro and Tapajós rivers

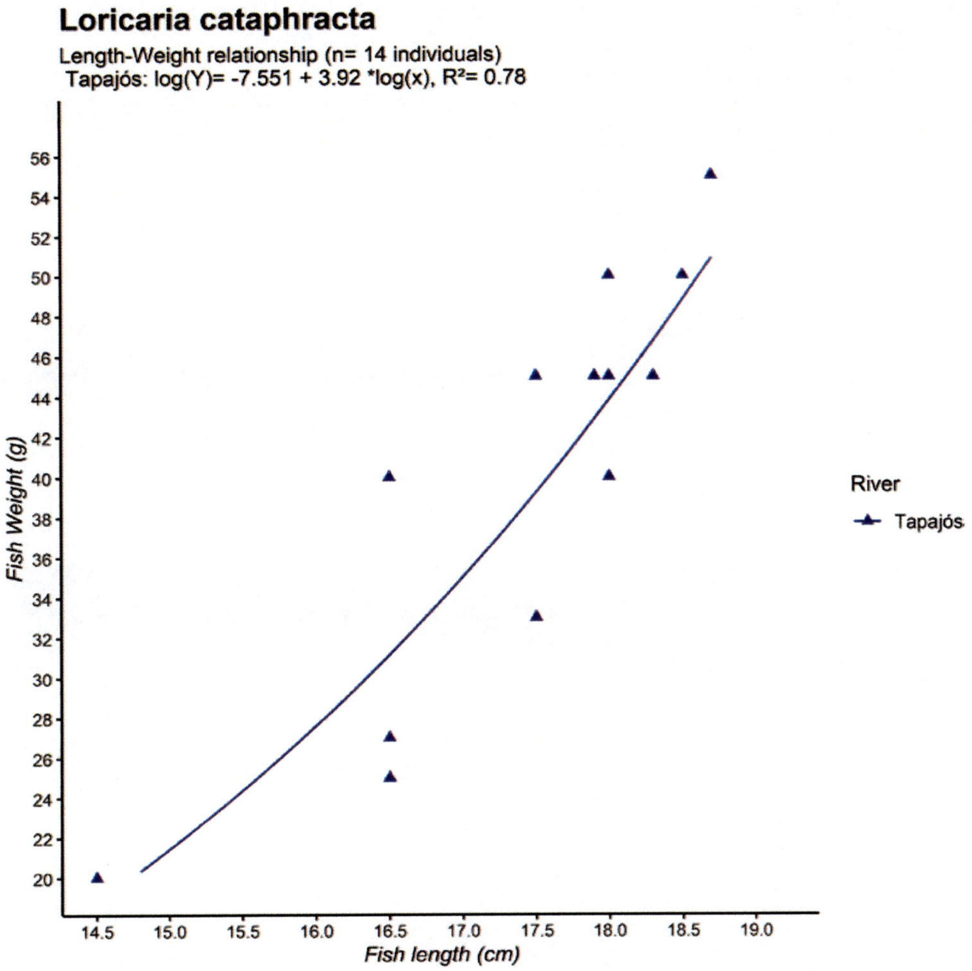

Fig. 5.94 Length-weight relationship for *Loricaria cataphracta* sampled in the Tapajós River

Fish species: ***Loricariichthys acutus*** **(Valenciennes 1840)**

Common name(s): acari-cachimbo

Size of fish sampled (standard length, both rivers): mean and median = 19 cm

Fisheries: not relevant to fisheries, due to its small size.

Fig. 5.95 *Loricariichthys acutus* sampled in the Tapajós River: **a** lateral view and **b** ventral view. (Photo: Renato Silvano)

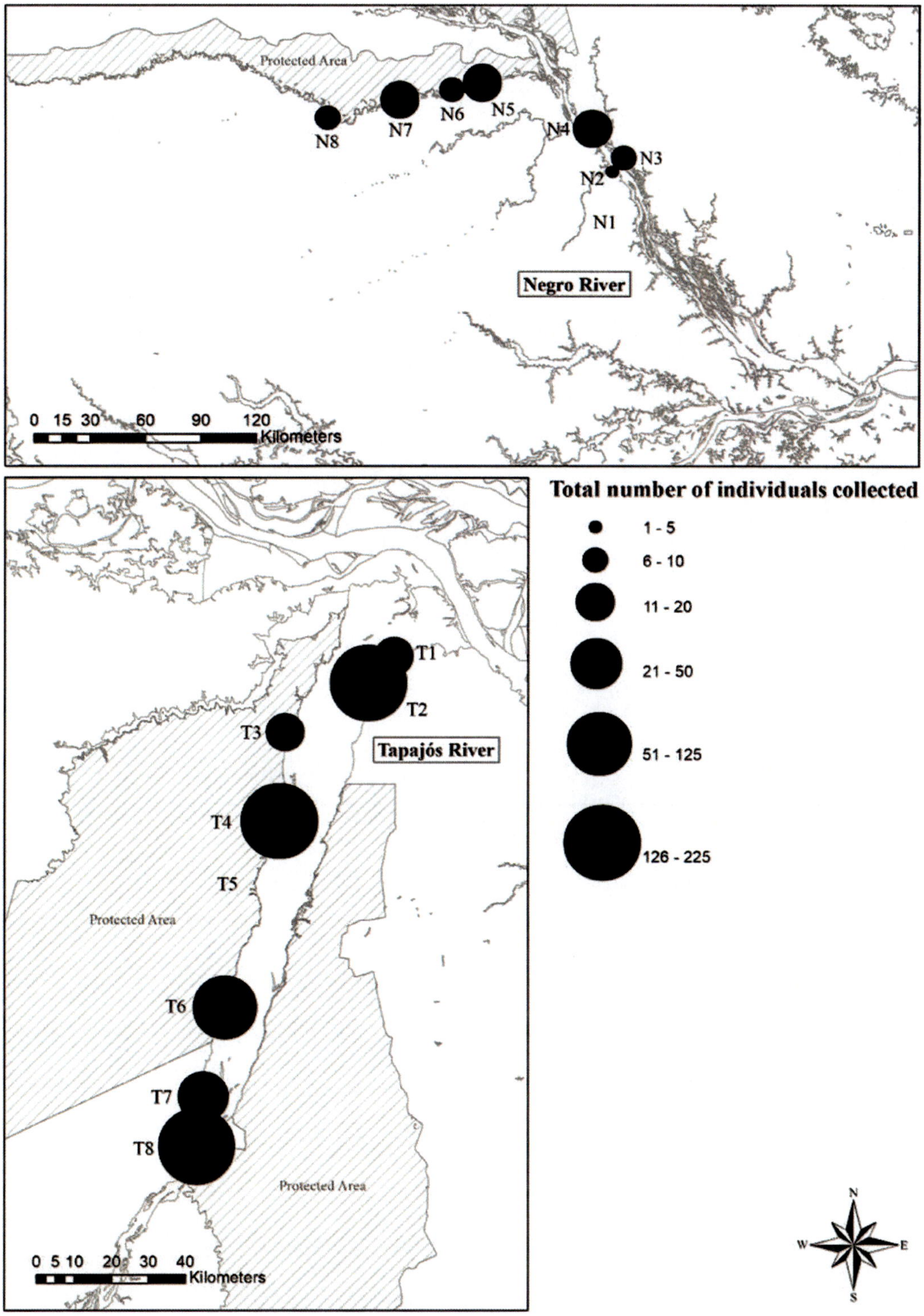

Fig. 5.96 Distribution of individuals of *Loricariichthys acutus* sampled in the Negro and Tapajós rivers

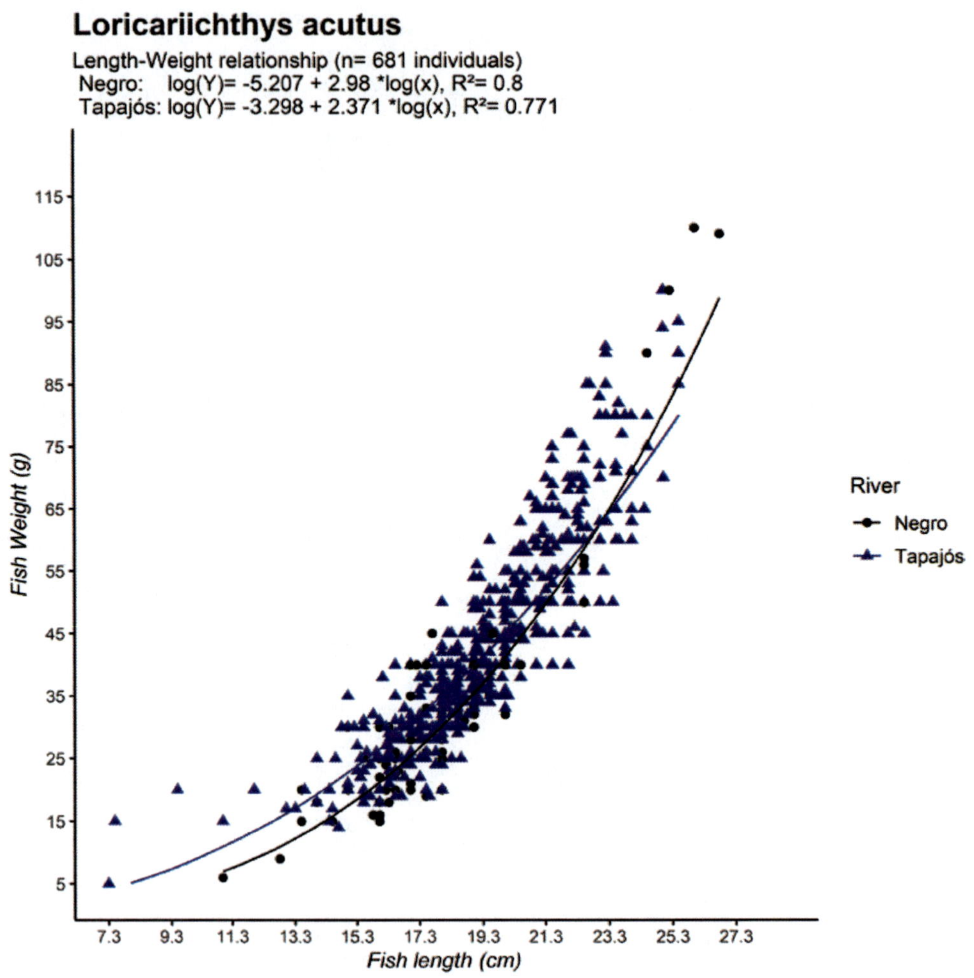

5

Fig. 5.97 Length-weight relationship for *Loricariichthys acutus* sampled in Negro and Tapajós rivers

Fish species: ***Pseudoloricaria laeviuscula*** **(Valenciennes 1840)**

Common name(s): acari-cachimbo

Size of fish sampled (standard length, both rivers): mean = 18.2 cm, median = 18.4 cm

Fisheries: not relevant to fisheries, due to its small size.

Fig. 5.98 *Pseudoloricaria laeviuscula* sampled in the Tapajós River: **a** lateral view and **b** ventral view. (Photo: Renato Silvano)

5

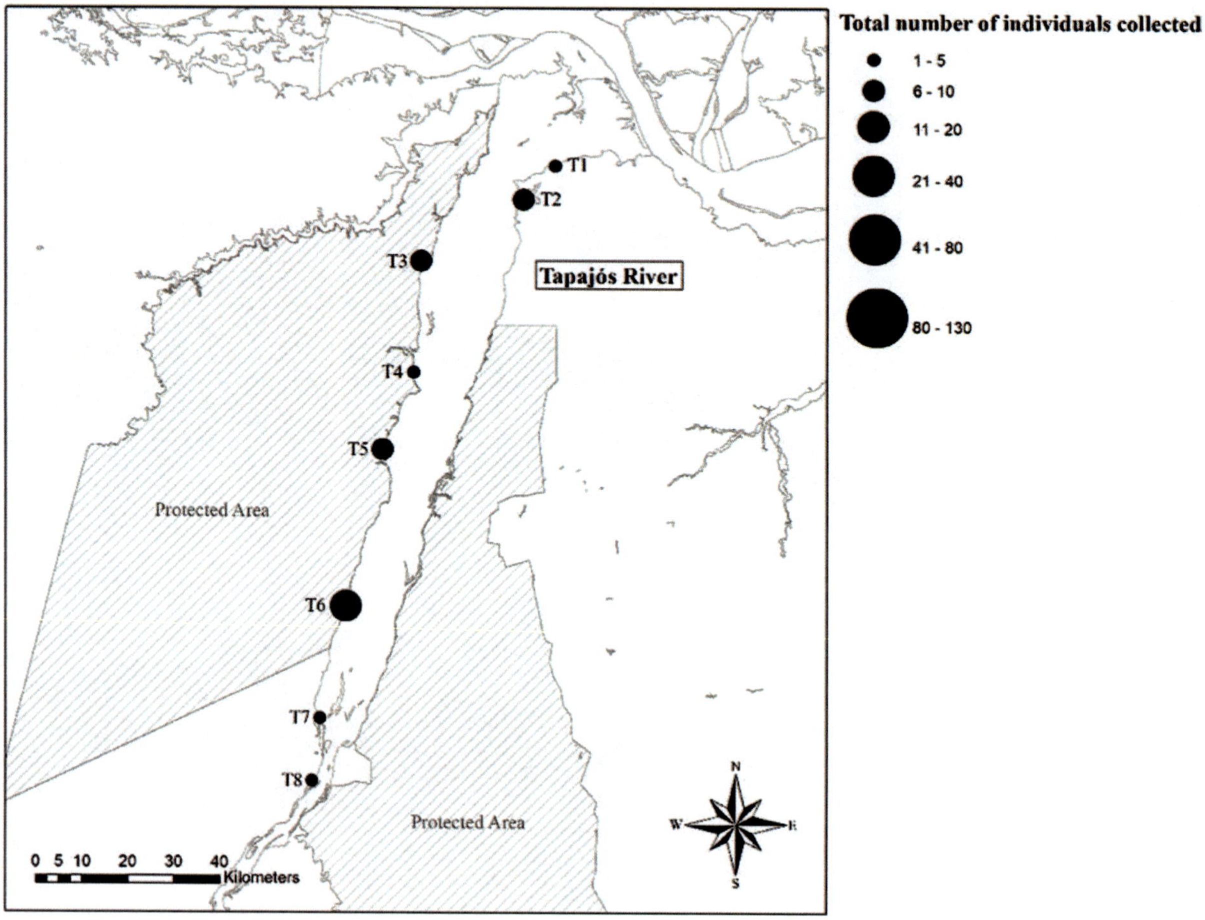

Fig. 5.99 Distribution of individuals of *Pseudoloricaria laeviuscula* sampled in the Tapajós River

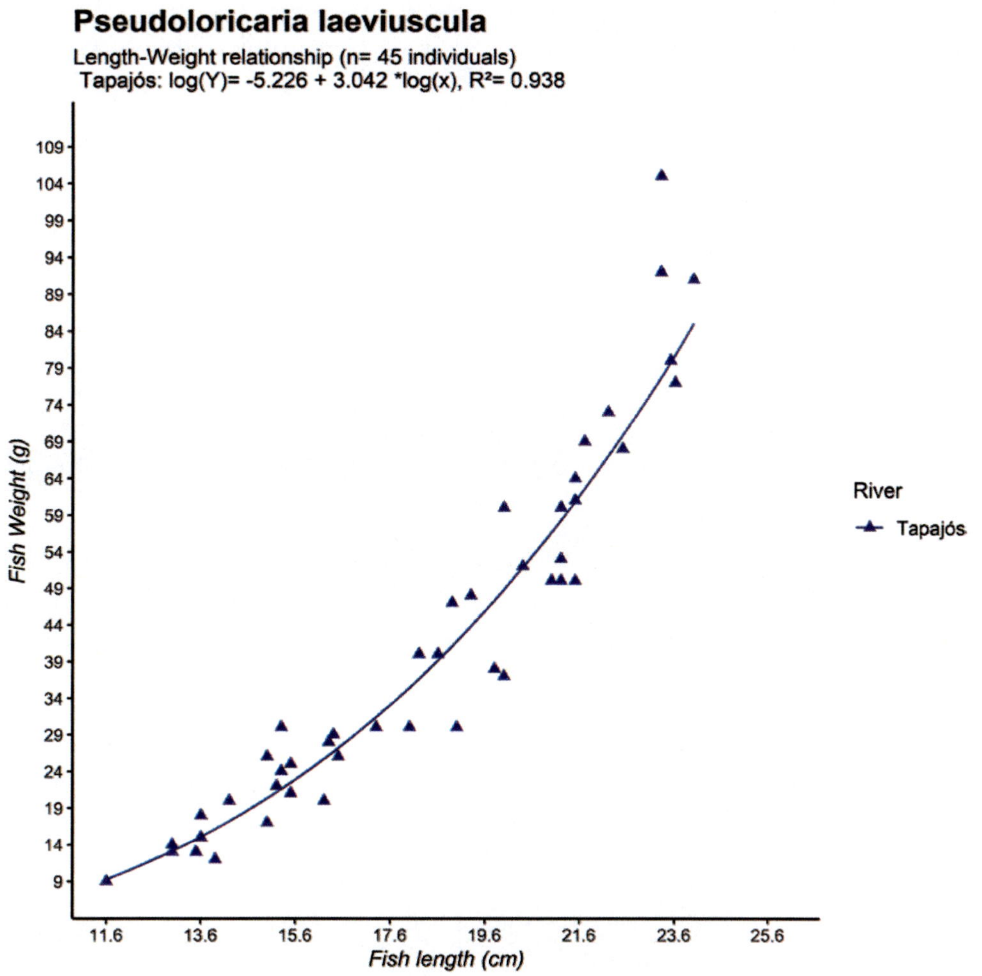

Fig. 5.100 Length-weight relationship for *Pseudoloricaria laeviuscula* sampled in Tapajós River

Fish species: *Pterygoplichthys gibbiceps* (Kner 1854)

Common name(s): acari, bodó, bodó-pedra, acari-bodó

Size of fish sampled (standard length, both rivers): mean = 24.3 cm, median = 27.5 cm

Fisheries: acaris (Loricariidae) are occasionally exploited by fisheries in both rivers. Biomass caught is in ◘ Table 3.3 (► Chap. 3).

◘ **Fig. 5.101** *Pterygoplichthys gibbiceps* sampled in the Tapajós River: **a** lateral view and **b** ventral view. (Photo: Renato Silvano)

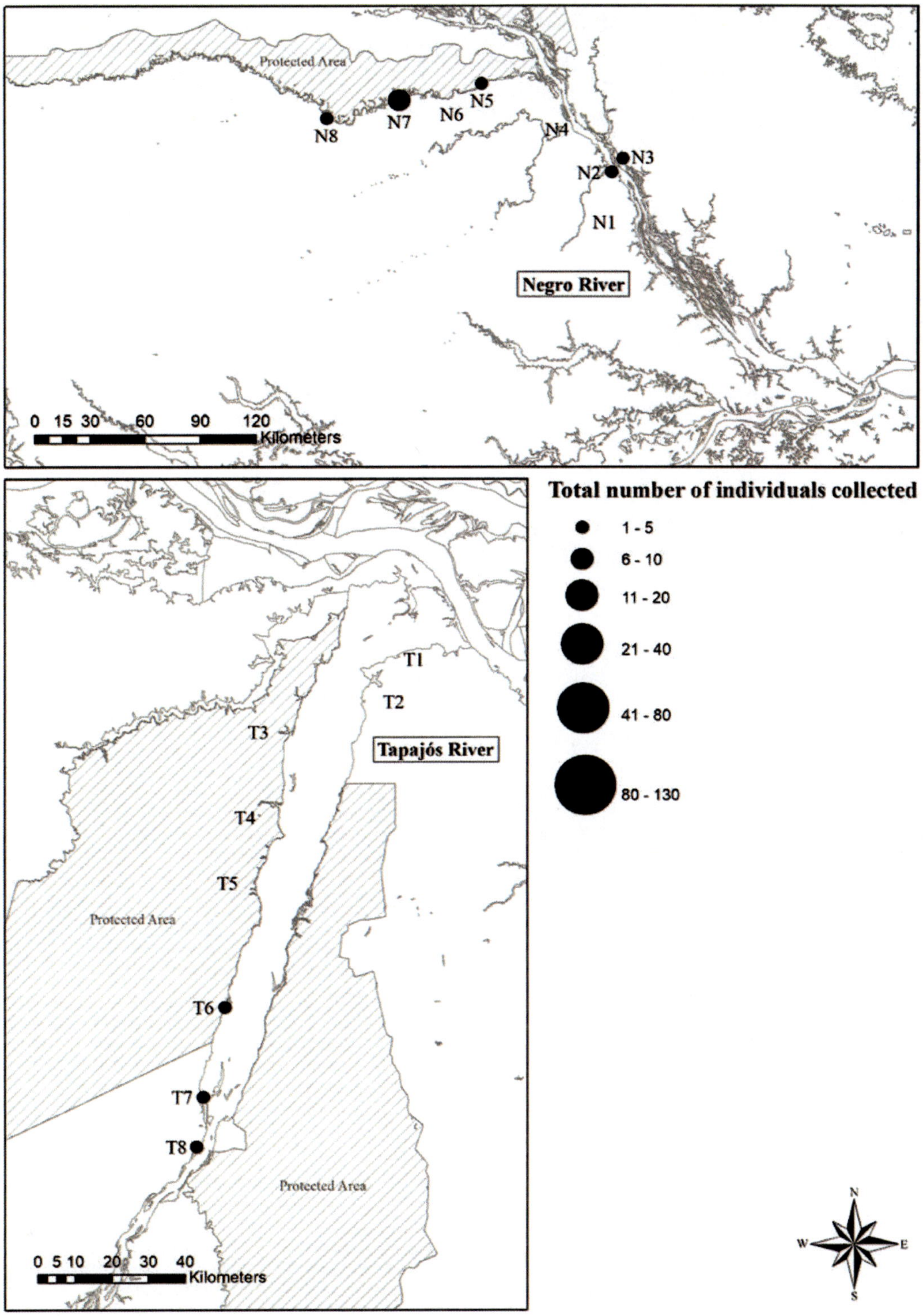

Fig. 5.102 Distribution of individuals of *Pterygoplichthys gibbiceps* sampled in the Negro and Tapajós rivers

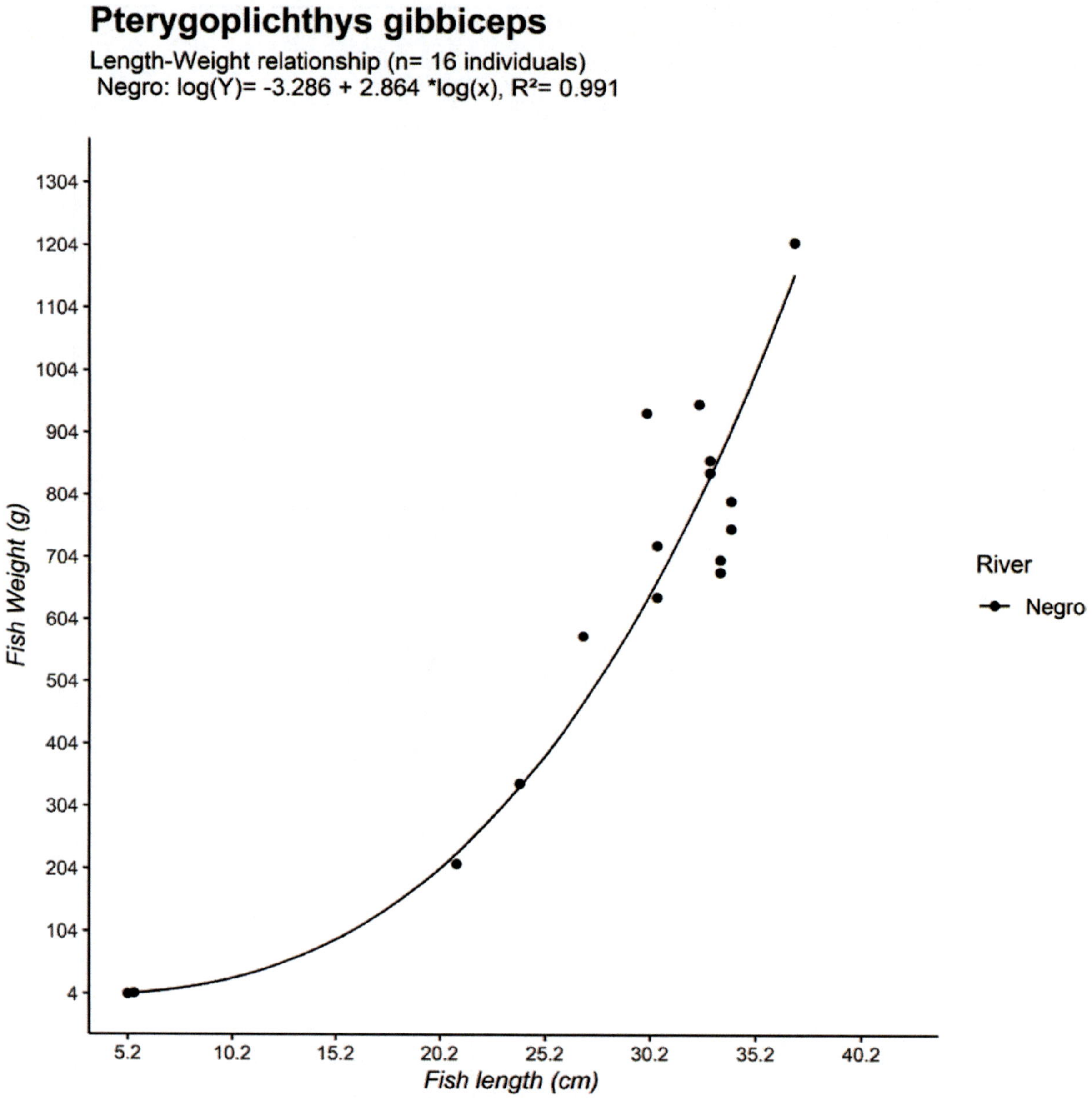

Fig. 5.103 Length-weight relationship for *Pterygoplichthys gibbiceps* sampled in Negro River

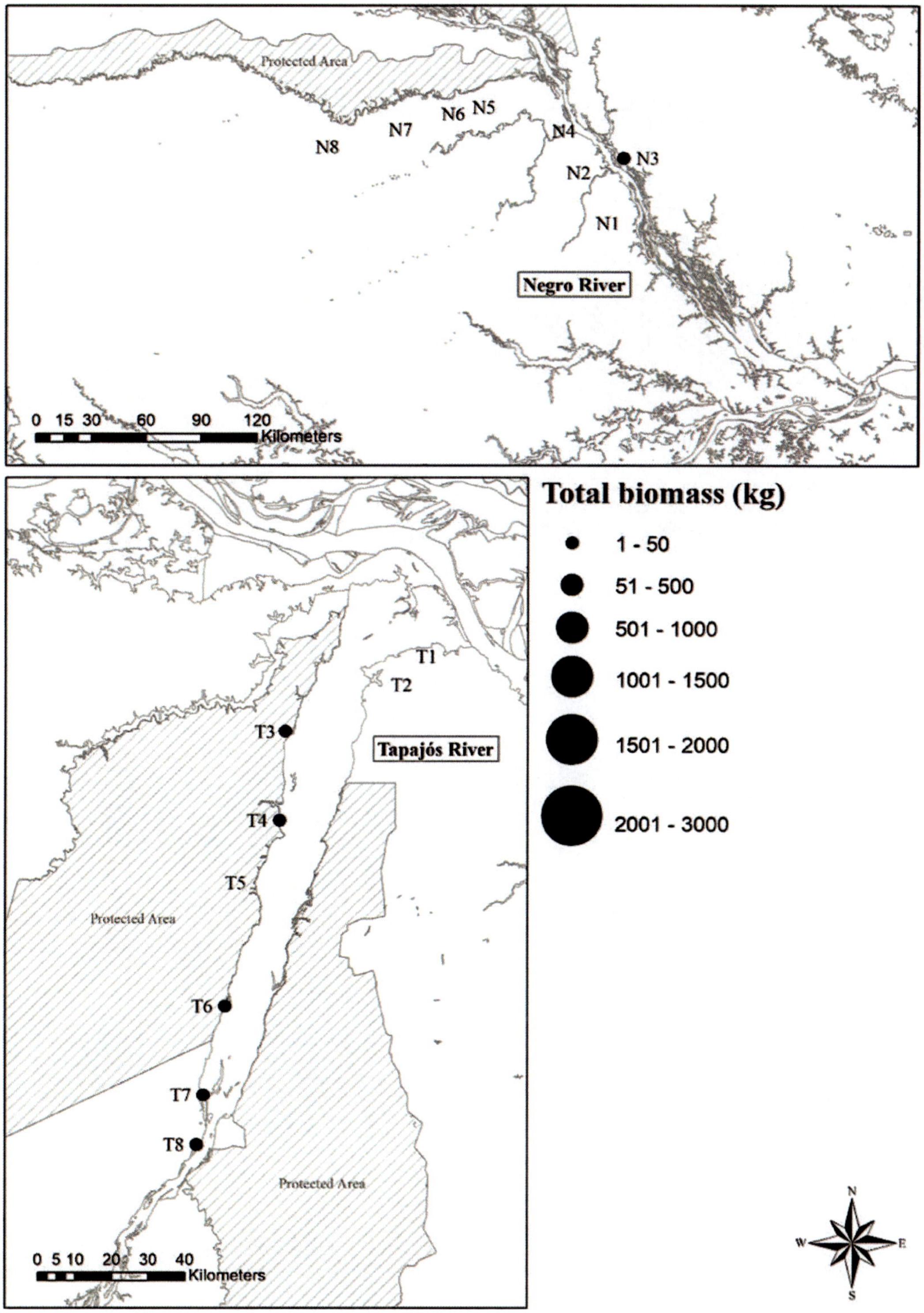

Fig. 5.104 Distribution of biomass of acaris (Loricariidae) caught in small-scale fisheries in the Negro and Tapajós rivers

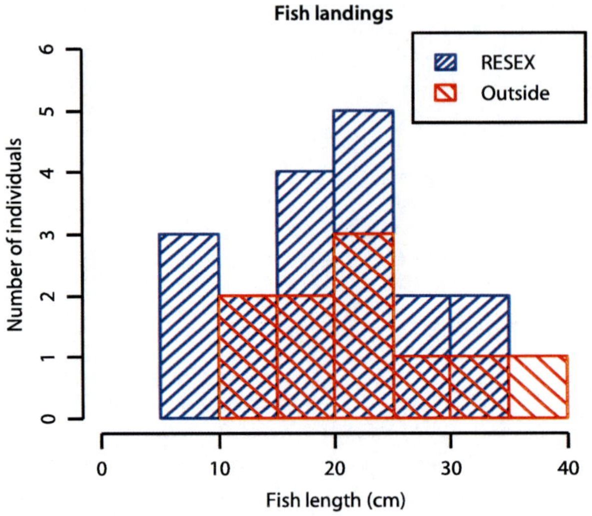

Fig. 5.105 Size distribution of acaris (Loricariidae) caught in small-scale fisheries in eight communities inside and outside the Extractive Reserves (RESEX) in the Tapajós River, N Resex = 18, N outside = 10

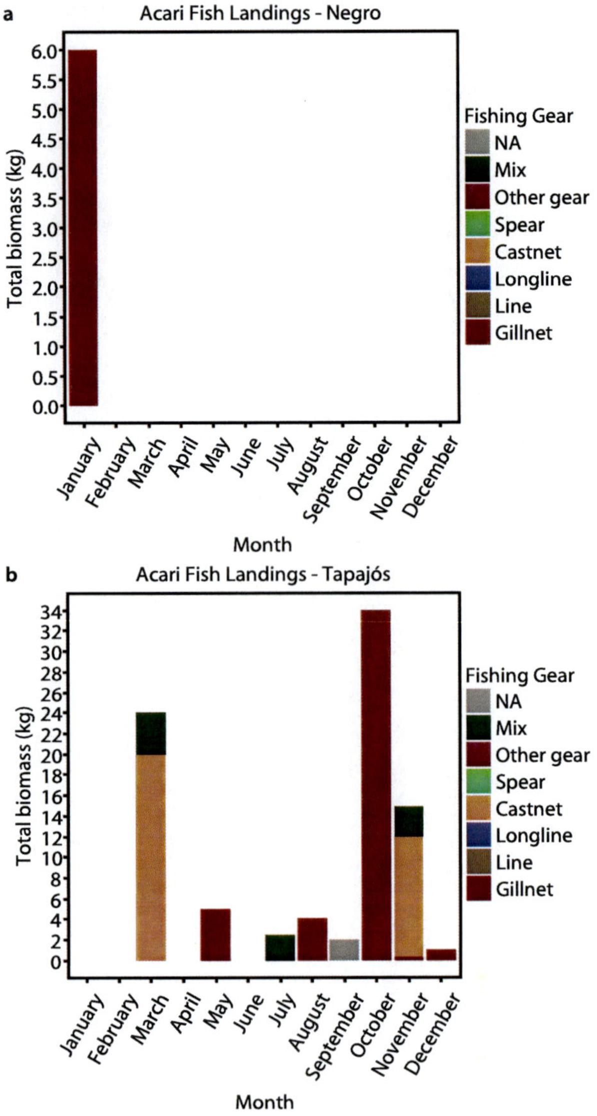

Fig. 5.106 Seasonal variation of biomass of acaris (Loricariidae) caught per fishing gear in small-scale fisheries in 16 communities in the rivers **a** Negro and **b** Tapajós. NA: fishing gear not informed; Mix: more than one gear was used

Fish species: *Rineloricaria daraha* Rapp Py-Daniel & Fichberg 2008

Common name(s): acari-cachimbo

Size of fish sampled (standard length, both rivers): 10.6 cm

Fisheries: not relevant to fisheries, due to its small size.

Fig. 5.107 *Rineloricaria daraha* sampled in the Negro River: **a** lateral view and **b** ventral view. (Photo: Renato Silvano)

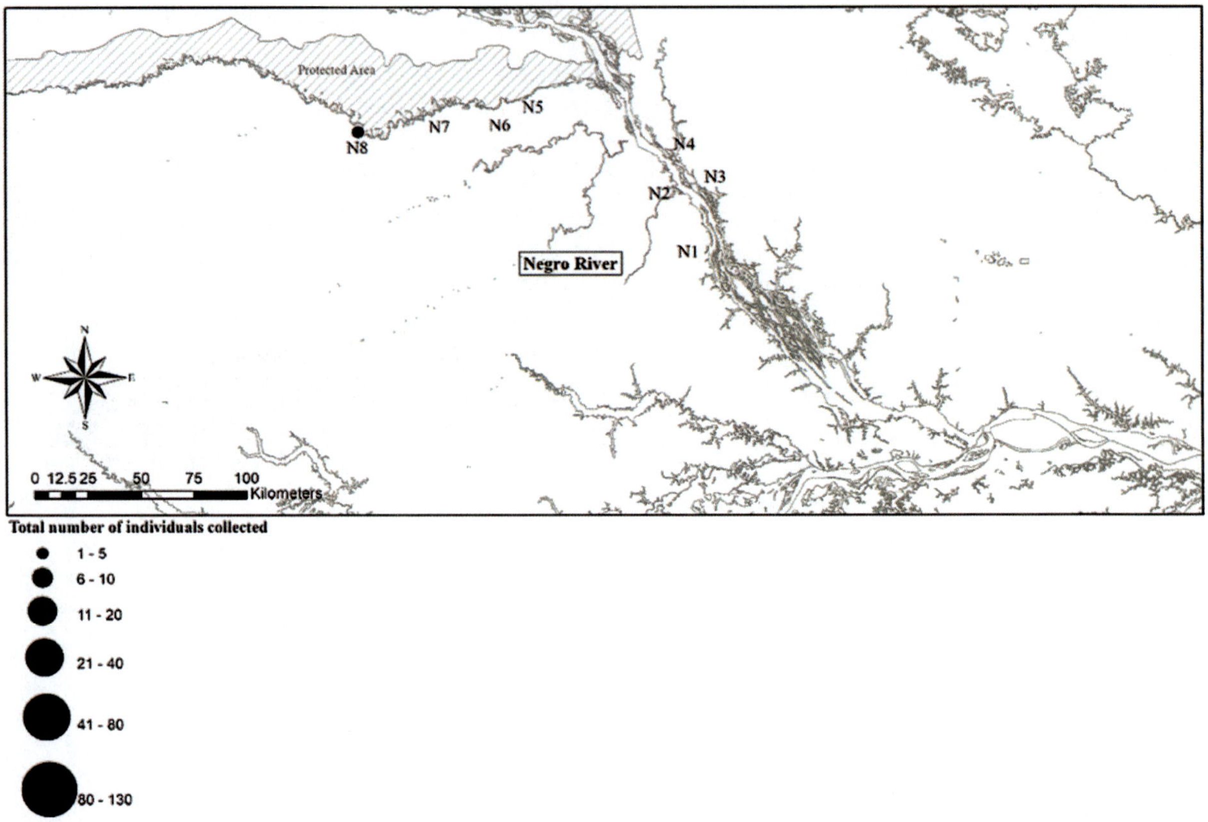

Fig. 5.108 Distribution of individuals of *Rineloricaria daraha* sampled in the Negro River

5.2.6 Family Heptapteridae

5.2.6.1 Summary of Biology and Ecology (Literature)

Diversity: Heptapteridae comprises a family of 214 species divided into 24 genera. The family is poorly known with many more species pending formal description.

Habitat: Heptapteridae is an endemic group of the Neotropical region; fishes of this family are benthic dwellers with exception to species of the genera *Brachyrhamdia* and *Pimelodella,* which frequents the water column. Most of the species from this family occupy small rainforest streams and are used as a bioindicator of environmental quality, since they inhabit shallow streams with dense canopy cover and good water quality.

Ecology and feeding: knowledge on feeding habits of the species of this family is scarce, but available studies indicate that most species are carnivorous or omnivorous. Crustaceans, mollusks, nematodes, and small fishes are the most common food items with algae and organic debris complementing the diet.

Morphology: fishes of this family are very difficult to separate from Pimelodidae and Pseudopimelodidae with most of its characters being internal features. The group can be separated from other Neotropical siluriforms by few external characters such as three pairs of barbels (maxillary, inner, and outer mentals); nares separated and lacking barbels; small adult body size (<20 cm SL, with 60% of species not reaching 10 cm SL); naked skin; cephalic laterosensory canals simple (one pore opening at skin surface); adipose fin with long base; first ray of dorsal and pectoral fins oscillate from soft to almost rigid (pungent spine); gill membranes free.

Reproduction: reproduction cycle remains poorly known for this family, but studies in other river basin indicate multiple spawning periods, synchronized with the rainy season.

Economic importance: species of this family don't have commercial importance to fisheries, although some species can be exploited for the aquarium trade.

References: Amaral et al. (1998), Eschmeyer (1998), Bockmann & Guazzelli (2003).

Fish species: *Goeldiella eques* (Müller & Troschel 1849)

Common name(s): mandi, jandiá

Size of fish sampled (standard length, both rivers): 9.5 cm

Fisheries: not relevant to fisheries, due to its small size.

Fig. 5.109 *Goeldiella eques* sampled in the Negro River. (Photo: Renato Silvano)

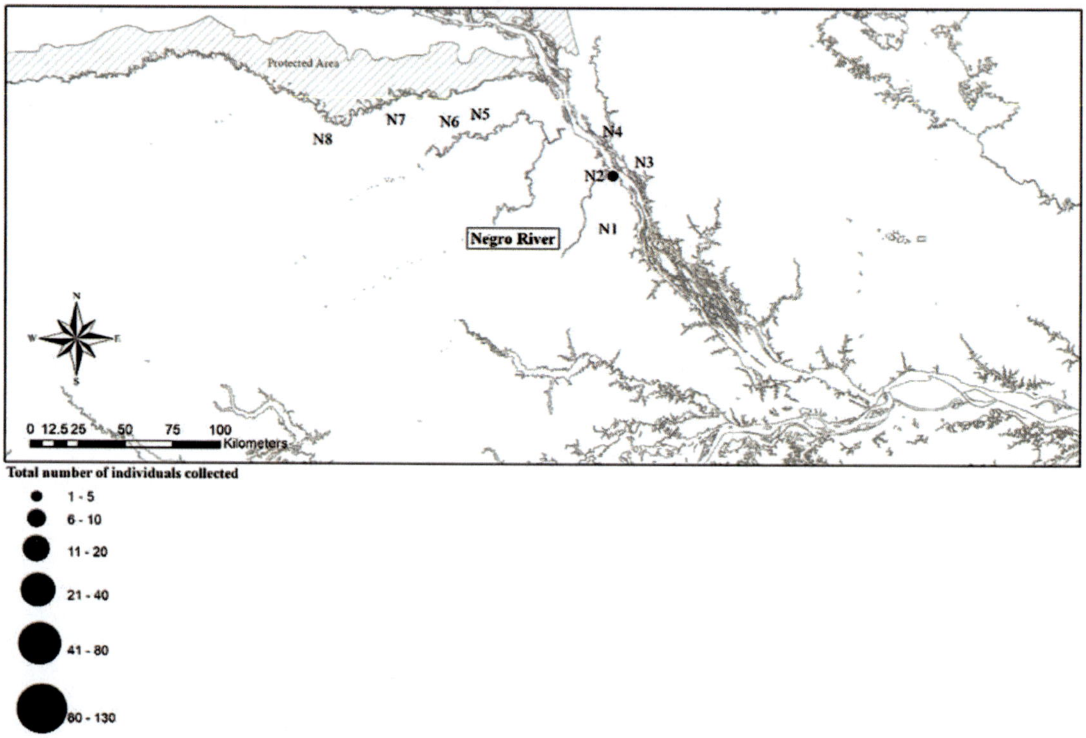

Fig. 5.110 Distribution of individuals of *Goeldiella eques* sampled in the Negro River

Fish species: *Pimelodella cristata* (Müller & Troschel 1849)

Common name(s): mandi

Fig. 5.111 *Pimelodella cristata* sampled in the Negro River. (Photo: Renato Silvano)

Fish species: *Pimelodella* sp.

Common name(s): mandi

Fig. 5.112 *Pimelodella* sp. sampled in the Tapajós River. (Photo: Renato Silvano)

Observation: abundance and size data were grouped for these two species, due to the difficulty in distinguishing them during sampling: *Pimelodella cristata* was collected in Negro River and *Pimelodella* sp. was collected in the Tapajós River; see Table 1.1 (▶ Chap. 1).

Size of fish sampled (standard length, both rivers): min. = 6.5 cm, max. = 23 cm, mean = 15.1 cm, median = 13.1 cm

Fisheries: The fura-calça and mandi (*Pimelodina flavipinnis, Pimelodella* spp.) were occasionally caught in the Tapajós River. Biomass caught is in Table 3.3 (▶ Chap. 3).

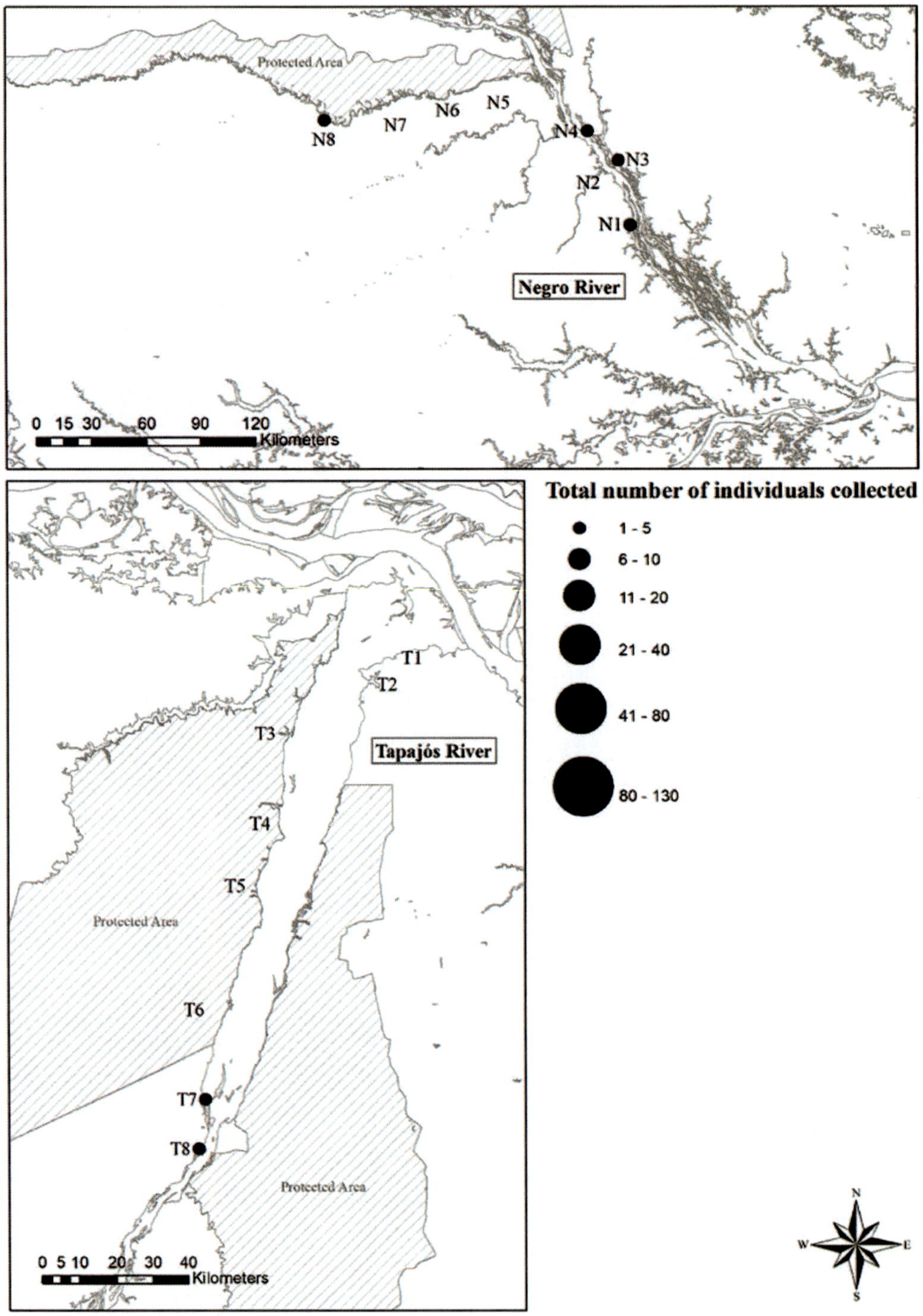

Fig. 5.113 Distribution of individuals of *Pimelodella* spp. sampled in the Negro and Tapajós rivers

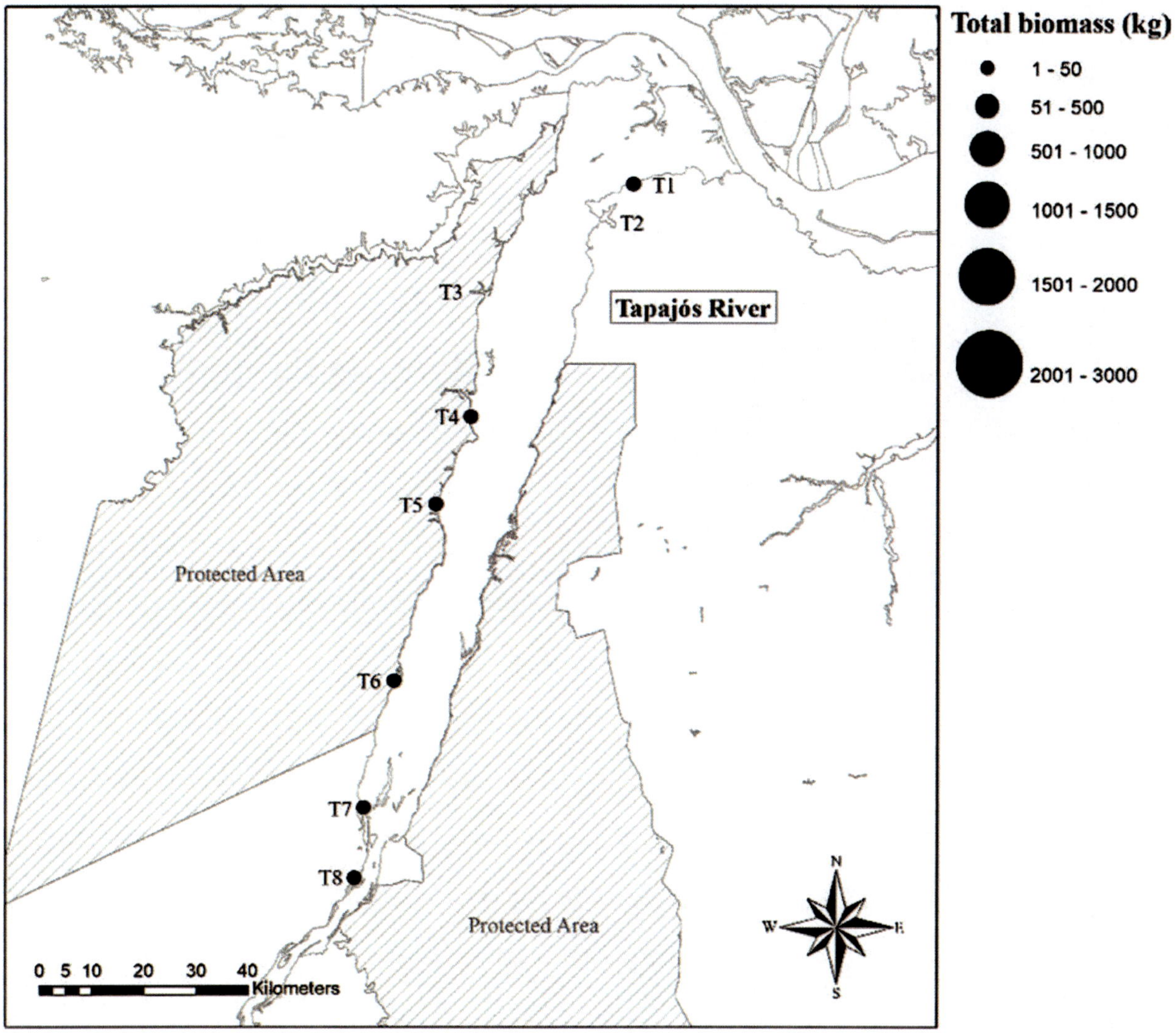

Fig. 5.114 Distribution of biomass of fura-calça and mandi (*Pimelodina flavipinnis, Pimelodella* spp.) caught in small-scale fisheries in the Tapajós River

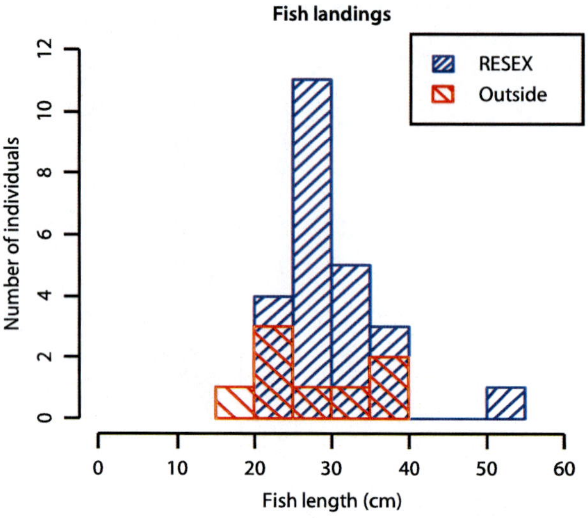

Fig. 5.115 Size distribution of fura-calça (*Pimelodina flavipinnis, Pimelodella* spp.) caught in small-scale fisheries in eight communities inside and outside the Extractive Reserves (RESEX) in the Tapajós River, N Resex = 24, N outside = 8

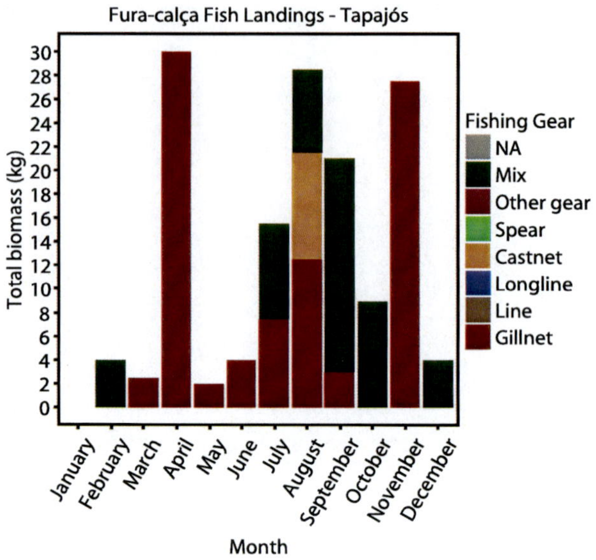

Fig. 5.116 Seasonal variation of biomass of fura-calça (*Pimelodina flavipinnis, Pimelodella* spp.) caught per fishing gear in small-scale fisheries in eight communities in the Tapajós River. NA: fishing gear not informed; Mix: more than one gear was used

5.2.7 Family Pimelodidae

5.2.7.1 Summary of Biology and Ecology (Literature)

Diversity: 114 valid species, distributed in 31 genera.

Habitat: endemic to the Neotropical region. The highest diversity is found in the Amazon River, Paraná, Orinoco, and large rivers in the Guianas.

Ecology and feeding: most pimelodids (also known as catfishes) have a carnivorous habit, but some species are omnivorous and may include fruits in their diets. Some of the largest catfishes, including those from the genus *Brachyplatystoma*, are considered top predators. Because of their higher position in the trophic chain, some of these large catfishes have potential to accumulate contaminants, such as heavy metals and mercury.

Morphology: several species can reach sizes larger than a meter. These fishes have a naked body (without bony plates neither scales) with an adipose fin relatively large and three pairs of long barbells (one maxillary pair and two mentonian). The color of the body can vary from uniform gray to elaborate strip, mole, and stain patterns.

Reproduction: some species are seasonal, spawning during the raising water period. The large catfish, especially those from the genus *Brachyplatystoma*, make extensive reproductive and feeding migrations of thousands of kilometers from the Amazon estuary to the headwaters of tributaries of the main Amazon River.

Economic importance: they are among the freshwater fishes with the highest economic importance for commercial and artisanal fisheries in South America and in the Brazilian Amazon. Although people living in the Amazon have some food taboos and cultural restrictions regarding consumption of large catfish, these fishes have been caught and sold to other Brazilian regions and even to other South American countries.

References: Barthem et al. (1991, 2017), Barthem and Goulding (1997), Ferreira et al. (1998), Boischio and Henshel (2000), Begossi et al. (2004), Petrere et al. (2004), Santos et al. (2006), Eschmeyer (1998), Queiroz et al. (2013a, b), Rocha and Zuanon (2013), Hallwass and Silvano (2016), Hallwass et al. (2020), Nunes et al. (2019).

Fish species: ***Brachyplatystoma capapretum*** **Lundberg & Akama 2005**

Common name(s): Filhote, piraíba

Size of fish sampled (standard length, both rivers): min. = 25.5 cm, max. = 62 cm, mean = 48.1 cm, median = 54 cm

Fisheries: the filhote (*Brachyplatystoma* spp.) is an important commercial fish exploited in fisheries, especially in the Tapajós River; see *Brachyplatystoma filamentosum* for filhote fisheries data.

Fig. 5.117 *Brachyplatystoma capapretum* sampled in the Negro River. (Photo: Renato Silvano)

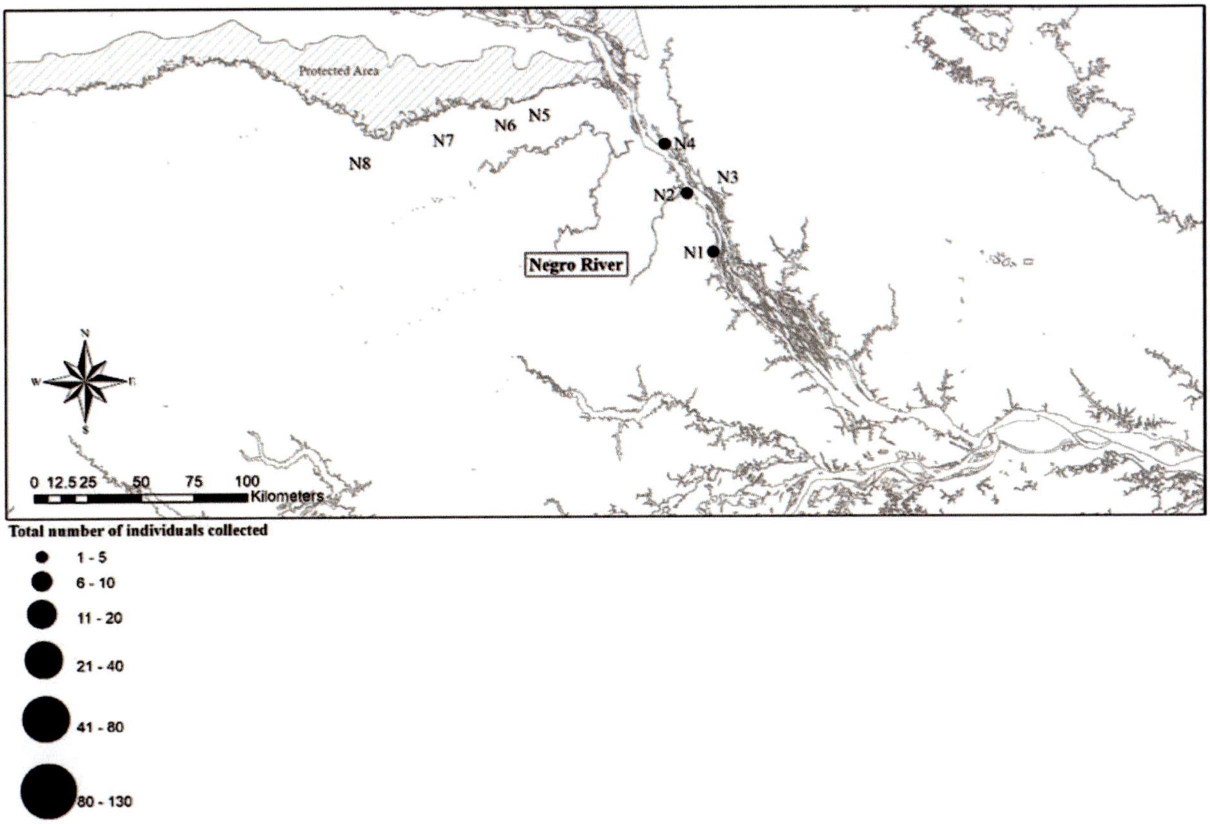

Fig. 5.118 Distribution of individuals of *Brachyplatystoma capapretum* sampled in the Negro River

Fish species: ***Brachyplatystoma filamentosum*** **(Lichtenstein 1819)**

Common name(s): Filhote

Size of fish sampled (standard length, both rivers): min. = 17.8 cm, max. = 55 cm, mean = 34.2 cm, median = 30.5 cm

Fisheries: The filhote (*Brachyplatystoma* spp.) is one of the main commercial fish caught, especially in the Tapajós River. Biomass caught is given in ▪ Table 3.3 (▶ Chap. 3).

5

▪ **Fig. 5.119** *Brachyplatystoma filamentosum* sampled in the Tapajós River. (Photo: Renato Silvano)

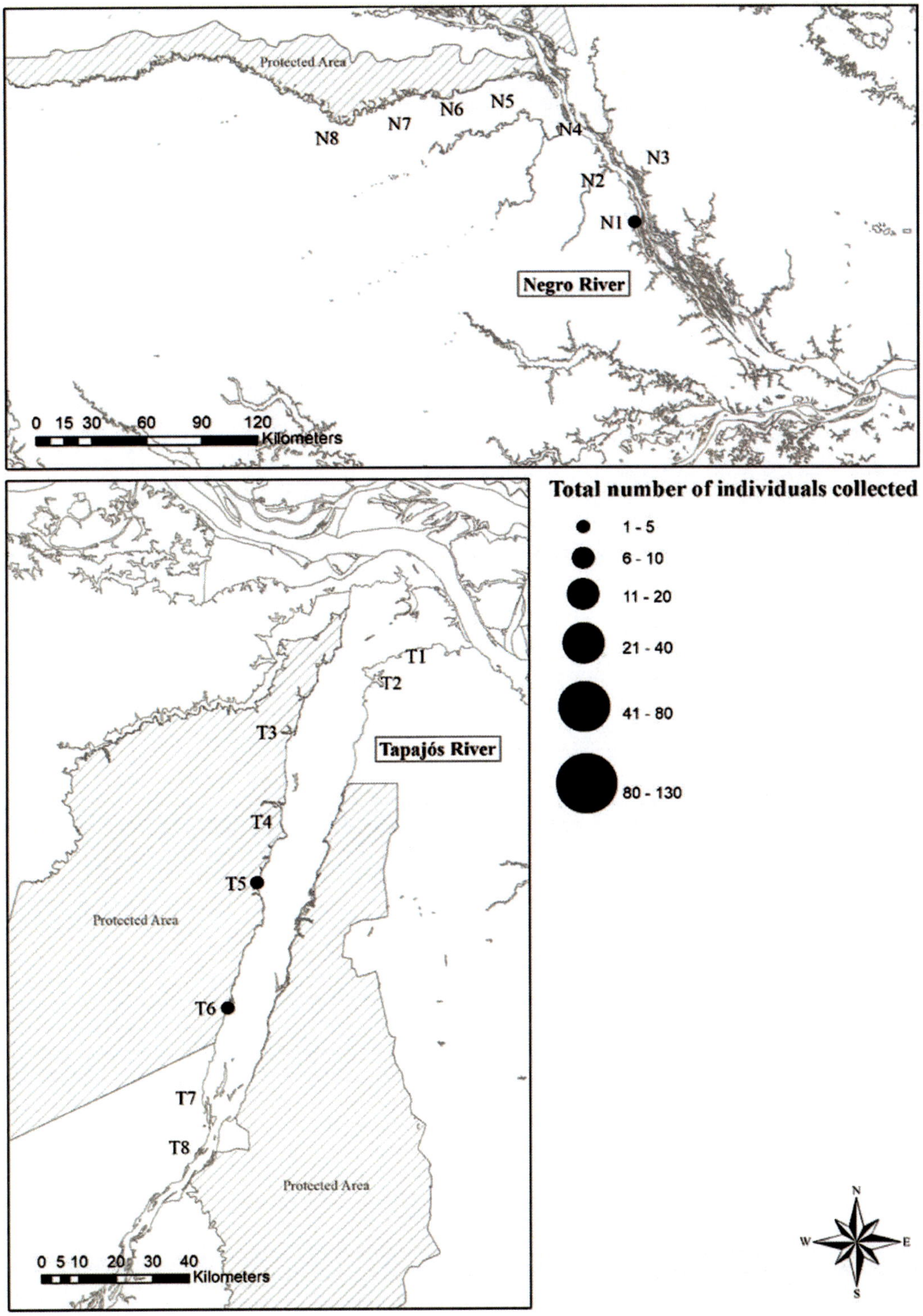

Fig. 5.120 Distribution of individuals of *Brachyplatystoma filamentosum* sampled in the Negro and Tapajós rivers

5

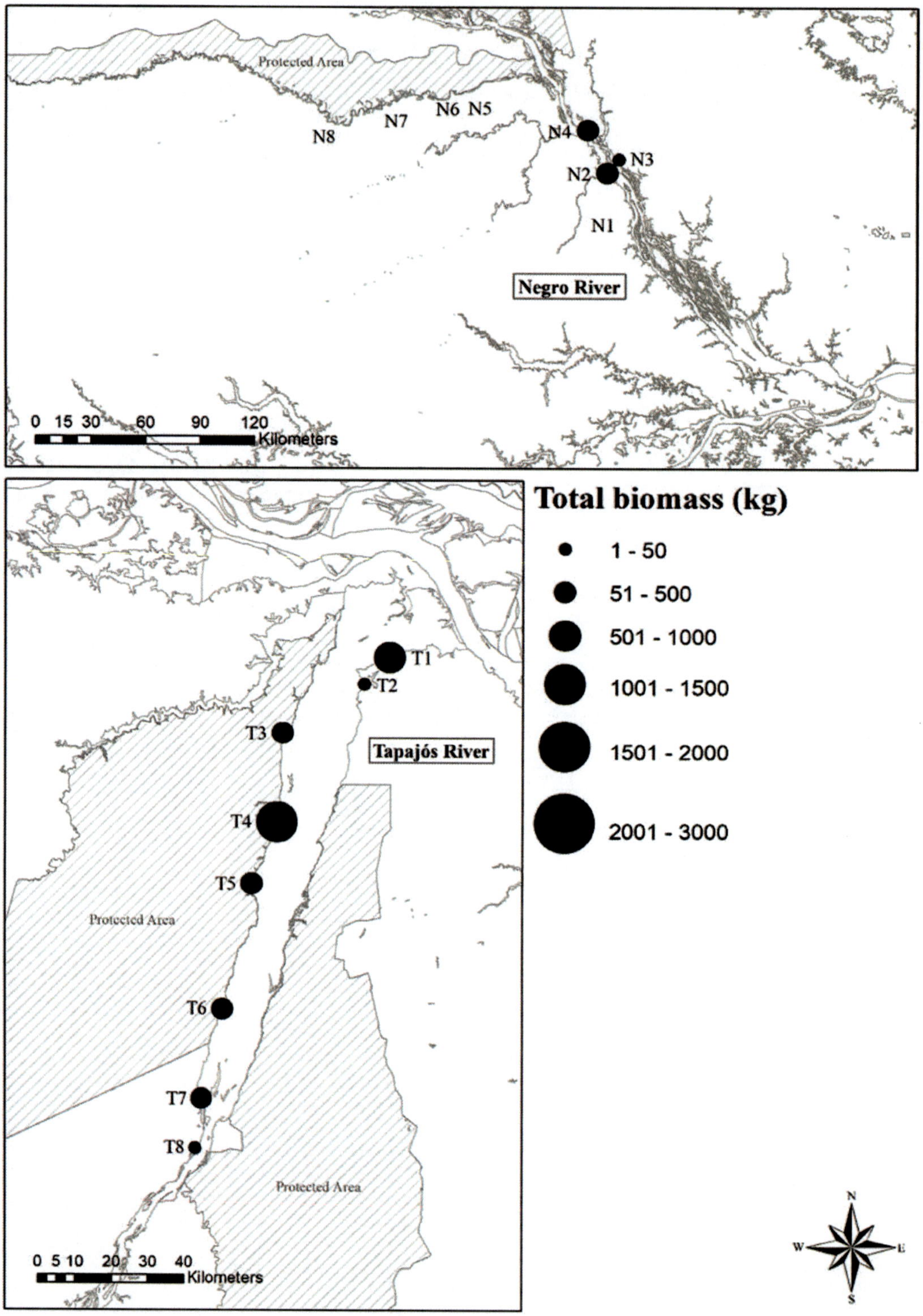

Fig. 5.121 Distribution of biomass of filhote (*Brachyplatystoma* spp.) caught in small-scale fisheries in the Negro and Tapajós rivers

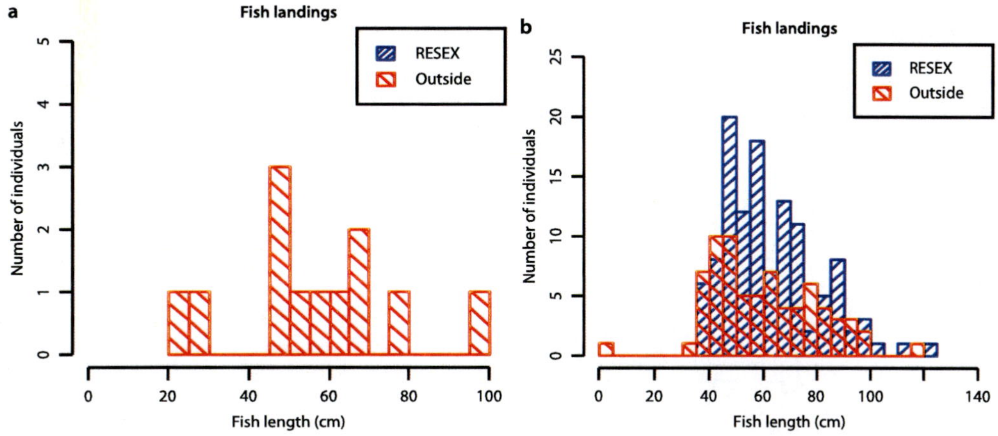

Fig. 5.122 Size distribution of filhote (*Brachyplatystoma* spp.) caught in small-scale fisheries in 16 communities inside and outside the Extractive Reserves (RESEX) in the rivers **a** Negro, N Resex = 0, N outside = 12 and **b** Tapajós, N Resex = 118, N outside = 73

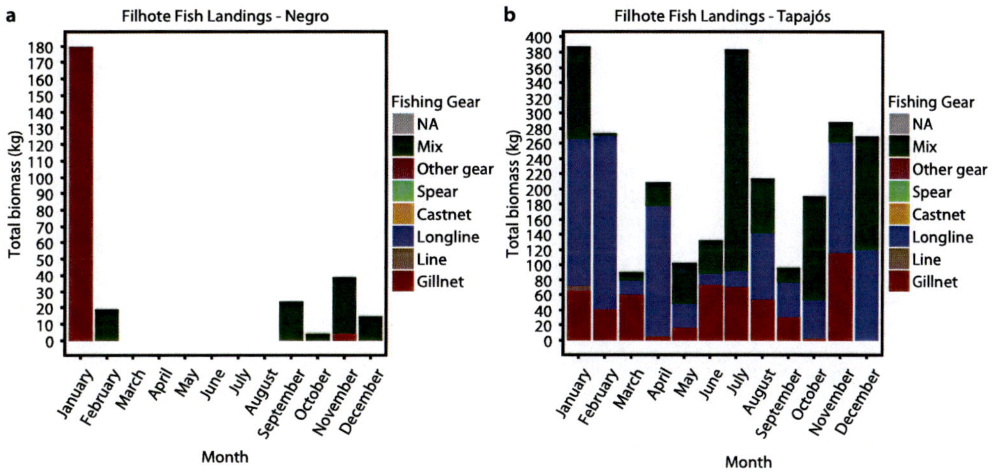

Fig. 5.123 Seasonal variation of biomass of filhote (*Brachyplatystoma* spp.) caught per fishing gear in small-scale fisheries in 16 communities in the rivers **a** Negro and **b** Tapajós. NA: fishing gear not informed; Mix: more than one gear was used

Fish species: *Calophysus macropterus* (Lichtenstein 1819)

Common name(s): piracatinga

Size of fish sampled (standard length, both rivers): mean = 15.6 cm, median = 13.5 cm

Fisheries: The piracatinga (*Calophysus macropterus*) was occasionally caught in the Tapajós River. Biomass caught is in Table 3.3 (► Chap. 3).

Fig. 5.124 *Calophysus macropterus* sampled in the Negro River. (Photo: Renato Silvano)

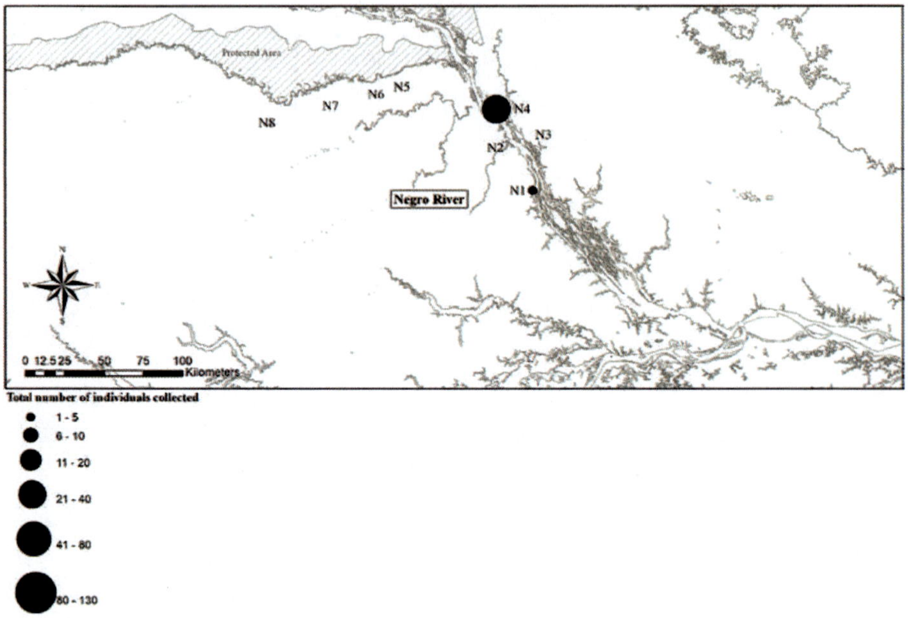

Fig. 5.125 Distribution of individuals of *Calophysus macropterus* sampled in the Negro River

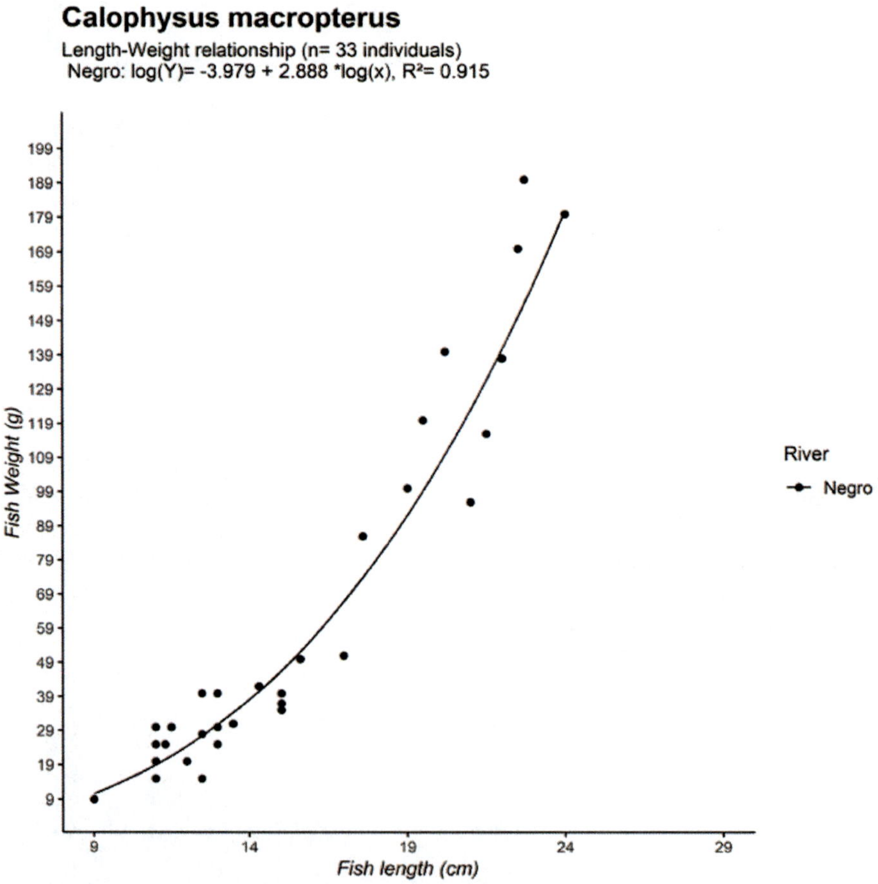

Fig. 5.126 Length-weight relationship for *Calophysus macropterus* sampled in the Negro River

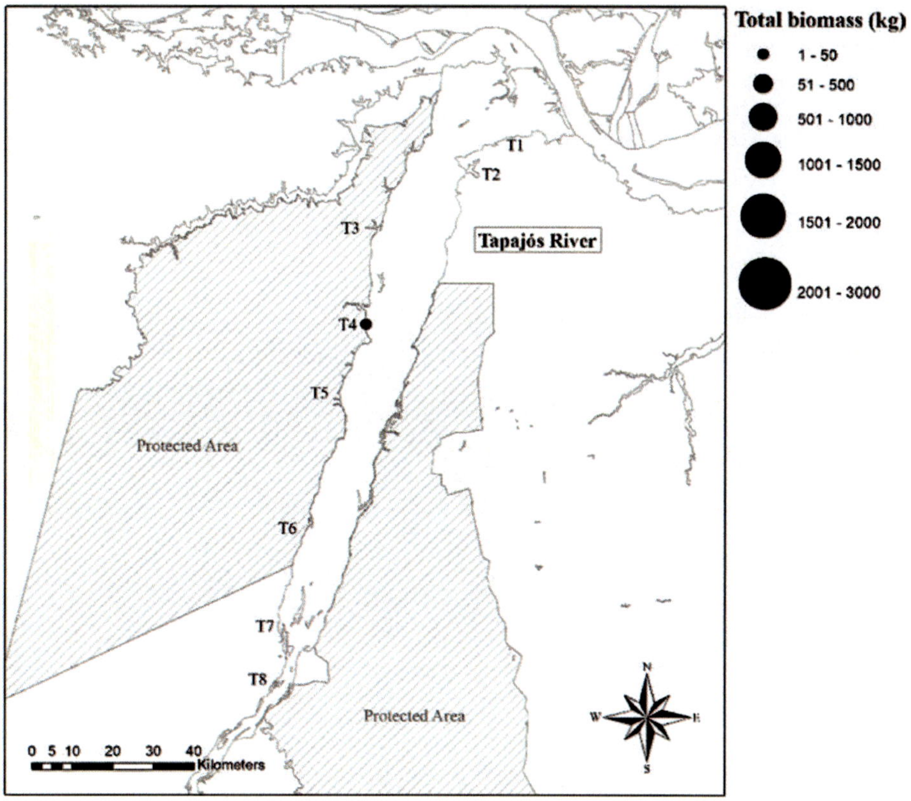

Fig. 5.127 Distribution of biomass of piracatinga (*Calophysus macropterus*) caught in small-scale fisheries in the Tapajós River

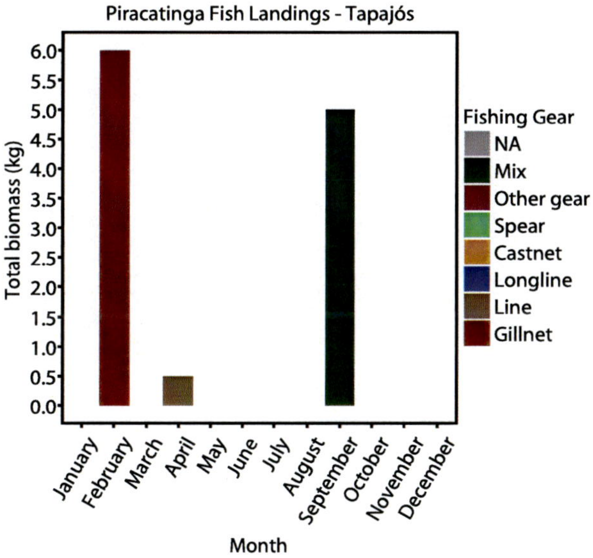

Fig. 5.128 Seasonal variation of biomass of piracatinga (*Calophysus macropterus*) caught per fishing gear in small-scale fisheries in eight communities in the Tapajós River. NA: fishing gear not informed; Mix: more than one gear was used

Fish species: *Hypophthalmus edentatus* Spix & Agassiz 1829

Common name(s): mapará

Size of fish sampled (standard length, both rivers): mean = 20.1 cm, median = 19.5 cm

Fisheries: The mapará (*Hypophthalmus* spp.) is among the most important fish for food and commerce, especially in the Tapajós River; see *Hypophthalmus marginatus* for mapará fisheries data.

5

Fig. 5.129 *Hypophthalmus edentatus* sampled in the Negro River. (Photo: Renato Silvano)

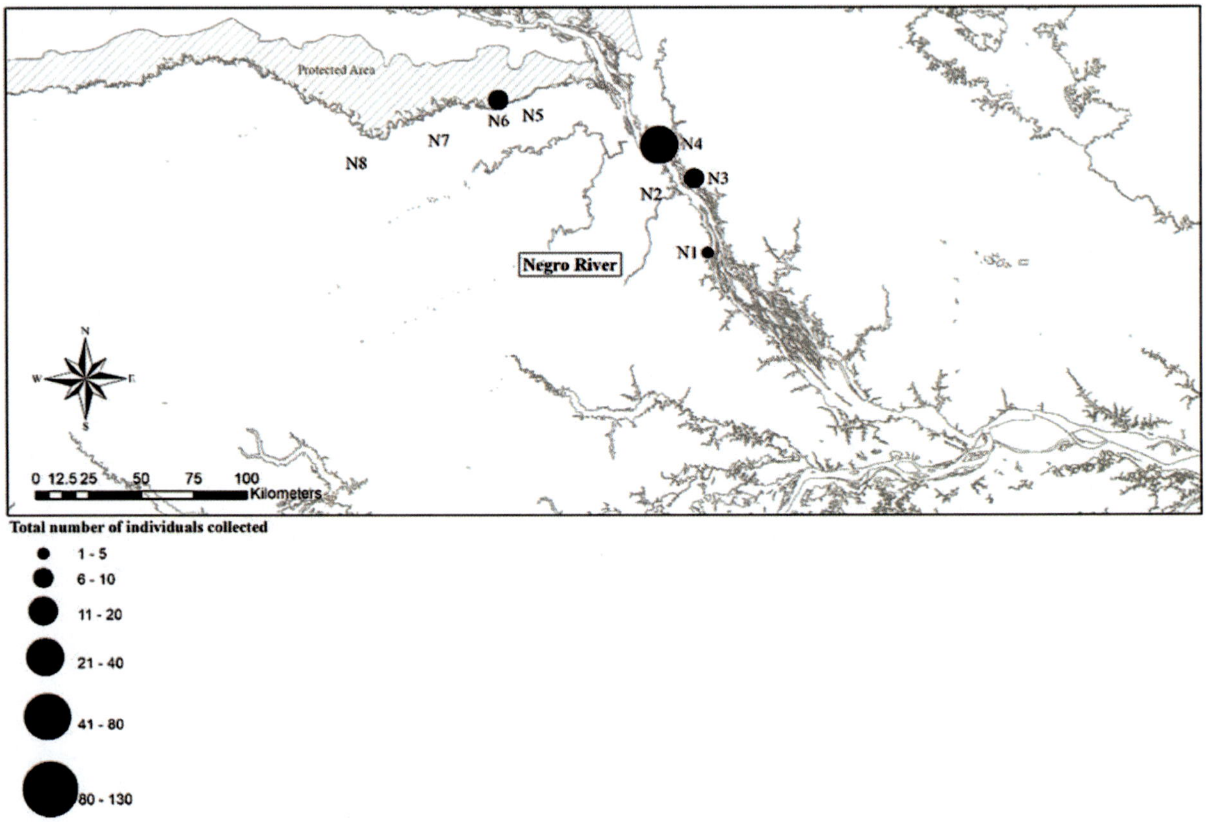

Fig. 5.130 Distribution of individuals of *Hypophthalmus edentatus* sampled in the Negro River

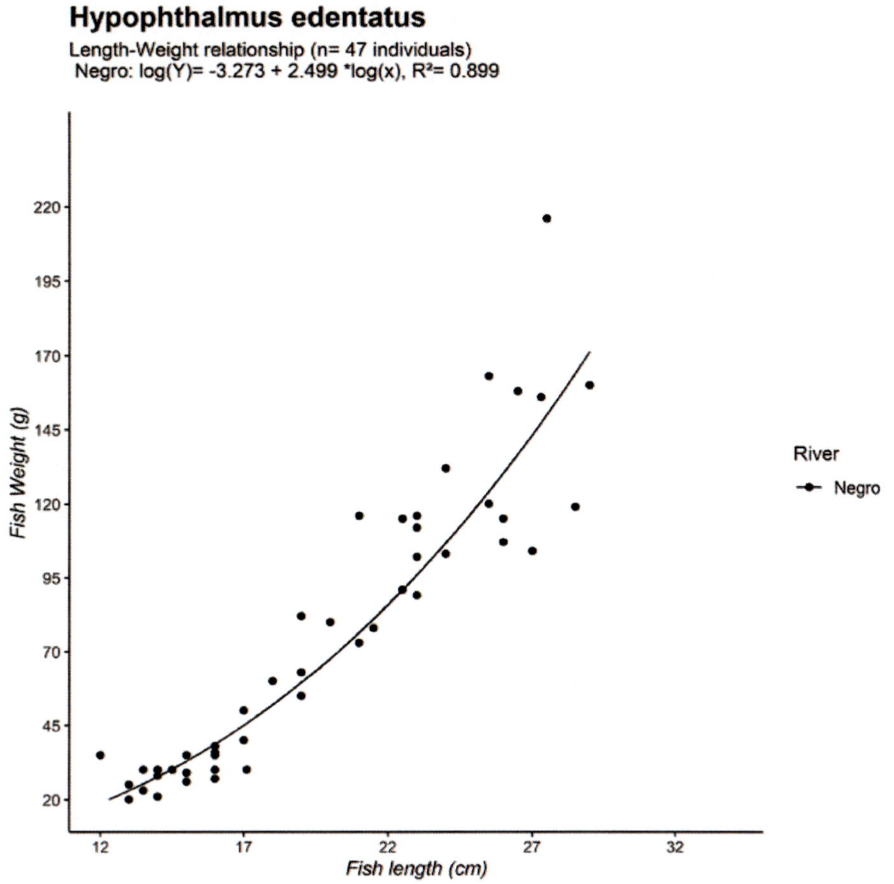

Fig. 5.131 Length-weight relationship for *Hypophthalmus edentatus* sampled in the Negro River

---

Fish species: ***Hypophthalmus fimbriatus*** **Kner 1858**

Common name(s): mapará

Size of fish sampled (standard length, both rivers): mean and median = 23.5 cm

Fisheries: The mapará (*Hypophthalmus* spp.) is among the most important fish for food and commerce, especially in the Tapajós River; see *Hypophthalmus marginatus* for mapará fisheries data.

Fig. 5.132 *Hypophthalmus fimbriatus* sampled in the Negro River. (Photo: Renato Silvano)

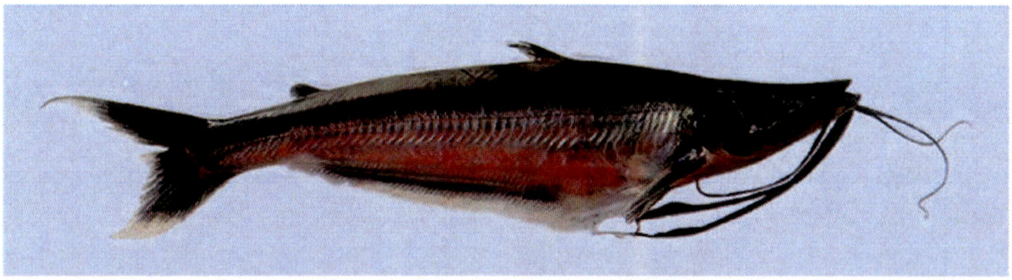

5

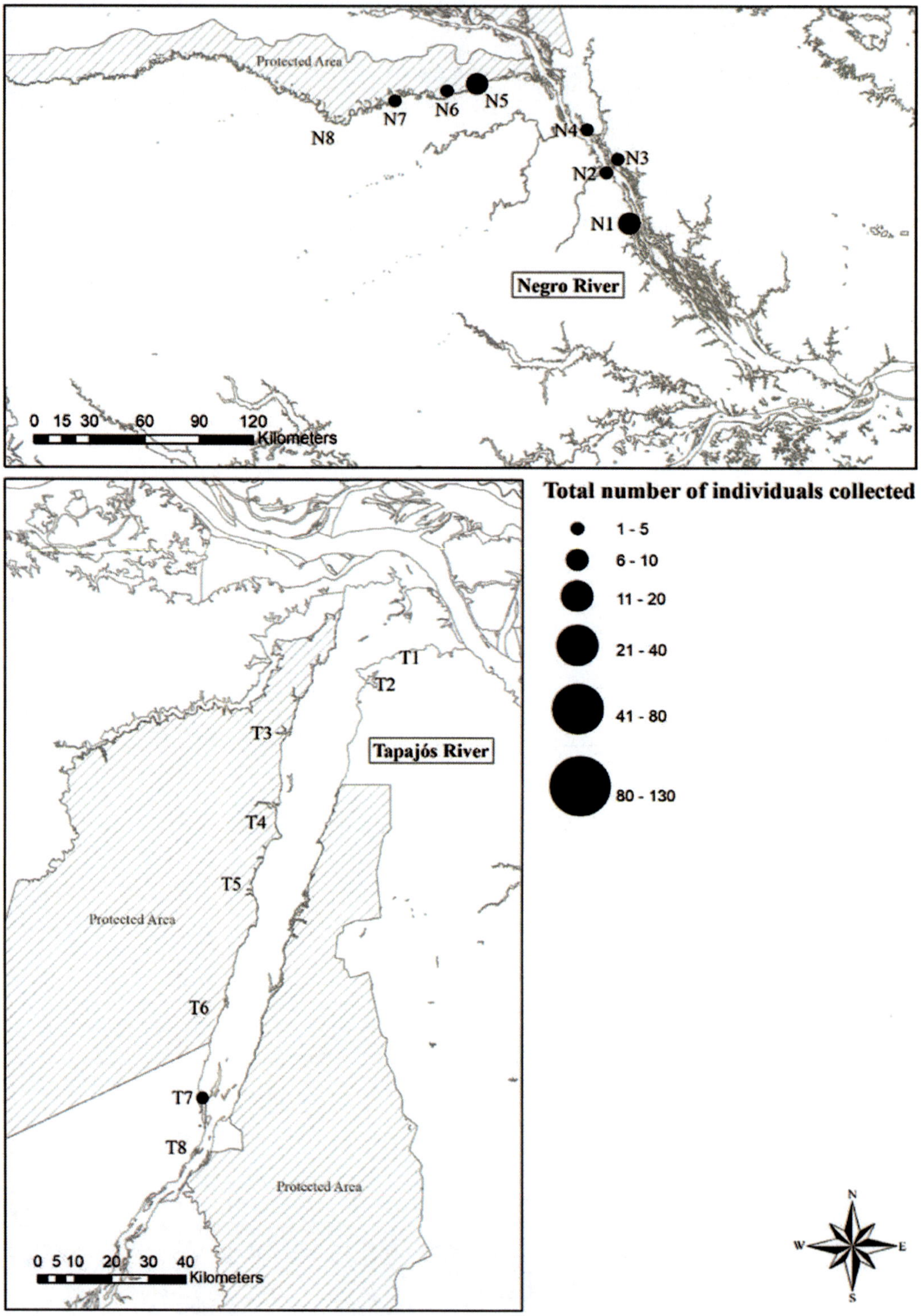

Fig. 5.133 Distribution of individuals of *Hypophthalmus fimbriatus* sampled in the Negro and Tapajós rivers

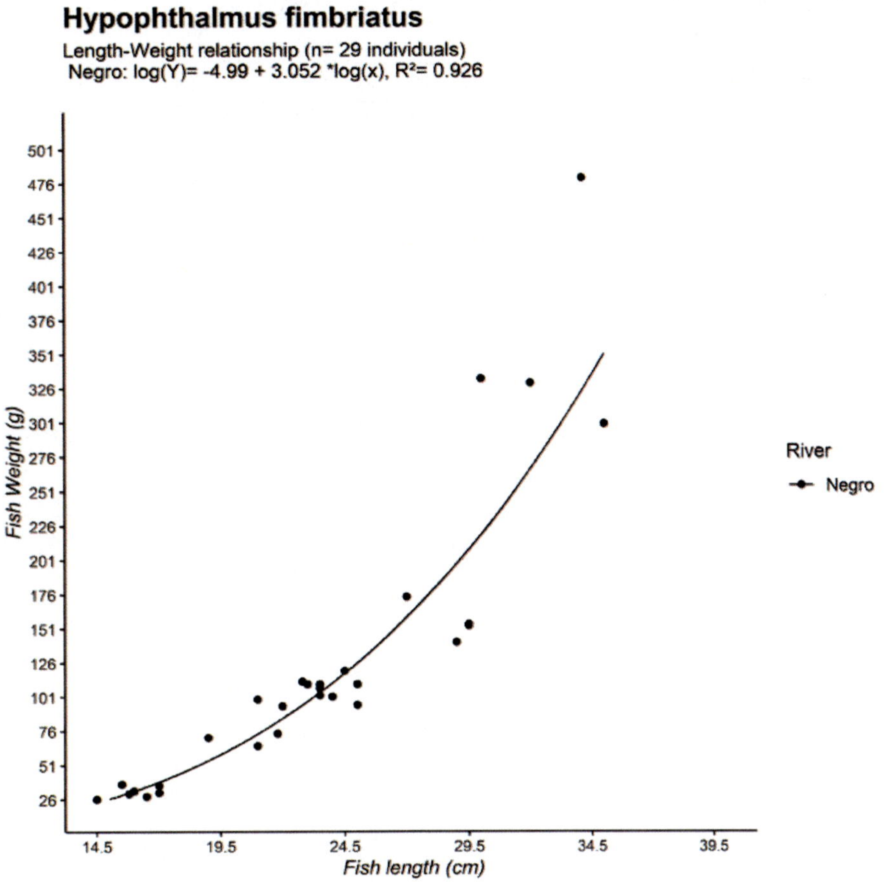

■ **Fig. 5.134** Length-weight relationship for *Hypophthalmus fimbriatus* sampled in the Negro River

---

Fish species: ***Hypophthalmus marginatus*** **Valenciennes 1840**

Common name(s): mapará

Size of fish sampled (standard length, both rivers): mean = 25.5 cm, median = 24.8 cm

Fisheries: The mapará (*Hypophthalmus* spp.) is among the most important fish for food and commerce, especially in the Tapajós River. Biomass caught is in ■ Table 3.3 (▶ Chap. 3).

■ **Fig. 5.135** *Hypophthalmus marginatus* sampled in the Tapajós River. (Photo: Renato Silvano)

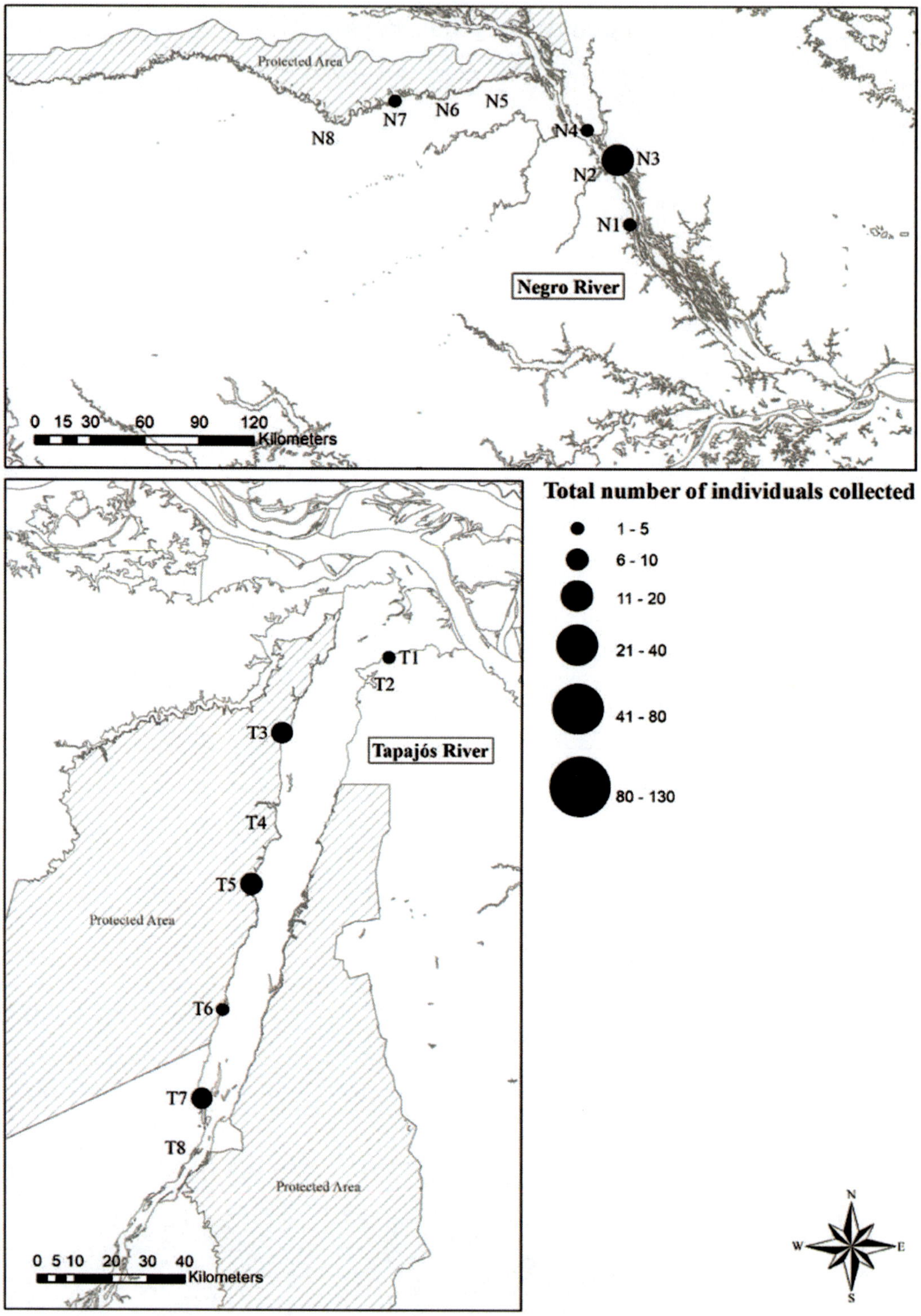

Fig. 5.136 Distribution of individuals of *Hypophthalmus marginatus* sampled in the Negro and Tapajós rivers

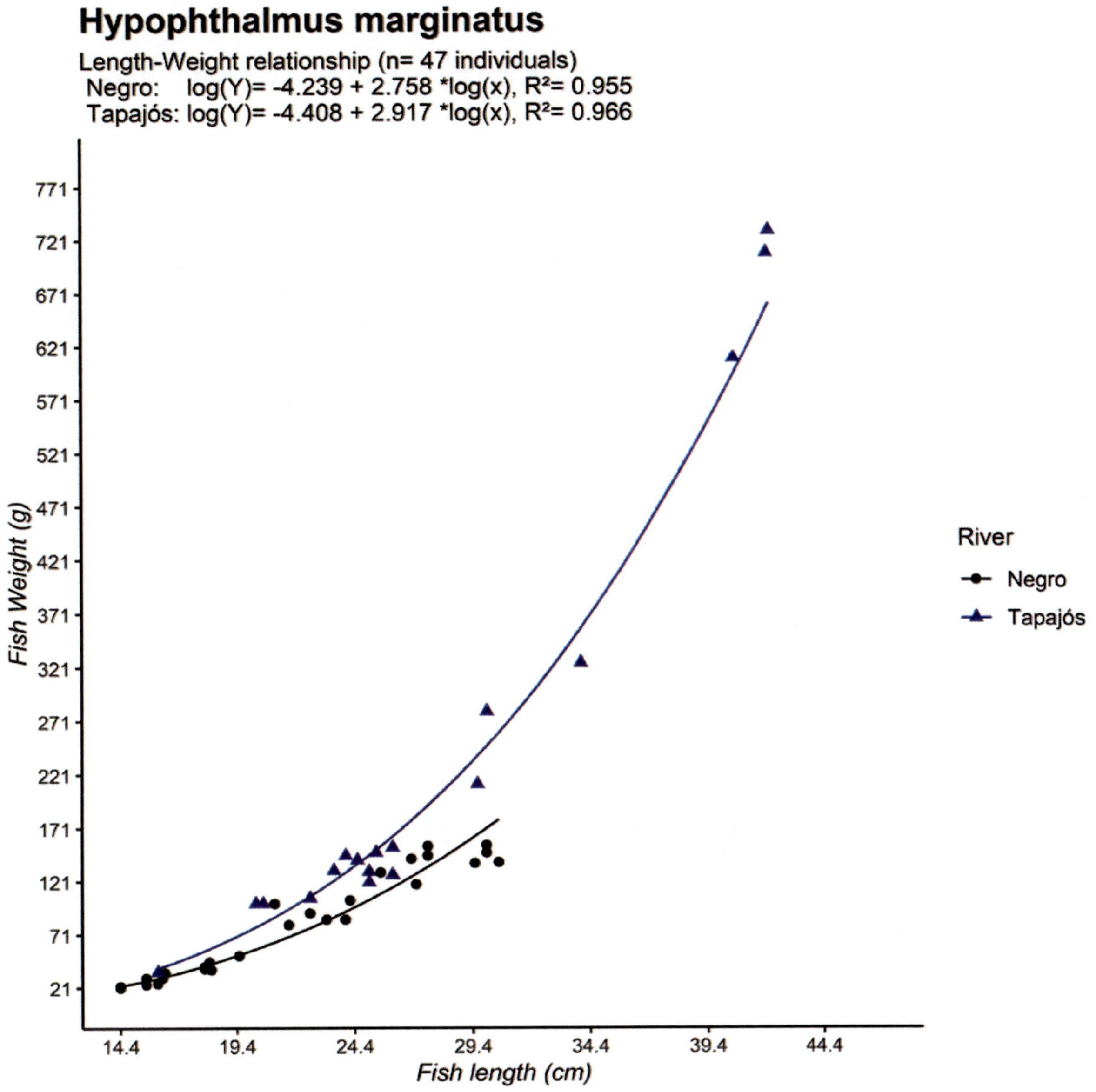

Fig. 5.137 Length-weight relationship for *Hypophthalmus marginatus* sampled in Tapajós and Negro rivers

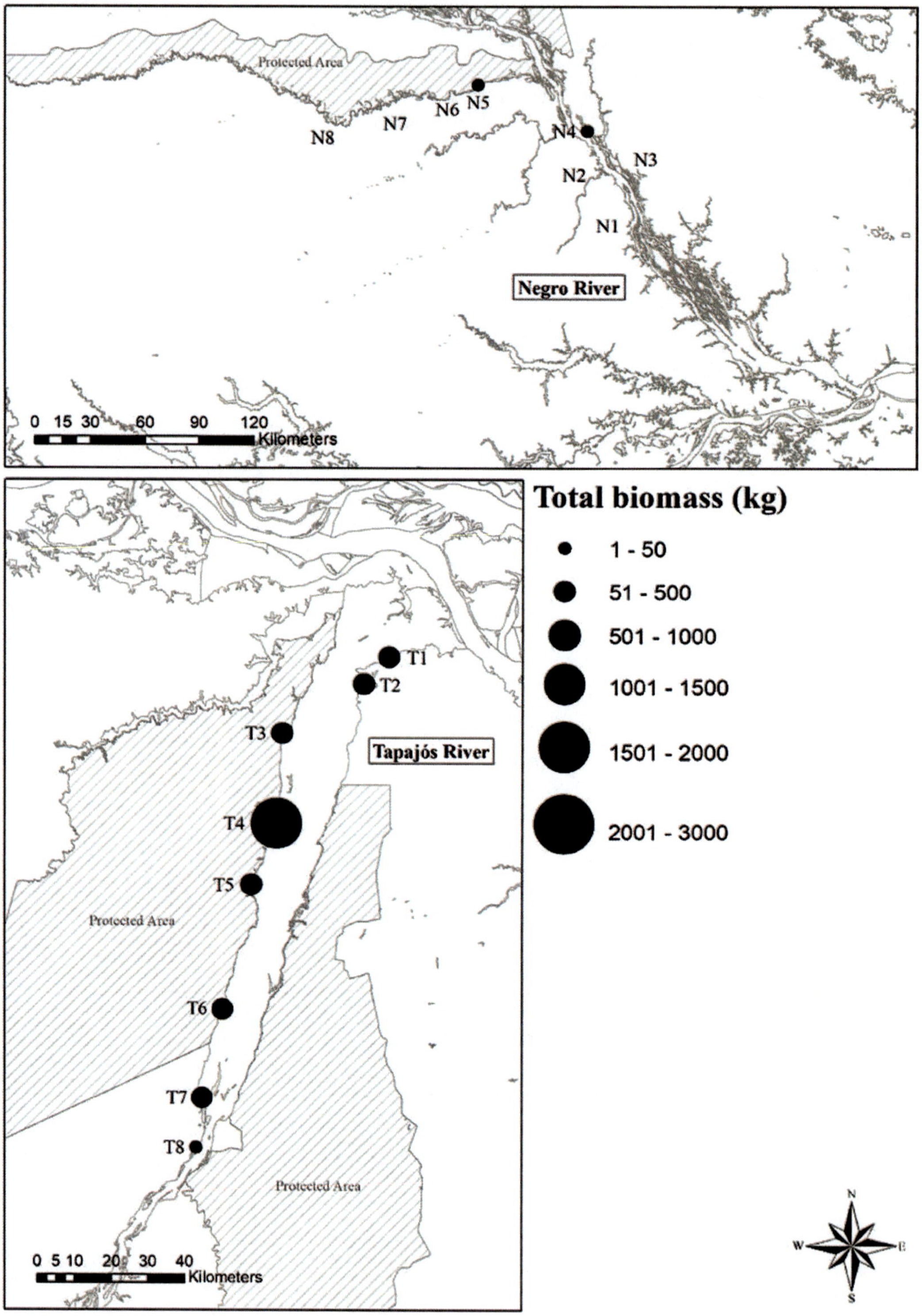

Fig. 5.138 Distribution of biomass of mapará (*Hypophthalmus* spp.) caught in small-scale fisheries in the Negro and Tapajós rivers

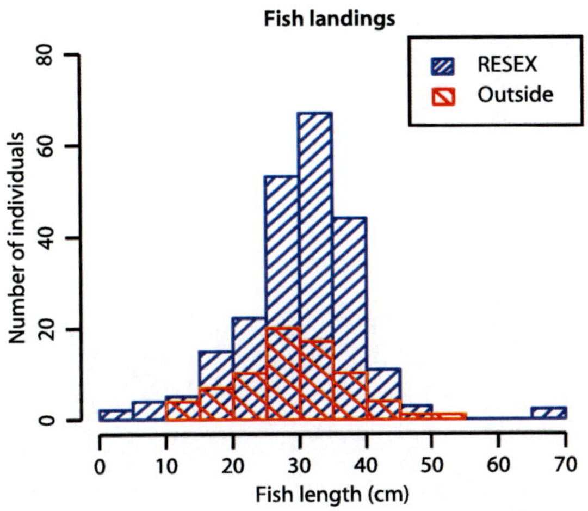

Fig. 5.139 Size distribution of mapará (*Hypophthalmus* spp.) caught in small-scale fisheries in eight communities inside and outside the Extractive Reserves (RESEX) in the Tapajós River, N Resex = 228, N outside = 74

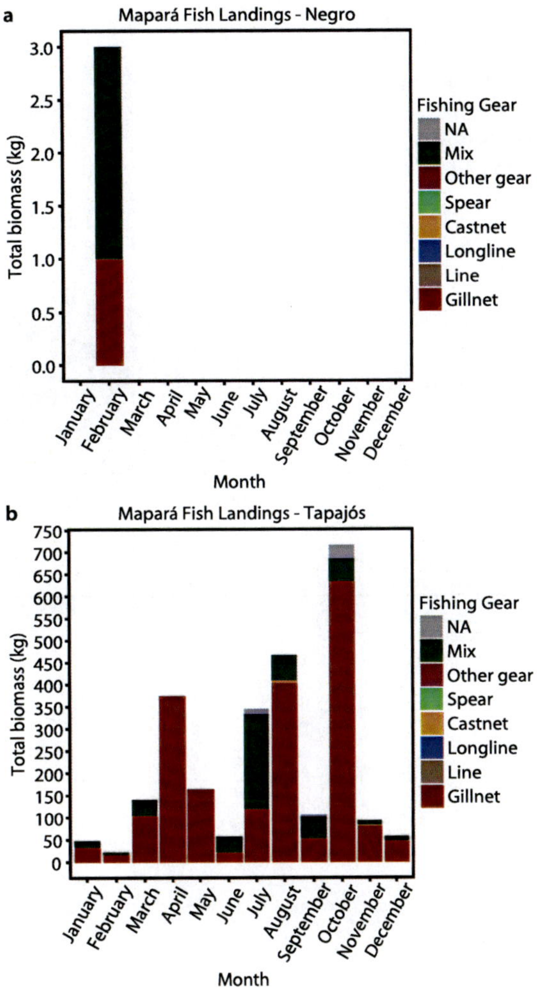

Fig. 5.140 Seasonal variation of biomass of mapará (*Hypophthalmus* spp.) caught per fishing gear in small-scale fisheries in 16 communities in the rivers **a** Negro and **b** Tapajós. NA: fishing gear not informed; Mix: more than one gear was used

Fish species: *Pimelodina flavipinnis* Steindachner 1876

Common name(s): Fura-calça, jandiá

Size of fish sampled (standard length, both rivers): min. = 23.3 cm, max. = 32.5 cm, mean = 26.4 cm, median = 24.9 cm

Fisheries: The fura-calça, mandi or jandiá (*Pimelodina flavipinnis, Pimelodella* spp.) was occasionally caught in the Tapajós River; see *Pimelodella cristata* for fura-calça fisheries data.

Fig. 5.141 *Pimelodina flavipinnis* sampled in the Negro River. (Photo: Renato Silvano)

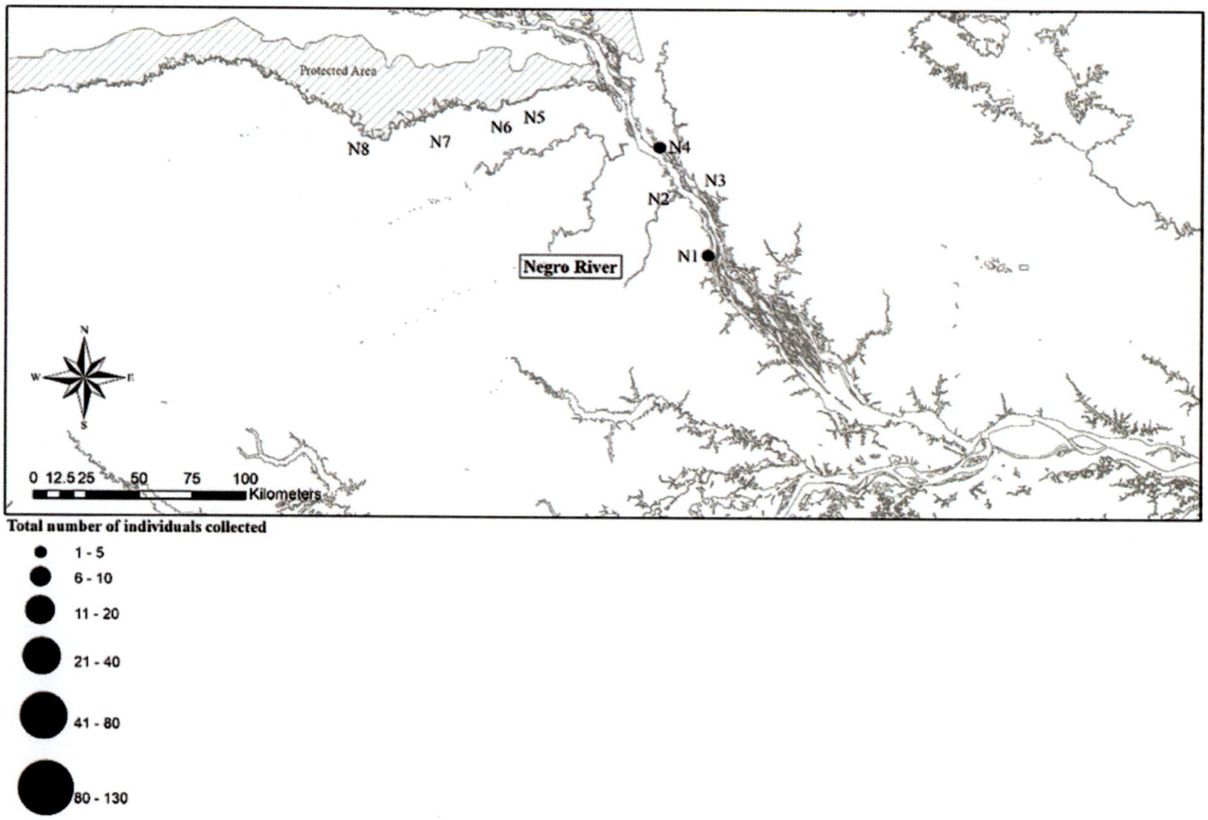

Fig. 5.142 Distribution of individuals of *Pimelodina flavipinnis* sampled in the Negro River

Fish species: ***Pimelodus blochii*** **Valenciennes 1840**

Common name(s): mandi

Size of fish sampled (standard length, both rivers): mean = 9.6 cm, median = 9.5 cm

Fisheries: The mandis (species from Pimelodidae with small to medium sizes) are occasionally caught in the fisheries of Negro River. Biomass caught is in ◘ Table 3.3 (► Chap. 3).

◘ **Fig. 5.143** *Pimelodus blochii* sampled in the Negro River. (Photo: Renato Silvano)

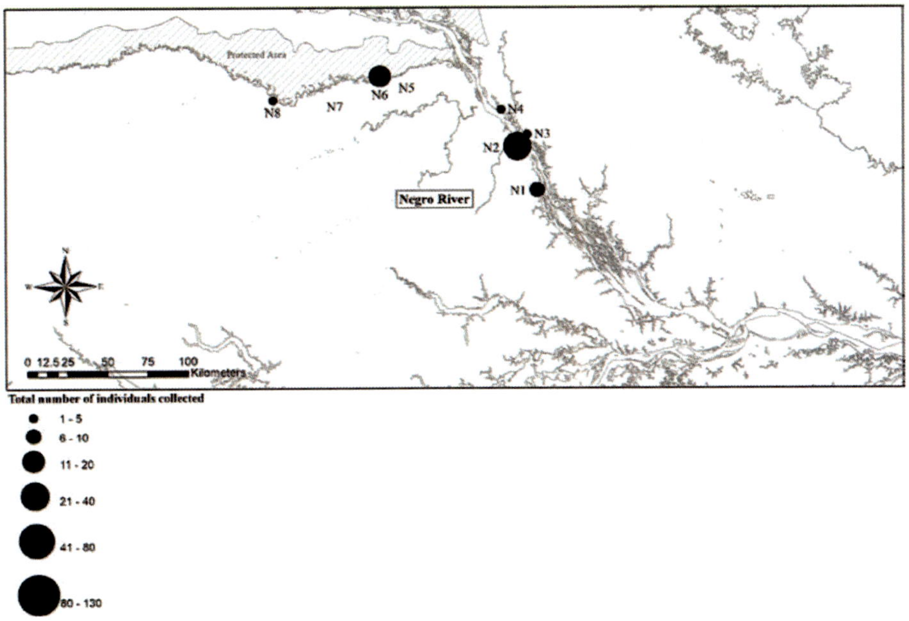

◘ **Fig. 5.144** Distribution of individuals of *Pimelodus blochii* sampled in the Negro River

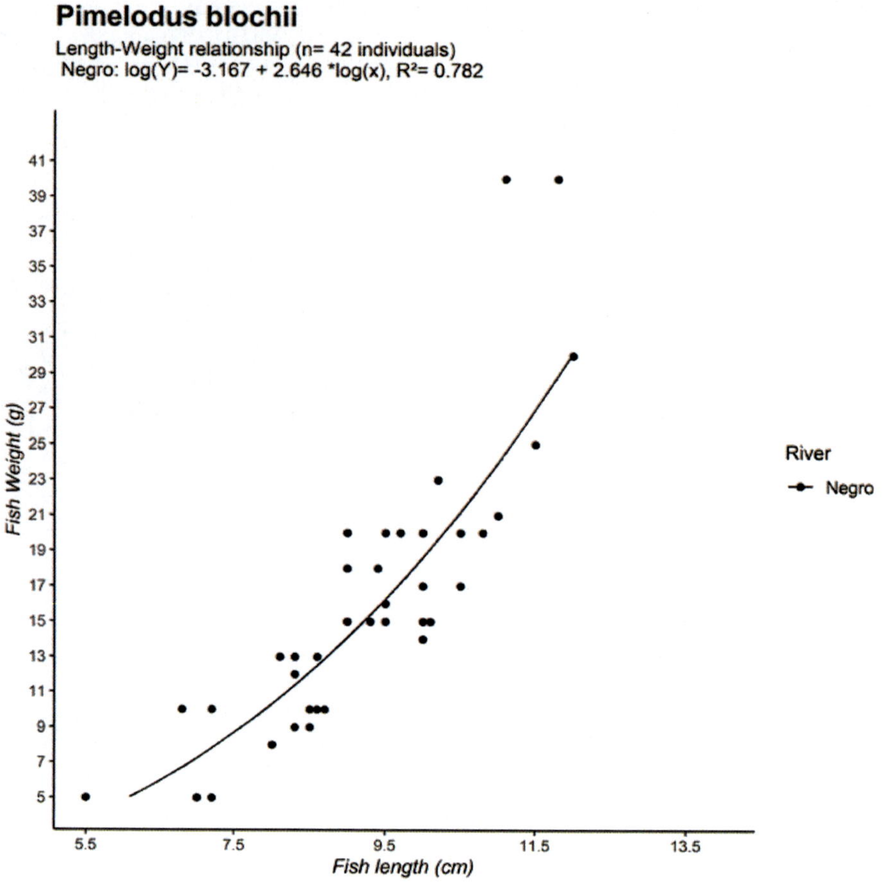

Fig. 5.145 Length-weight relationship for *Pimelodus blochii* sampled in Negro River

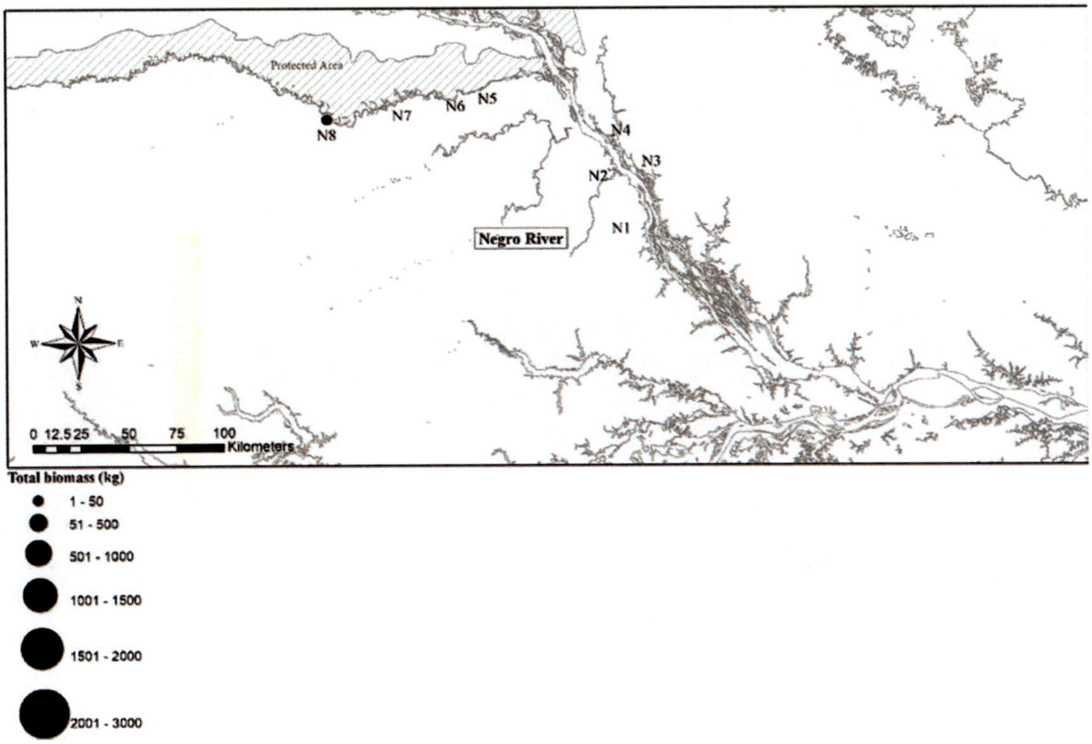

Fig. 5.146 Distribution of biomass of mandis (Pimelodidae) caught in small-scale fisheries in the Negro River

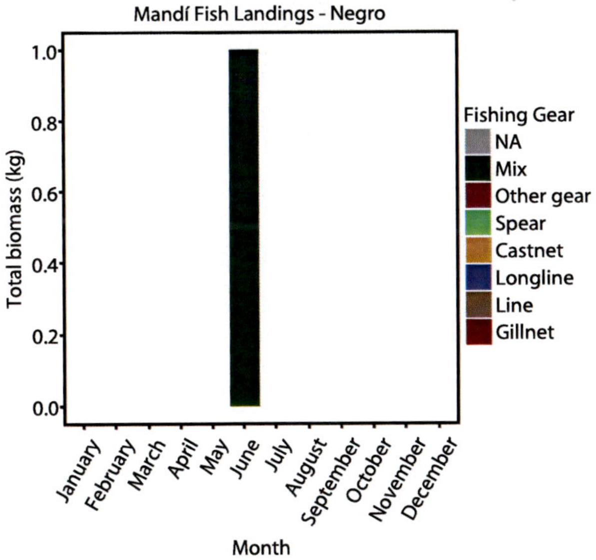

Fig. 5.147 Seasonal variation of biomass of mandis (Pimelodidae) caught per fishing gear in small-scale fisheries in eight communities in the Negro River. NA: fishing gear not informed; Mix: more than one gear was used

Fish species: *Pimelodus tetramerus* Ribeiro & Lucena 2006

Common name(s): mandi, mandi-pintado

Size of fish sampled (standard length, both rivers): mean = 10.3 cm, median = 10 cm

Fisheries: The mandis (species from Pimelodidae with small to medium sizes) are occasionally caught in the fisheries of Negro River; see *Pimelodus blochii* for mandi fisheries data.

Fig. 5.148 *Pimelodus tetramerus* sampled in the Tapajós River. (Photo: Renato Silvano)

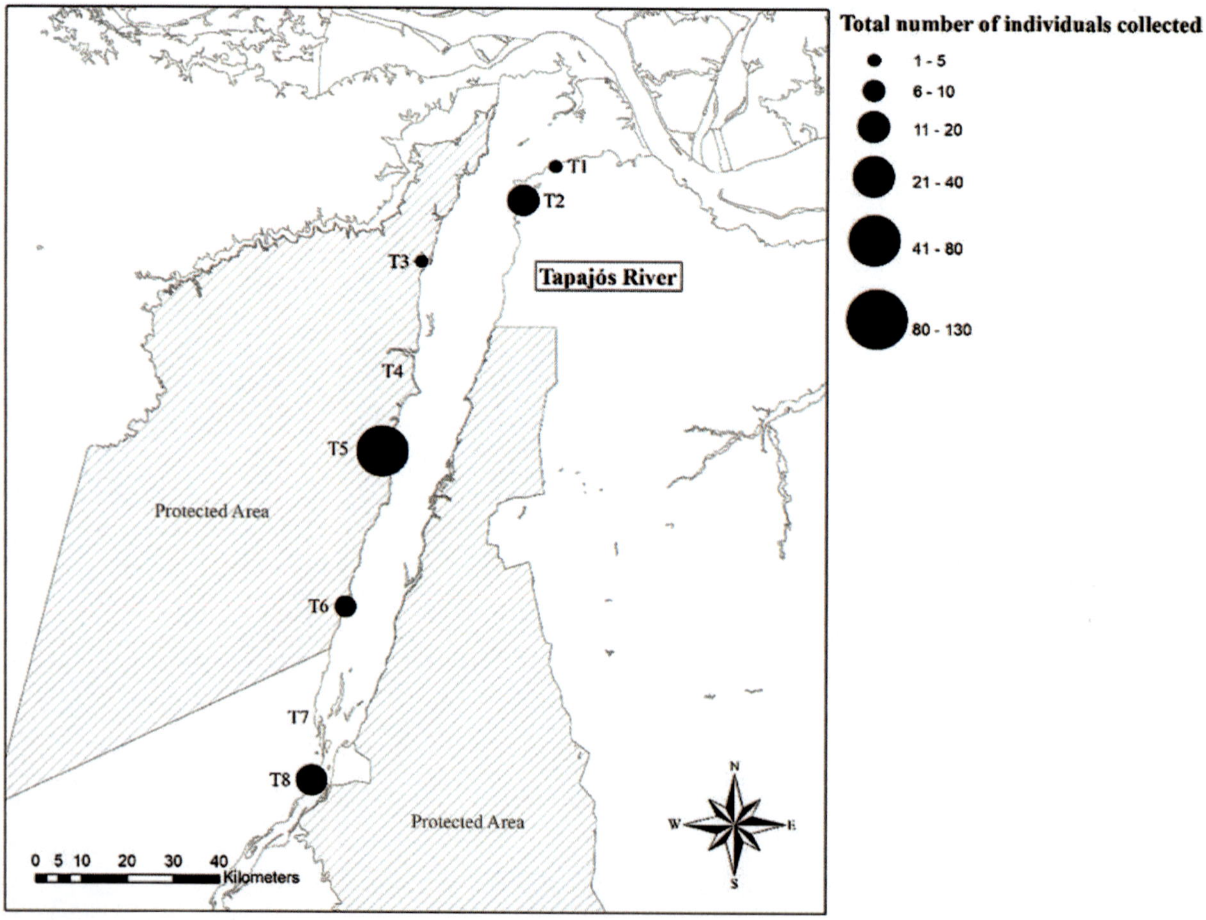

Fig. 5.149 Distribution of individuals of *Pimelodus tetramerus* sampled in the Tapajós River

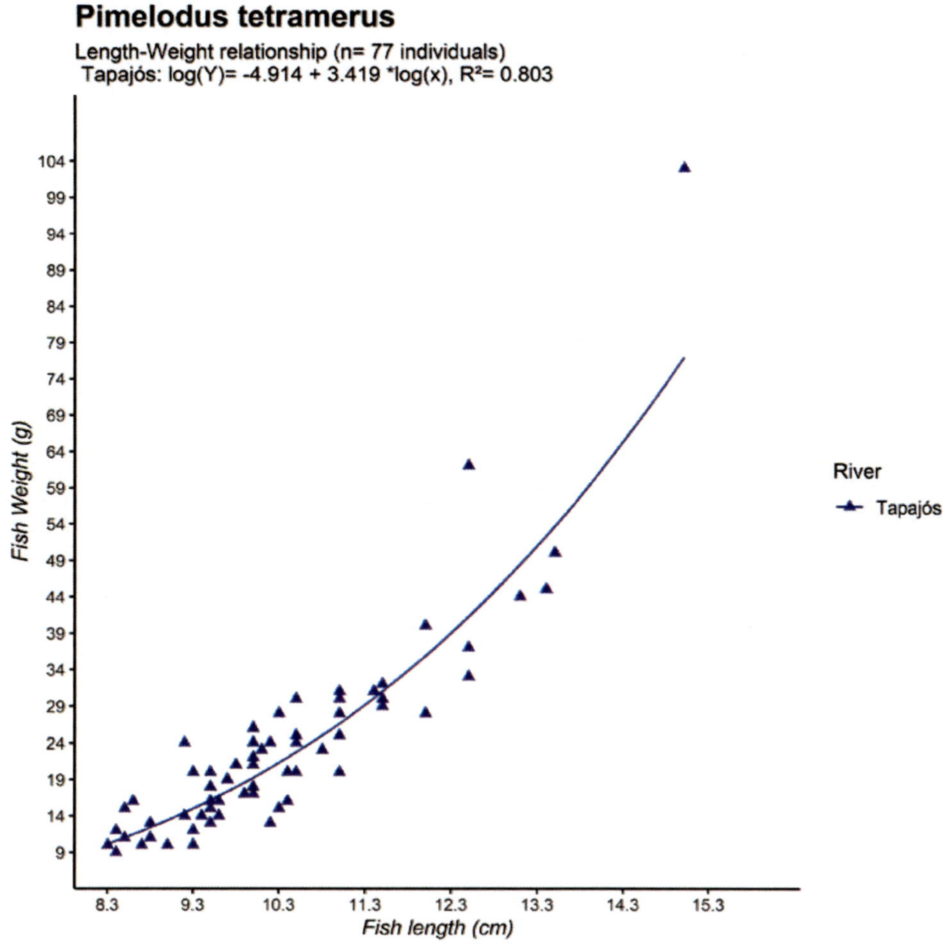

Fig. 5.150 Length-weight relationship for *Pimelodus tetramerus* sampled in the Tapajós River

---

Fish species: *Pimelodus ornatus* **Kner 1858**

Common name(s): mandi

Size of fish sampled (standard length, both rivers): min. = 11 cm, max. = 17 cm, mean and median = 14 cm

Fisheries: The mandis (species from Pimelodidae with small to medium sizes) are occasionally caught in the fisheries of Negro River; see *Pimelodus blochii* for mandi fisheries data.

Fig. 5.151 *Pimelodus ornatus* sampled in the Negro River. (Photo: Renato Silvano)

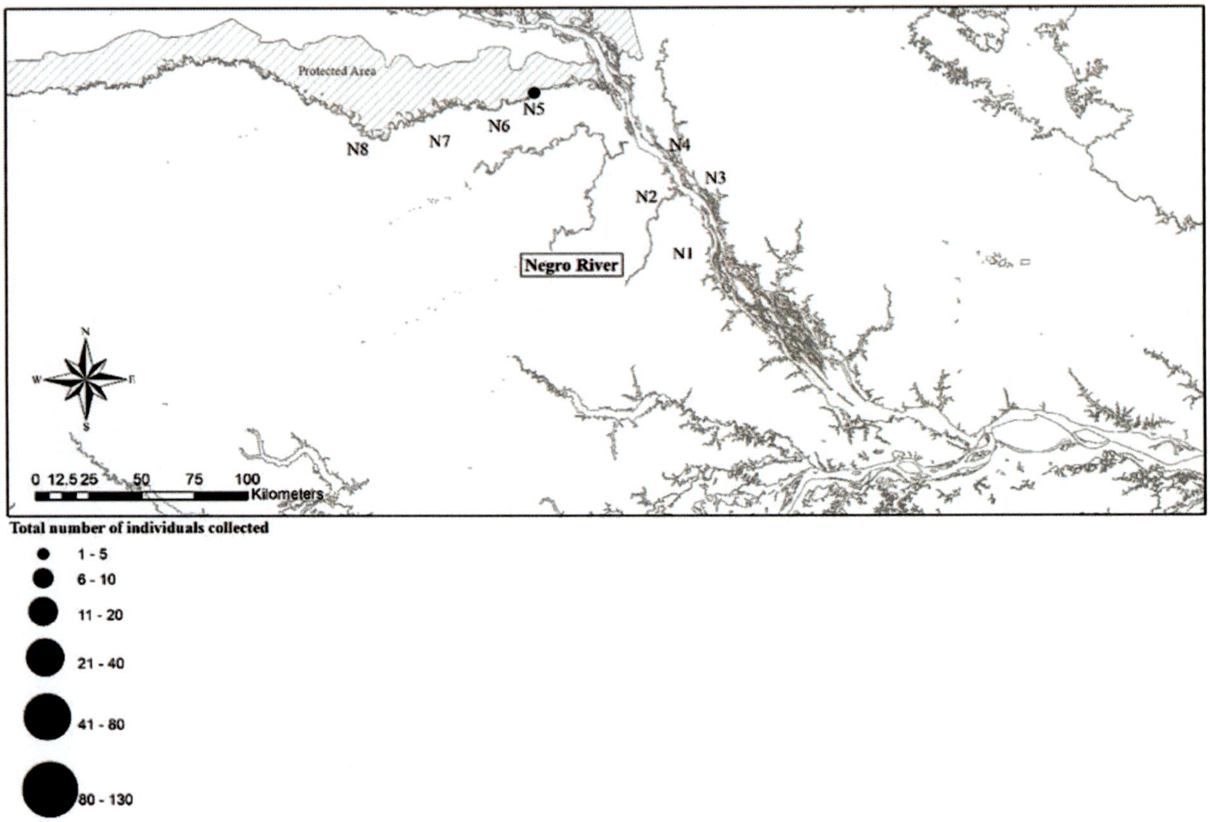

Fig. 5.152 Distribution of individuals of *Pimelodus ornatus* sampled in the Negro River

Fish species: *Pinirampus pirinampu* (Spix & Agassiz 1829)

Common name(s): barbado, barba-chata

Size of fish sampled (standard length, both rivers): mean = 22.7 cm, median = 22.5 cm

Fisheries: The barbado (*Pinirampus pirinampu*) is occasionally caught in both rivers. Biomass caught is in Table 3.3 (▶ Chap. 3).

Fig. 5.153 *Pinirampus pirinampu* sampled in the Negro River. (Photo: Renato Silvano)

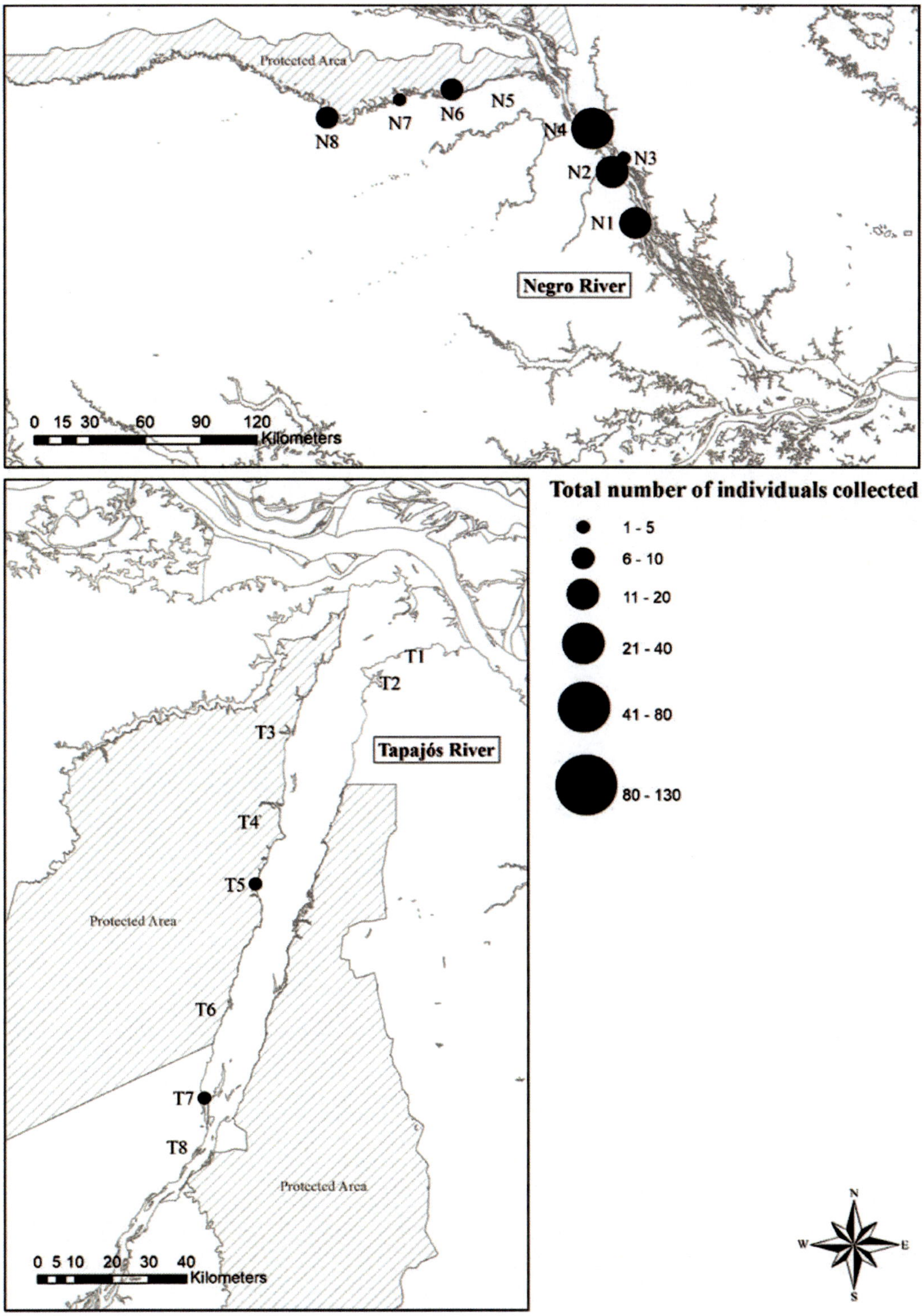

Fig. 5.154 Distribution of individuals of *Pinirampus pirinampu* sampled in the Negro and Tapajós rivers

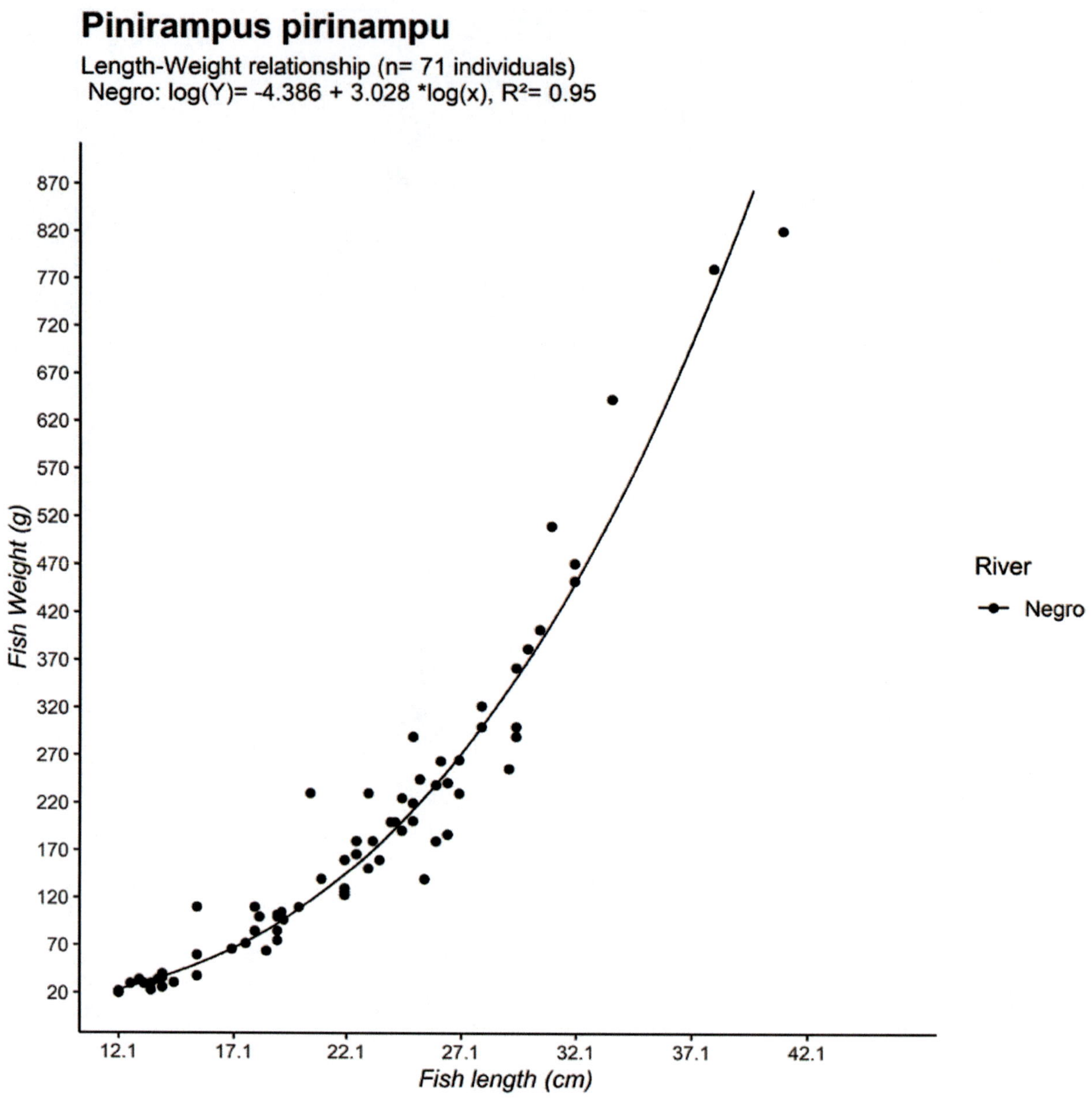

Fig. 5.155 Length-weight relationship for *Pinirampus pirinampu* sampled in the Negro River

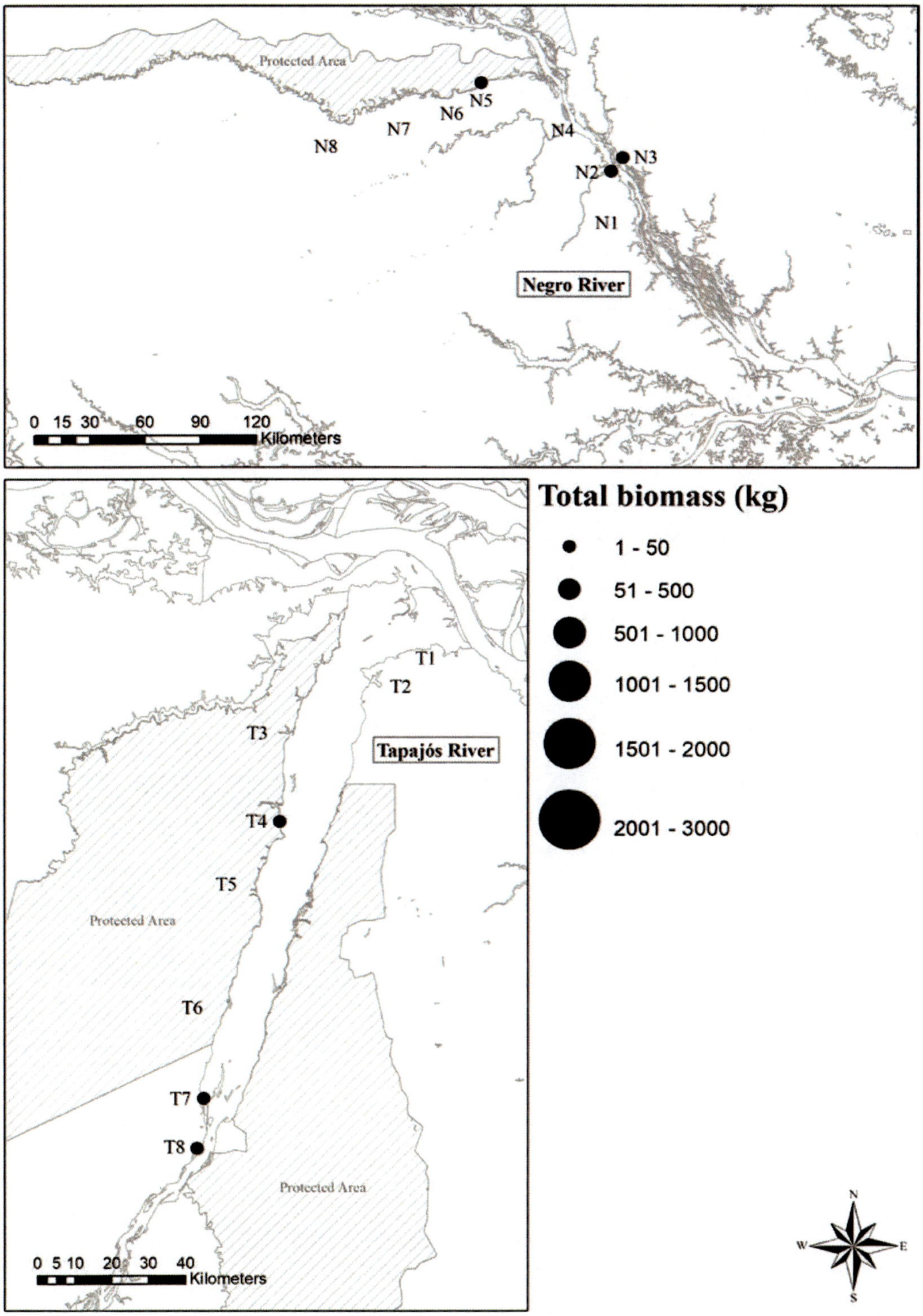

Fig. 5.156 Distribution of biomass of barbado (*P. pirinampu*) caught in small-scale fisheries in the Negro and Tapajós Rivers

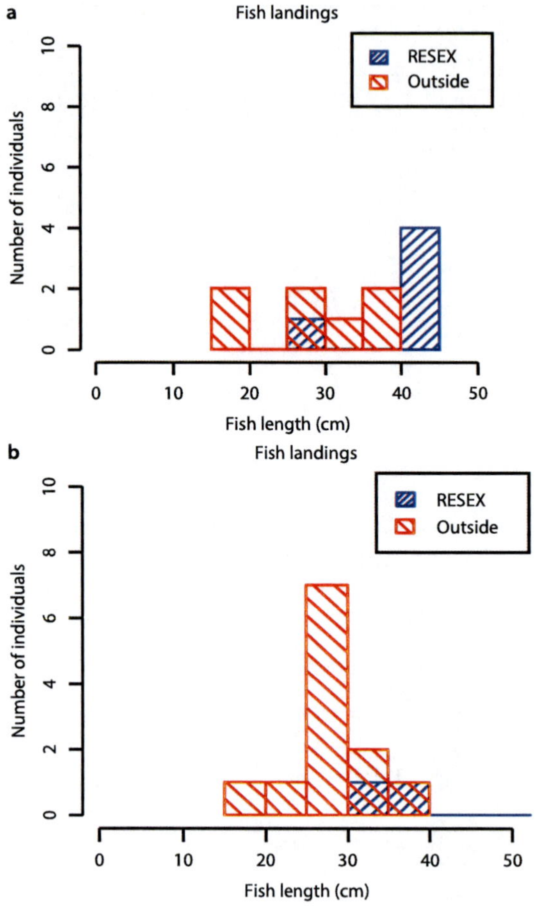

Fig. 5.157 Size distribution of barbado (*P. pirinampu*) caught in small-scale fisheries in 16 communities inside and outside the Extractive Reserves (RESEX) in the rivers **a** Negro, N Resex = 5, N outside = 7 and **b** Tapajós, N Resex = 2, N outside = 12

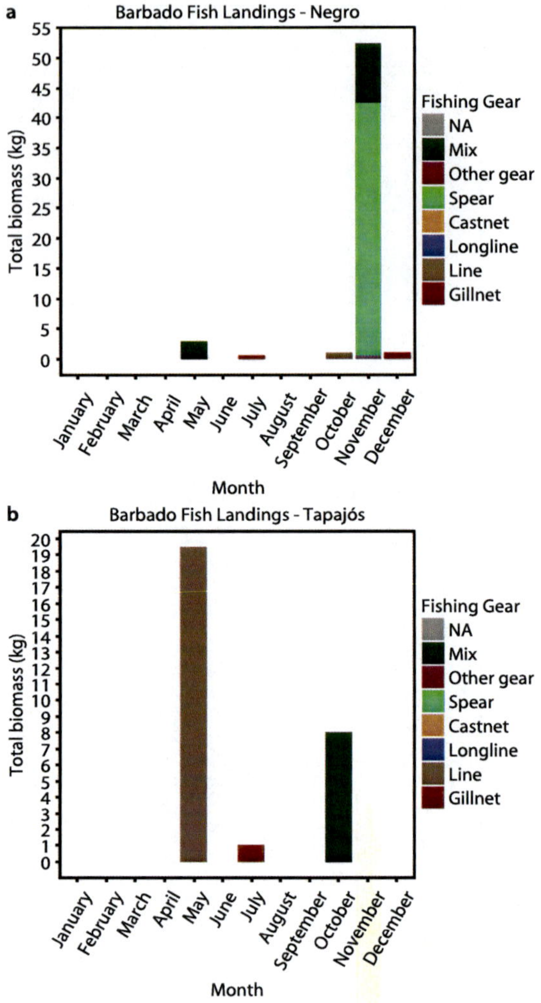

Fig. 5.158 Seasonal variation of biomass of barbado (*P. pirinampu*) caught per fishing gear in small-scale fisheries in 16 communities in the rivers **a** Negro and **b** Tapajós. NA: fishing gear not informed; Mix: more than one gear was used

Fish species: *Platynematichthys notatus* **(Jardine 1841)**

Common name(s): cara-de-gato, piranambu

Size of fish sampled (standard length, both rivers): mean = 20.5 cm, median = 18.4 cm

Fisheries: The cara de gato (*P. notatus*) is occasionally caught in fisheries in Tapajós River. Biomass caught is in ■ Table 3.3 (► Chap. 3).

■ **Fig. 5.159** *Platynematichthys notatus* sampled in the Tapajós River. (Photo: Renato Silvano)

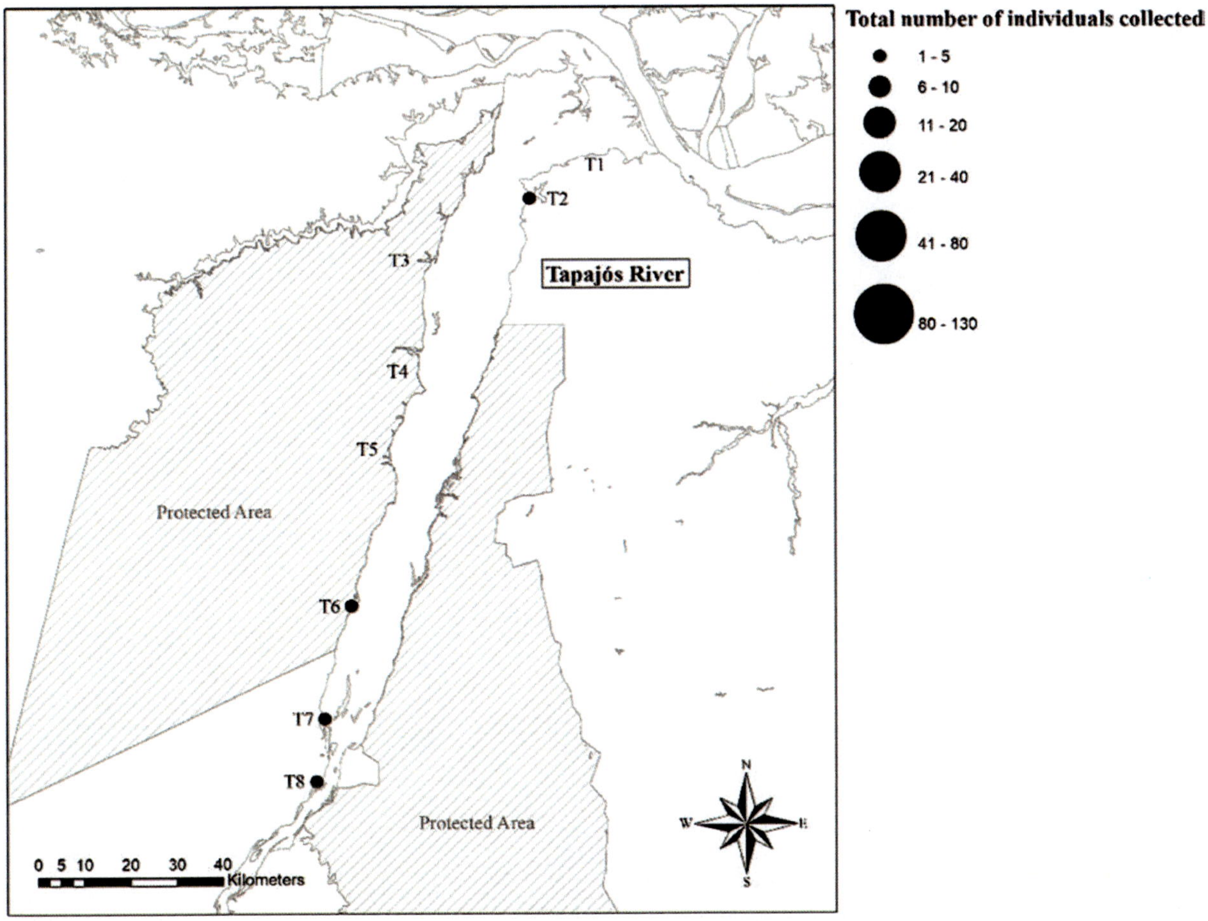

■ **Fig. 5.160** Distribution of individuals of *Platynematichthys notatus* sampled in the Tapajós River

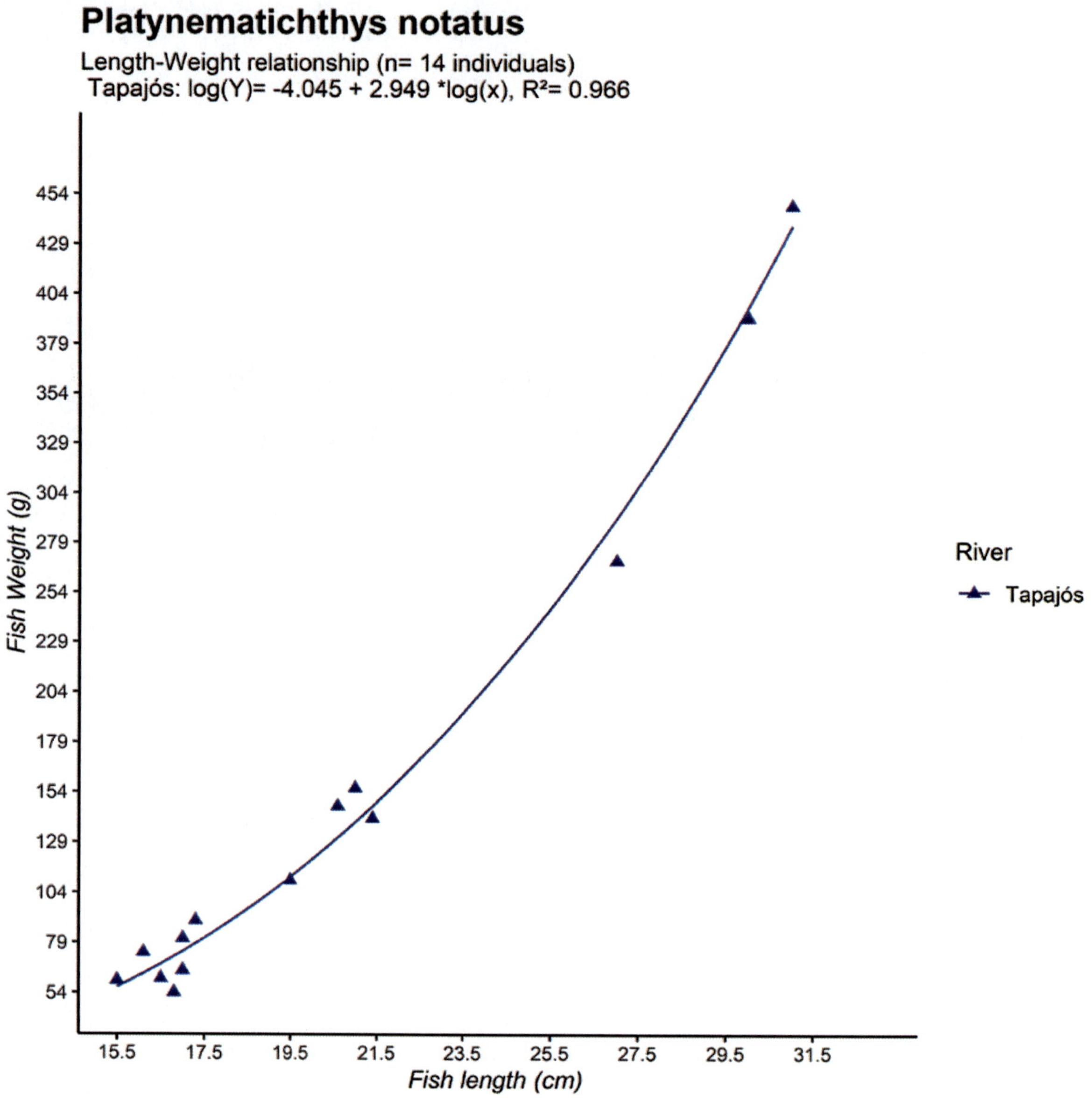

Fig. 5.161 Length-weight relationship for *Platynematichthys notatus* sampled in Tapajós River

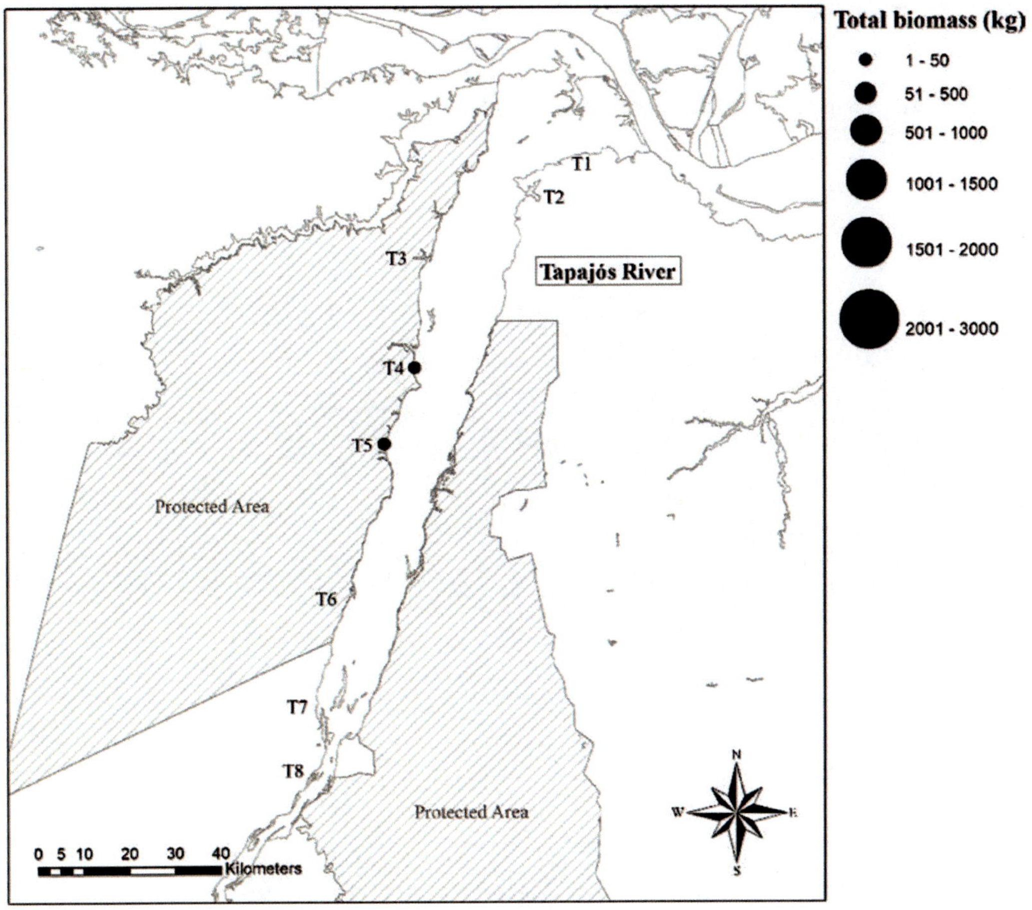

Fig. 5.162 Distribution of biomass of the cara de gato (*P. notatus*) caught in small-scale fisheries in the Tapajós River

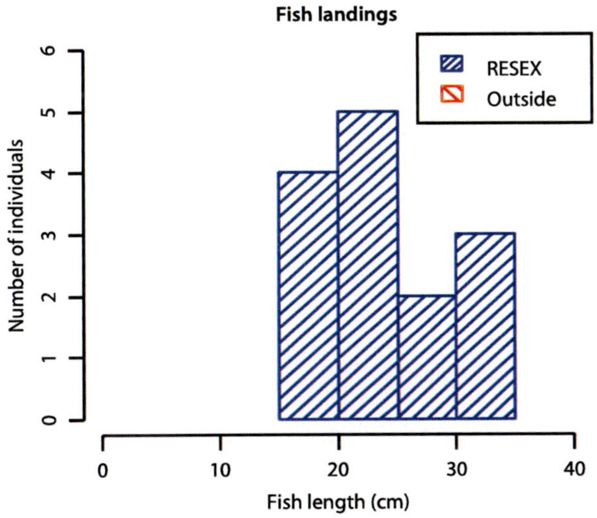

Fig. 5.163 Size distribution of cara de gato (*P. notatus*) caught in small-scale fisheries in eight communities inside and outside the Extractive Reserves (RESEX) in the Tapajós River, N Resex = 14, N outside = 0

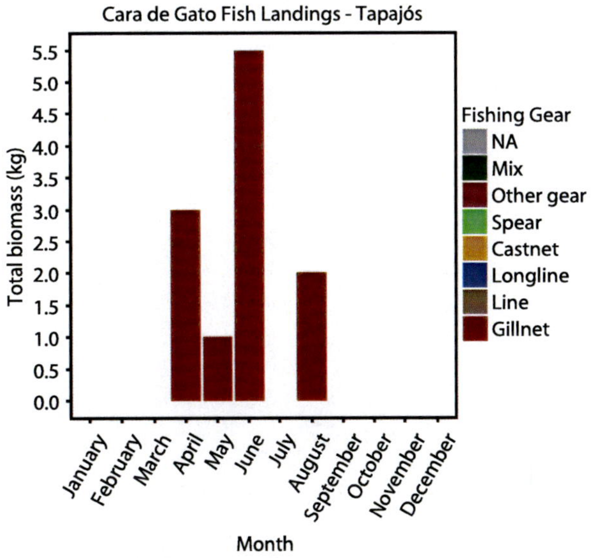

Fig. 5.164 Seasonal variation of biomass of the cara de gato (*P. notatus*) caught per fishing gear in small-scale fisheries in eight communities in the Tapajós River. NA: fishing gear not informed; Mix: more than one gear was used

5

Fish species: *Platystomatichthys sturio* (Kner 1858)

Common name(s): bico-de-pato

Size of fish sampled (standard length, both rivers): mean = 24 cm, median = 24.4 cm

Fisheries: Not relevant to fisheries in both rivers, sometimes may be occasionally caught together with mandis and fura-calça. See *Pimelodus blochii* for mandi fisheries data.

Fig. 5.165 *Platystomatichthys sturio* sampled in the Tapajós River. (Photo: Renato Silvano)

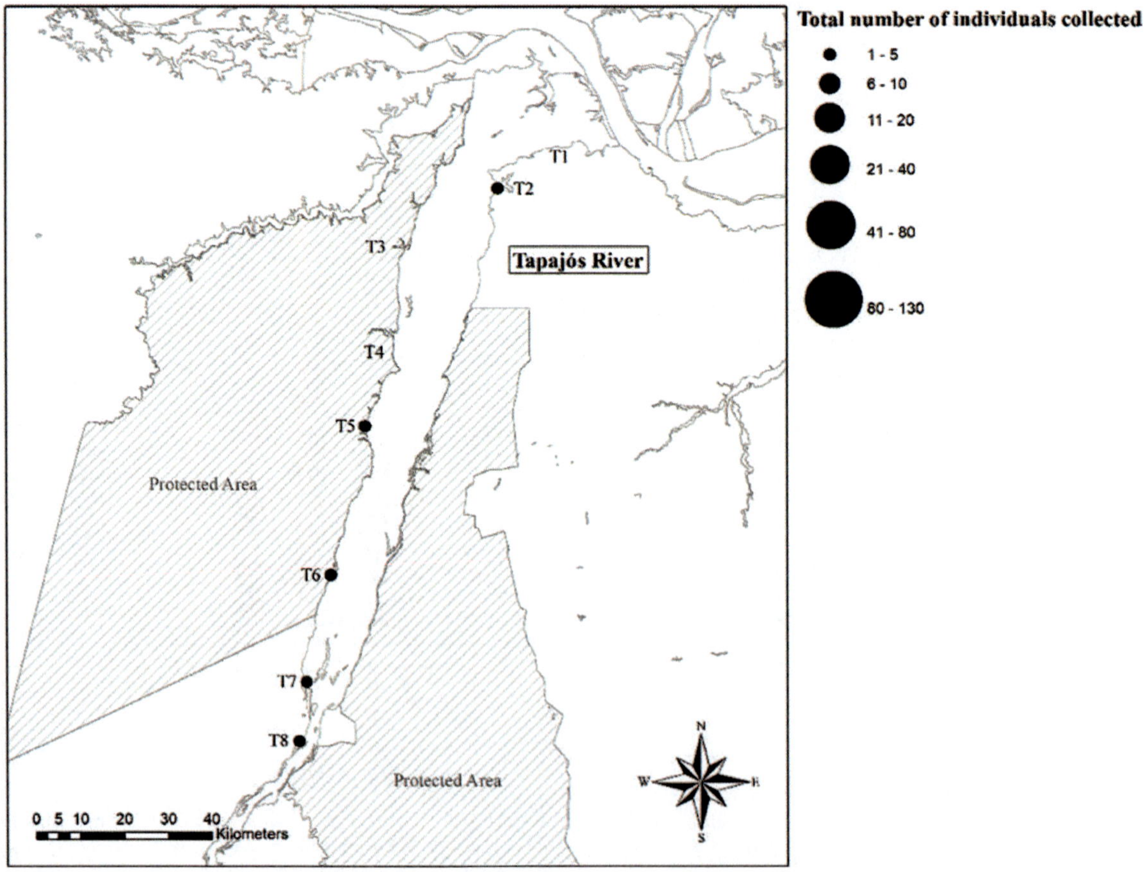

Fig. 5.166 Distribution of individuals of *Platystomatichthys sturio* sampled in the Tapajós River

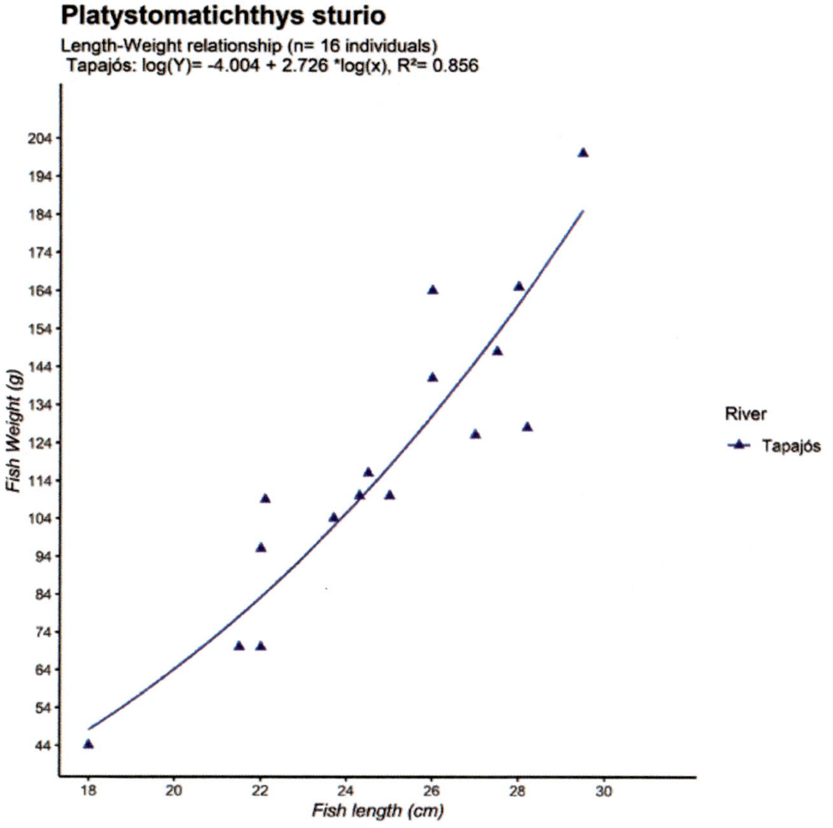

Fig. 5.167 Length-weight relationship for *Platystomatichthys sturio* sampled in the Tapajós River

Fish species: *Pseudoplatystoma punctifer* (Castelnau 1855)

Common name(s): surubim

Size of fish sampled (standard length, both rivers): min. = 24.5 cm, max. = 26.5 cm, mean and median = 25.5 cm

Fisheries: Although not among the most caught fish, the surubim (*Pseudoplatystoma* spp.) is an important commercial fish in both rivers. Biomass caught is in Table 3.3 (▶ Chap. 3); see *Pseudoplatystoma tigrinum* for surubim fisheries data.

Fig. 5.168 *Pseudoplatystoma punctifer* sampled in the Negro River. (Photo: Renato Silvano)

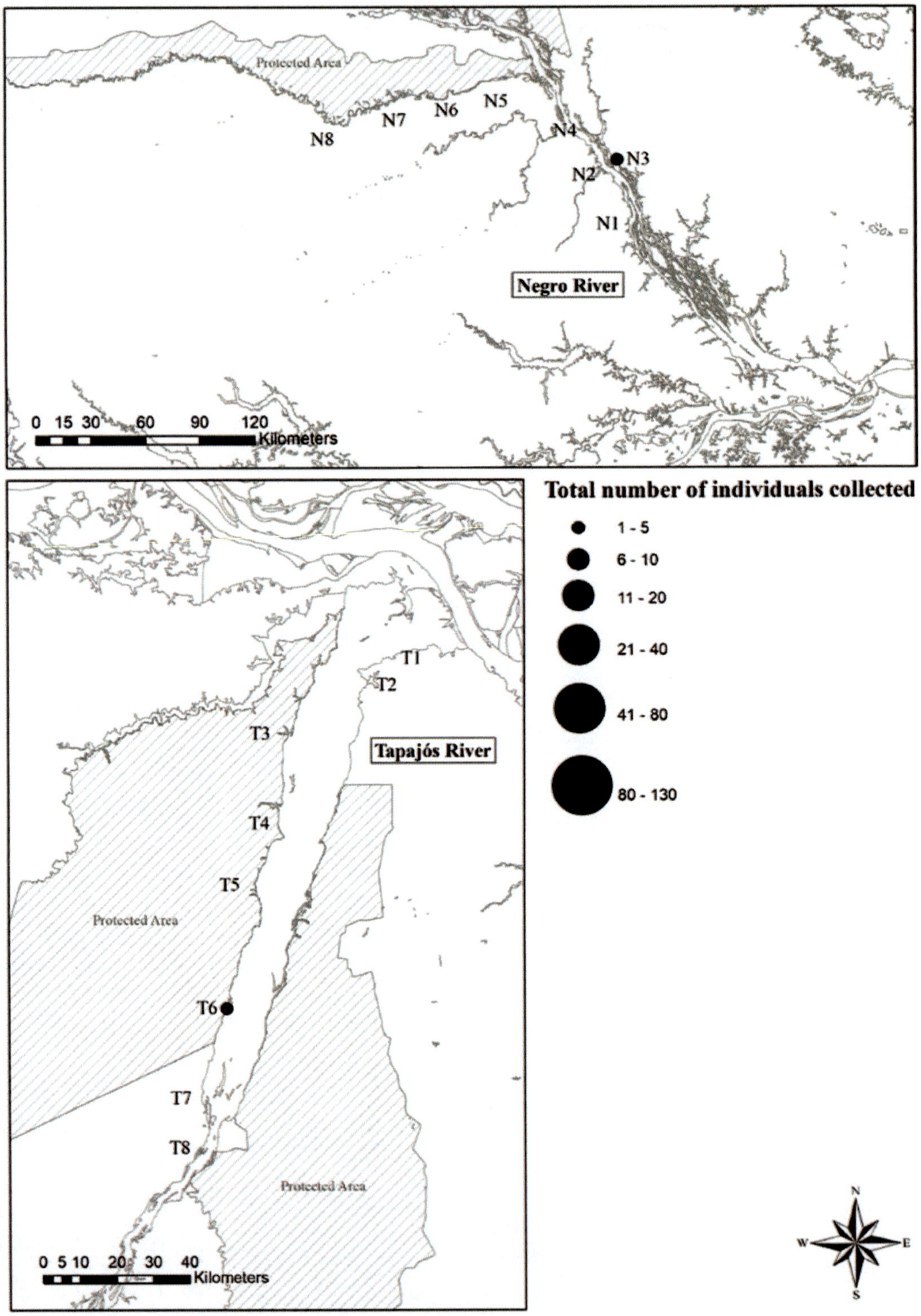

Fig. 5.169 Distribution of individuals of *Pseudoplatystoma punctifer* sampled in the Negro and Tapajós rivers

Fish species: ***Pseudoplatystoma tigrinum* (Valenciennes 1840)**

Common name(s): surubim, surubim-pintado, caparari

Size of fish sampled (standard length, both rivers): mean = 27.7 cm, median = 27 cm

Fisheries: Although not among the most caught fish, the surubim (*Pseudoplatystoma* spp.) is an important commercial fish in both rivers. Biomass caught is in ◘ Table 3.3 (► Chap. 3).

◘ **Fig. 5.170** *Pseudoplatystoma tigrinum* sampled in the Tapajós River. (Photo: Renato Silvano)

5

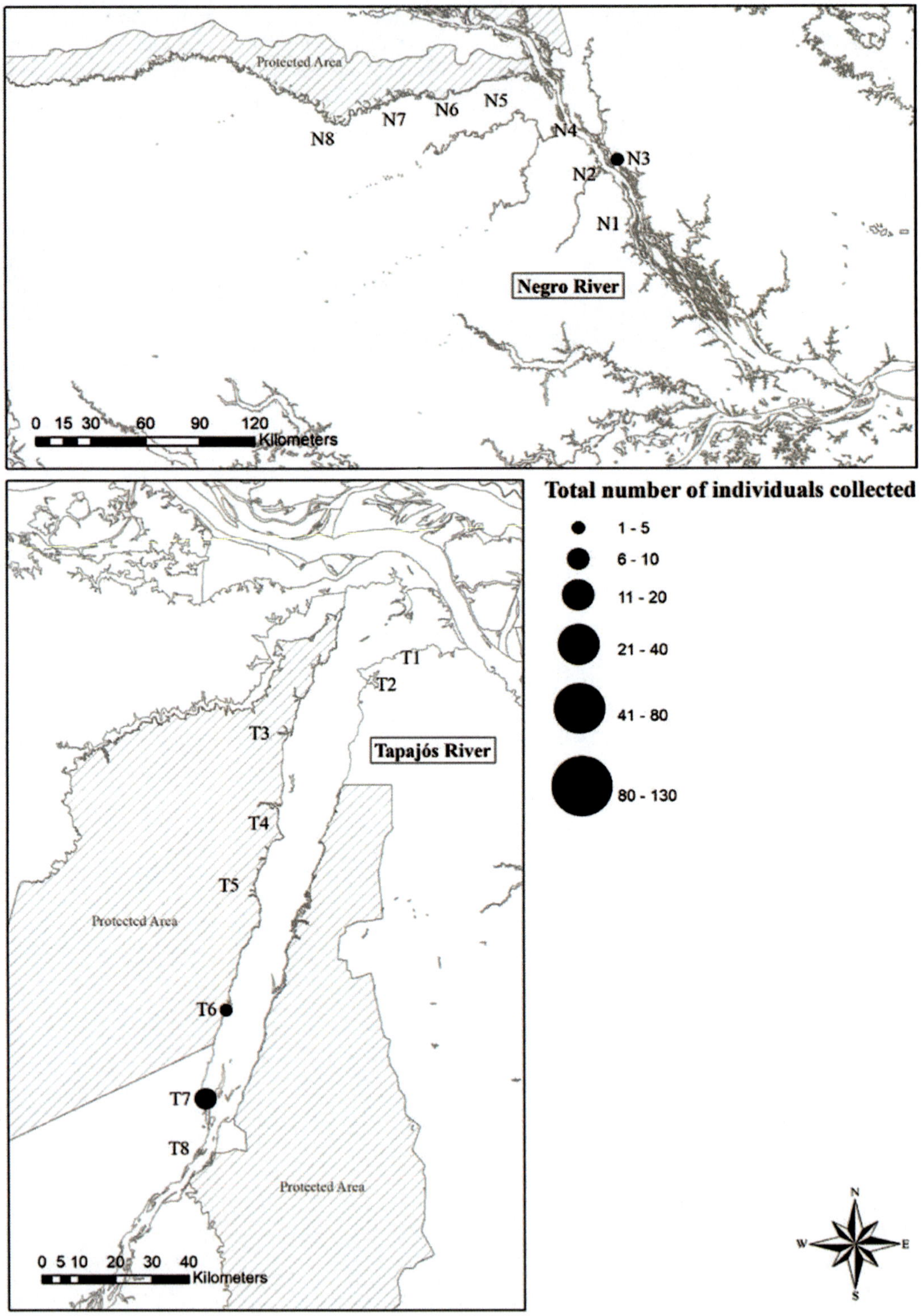

Fig. 5.171 Distribution of individuals of *Pseudoplatystoma tigrinum* sampled in the Negro and Tapajós rivers

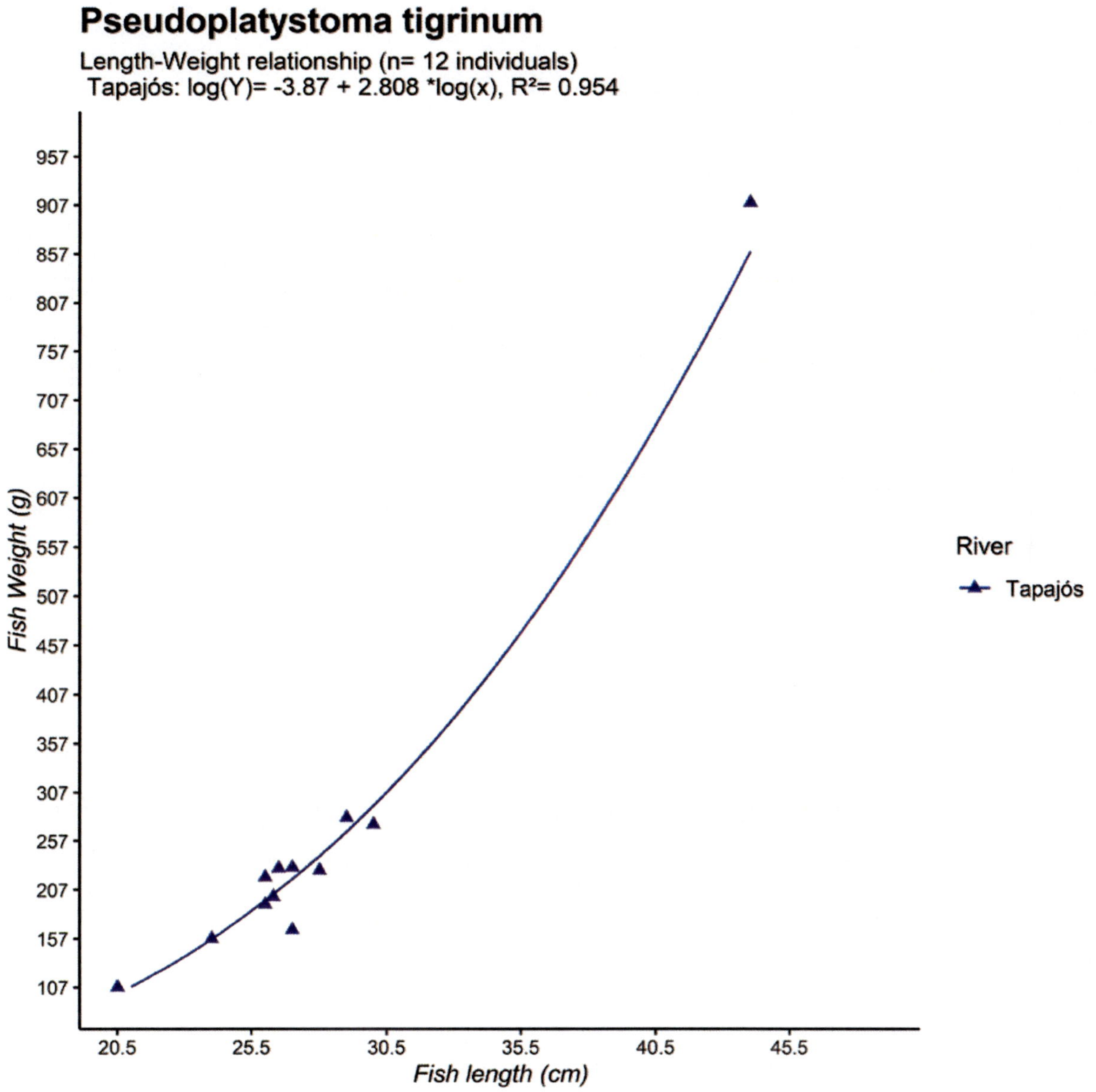

Fig. 5.172 Length-weight relationship for *Pseudoplatystoma tigrinum* sampled in the Tapajós River

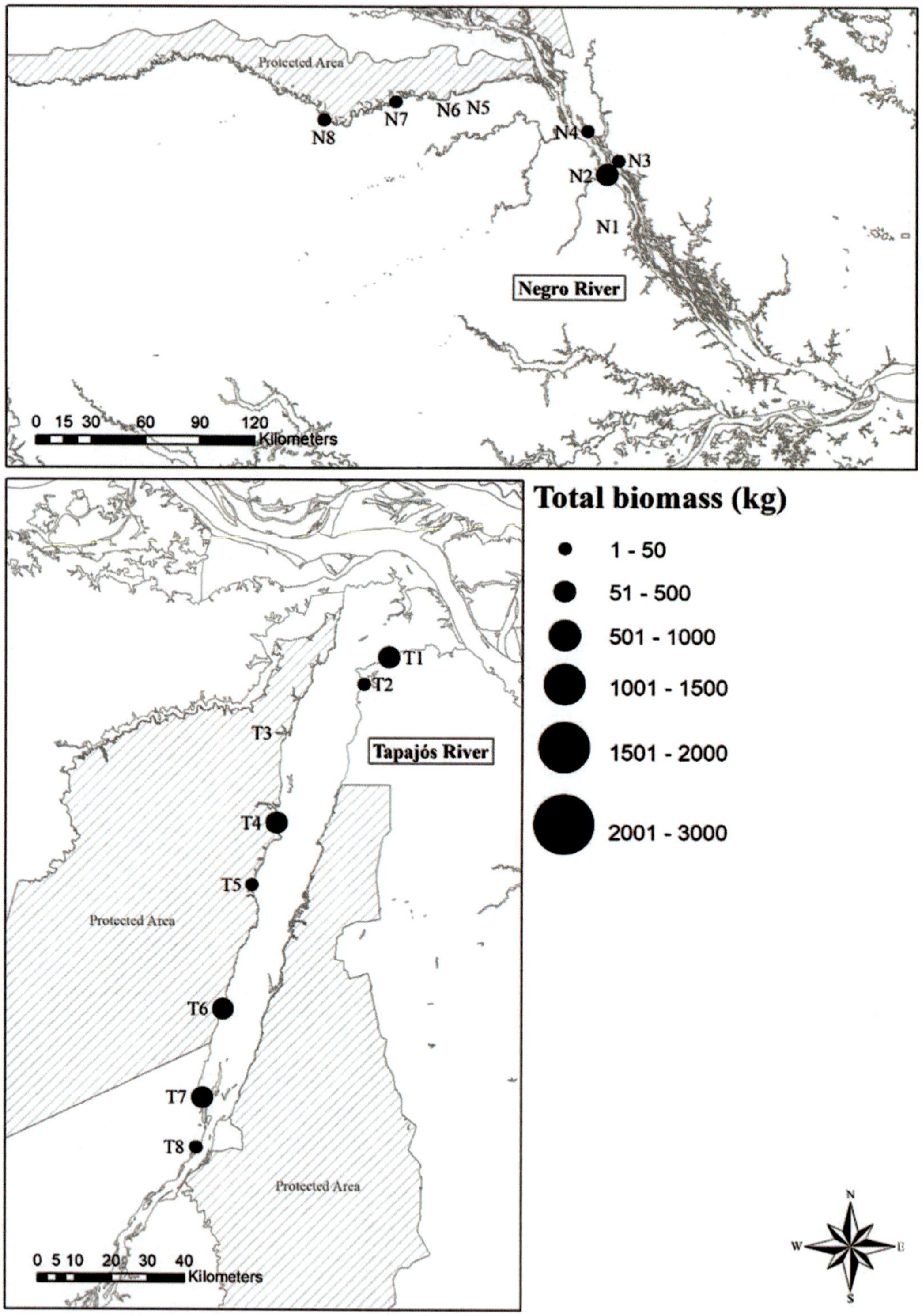

■ **Fig. 5.173** Distribution of biomass of surubim (*Pseudoplatystoma* spp.) caught in small-scale fisheries in the Negro and Tapajós rivers

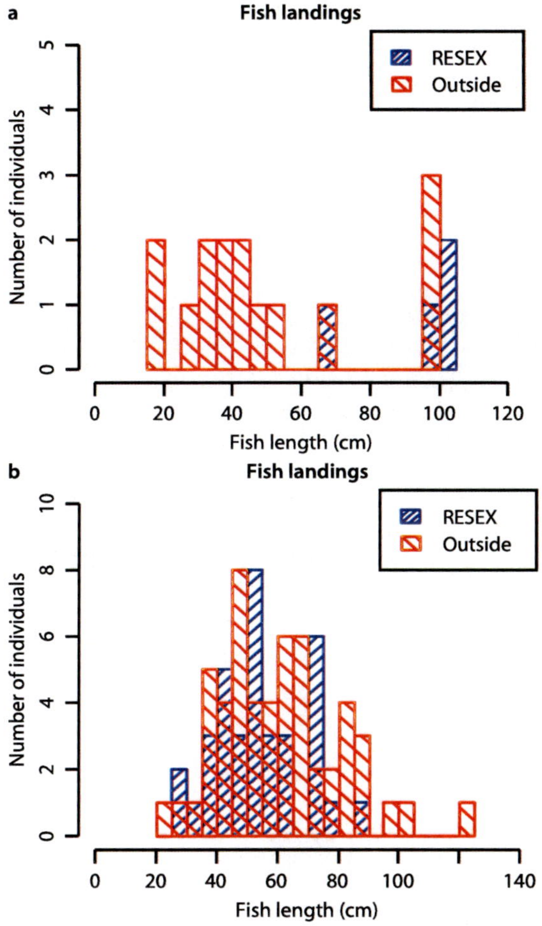

Fig. 5.174 Size distribution of surubim (*Pseudoplatystoma* spp.) caught in small-scale fisheries in 16 communities inside and outside the Extractive Reserves (RESEX) in the rivers **a** Negro, N Resex = 4, N outside = 15 and **b** Tapajós, N Resex = 36, N outside = 54

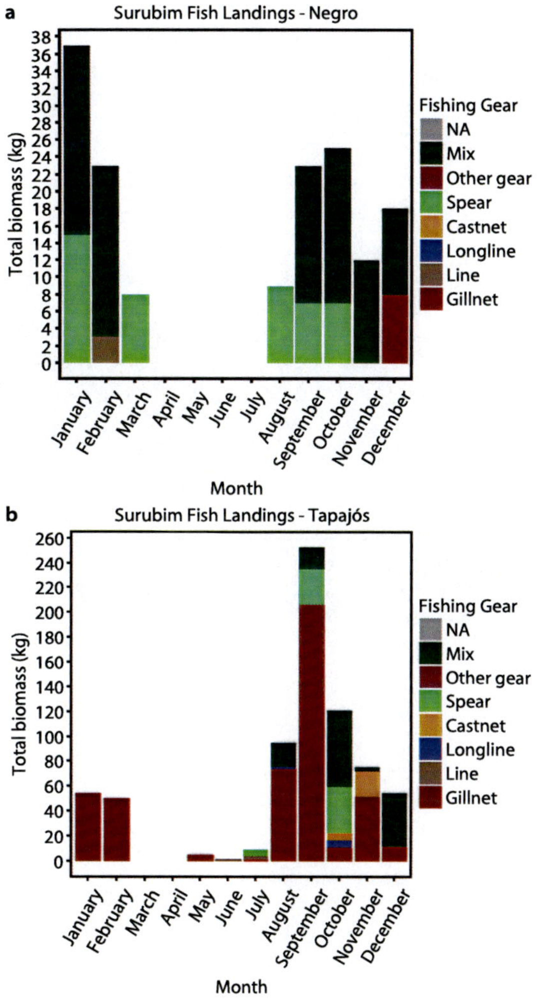

Fig. 5.175 Seasonal variation of biomass of surubim (*Pseudoplatystoma* spp.) caught per fishing gear in small-scale fisheries in 16 communities in the rivers **a** Negro and **b** Tapajós. NA: fishing gear not informed; Mix: more than one gear was used

Fish species: *Sorubim elongatus* **Littmann, Burr, Schmidt & Isern 2001**
Common name(s): maparaí

Size of fish sampled (standard length, both rivers): min. = 19.6 cm, max. = 22.2 cm, mean = 20.7 cm, median = 20.5 cm

Fisheries: Not relevant to fisheries in both rivers, sometimes may be occasionally caught together with mandis and fura-calça. Biomass caught is in ◘ Table 3.3 (► Chap. 3); see *Pimelodus blochii* for mandi fisheries data.

5

◘ **Fig. 5.176** *Sorubim elongatus* sampled in the Tapajós River. (Photo: Renato Silvano)

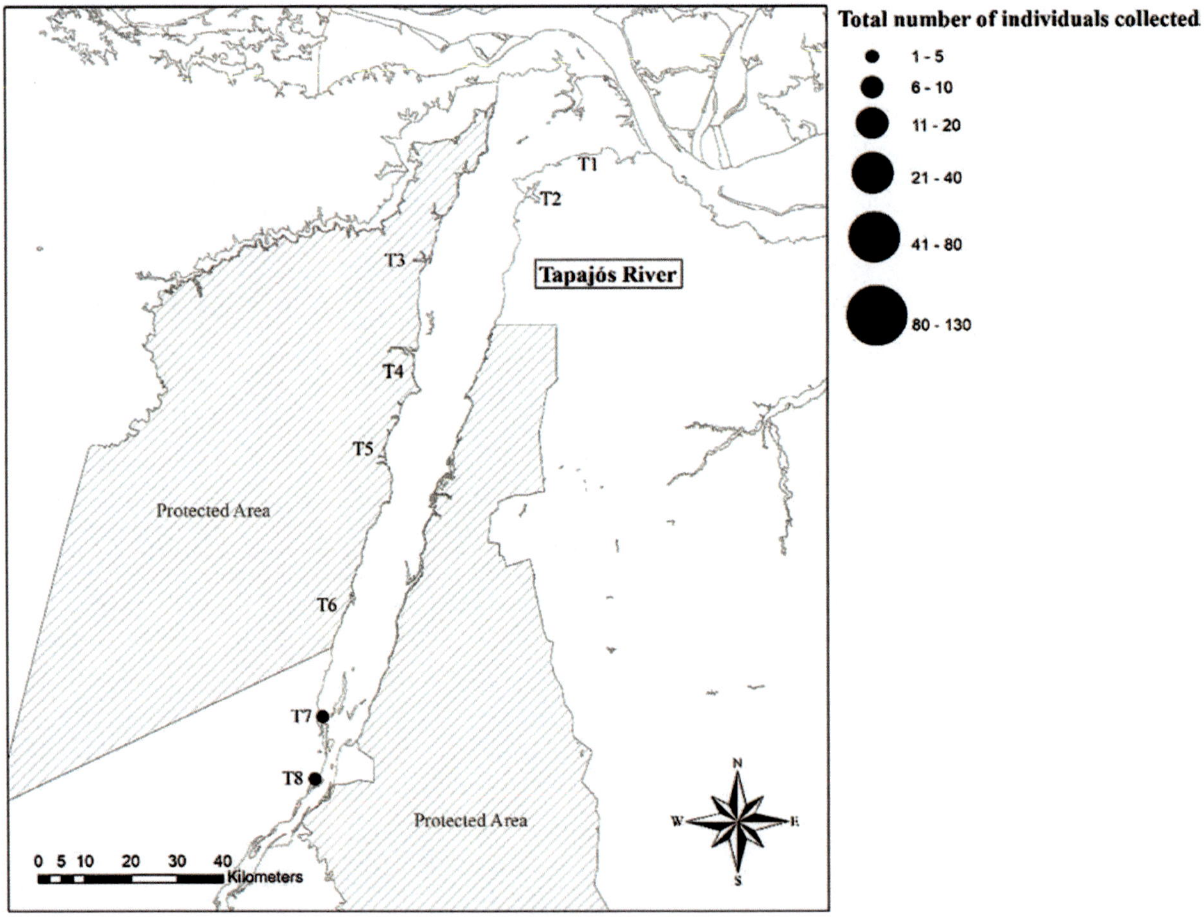

◘ **Fig. 5.177** Distribution of individuals of *Sorubim elongatus* sampled in the Tapajós River

References

Akama A, Ribeiro FRV (2013) Auchenipteridae. In: Queiroz LJ, Vilara GT, Ohara WM, THS P, Zuanon J, Doria CR (eds) Peixes do Rio Madeira, vol 3. Dialeto, São Paulo, pp 172–205

Albert JS (2001) Species diversity and phylogenetic systematics of American knifefishes (Gymnotiformes, Teleostei). Misc Publ Mus Zool Univ Mic 190:1–129

Albert JS (2003) Sternopygidae (Glass knifefishes, Rattail knifefishes). In: Reis RE, Kullander SO, Ferraris CJ Jr (eds) Checklist of the freshwater fishes of South and Central America. EDIPUCRS, Porto Alegre, pp 487–491

Amaral MF, Aranha JMR, Menezes MS (1998) Reproduction of the freshwater catfish *Pimelodella pappenheimi* in Southern Brazil. Stud Neotropical Fauna Environ 33:2–10

Barthem R, Goulding M (1997) The catfish connection. In: Ecology, migration and conservation of Amazonian predators. Columbia University Press, New York, 144pp

Barthem RB, Ribeiro MCLB, Petrere M (1991) Life strategies of some long-distance migratory catfish in relation to hydroelectric dams in the Amazon Basin. Biol Conserv 55:339–345

Barthem R, Goulding RG, Leite C, Cañas B, Forsberg E, Venticinque P, Petry MLB, Ribeiro J, Chuctaya MA (2017) Goliath catfish spawning in the far western Amazon confirmed by the distribution of mature adults, drifting larvae and migrating juveniles. Nat Sci Rep 7:41–784

Baskin JN, Zaret TM, Leccia FM (1980) Feeding of reportedly parasitic catfishes (Trichomycteridae and Cetopsidae) in the Rio Portuguesa Basin, Venezuela. Biotropica 12(3):182–186

Begossi A, Hanazaki N, Ramos R (2004) Food chain and the reasons for food taboos in the Amazon and on the Atlantic Forest coast. Ecol Appl 14:1334–1343

Birindelli JLO (2010) Relações filogenéticas da superfamília Doradoidea (Ostariophysi, Siluriformes). Tese de doutorado, Universidade de São Paulo, São Paulo. pp. 376

Birindelli JL, Sousa LM (2013) Doradidae. In: Queiroz LJ, Vilara GT, Ohara WM, THS P, Zuanon J, Doria CR (eds) Peixes do Rio Madeira, vol 3. Dialeto, São Paulo, pp 130–171

Bockmann FA, Guazzelli GM (2003) Family Heptapteridae (Heptapterids). In: Reis RE, Kullander SO, Ferraris CJ Jr (eds) Check list of the freshwater fishes of South and Central America. Editora EDIPUCRS, Porto Alegre, pp 406–431

Boischio AAP, Henshel D (2000) Fish consumption, fish lore, and mercury pollution: risk communication for the Madeira River people. Environ Res A 84:108–126

Burgess WE (1989) An atlas of freshwater and marine catfishes. A preliminary survey of Siluriformes. T.F.H. Publications, Neptune City

Carvalho TP, Ramos CS, Albert JS (2011) A new species of *Gymnorhamphichthys* (Gymnotiformes: Rhamphichthyidae) from the Paraná-Paraguay basin. Copeia 2011(3):400–406

Chamon CC, Silva TC (2018) *Pseudacanthicus major*: description of one of the largest known Loricariidae. Zootaxa 3:499–510

Crampton WGR, Thorsen DH, Albert JS (2004) *Steatogenys ocellatus*: a new species of neotropical electric fish (Gymnotiformes: Hypopomidae) from the Lowland Amazon Basin. Copeia 2004(1):78–91

Daniel LHRP, Fernandes CC (2005) Dimorfismo sexual em Siluriformes e Gymnotiformes (Ostariophysi) da Amazônia. Acta Amazon 35:97–110

Daniel LHRP, de Deus CP, Henriques AL, Pimpão DM, Ribeiro OM (2007) Biodiversidade do médio Madeira: bases científicas para propostas de conservação. Instituto Nacional de Pesquisas da Amazônia-INPA, Manaus

de Pinna MCC, Ferraris CJ, Vari RP (2007) A phylogenetic study of the neotropical catfishes family Cetopsidae (Osteichthyes, Ostariophysi, Siluriformes), with a new classification. Zool J Linnean Soc 150:755–813

Doria CRC, Lima MAL (2015) Rio Madeira: seus peixes e sua pesca. Editora Rima, São Paulo

Eschmeyer WN (1998) Catalog of fishes. Special publication, vol 3. California Academy of Sciences, San Francisco, p 2905. Available at: research.calacademy.org/research/ichthyology/catalog/fishcatmain.asp. Accessed July 2019

Ferraris CJ Jr (2003) Family Auchenipteridae. In: Reis RE, Kullander SO (eds) Check list of the freshwater fishes of South and Central America. EDIPUCRS, Porto Alegre, pp 470–482

Ferraris CJ Jr, Brown BA (1991) A new species of *Pseudocetopsis* from the Río Negro drainage of Venezuela (Siluriformes: Cetopsidae). Copeia 1991(1):161–165

Ferraris CJ Jr, de Santana CD, Vari RP (2017) Checklist of Gymnotiformes (Osteichthyes: Ostariophysi) and catalogue of primary types. Neotrop Ichthyol 15(1):e160067

Ferreira EJG, Zuanon JA, dos Santos GM (1998) Peixes comerciais do médio Amazonas: região de Santarém, Pará. IBAMA, Brasília

Gee JH, Graham JB (1978) Respiratory and hydrostatic functions of the intestine of the catfishes *Hoplosternum thoracatum* and *Brochis splendens* (Callichthyidae). J Exp Biol 74:1–16

Giora J, Fialho CB, Dufech APS (2005) Feeding habit of *Eigenmannia trilineata* Lopez & Castello, 1966 (Teleostei: Sternopygidae) of Parque Estadual de Itapuã, RS, Brazil. Neotrop Ichthyol 3(2):291–298

Goulding M (1979) Ecologia da Pesca do Rio Madeira. Instituto Nacional de Pesquisas da Amazônia (INPA), CNPq, Manaus, 172pp

Goulding M (1980) The fishes and the forest: explorations in Amazonian natural history. University of California Press, Berkeley

Goulding M (1981) Man and fisheries on an Amazon Frontier. W. Junk Publishers, The Hague, Boston, London

Hallwass G, Silvano RAM (2016) Patterns of selectiveness in the Amazonian freshwater fisheries: implications for management. J Environ Plan Manag 59:1537–1559

Hallwass G, Schiavetti A, Silvano RAM (2020) 'Fishers' knowledge indicates temporal changes in composition and abundance of fishing resources in Amazon protected areas. Anim Conserv. 23:36–47. https://doi.org/10.1111/acv.12504

Hostache G, Mol JH (1998) Reproductive biology of the neotropical armoured catfish *Hoplosternum littorale* (Siluriformes – Callichthyidae): a synthesis stressing the role of the floating bubble nest. Aquat Living Resour 11:173–185

Kohda M, Tanimura M, Nakamura MK, Yamagishi S (1995) Sperm drinking by female catfishes: a novel mode of insemination. Environ Biol Fish 42:1–6

Lundberg JG, Stager JC (1985) Microgeographic diversity in the neotropical knife-fish *Eigenmannia macrops* (Gymnotiformes, Sternopygidae). Environ Biol Fish 13(3):173–181

Marrero C, Winemiller KO (1993) Tube-snouted gymnotiform and mormyriform fishes: convergence of a specialized foraging mode in teleosts. Environ Biol Fish 38(4):299–309

Mayer F (1950) Callichthys callichthys. Aqua J 21(1):1–3

Nelson JS (1994) Fishes of the world, 3rd edn. Wiley, New York

Nunes MUS, Hallwass G, Silvano RAM (2019) 'Fishers' local ecological knowledge indicate migration patterns of tropical freshwater fish in an Amazonian river. Hydrobiologia 833(1): 197–215

Ohara WM, Lima FCT, Salvador GN, Andrade MC (2017) Peixes do Rio Teles Pires: diversidade e guia de identificação. Museu de

Zoologia da Universidade de São. Paulo/Universidade Federal do Pará/Museu de Zoologia da Universidade Estadual de Campinas, São Paulo

Petrere M, Barthem RB, Córdoba EA, Gómez BC (2004) Review of the large catfish fisheries in the upper Amazon and the stock depletion of piraíba (*Brachyplatystoma filamentosum* Lichtenstein). Rev Fish Biol Fish 14:403–414

Queiroz LJ, Vilara GT, Ohara WM, Pires THS, Zuanon J, Doria CR (2013a) Peixes do rio madeira, vol 2. Editora Dialeto, São Paulo

Queiroz LJ, Vilara GT, Ohara WM, Pires THS, Zuanon J, Doria CR (2013b) Peixes do rio madeira, vol 3. Editora Dialeto, São Paulo

Reis RE (1997) Revision of the Neotropical genus *Hoplosternum* (Ostariophysi: Siluriformes: Callichthyidae) with the description of two new genera and three new species. Ichthyol Explor Fres 7:299–326

Reis RE (1998) Anatomy and phylogenetic analysis of the Neotropical callichthyid catfish (Ostariophysi, Siluriformes). Zool J Linnaean Soc 124:105–168

Ribeiro AC, Vilara GT, Filho JAL, Doria CRC (2016) Ecologia e Biologia de Peixes do rio Madeira. Editora EDUFRO, Porto Velho

Rocha MS, Zuanon J (2013) Pimelodidae. In: Queiroz LJ, Vilara GT, Ohara WM, THS P, Zuanon J, Doria CR (eds) Peixes do rio Madeira, vol 3. Dialeto, São Paulo, pp 78–129

Santos GM, de Mérona B, Juras AA, Jégu M (2004) Peixes do baixo rio Tocantins: 20 anos depois da Usina Hidrelétrica Tucuruí. Eletronorte, Brasília

Santos GM, Ferreira EJG, Zuanon J (2006) Peixes Comerciais de Manaus. ProVárzea, IBAMA/AM, Manaus

Saul WG (1975) An ecological study of fishes at a site in upper Amazonian Ecuador. Proc Acad Natl Sci Phila 127(12): 93–134

Sazima I, Zuanon J, Haddad V Jr (2005) Puncture wounds by driftwood catfish during bucket baths: local habits of riverside people and fish natural history in the Amazon. Wilderness Environ Med 16(204):208

Schaefer SA (2003) Subfamily Hypoptopomatinae. In: Reis RE, Kullander SO (eds) Check list of the freshwater fishes of South and Central America. EDIPUCRS, Porto Alegre, pp 321–329

Vari RP, Ortega H (1986) The catfishes of the neotropical family Helogenidae (Ostariophysi: Siluroidei). Smithsonian Contrib Zool 442:i–iv. + 1–20

Vari RP, Ferraris CJ, de Pinna MCC (2005) The neotropical whale catfishes (Siluriformes, Cetopsidae, Cetopsinae), a revisionary study. Neotrop Ichthyol 3(2):127–238

Winemiller KO, Adite A (1997) Convergent evolution of weakly electric fishes from floodplain habitats in Africa and South America. Environ Biol Fish 49(2):175–186

5

Atlas of Fish of Tapajós and Negro Rivers III: Perciformes and Other Fish Groups

Renato A. M. Silvano, Pedro P. Nitschke, Kaluan C. Vieira, Paula Nagl, Astrid T. R. Martínez, Márcia C. F. Dutra, Friedrich W. Keppeler, Junior A. Chuctaya, and Marcelo C. Andrade

Contents

Renato A. M. Silvano` (ed.), *Fish and Fisheries in the Brazilian Amazon*, https://doi.org/10.1007/978-3-030-49146-8_6

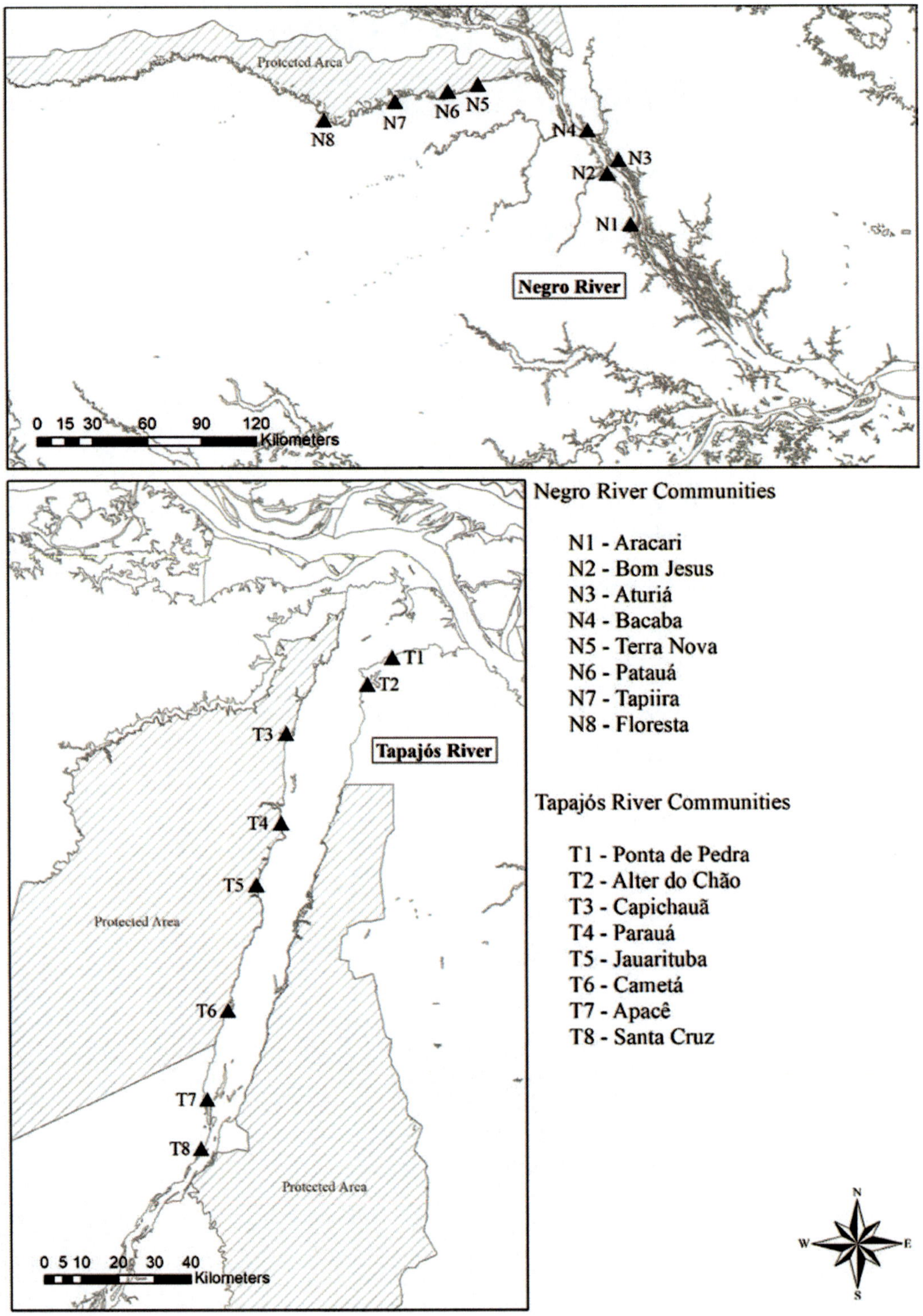

Fig. 6.1 Map of the studied region, showing the 16 studied communities with codes for community localization in maps showing the occurrence of fish species

6.1 Order Cichliformes

6.1.1 Family Cichlidae

6.1.1.1 Summary of Biology and Ecology (Literature)

Diversity: Cichlidae family is currently composed of 1720 valid species and 202 genera, which makes it the largest non-otophysii family of freshwater fish on the planet. This family was considered for a long time to belong to the order Perciformes, but recent molecular studies replaced the family in an exclusive new order, called Cichliformes. Molecular studies recommended provisional changes of species within the genus *Cichla* (Willis et al. 2012), but due to the lack of a recent revision of diagnostic characters, the identification of species of *Cichla* in this book was based on Kullander and Ferreira (2006).

Habitat: currently, the cichlids are distributed in freshwaters and occasionally in brackish waters of Central and South America, the southern regions of North America (Mexico and Texas), West Indies, Africa, Madagascar, Israel, Syria, coastal regions of India, and Sri Lanka. These fishes are commonly found in lentic environments, such as lakes, ponds, slow-flowing rivers, and reservoirs, but some species have adapted to occur in fast-flowing waters of rapids. These fishes are usually associated with vegetation areas, such as root banks and submerged vegetation. The majority of species have a benthic-pelagic habit.

Ecology and feeding: diet varies according to the species and ontogeny. Most species are omnivores or invertivorous (e.g., *Satanoperca, Geophagus*), some can be strictly herbivorous (e.g., *Uaru amphiacanthoides*), and some are predators or piscivorous (e.g., *Cichla* spp.).

Morphology: fishes with spines in the dorsal, anal, and pelvic fins. The latter is located in the thorax region, close to the pectoral fins, rather than in the abdominal region. Cichlids do not possess adipose fin and lateral line is not continuous. This group is very diverse in terms of body shape and color patterns. Predators usually have a fusiform body shape, while omnivores and herbivores tend to possess a more compressed shape. A great part of the diversity found in this family is linked with events of fast adaptive radiation, especially in the old aged and large tropical African lakes.

Reproduction: most cichlid species are equilibrium strategists, which mean that they have parental care, produce few large eggs per batch, have multiple batches per year, and present low growth rates. These fish tend to be aggressive, especially during the reproductive period. Polygamic species tend to protect the eggs inside their mouths, while monogamic species lay their eggs on a nest and guard them. A mix of both strategies can also occur, such as in cases when the cichlid lay eggs on a nest and, after hatching, keep the offspring inside its mouth.

Economic importance: medium and large bodied species are very important to small-scale fisheries in the Amazon, either as a source of protein for riverine communities or as relevant commercial fish (e.g., *Geophagus* spp., *Cichla* spp.). Fishes of this family are really appreciated by aquarists worldwide given their colorful patterns, morphological diversity, and complex behavior. The tucunarés (*Cichla* spp.) are also highly appreciated by recreational fishers due to their large size, voracity, and fighting behavior when hooked.

References: Berra (2001), Nelson (2006), Turner (2007), Hallwass et al. (2013), Hallwass and Silvano (2016), Keppeler et al. (2017), Froese and Pauly (2018), Bower and Winemiller (2019), Fricke et al. (2019), Willis et al. (2012), Kullander and Ferreira (2006).

Fish species: *Acarichthys heckelii* (Müller & Troschel 1849)

Common name(s): cará, cará-papa-terra, papa-terra

Size of fish sampled (standard length, both rivers): mean = 7.9 cm, median = 7.4 cm

Fisheries: small size, usually not exploited; see *Satanoperca lilith* for data on cará and acará fisheries. This fish can be targeted by the aquarium trade.

Fig. 6.2 *Acarichthys heckelii* collected in Tapajós River. (Photo: Renato Silvano)

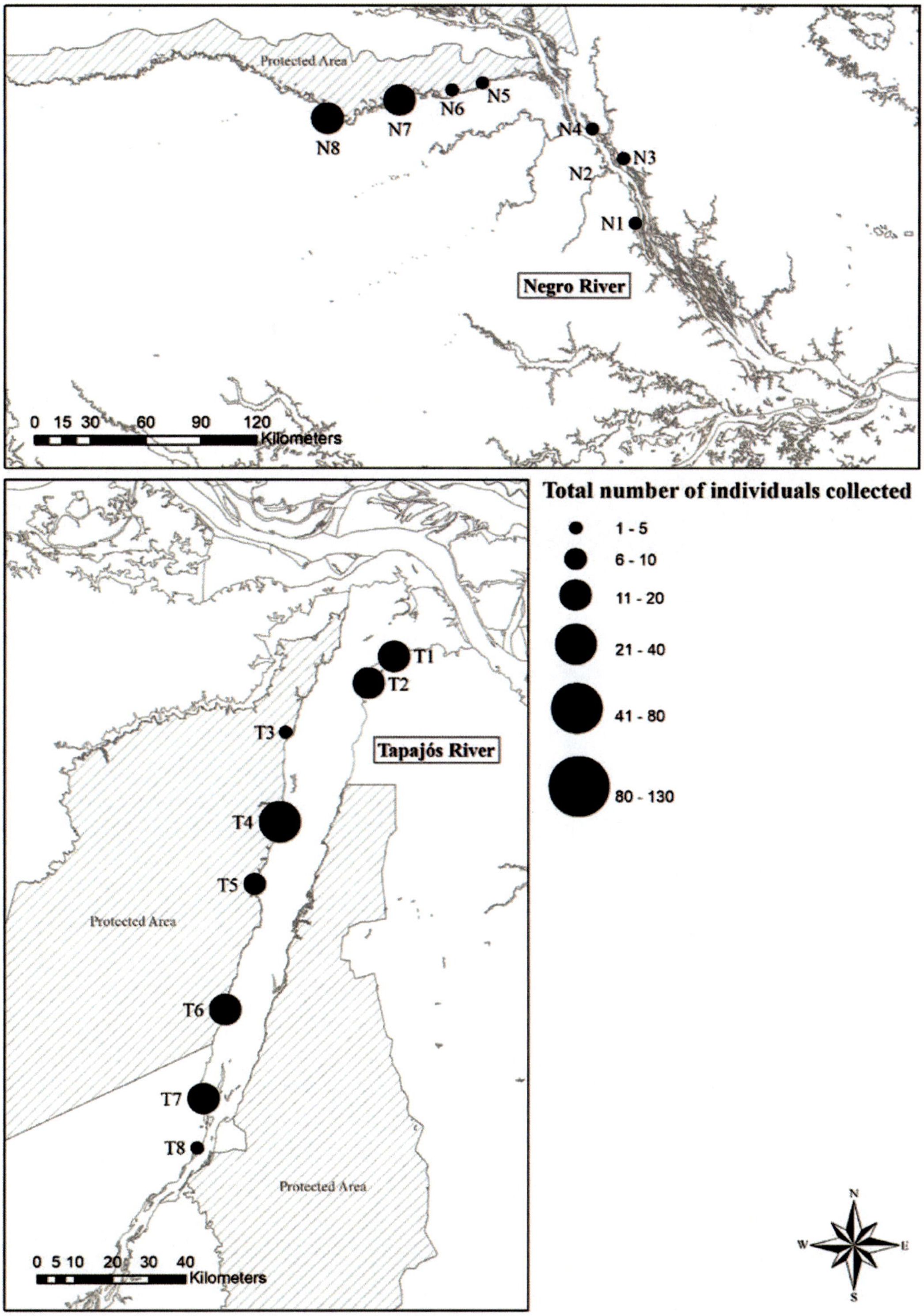

Fig. 6.3 Distribution of individuals of *Acarichthys heckelii* sampled in the Negro and Tapajós rivers

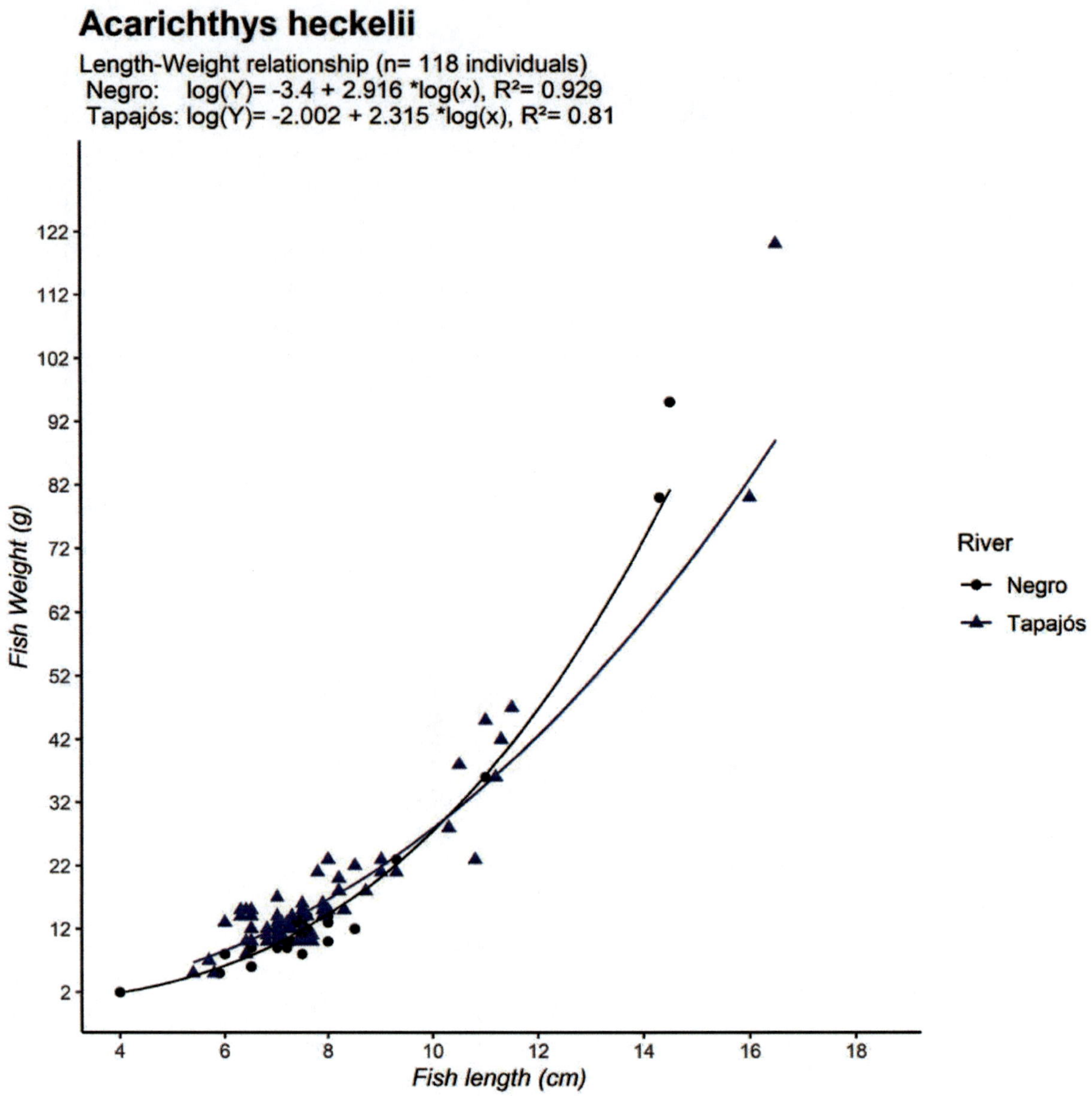

Fig. 6.4 Length-weight relationship for *Acarichthys heckelii* sampled in Tapajós and Negro rivers

Fish species: *Acaronia nassa* (Heckel 1840)

Common name(s): cará, cará-cascudo, cará-boca-de-pote, cará-foró

Size of fish sampled (standard length, both rivers): min. = 5.5 cm, max. = 12 cm, mean = 8.5 cm, median = 8.3 cm

Fisheries data: small size, usually not exploited.

Fig. 6.5 *Acaronia nassa* collected in Negro River. (Photo: Renato Silvano)

6

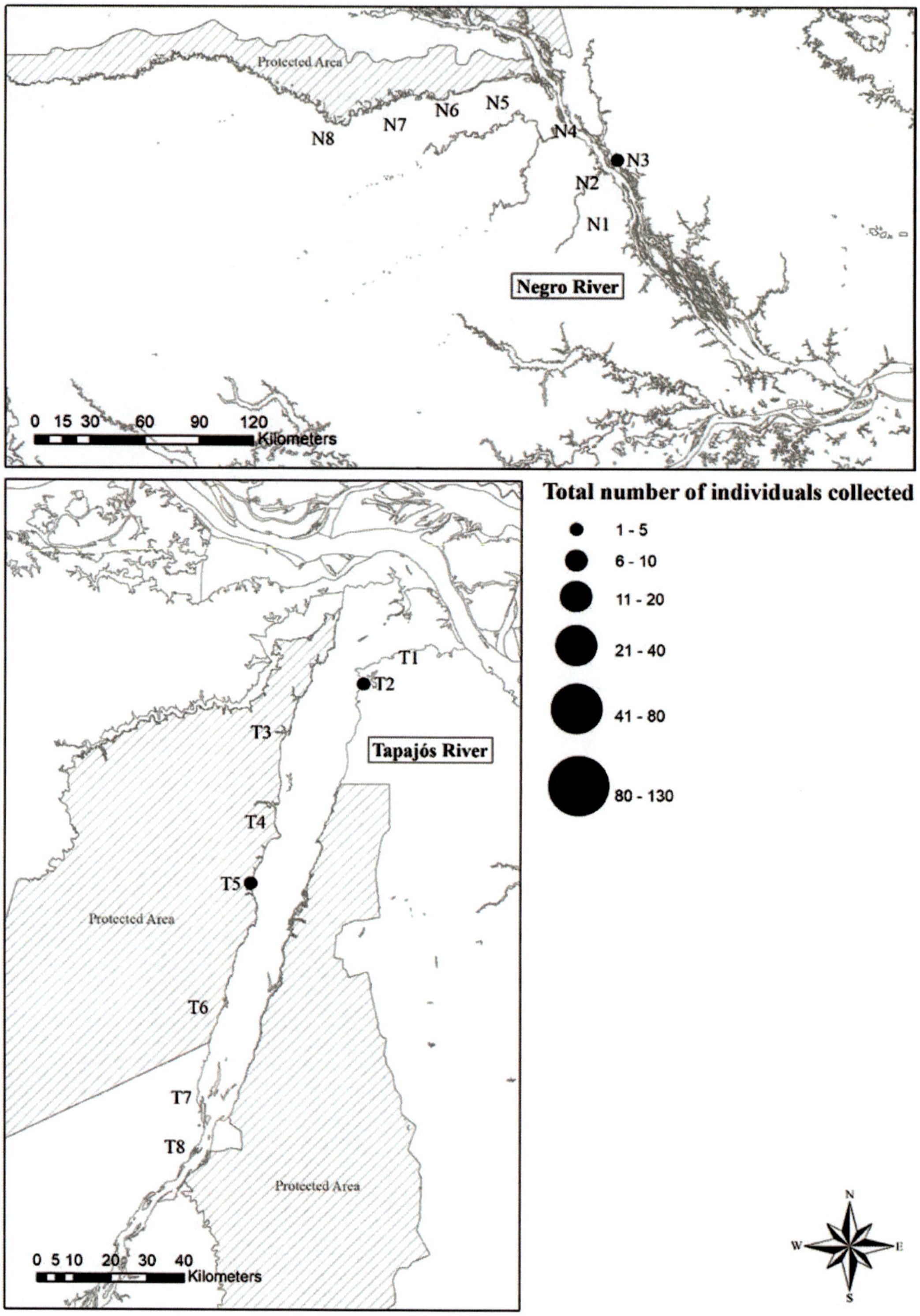

Fig. 6.6 Distribution of individuals of *Acaronia nassa* sampled in the Negro and Tapajós rivers

Fish species: ***Astronotus ocellatus*** **(Agassiz 1831)**

Common name(s): cará, cará-açu, carauaçu

Size of fish sampled (standard length, both rivers): min. = 14.5 cm, max. = 23 cm, mean = 17.4 cm, median = 16 cm

Fisheries: occasionally caught, usually with other carás, but cará-açu is larger and usually more valuable. Sizes of individuals caught are shown grouped with sizes of other carás caught below (see *Satanoperca lilith* for data on cará and acará fisheries). Biomass caught is in ■ Table 3.3 (▶ Chap. 3).

■ **Fig. 6.7** *Astronotus ocellatus* collected in the Tapajós River. (Photo: Pedro Nitschke)

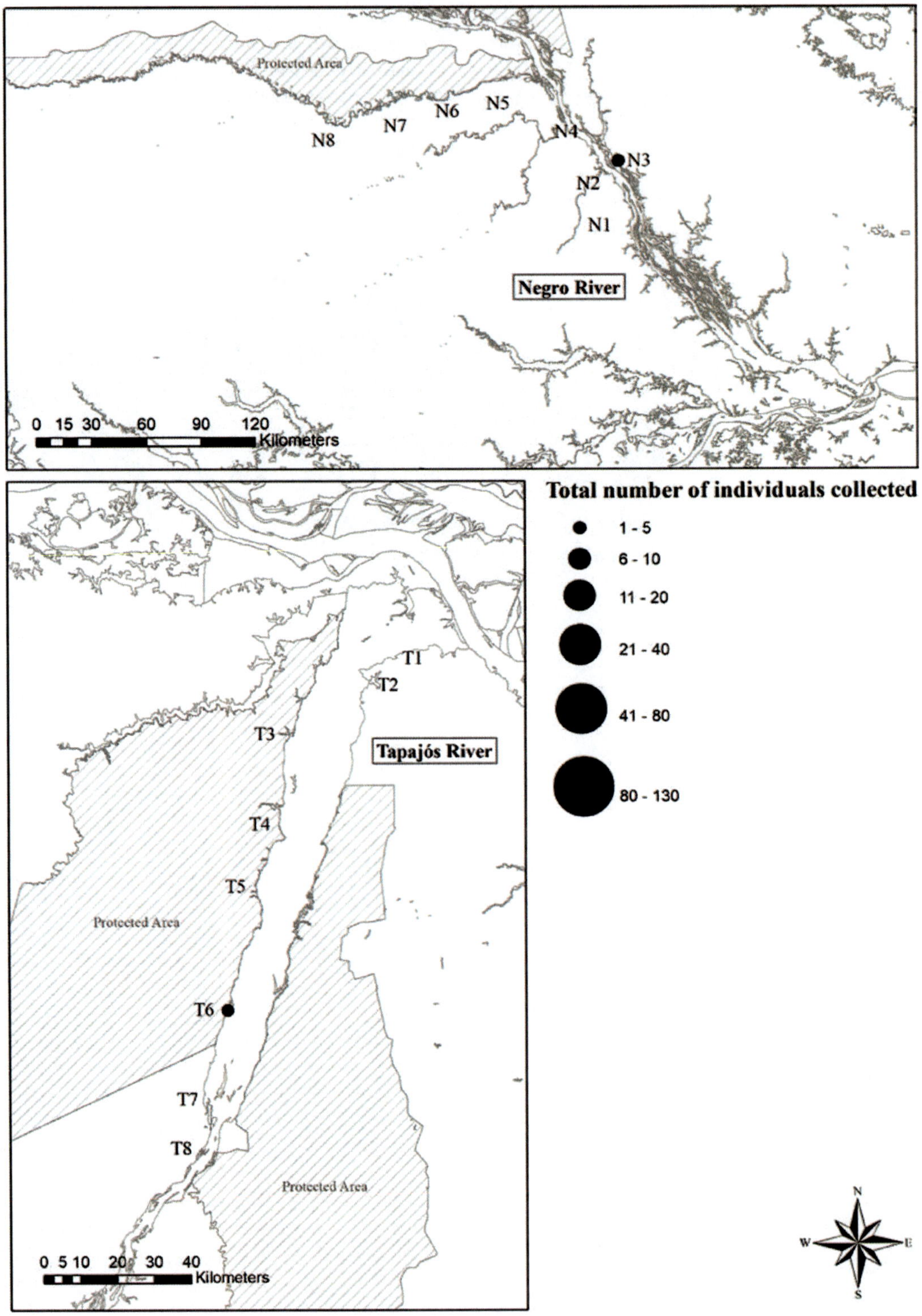

Fig. 6.8 Distribution of individuals of *Astronotus ocellatus* sampled in the Negro and Tapajós rivers

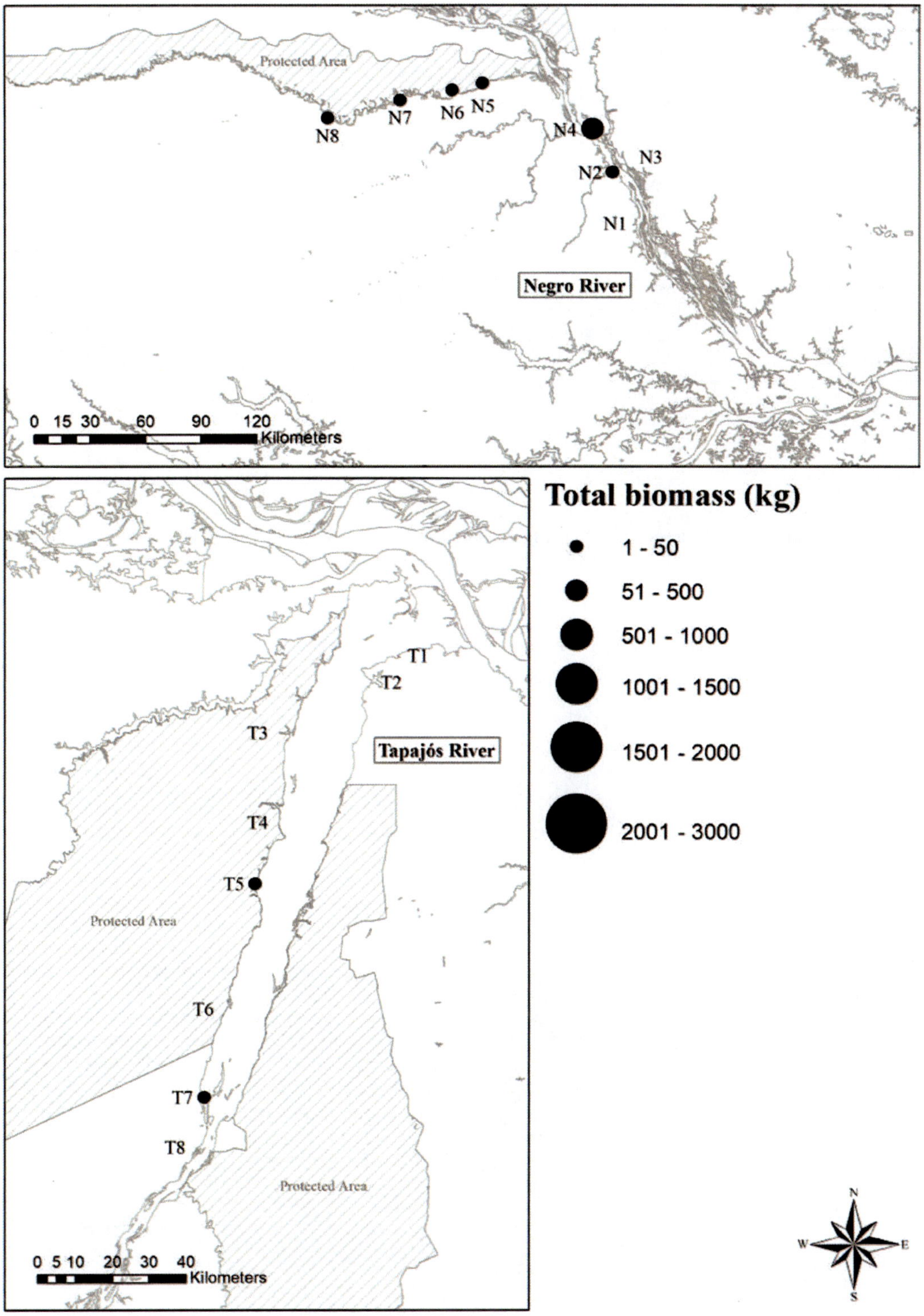

Fig. 6.9 Distribution of biomass of acará-açu (*A. ocellatus*) caught in small-scale fisheries in the Negro and Tapajós rivers

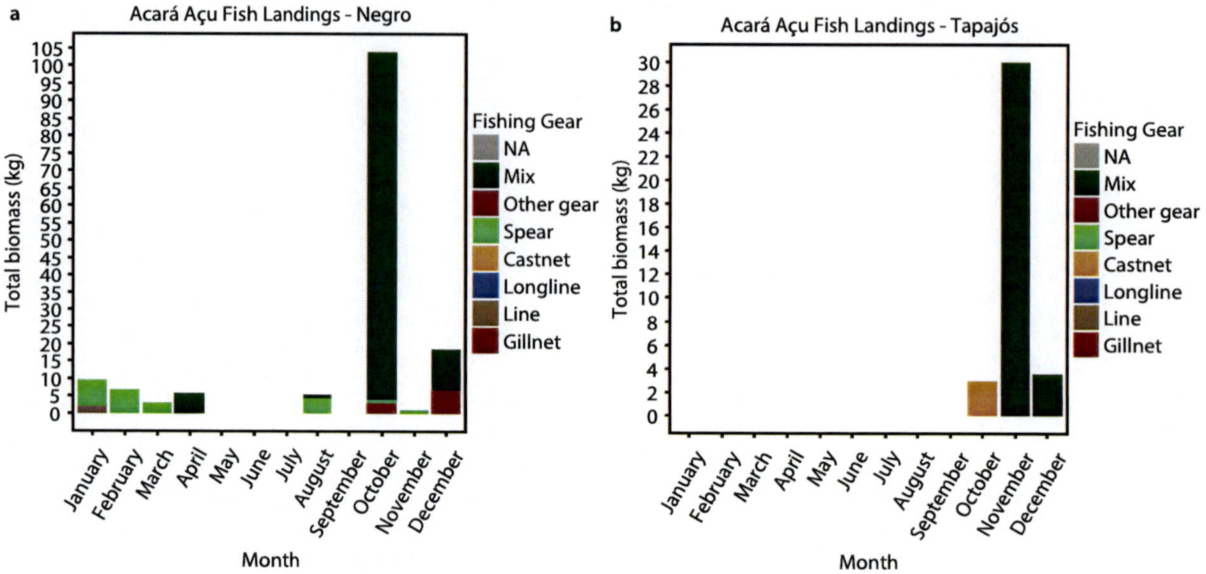

Fig. 6.10 Seasonal variation of biomass of acará-açu (*A. ocellatus*) caught per fishing gear in small-scale fisheries in 16 communities in the rivers **a** Negro and **b** Tapajós. NA: fishing gear not informed; Mix: more than one gear was used

Fish species: *Caquetaia spectabilis* (Steindachner 1875)

Common name(s): cará, cará-preto, cará-cascudo

Size of fish sampled (standard length, both rivers): min. = 11.7 cm, max. = 13 cm, mean and median = 12.4 cm

Fisheries: this species is not common and is usually not exploited; see *Satanoperca lilith* for data on cará and acará fisheries.

Fig. 6.11 *Caquetaia spectabilis* sampled in the Tapajós River. (Photo: Renato Silvano)

6

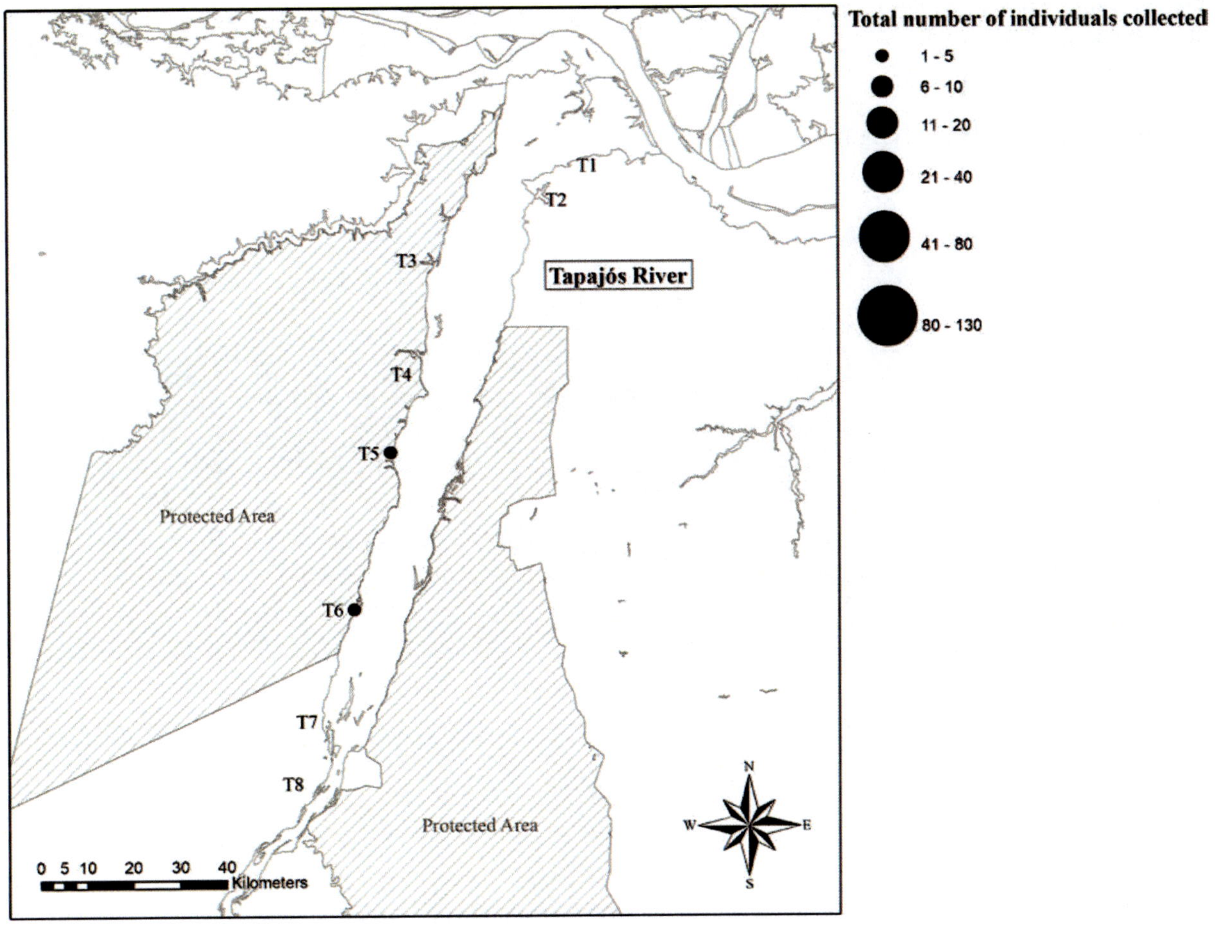

Fig. 6.12 Distribution of individuals of *Caquetaia spectabilis* sampled in the Tapajós River

Fish species: *Chaetobranchus flavescens* **Heckel 1840**

Common name(s): cará-prata

Size of fish sampled (standard length, both rivers): 15.3 cm

Fisheries: this species is not common and is usually not exploited; see *Satanoperca lilith* for data on cará and acará fisheries.

Fig. 6.13 *Chaetobranchus flavescens* sampled in the Negro River. (Photo: Renato Silvano)

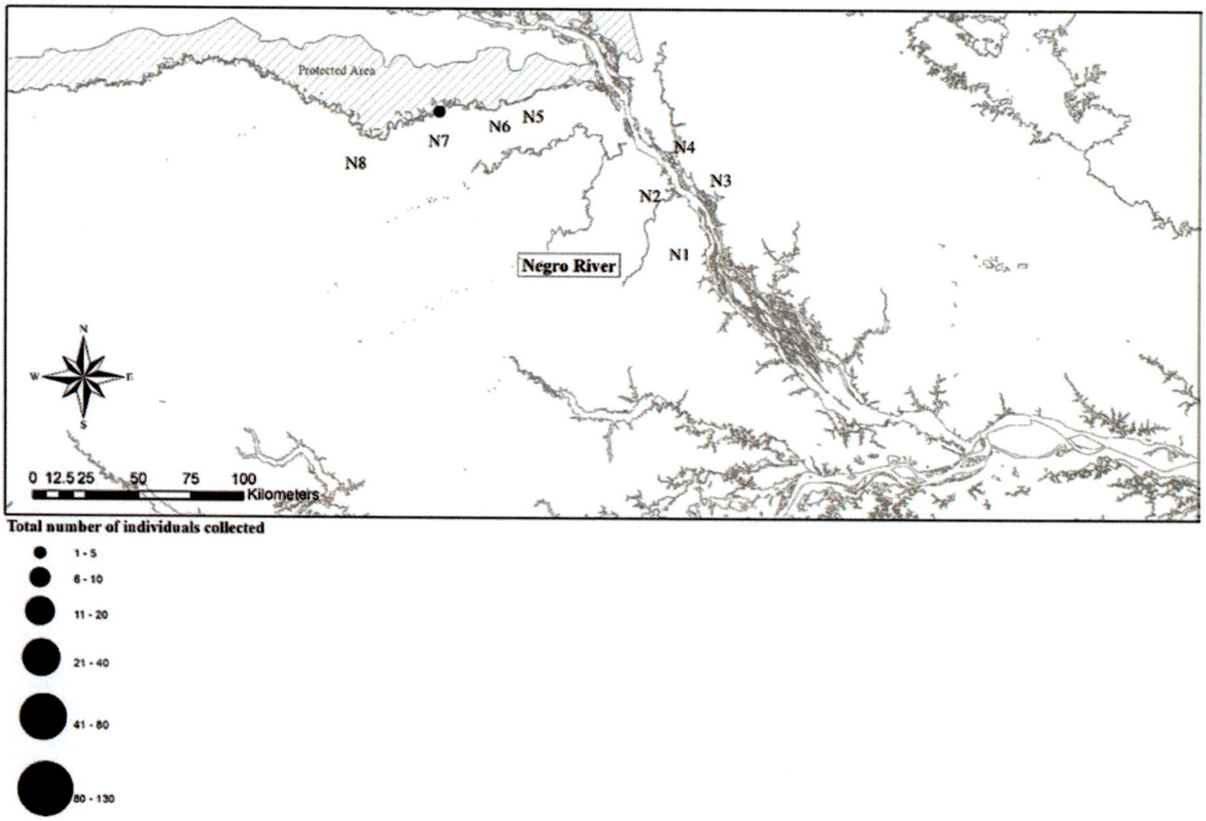

Fig. 6.14 Distribution of individuals of *Chaetobranchus flavescens* sampled in the Negro River

Fish species: *Cichla ocellaris* **Bloch & Schneider 1801**

Common name(s): tucunaré, tucunaré-açu, tucunaré-pitanga, tucunaré-tauá, tucunaré-botão

Size of fish sampled (standard length, both rivers): mean = 24.9 cm, median = 23.3 cm

Fisheries: tucunarés (*Cichla* spp.) are among the most commercially valuable and caught fish in both rivers. Biomass caught is in ◘ Table 3.3 (► Chap. 3).

◘ **Fig. 6.15** *Cichla ocellaris* sampled in the Tapajós River. (Photo: Renato Silvano)

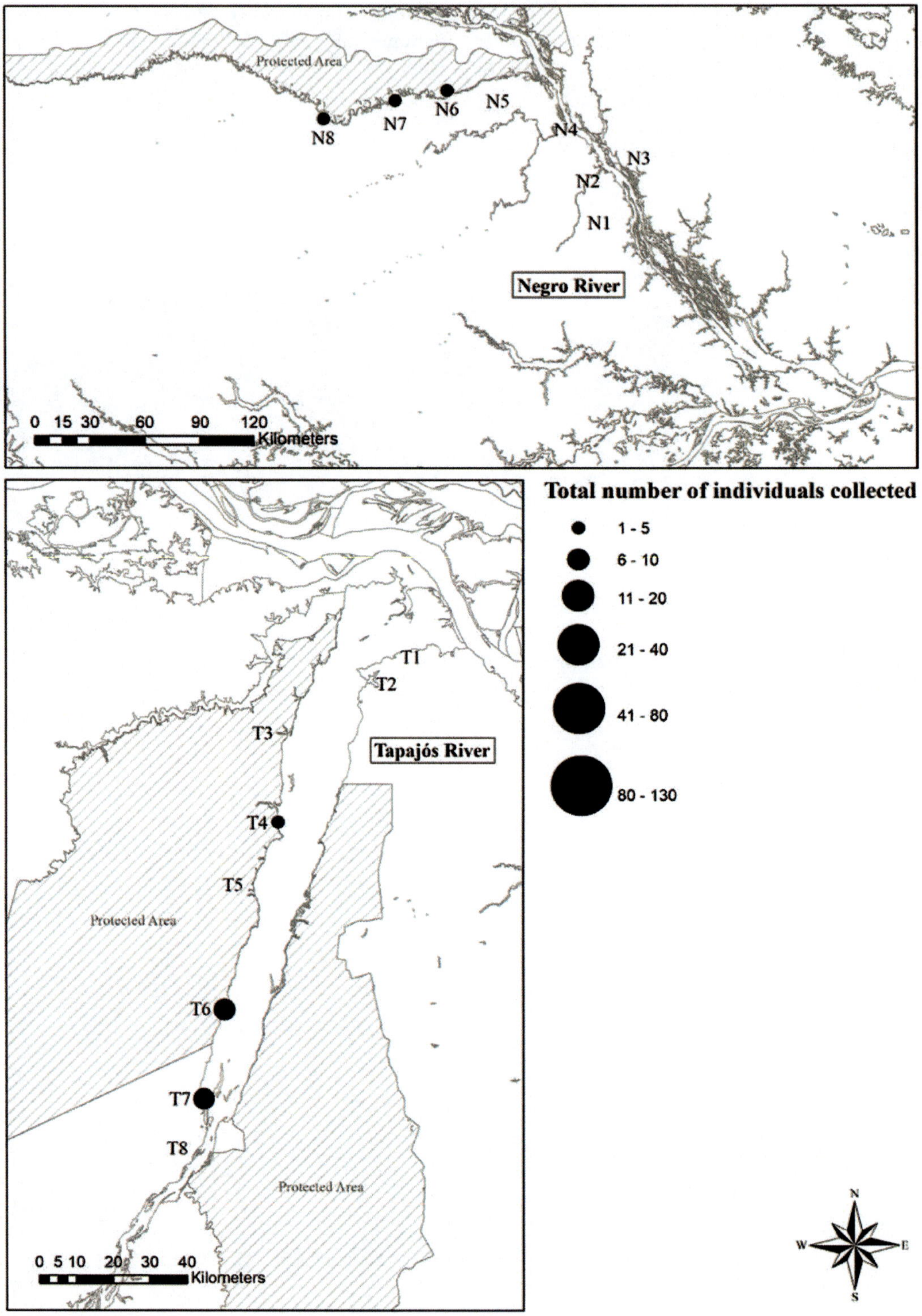

Fig. 6.16 Distribution of individuals of *Cichla ocellaris* sampled in the Negro and Tapajós rivers

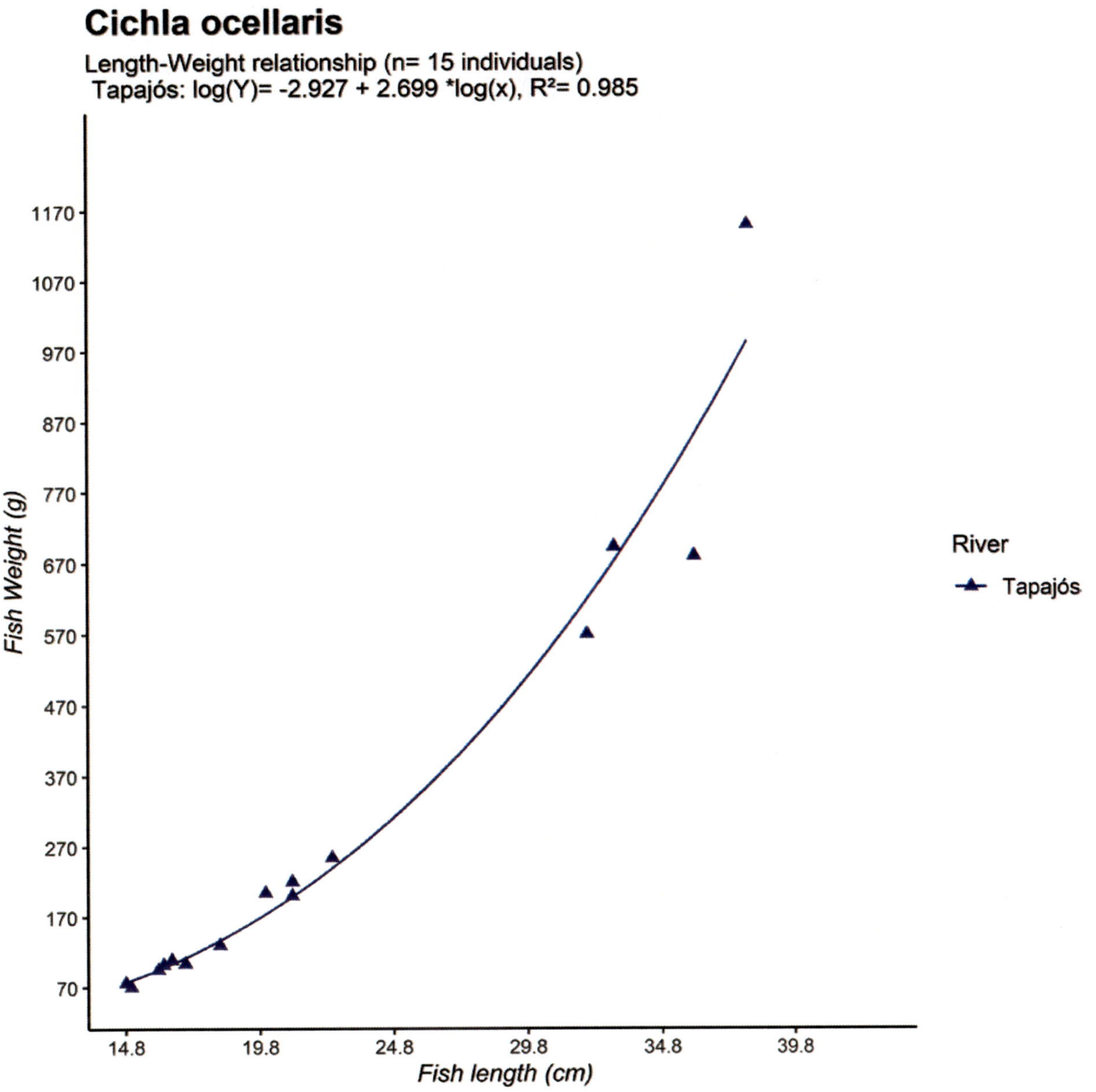

Fig. 6.17 Length-weight relationship for *Cichla ocellaris* sampled in the Tapajós River

6

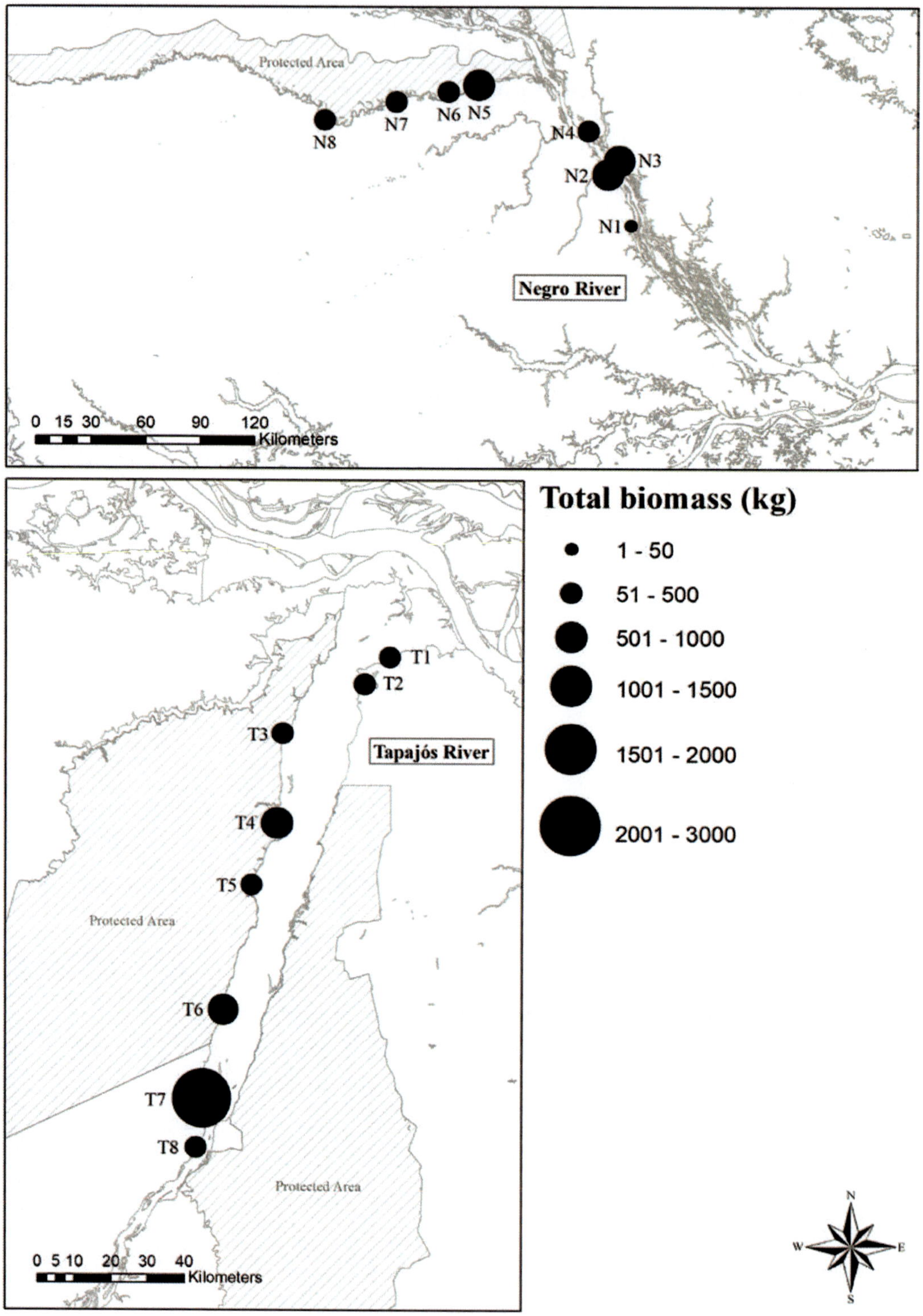

Fig. 6.18 Distribution of biomass of tucunarés (*Cichla* spp.) caught in small-scale fisheries in the Negro and Tapajós rivers

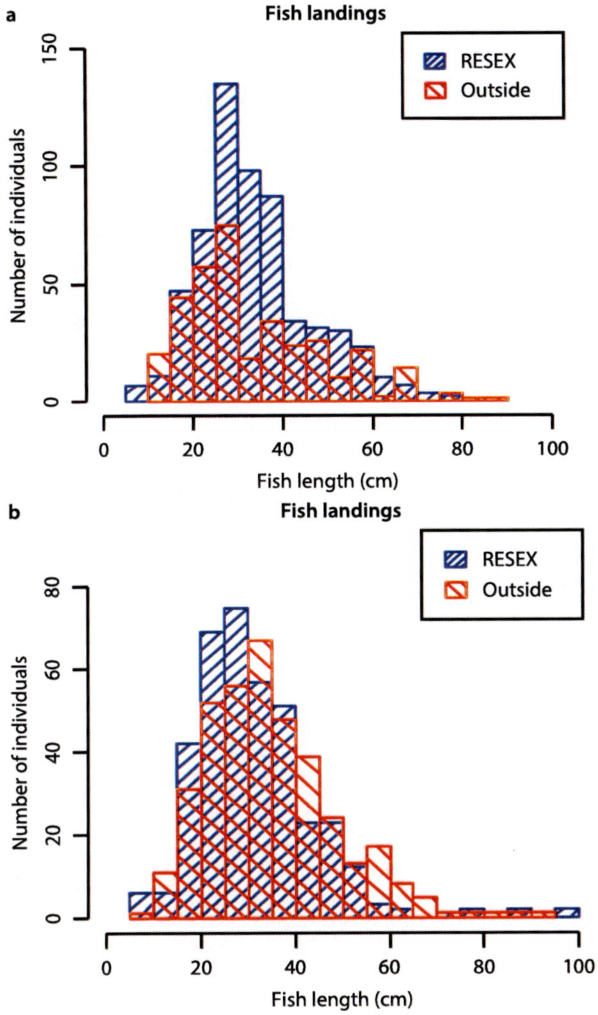

Fig. 6.19 Size distribution of tucunarés (*Cichla* spp.) caught in small-scale fisheries in 16 communities inside and outside the Extractive Reserves (RESEX) in the rivers **a** Negro, N Resex = 598, N outside = 351 and **b** Tapajós, N Resex = 376, N outside = 377

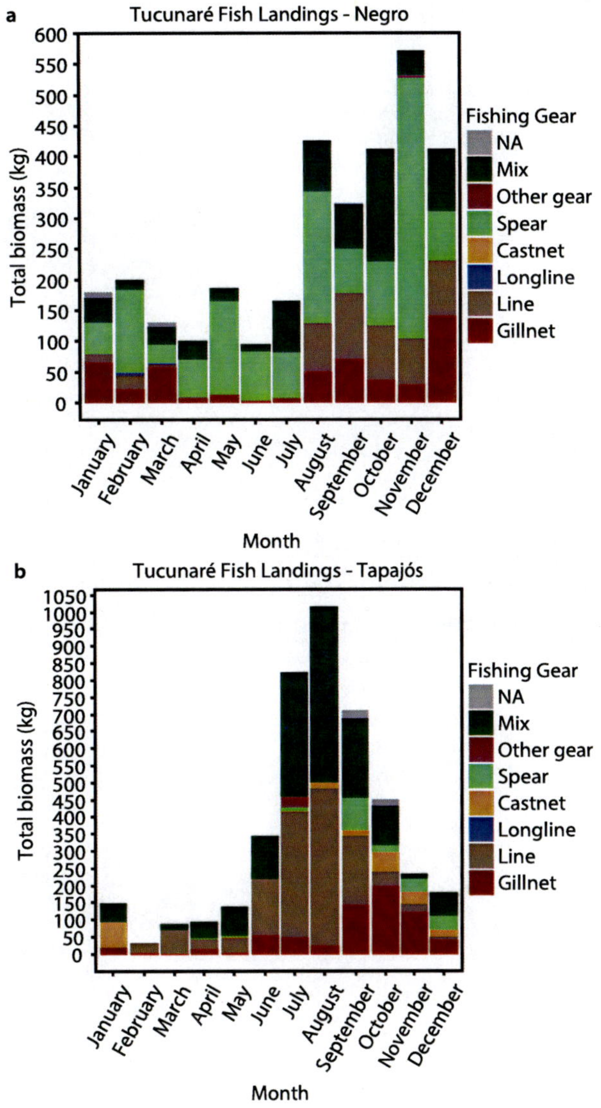

Fig. 6.20 Seasonal variation of biomass of tucunarés (*Cichla* spp.) caught per fishing gear in small-scale fisheries in 16 communities in the rivers **a** Negro and **b** Tapajós. NA: fishing gear not informed; Mix: more than one gear was used

Fish species: *Cichla* aff. *pinima*

Fig. 6.21 *Cichla* aff. *pinima* sampled in the Tapajós River. (Photo: Renato Silvano)

Fish species: *Cichla pinima* Kullander & Ferreira 2006

Common name(s): tucunaré, tucunaré-pinima, tucunaré-pintado, tucunaré-açu, tucunaré-amarelo

Fig. 6.22 *Cichla pinima* sampled in the Tapajós River. (Photo: Renato Silvano)

Observation: abundance and size data were grouped for these two species, due to the difficulty in distinguishing them during sampling.

Size of fish sampled (standard length, both rivers): min. = 8.9 cm, max. = 36.6, mean = 14.4 cm, median = 14.8 cm

Fisheries: tucunarés (*Cichla* spp.) are among the most commercially valuable and caught fish in both rivers; see figures in *Cichla ocellaris* for fisheries data of tucunarés.

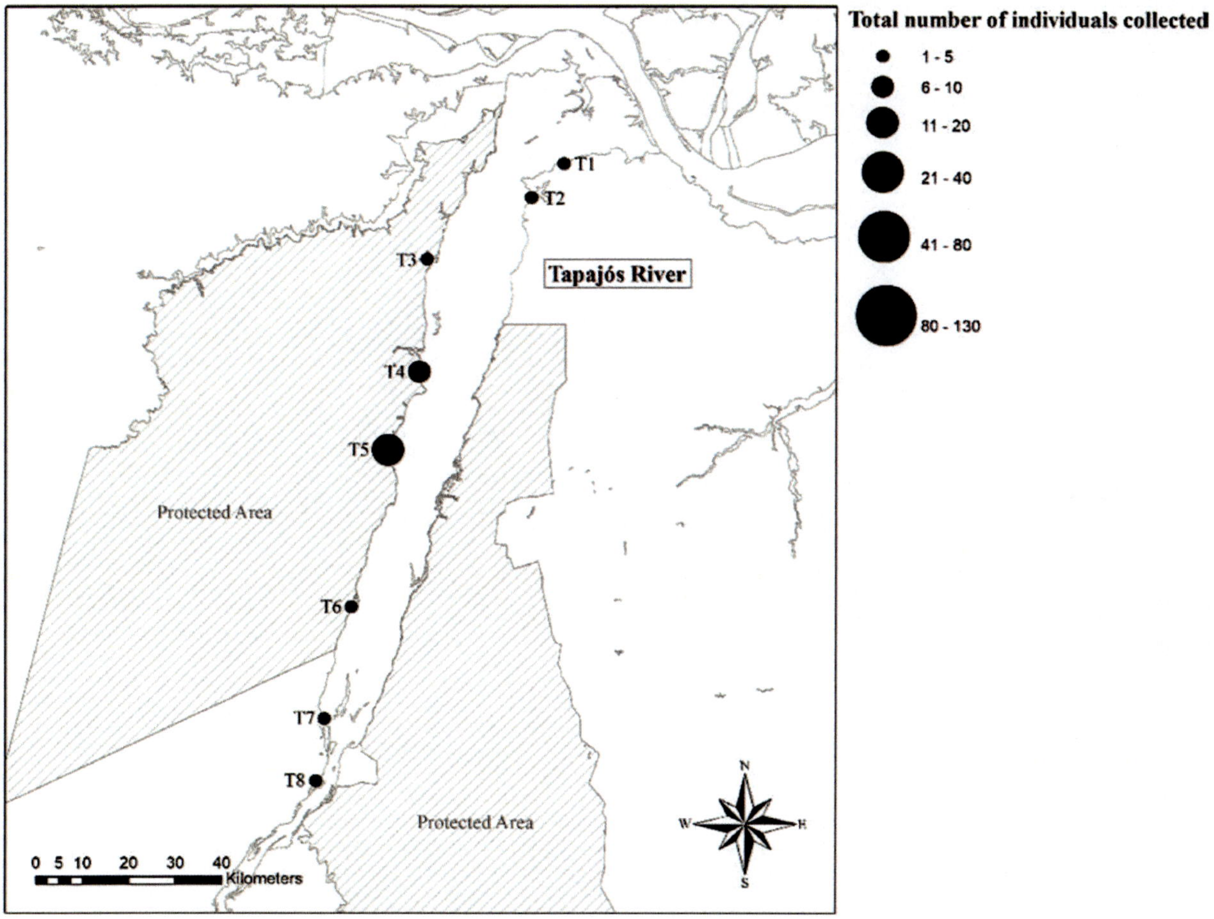

Fig. 6.23 Distribution of individuals of *C. pinima* and *Cichla* aff. *pinima* sampled in the Tapajós River

Fish species: *Cichla nigromaculata* Jardine & Schomburgk 1843

Common name(s): tucunaré, tucunaré-tauá, tucunaré-sarabiano

Size of fish sampled (standard length, both rivers): 16.5 cm

Fisheries: tucunarés (*Cichla* spp.) are among the most commercially valuable and caught fish in both rivers; see figures in *Cichla ocellaris* for fisheries data of tucunarés.

Fig. 6.24 *Cichla nigromaculata* sampled in the Negro River. (Photo: Renato Silvano)

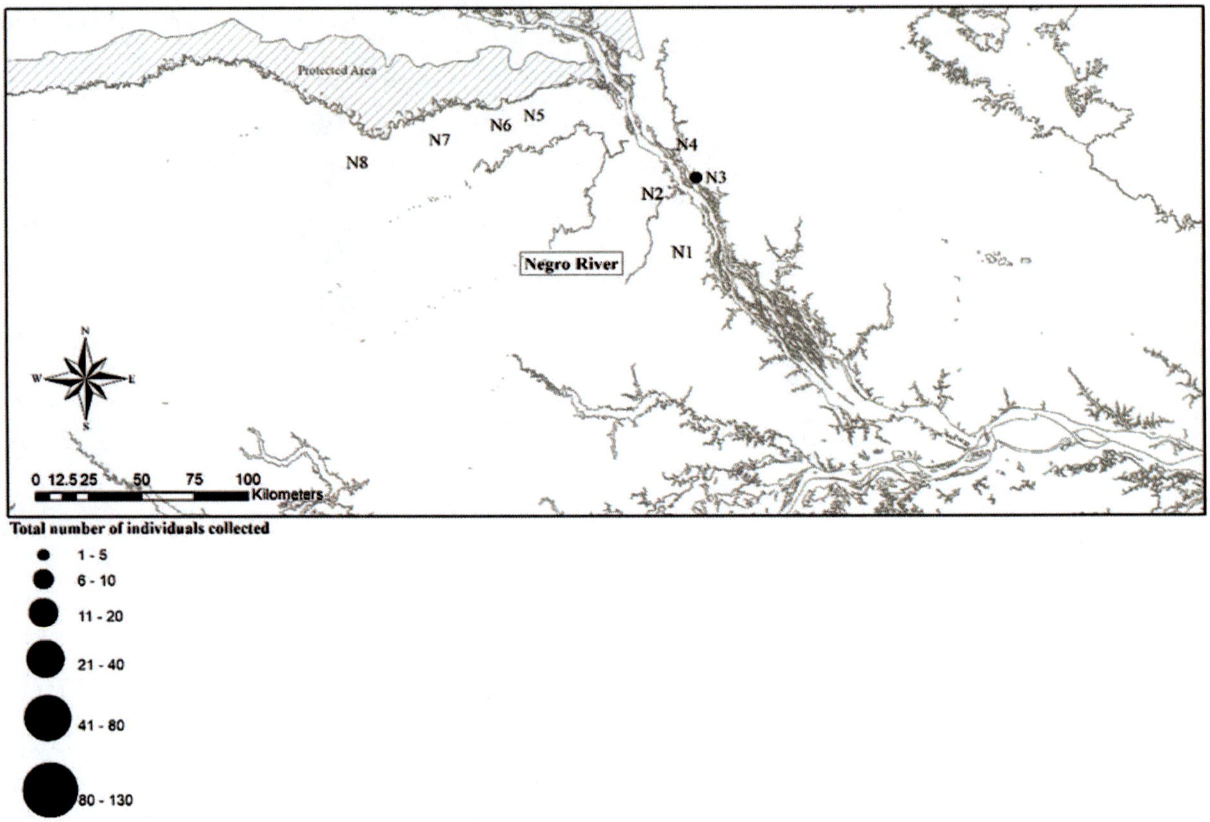

Fig. 6.25 Distribution of individuals of *Cichla nigromaculata* sampled in the Negro River

Fish species: *Cichla temensis* Humboldt 1821

Common name(s): tucunaré, tucunaré-paca, tucunaré-sarabiano

Size of fish sampled (standard length, both rivers): mean = 15.5 cm, median = 15.8 cm

Fisheries: tucunarés (*Cichla* spp.) are among the most commercially valuable and caught fish in both rivers; see figures in *Cichla ocellaris* for fisheries data of tucunarés.

Fig. 6.26 *Cichla temensis* sampled in the Negro River. (Photo: Renato Silvano)

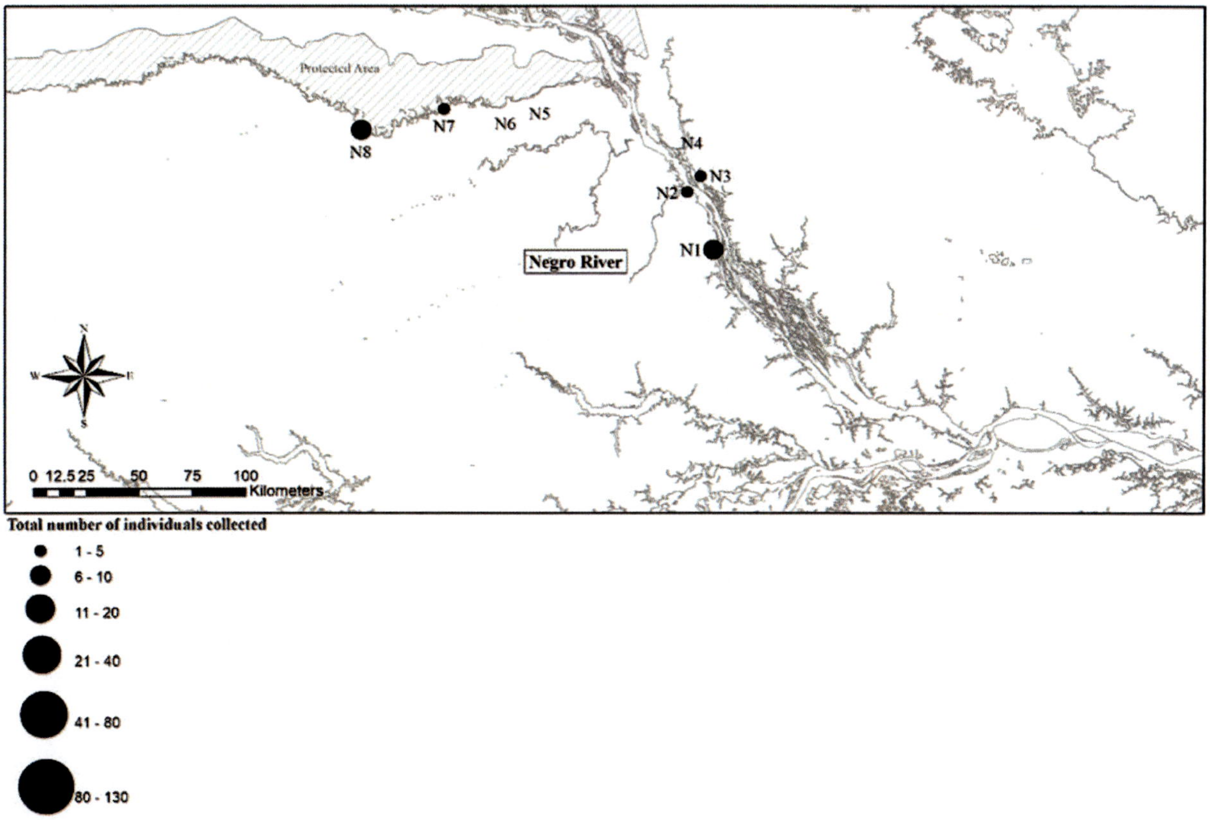

Fig. 6.27 Distribution of individuals of *Cichla temensis* sampled in the Negro River

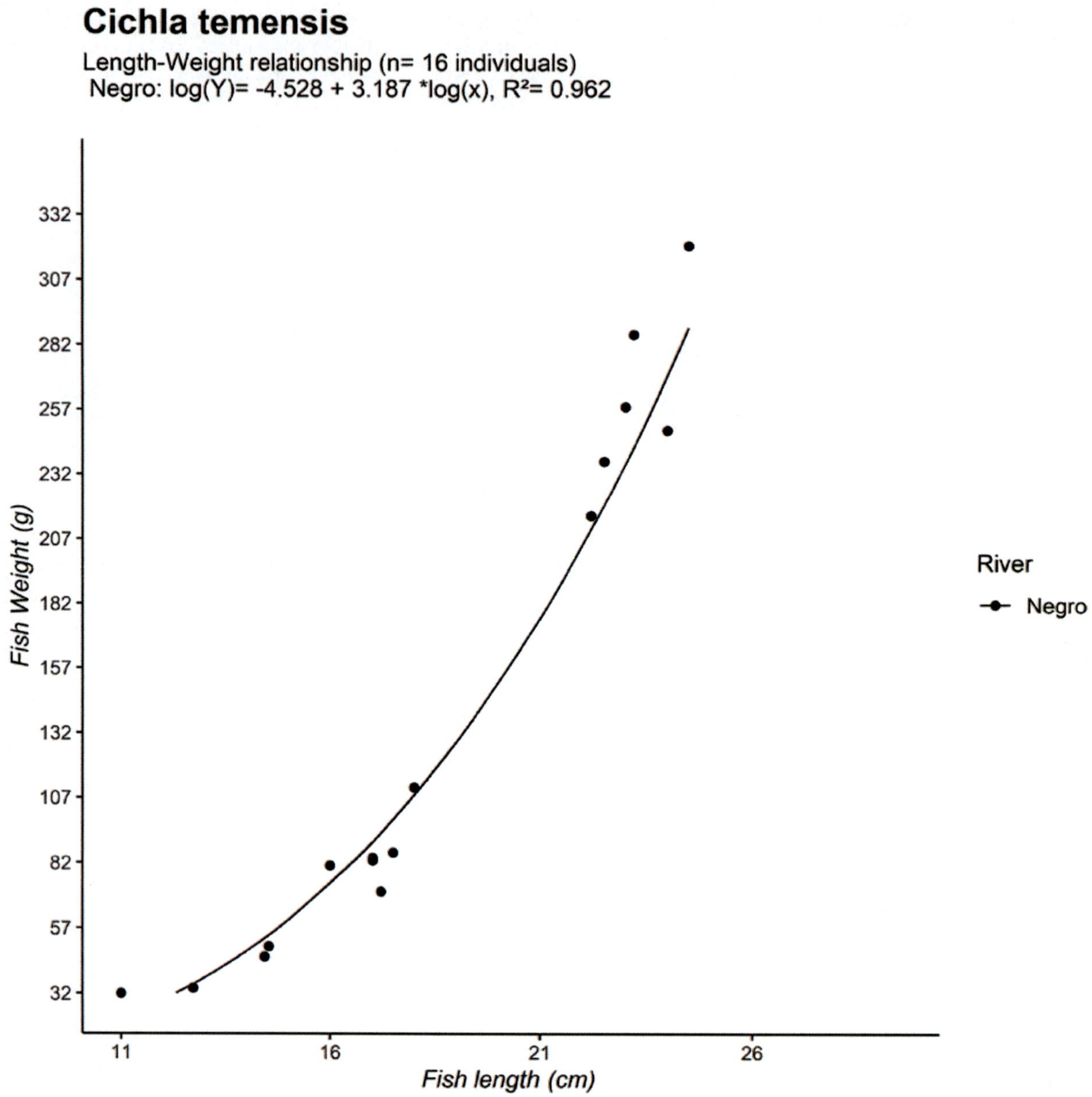

Fig. 6.28 Length-weight relationship for *Cichla temensis* sampled in the Negro River

Fish species: ***Crenicichla lenticulata*** **Heckel 1840**
Common name(s): jacundá

Size of fish sampled (standard length, both rivers): 21 cm

Fisheries: jacundás (*Crenicichla* spp.) are occasionally caught and are not among the most valuable fish in both rivers. Biomass caught is in ◘ Table 3.3 (▶ Chap. 3).

◘ **Fig. 6.29** *Crenicichla lenticulata* sampled in the Negro River. (Photo: Renato Silvano)

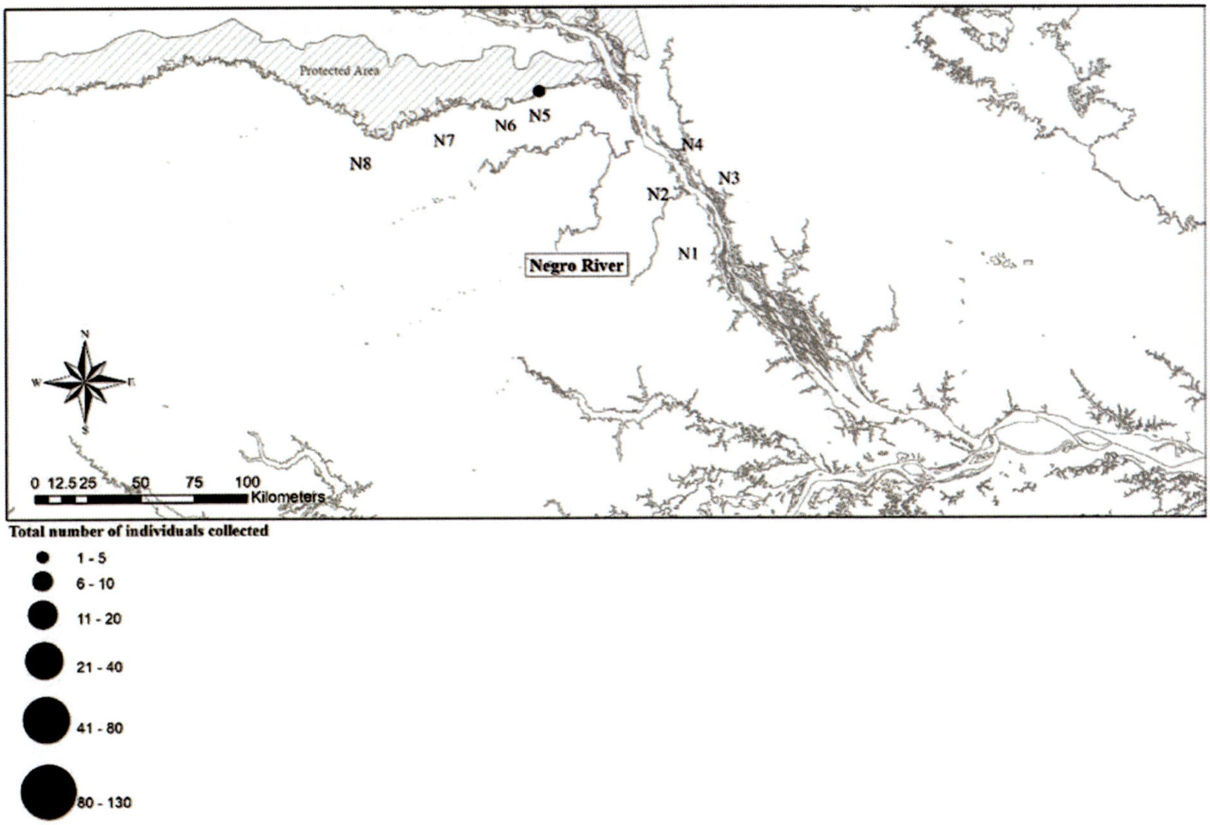

◘ **Fig. 6.30** Distribution of individuals of *Crenicichla lenticulata* sampled in the Negro River

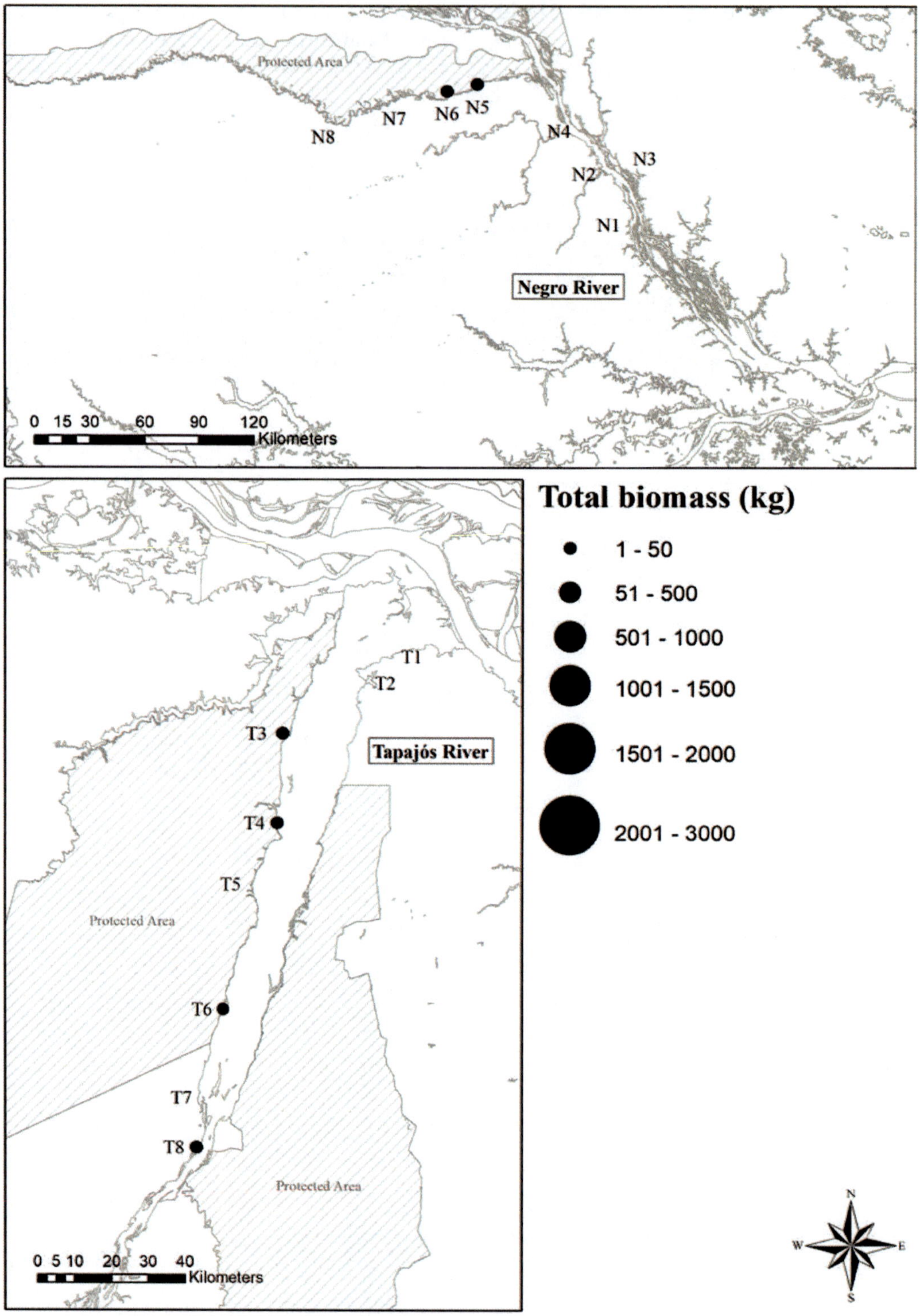

Fig. 6.31 Distribution of biomass of jacundás (*Crenicichla* spp.) caught in small-scale fisheries in the Negro and Tapajós rivers

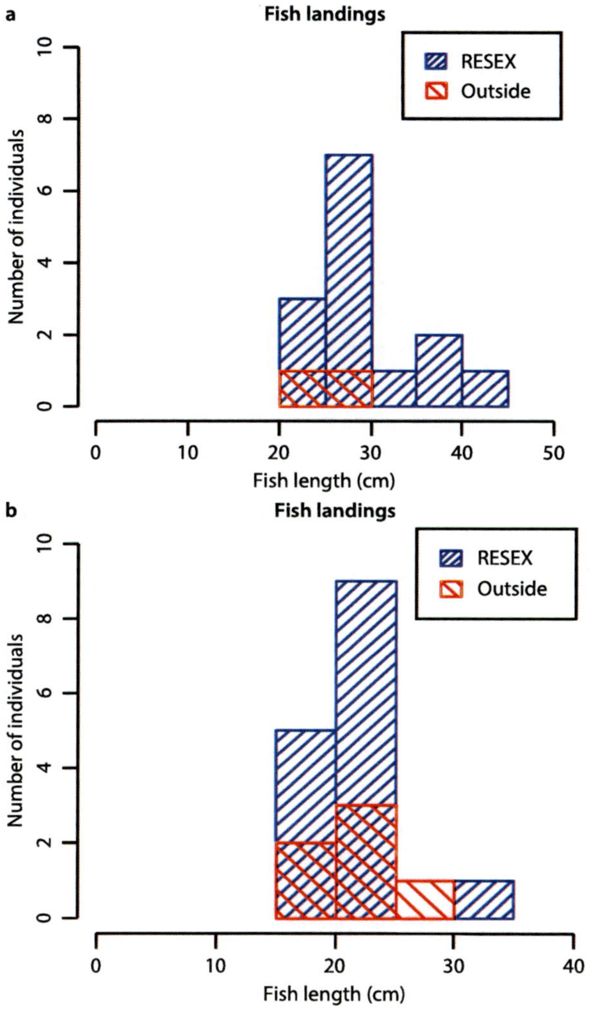

Fig. 6.32 Size distribution of jacundás (*Crenicichla* spp.) caught in small-scale fisheries in 16 communities inside and outside the Extractive Reserves (RESEX) in the rivers **a** Negro, N Resex = 14, N outside = 2 and **b** Tapajós, N Resex = 15, N outside = 6

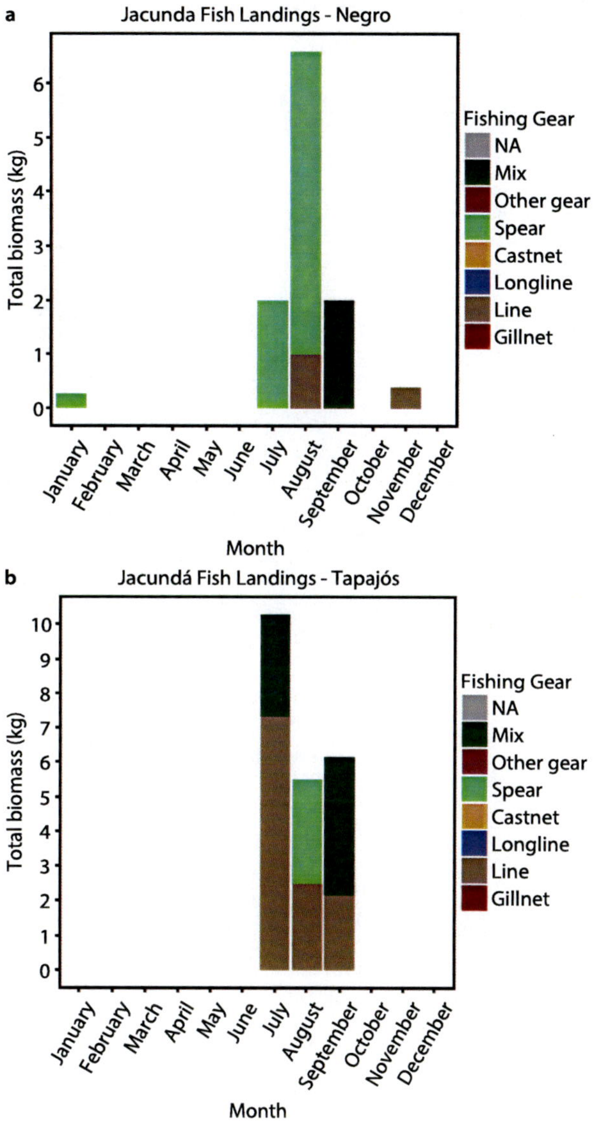

Fig. 6.33 Seasonal variation of biomass of jacundás (*Crenicichla* spp.) caught per fishing gear in small-scale fisheries in 16 communities in the rivers **a** Negro and **b** Tapajós. NA: fishing gear not informed; Mix: more than one gear was used

Fish species: *Crenicichla lugubris* Heckel 1840

Common name(s): jacundá

Size of fish sampled (standard length, both rivers): 15.2 cm

Fisheries: jacundás (*Crenicichla* spp.) are occasionally caught and are not among the most valuable fish in both rivers; see *Crenicichla lenticulata* for data on fisheries of jacundás (*Crenicichla* spp.).

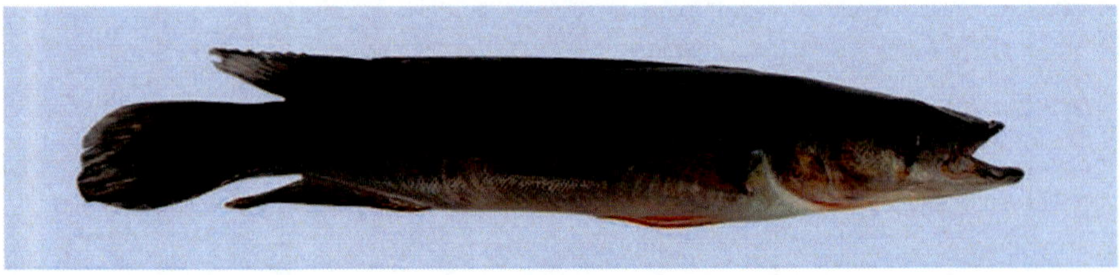

Fig. 6.34 *Crenicichla lugubris* sampled in the Negro River. (Photo: Renato Silvano)

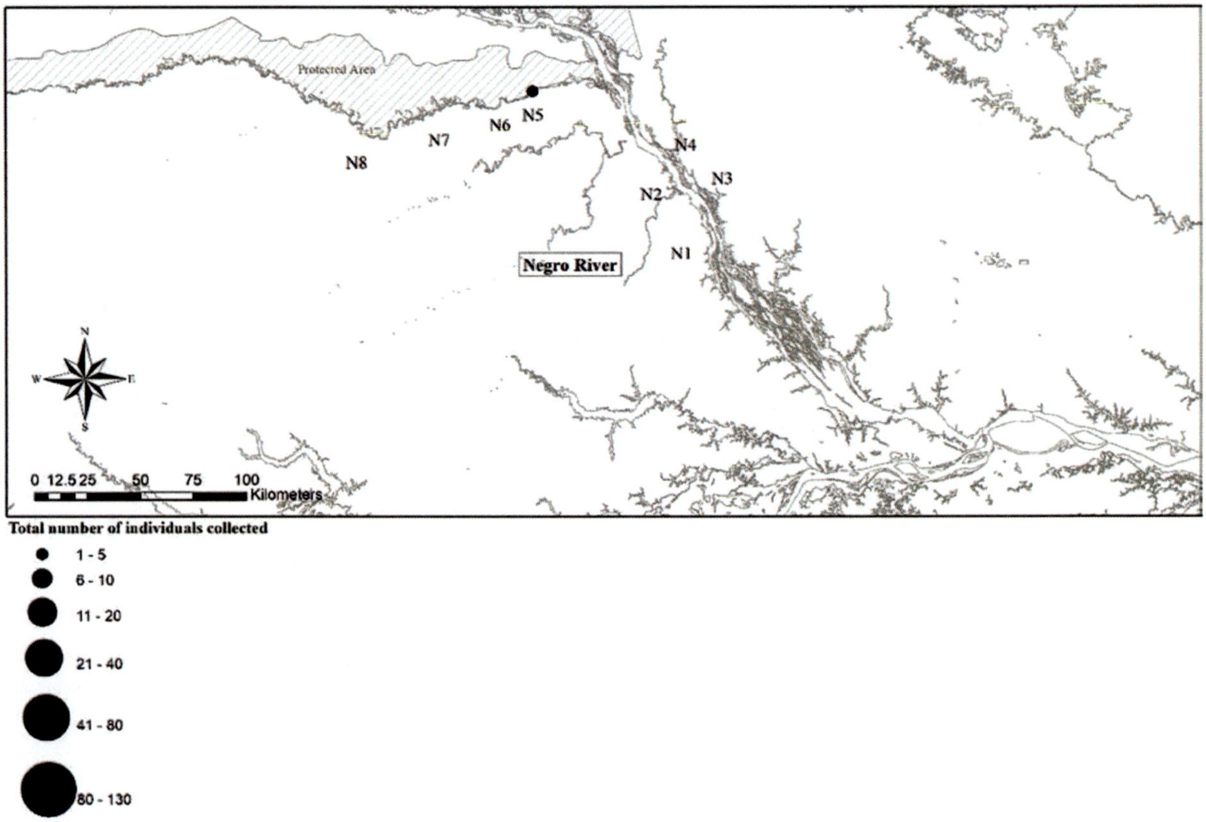

Fig. 6.35 Distribution of individuals of *Crenicichla lugubris* sampled in the Negro River

Fish species: *Geophagus altifrons* Heckel 1840

Common name(s): acaratinga, chaperema

Observation: abundance and size data were grouped for *Geophagus altifrons* and *Geophagus* cf. *altifrons* (no photo shown) due to the difficulty in distinguishing these possible two species during sampling.

Size of fish sampled (standard length, both rivers): min = 6.5 cm, max = 20 cm, mean = 13.1 cm, median = 13.5 cm

Fisheries: acaratingas (*Geophagus* spp.) are important for fisheries in the Tapajós River, being caught mostly for food. Biomass caught is in ◘ Table 3.3 (► Chap. 3).

◘ **Fig. 6.36** *Geophagus altifrons* sampled in the Tapajós River. (Photo: Renato Silvano)

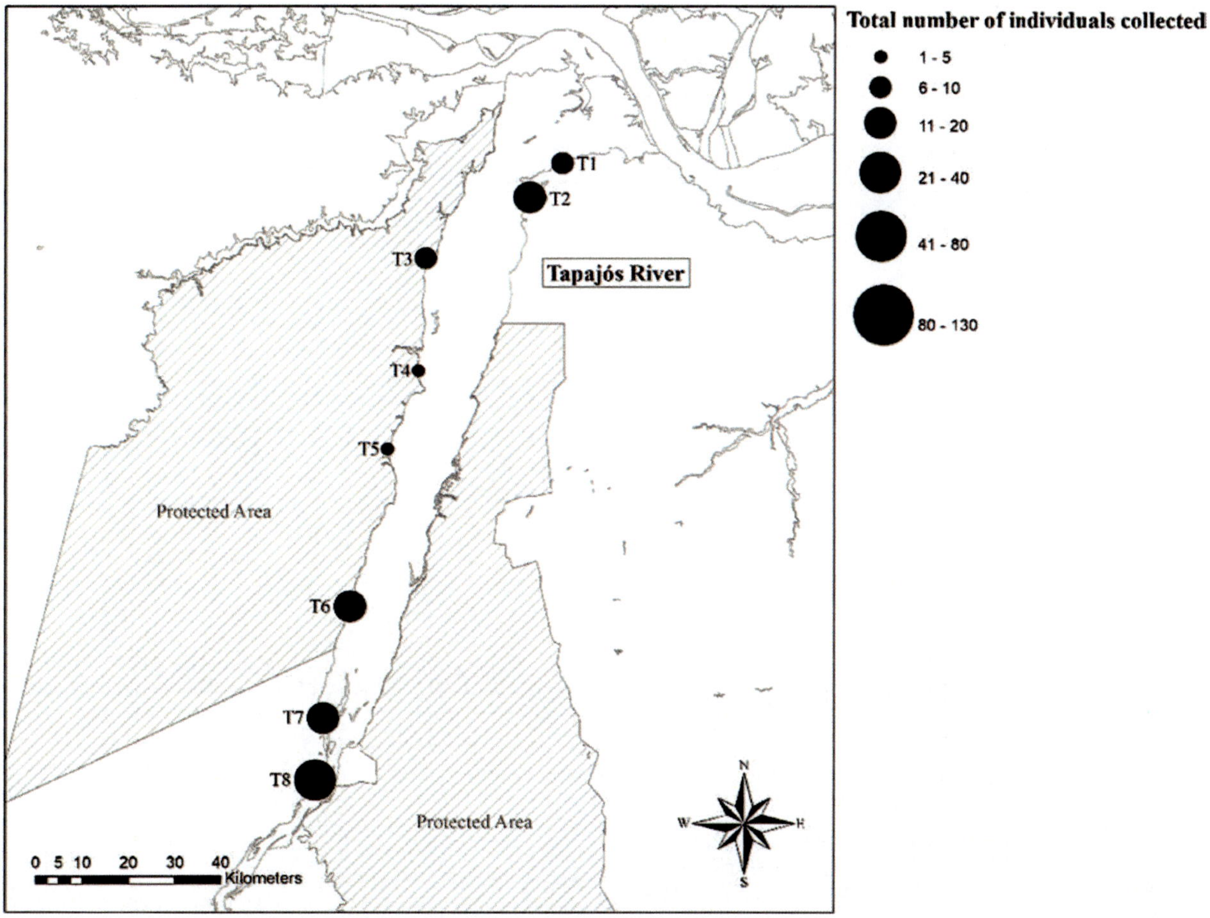

◘ **Fig. 6.37** Distribution of individuals of *G. altifrons* sampled in the Tapajós River

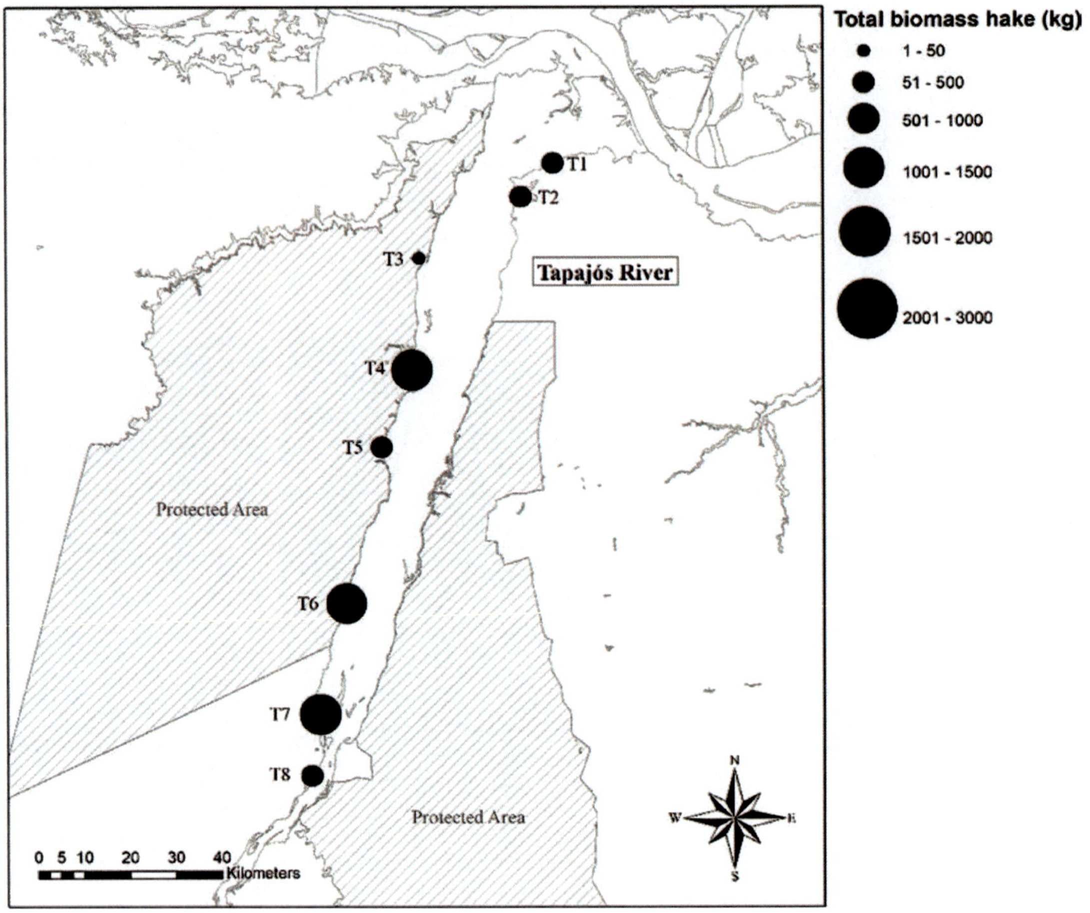

Fig. 6.38 Distribution of biomass of acaratingas (*Geophagus* spp.) caught in small-scale fisheries in the Tapajós River

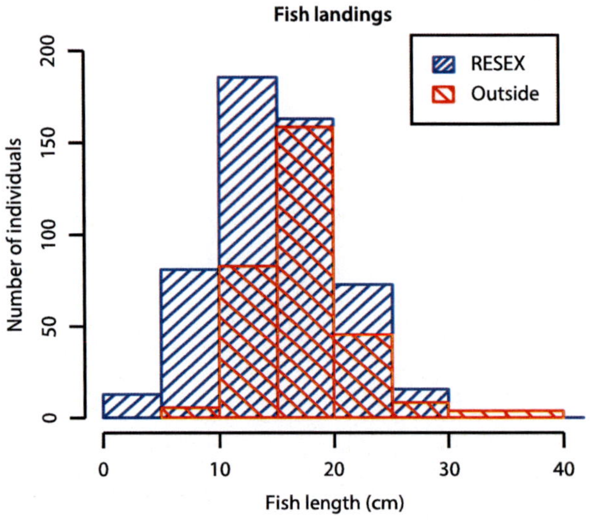

Fig. 6.39 Size distribution of acaratingas (*Geophagus* spp.) caught in small-scale fisheries in eight communities inside and outside the Extractive Reserves (RESEX) in the Tapajós River, N Resex = 531, N outside = 305

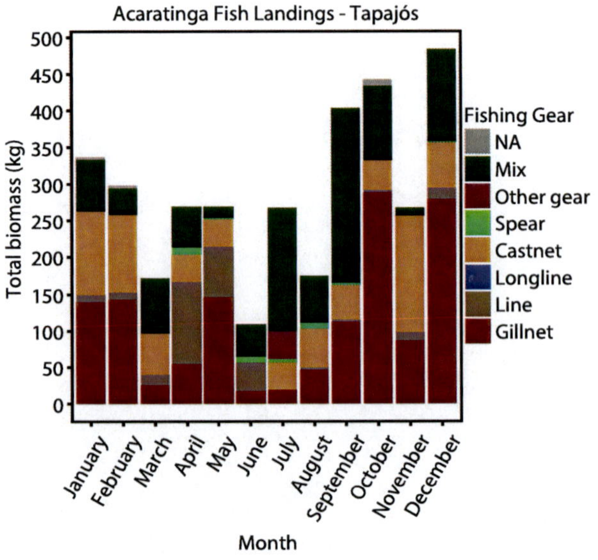

Fig. 6.40 Seasonal variation of biomass of acaratingas (*Geophagus* spp.) caught per fishing gear in small-scale fisheries in eight communities in the Tapajós River. NA: fishing gear not informed; Mix: more than one gear was used

Fish species: *Geophagus winemilleri* López-Fernández & Taphorn 2004

Common name(s): cará, cará-banda-de-remo, cará-crecréu, cará-rói-vidro

Size of fish sampled (standard length, both rivers): mean = 13.7cm, median = 13 cm

Fisheries data: acaratingas (*Geophagus* spp.) are important for fisheries in the Tapajós River, but these fish are not regularly caught in Negro River, being usually grouped with other species of cará (Cichlidae); see *Geophagus altifrons* for data on acaratinga fisheries and *Satanoperca lilith* for data on cará and acará fisheries.

Fig. 6.41 *Geophagus winemilleri* sampled in the Negro River. (Photo: Renato Silvano)

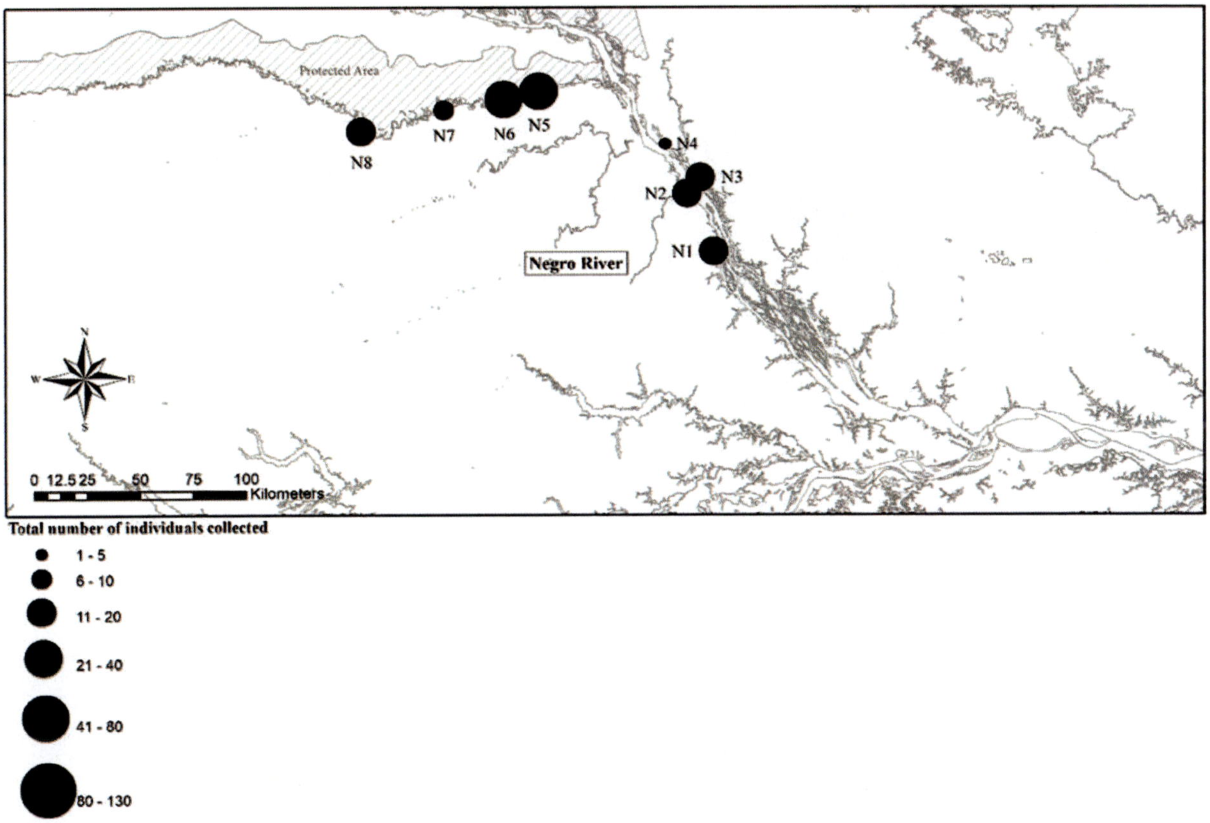

Fig. 6.42 Distribution of individuals of *Geophagus winemilleri* sampled in the Negro River

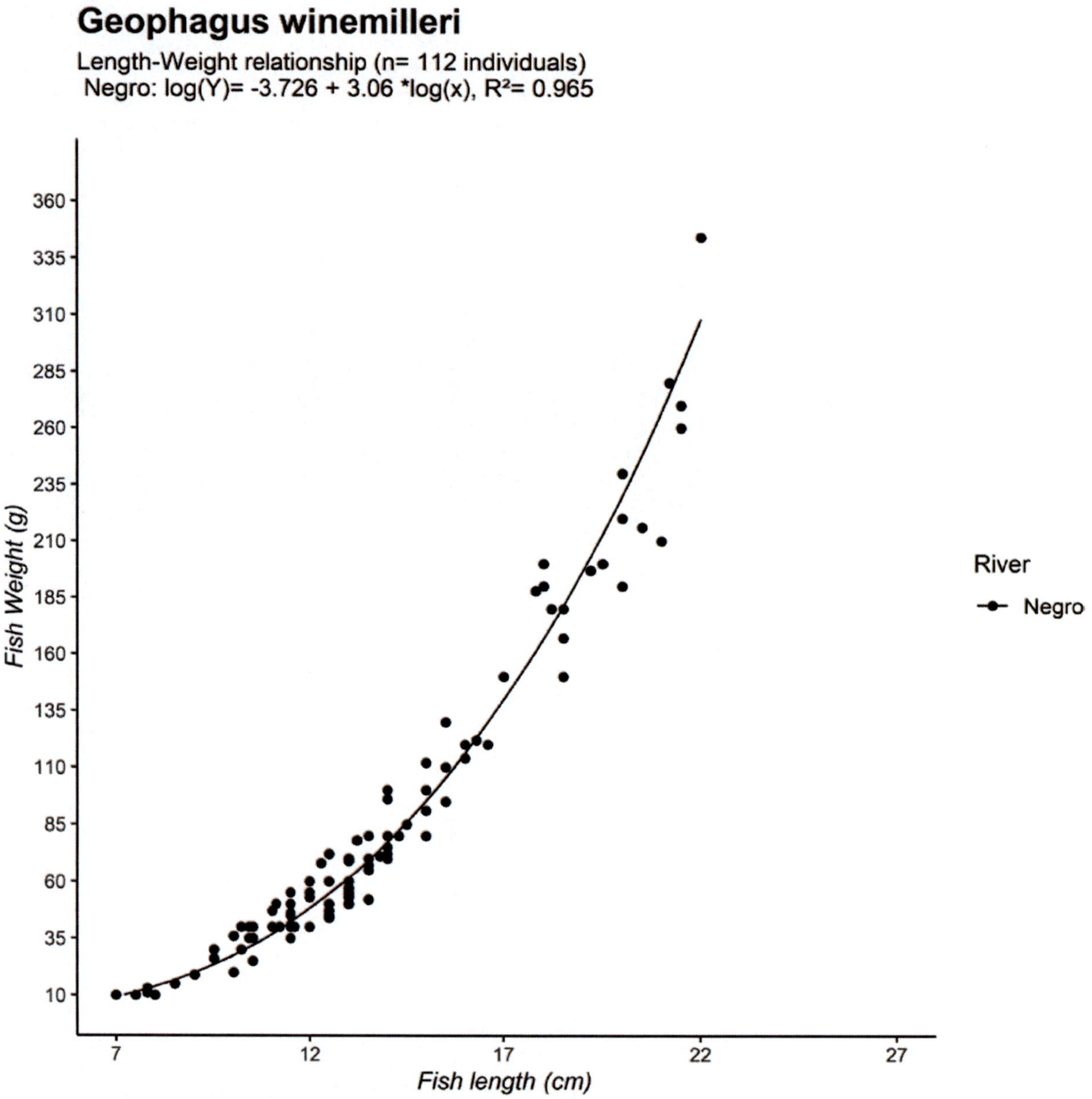

Fig. 6.43 Length-weight relationship for *G. winemilleri* sampled in the Negro River

Fish species: *Geophagus* sp. 'red head'

Common name(s): chaperema, chaperema-cabeça-vermelha

Size of fish sampled (standard length, both rivers): mean = 10.7 cm, median = 11 cm

Fisheries data: acaratingas (*Geophagus* spp.) are important for fisheries in the Tapajós River, being caught mostly for food; see *Geophagus altifrons* for data on acaratinga fisheries.

Fig. 6.44 *Geophagus* sp. 'red head' sampled in the Tapajós River. (Photo: Renato Silvano)

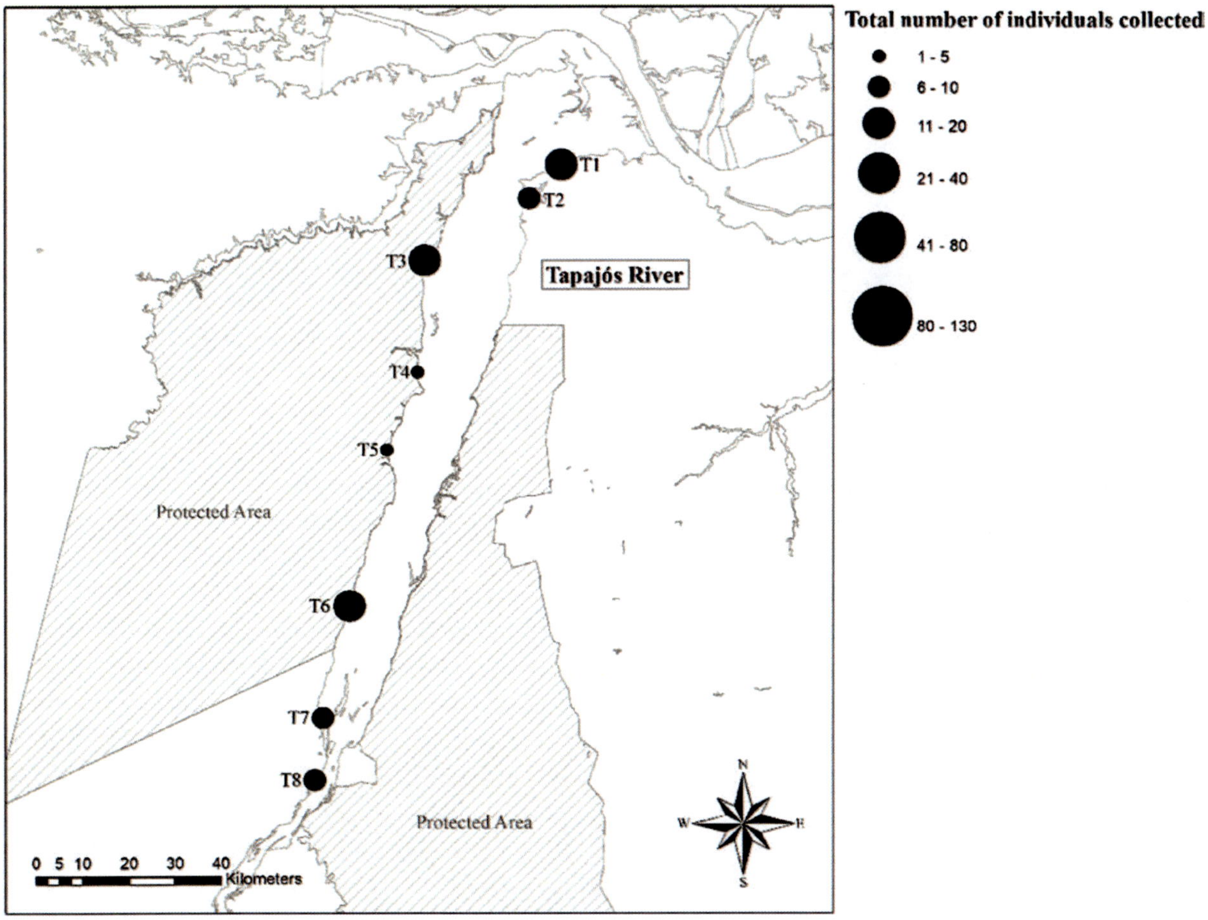

Fig. 6.45 Distribution of individuals of *Geophagus* sp. 'red head' sampled in the Tapajós River

6

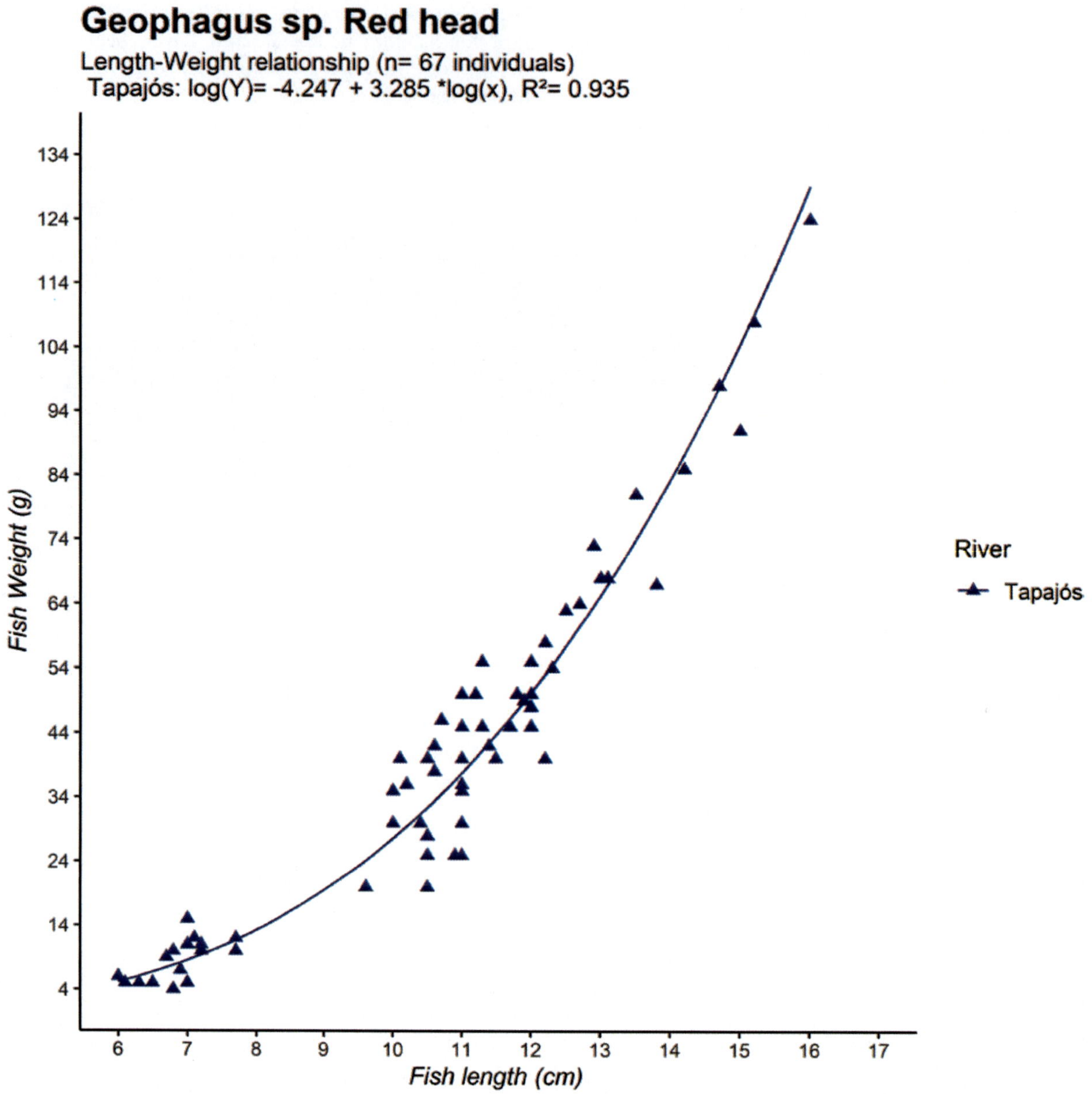

Fig. 6.46 Length-weight relationship for *Geophagus* sp. 'red head' sampled in the Tapajós River

Fish species: *Hoplarchus psittacus* **(Heckel 1840)**

Common name(s): cará-papagaio, cará-azulão

Size of fish sampled (standard length, both rivers): mean = 21.6 cm, median = 22 cm

Fisheries: may be occasionally caught with other carás or acarás in Negro River; see *Satanoperca lilith* for data on cará and acará fisheries.

Fig. 6.47 *Hoplarchus psittacus* sampled in the Negro River. (Photo: Renato Silvano)

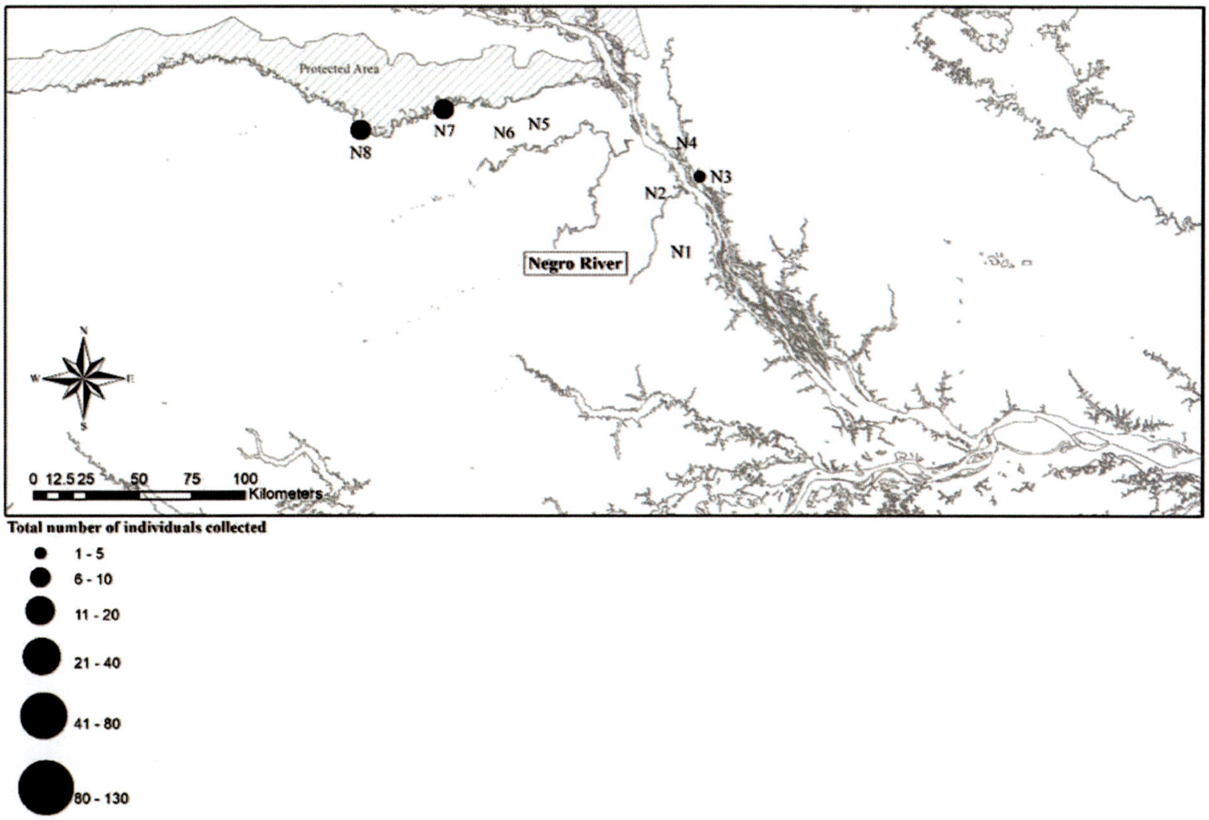

Fig. 6.48 Distribution of individuals of *Hoplarchus psittacus* sampled in the Negro River

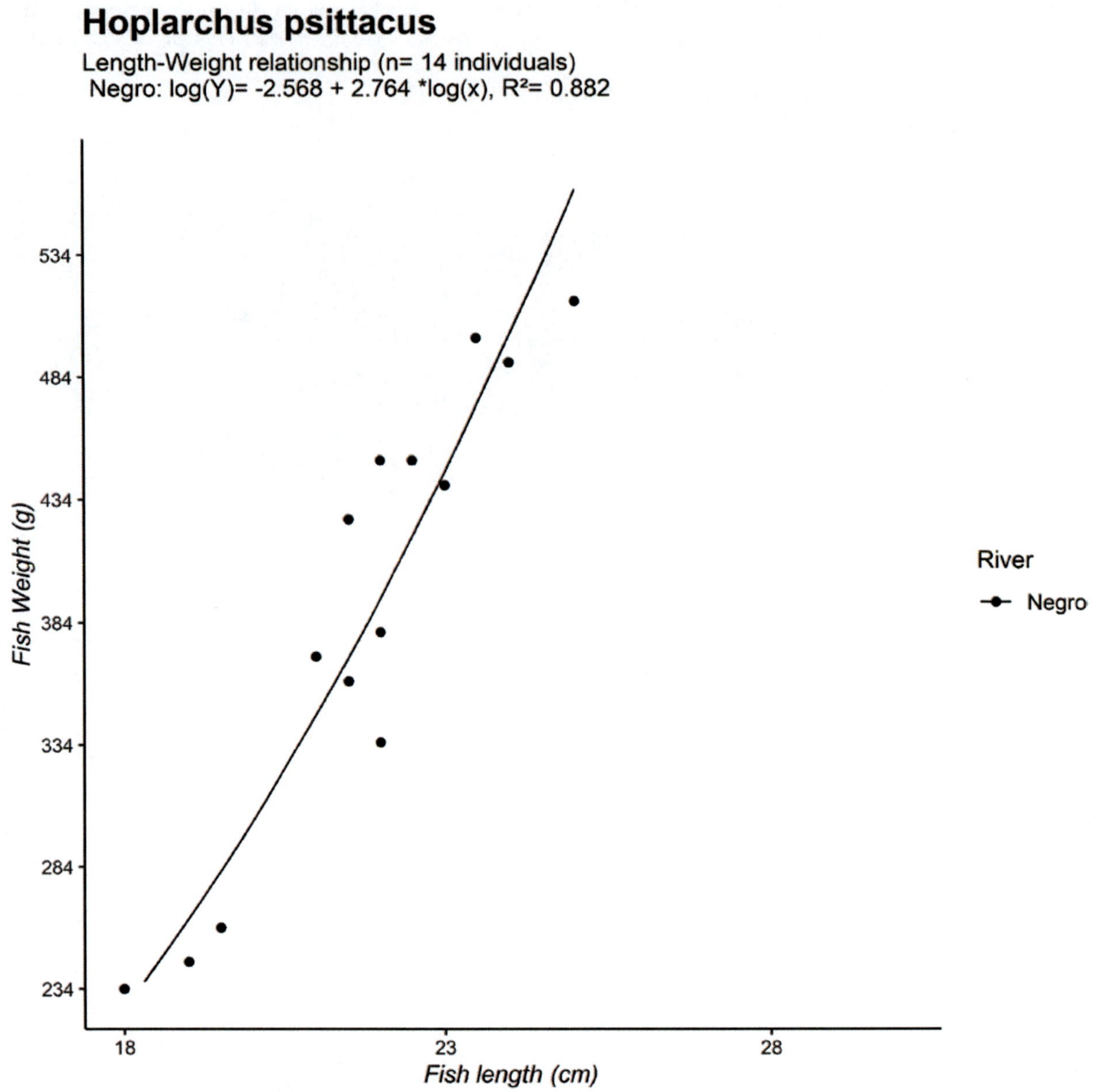

Fig. 6.49 Length-weight relationship for *Hoplarchus psittacus* sampled in the Negro River

Fish species: *Satanoperca acuticeps* **(Heckel 1840)**

Common name(s): cará, cará-bicudo, cará-pucu, cará-biquara

Size of fish sampled (standard length, both rivers): mean = 11.1 cm, median = 11.5 cm

Fisheries: carás and acarás (Cichlidae) are exploited in fisheries mostly for food, not being among the most caught fish; see *Satanoperca lilith* for data on cará and acará fisheries.

Fig. 6.50 *Satanoperca acuticeps* sampled in the Tapajós River. (Photo: Renato Silvano)

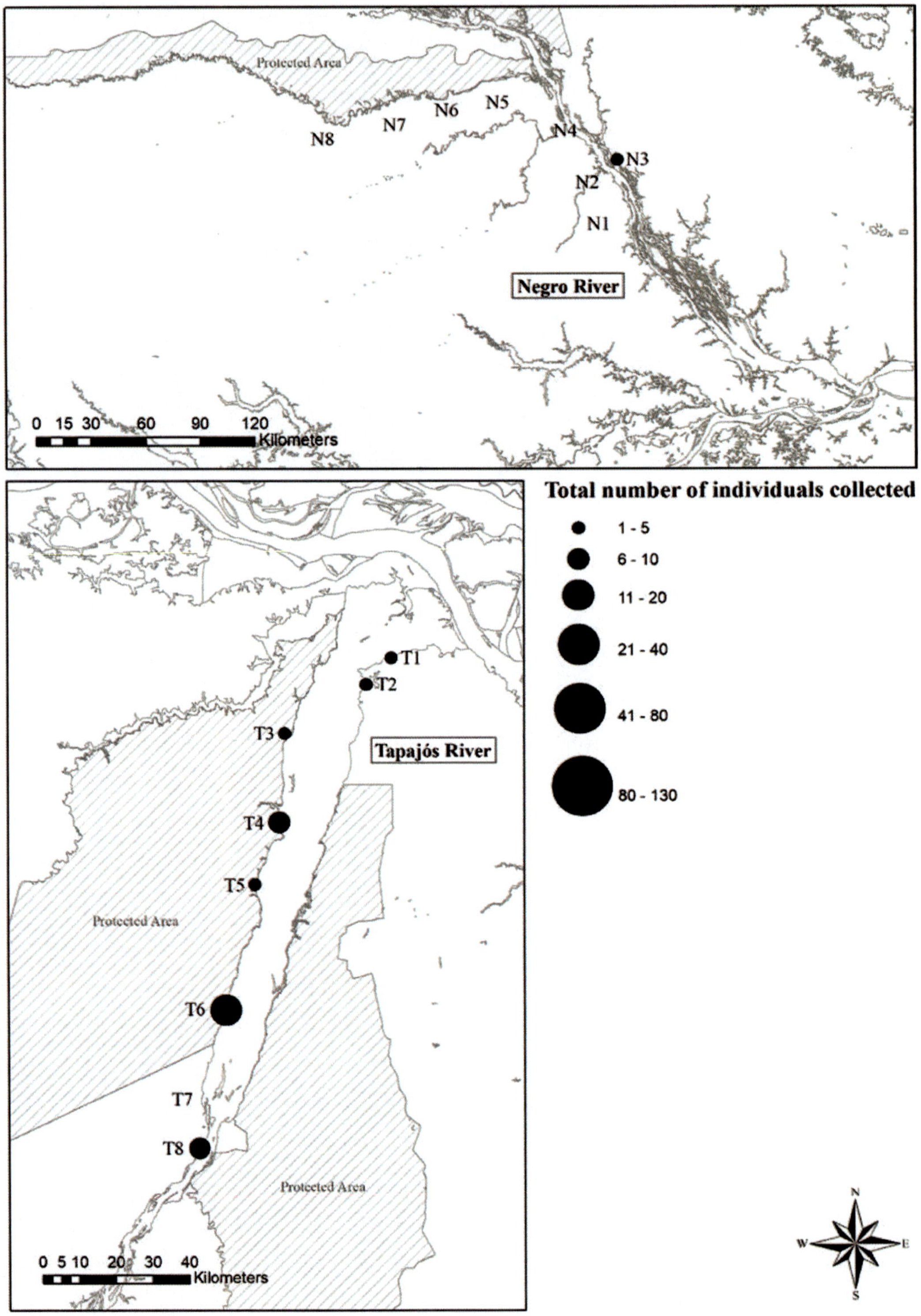

Fig. 6.51 Distribution of individuals of *Satanoperca acuticeps* sampled in the Negro and Tapajós rivers

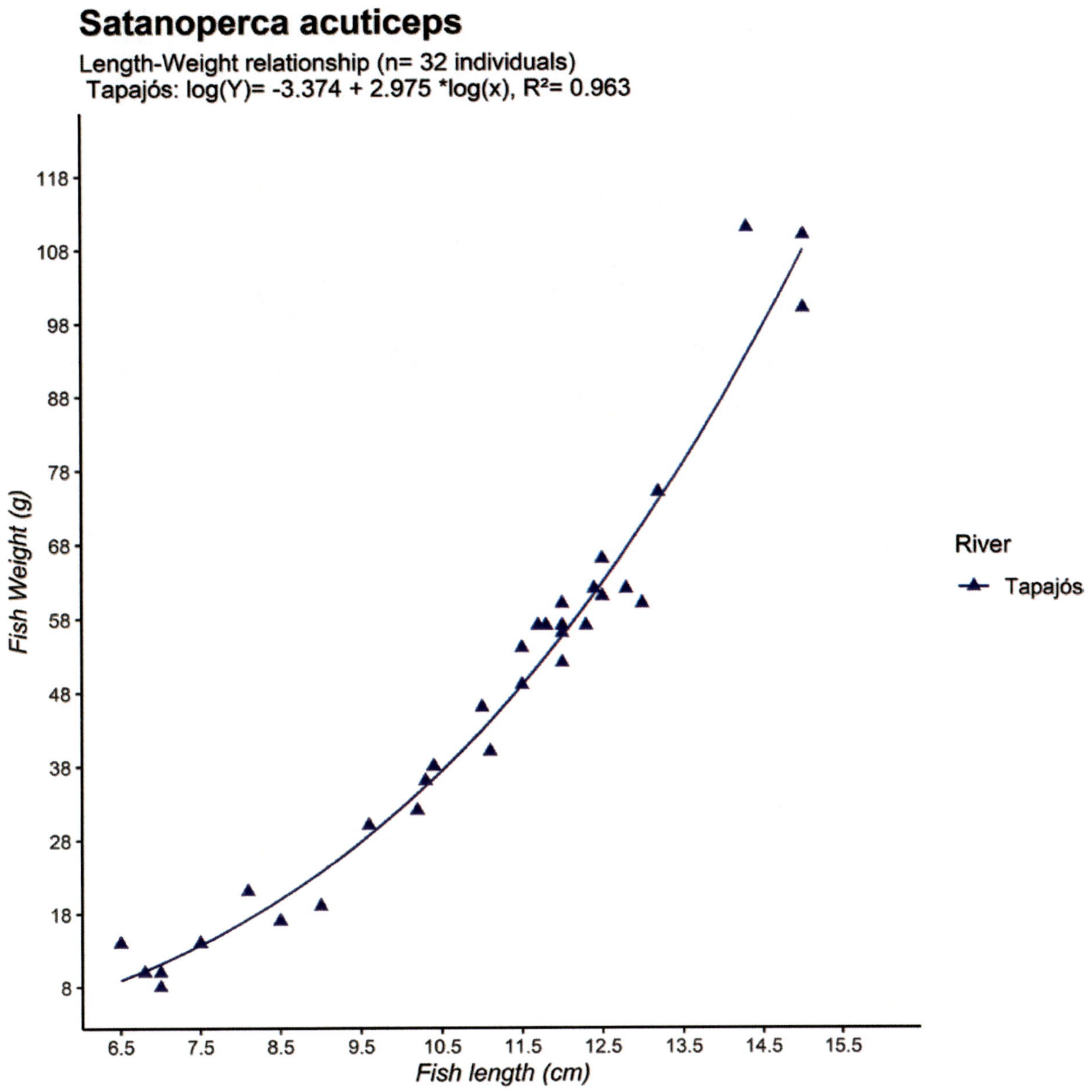

Fig. 6.52 Length-weight relationship for *Satanoperca acuticeps* sampled in Tapajós River

Fish species: *Satanoperca jurupari* **(Heckel 1840)**

Common name(s): cará, cará-barrão, cará-bicudo

Size of fish sampled (standard length, both rivers): min = 7 cm, max = 16.8 cm, mean = 11.5 cm, median = 11.4 cm

Fisheries: carás and acarás (Cichlidae) are exploited in fisheries mostly for food, not being among the most caught fish; see *Satanoperca lilith* for data on cará and acará fisheries.

Fig. 6.53 *Satanoperca jurupari* sampled in the Negro River. (Photo: Renato Silvano)

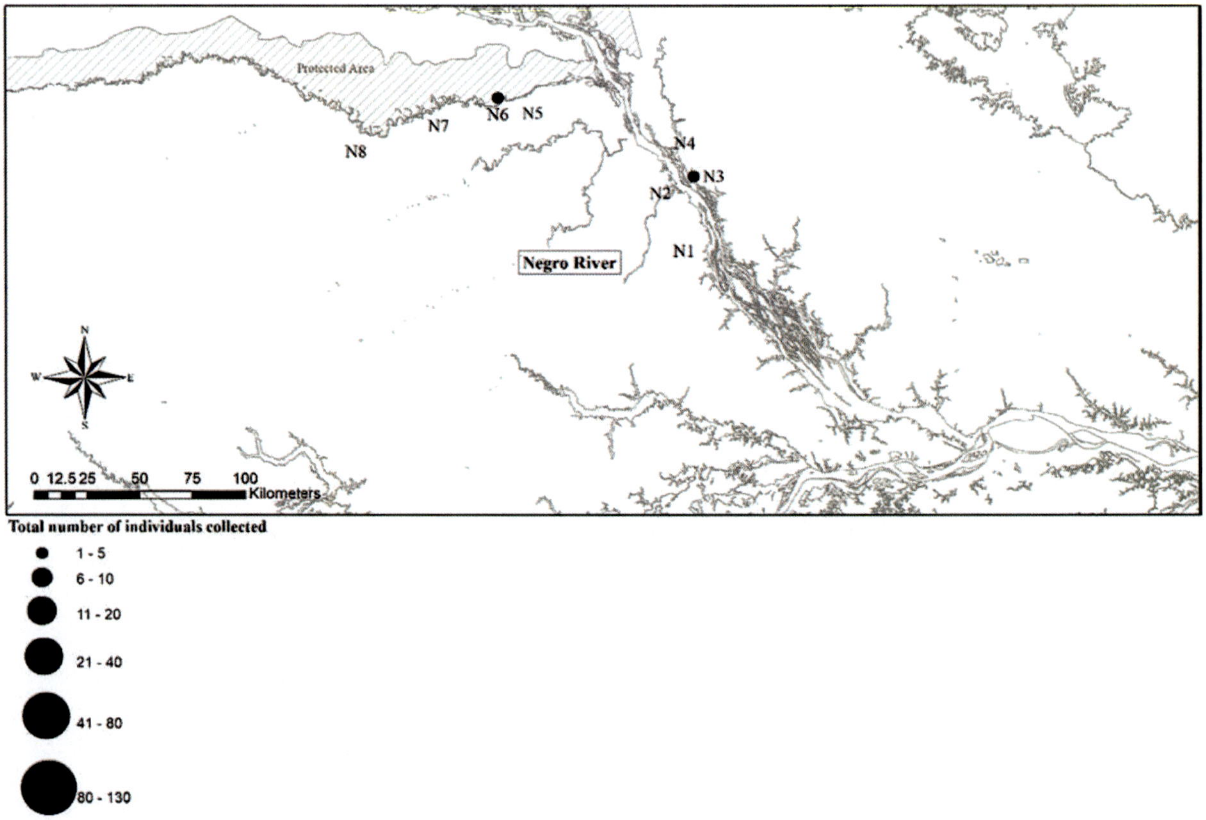

Fig. 6.54 Distribution of individuals of *Satanoperca jurupari* sampled in the Negro River

Fish species: *Satanoperca lilith* Kullander & Ferreira 1988

Common name(s): cará, cará-bicudo, cará-pucu

Size of fish sampled (standard length, both rivers): min. = 7.3 cm, max. = 22 cm, mean = 14 cm, median = 12.7 cm

Fisheries: carás and acarás (Cichlidae) are exploited in fisheries mostly for food, not being among the most caught fish. Biomass caught is in ◘ Table 3.3 (▸ Chap. 3).

◘ **Fig. 6.55** *Satanoperca lilith* sampled in the Negro River. (Photo: Renato Silvano)

6

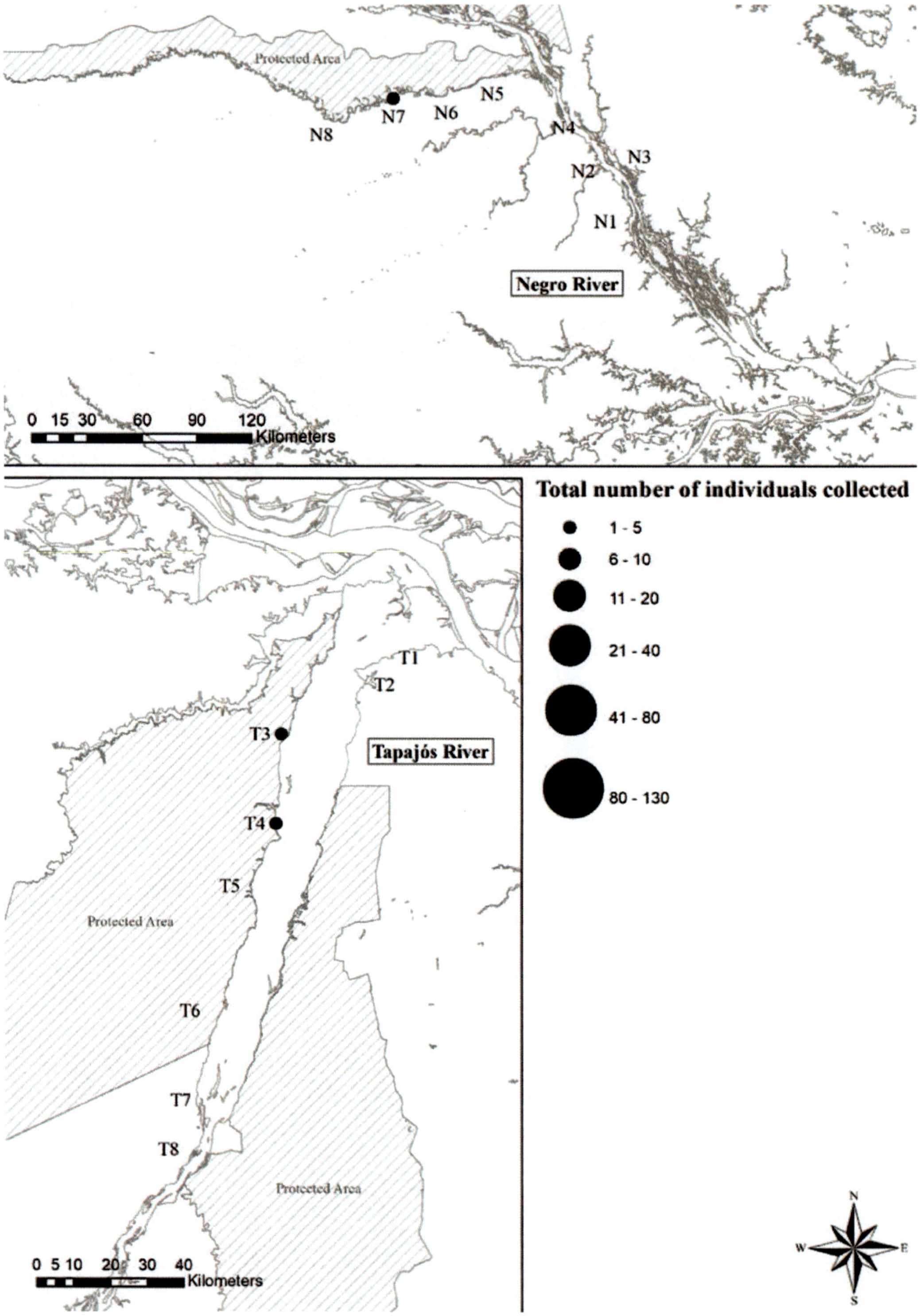

Fig. 6.56 Distribution of individuals of *Satanoperca lilith* sampled in the Negro and Tapajós rivers

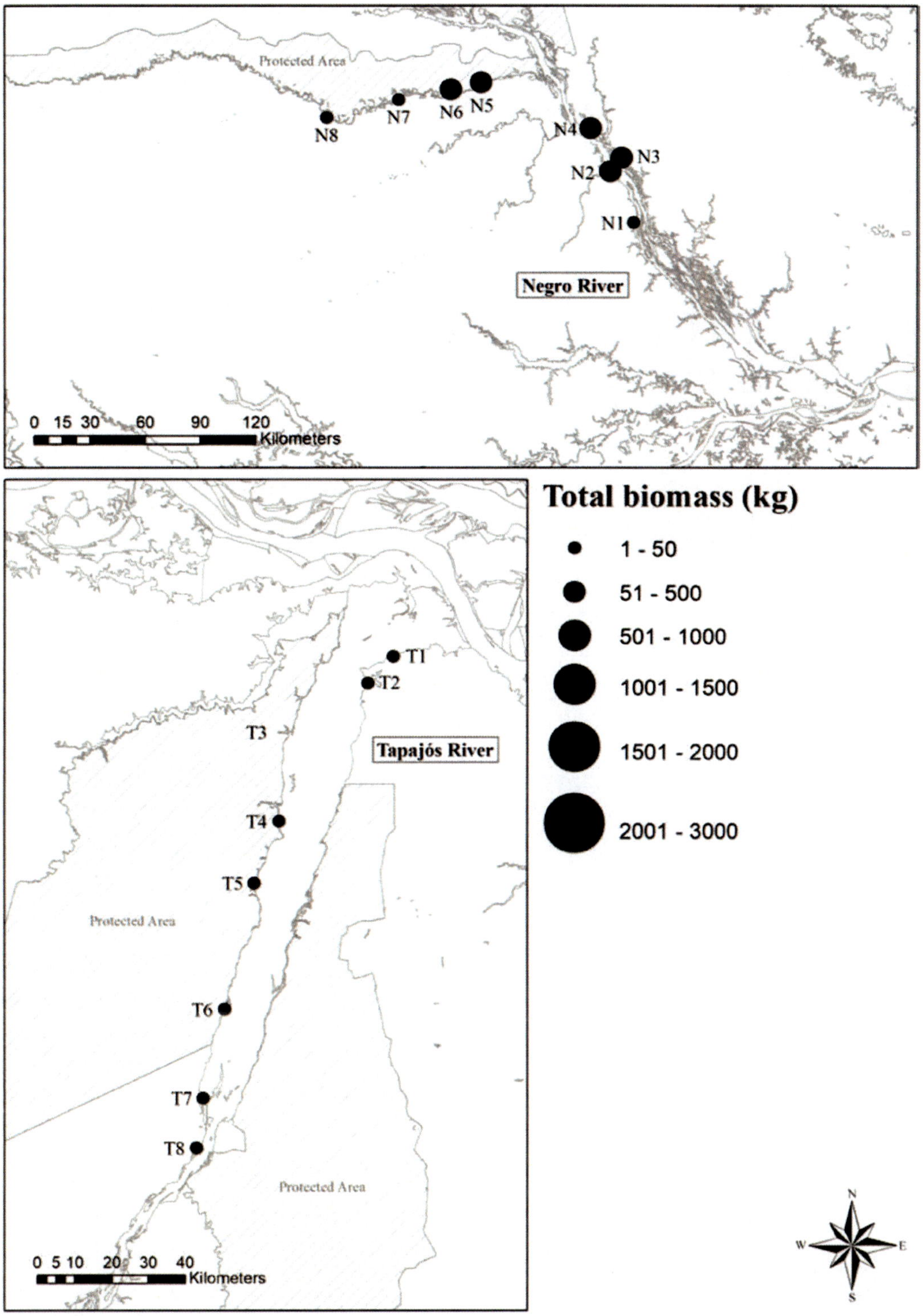

Fig. 6.57 Distribution of biomass of carás and acarás (Cichlidae) caught in small-scale fisheries in the Negro and Tapajós rivers

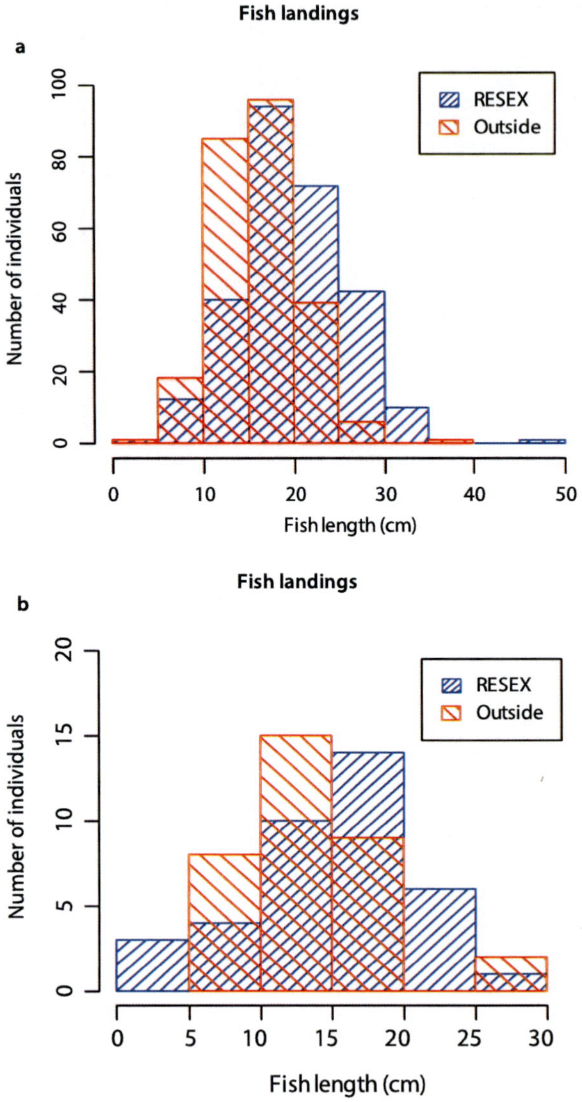

Fig. 6.58 Size distribution of carás and acarás (Cichlidae) caught in small-scale fisheries in 16 communities inside and outside the Extractive Reserves (RESEX) in the rivers **a** Negro, N Resex = 272, N outside = 276 and **b** Tapajós, N Resex = 38, N outside = 34

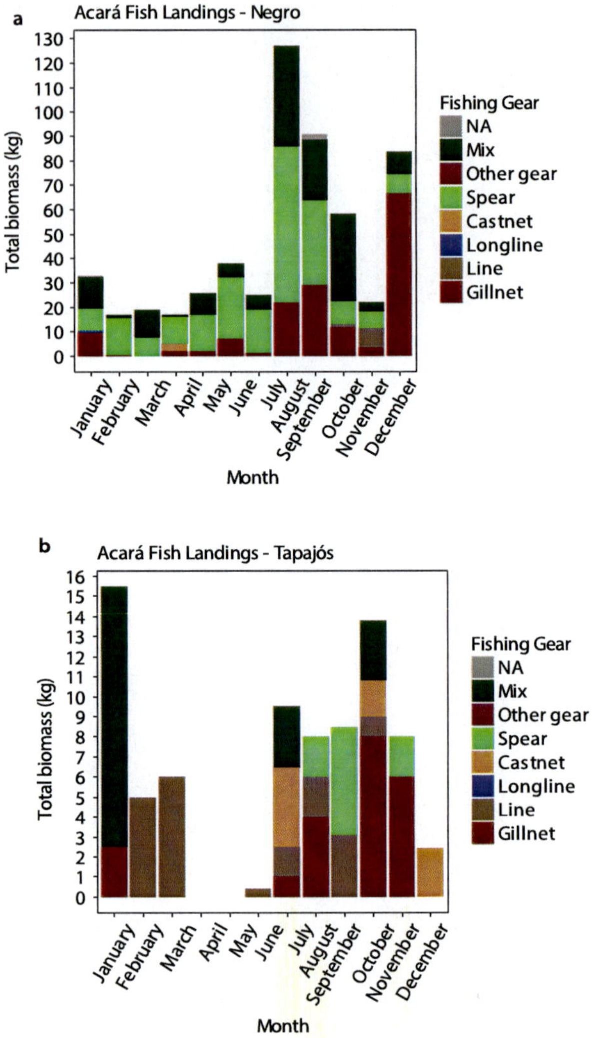

Fig. 6.59 Seasonal variation of biomass of carás and acarás (Cichlidae) caught per fishing gear in small-scale fisheries in 16 communities in the rivers **a** Negro and **b** Tapajós. NA: fishing gear not informed; Mix: more than one gear was used

Fish species: ***Uaru amphiacanthoides*** **Heckel 1840**

Common name(s): bararuá, cará-baru, baru

Size of fish sampled (standard length, both rivers): min. = 13 cm, max. = 18 cm, mean = 16.3 cm, median = 16.5 cm

Fisheries: barus or bararuás (*Uaru amphiacanthoides*) are exploited in fisheries mostly for food, not being among the most caught fish. Biomass caught is in ◘ Table 3.3 (► Chap. 3).

◘ **Fig. 6.60** *Uaru amphiacanthoides* sampled in the Tapajós River. (Photo: Renato Silvano)

6

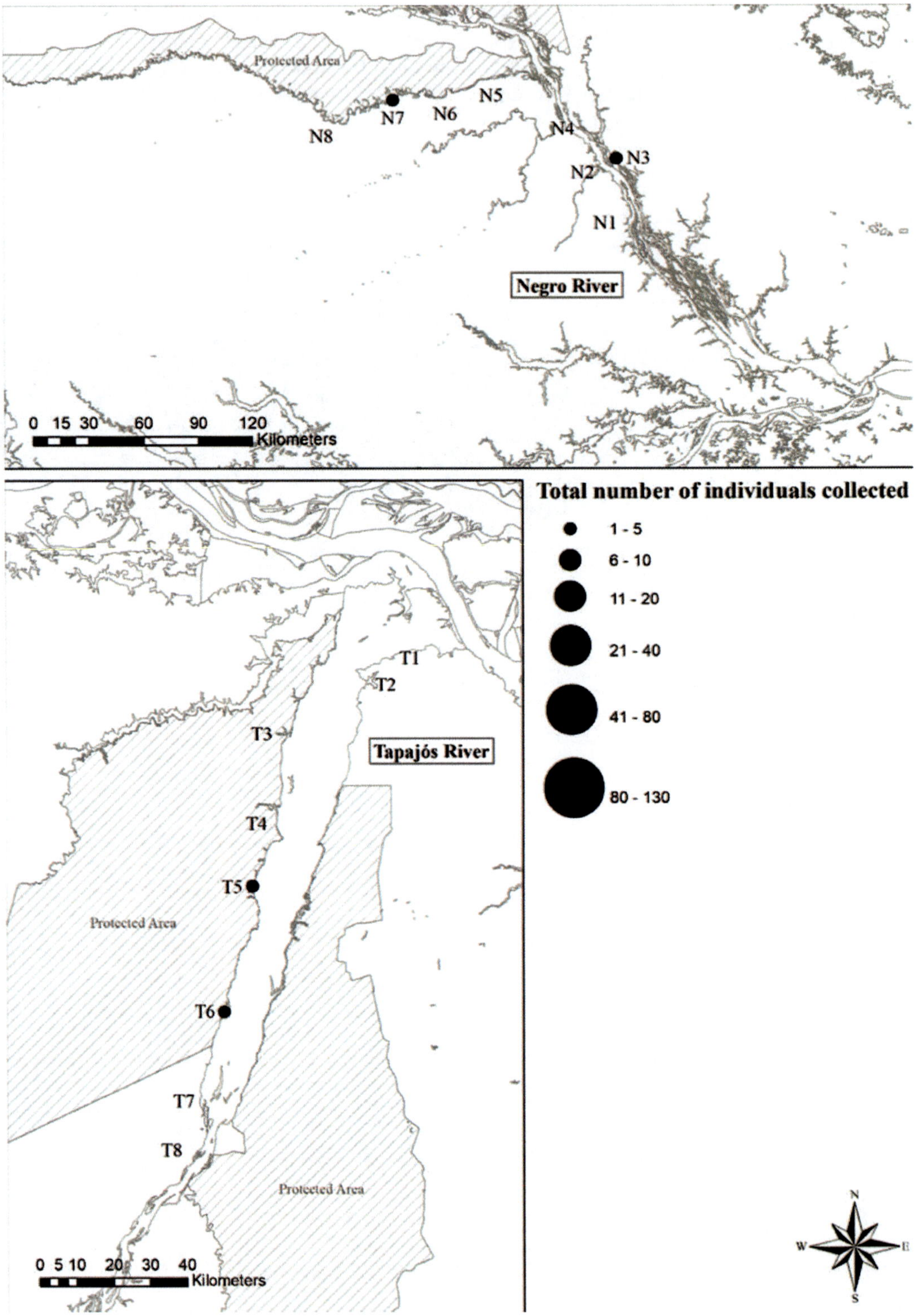

Fig. 6.61 Distribution of individuals of *Uaru amphiacanthoides* sampled in the Negro and Tapajós rivers

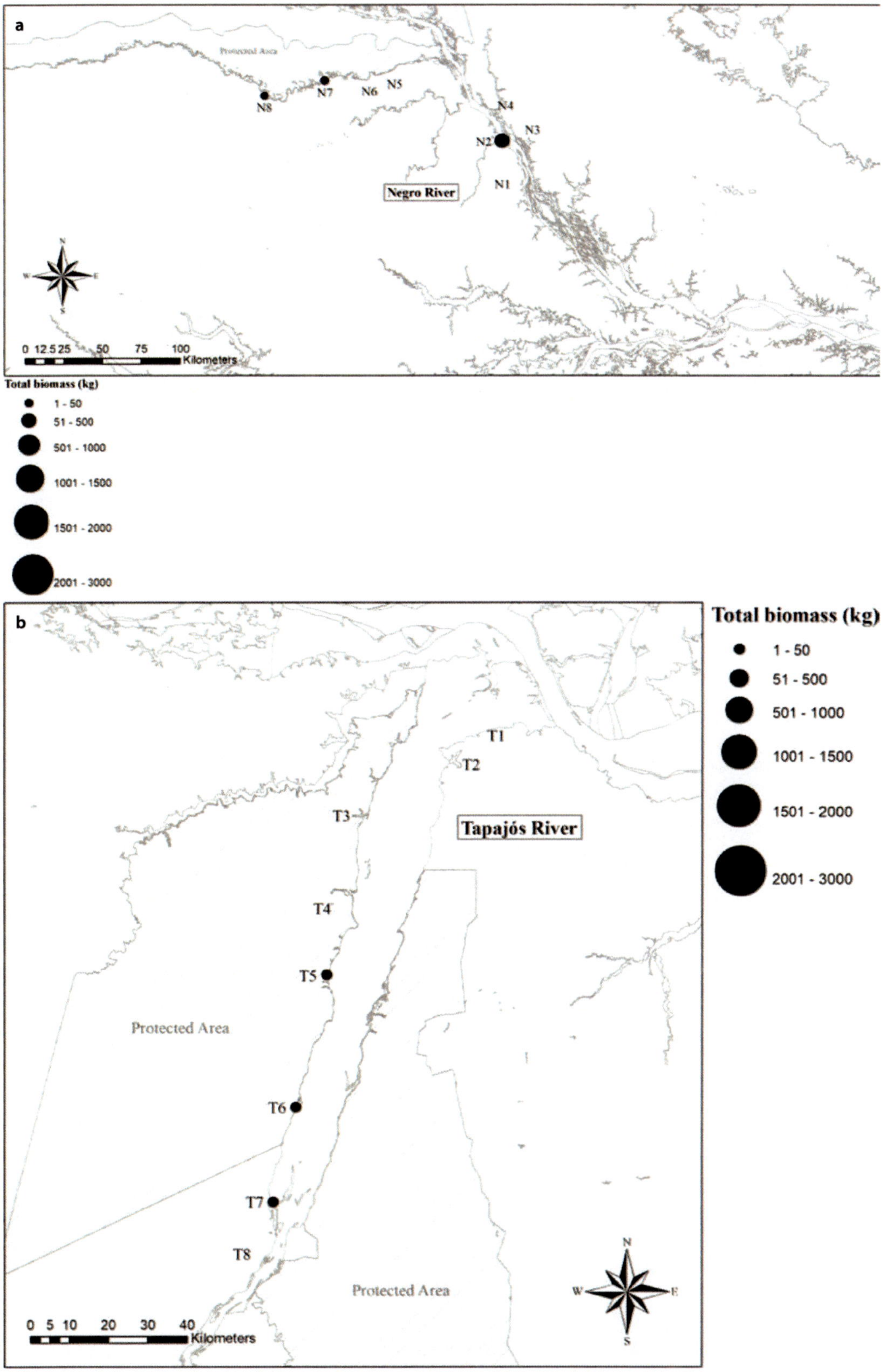

Fig. 6.62 Distribution of biomass of barus or bararuás (*Uaru amphiacanthoides*) caught in small-scale fisheries in the **a** Negro and **b** Tapajós rivers

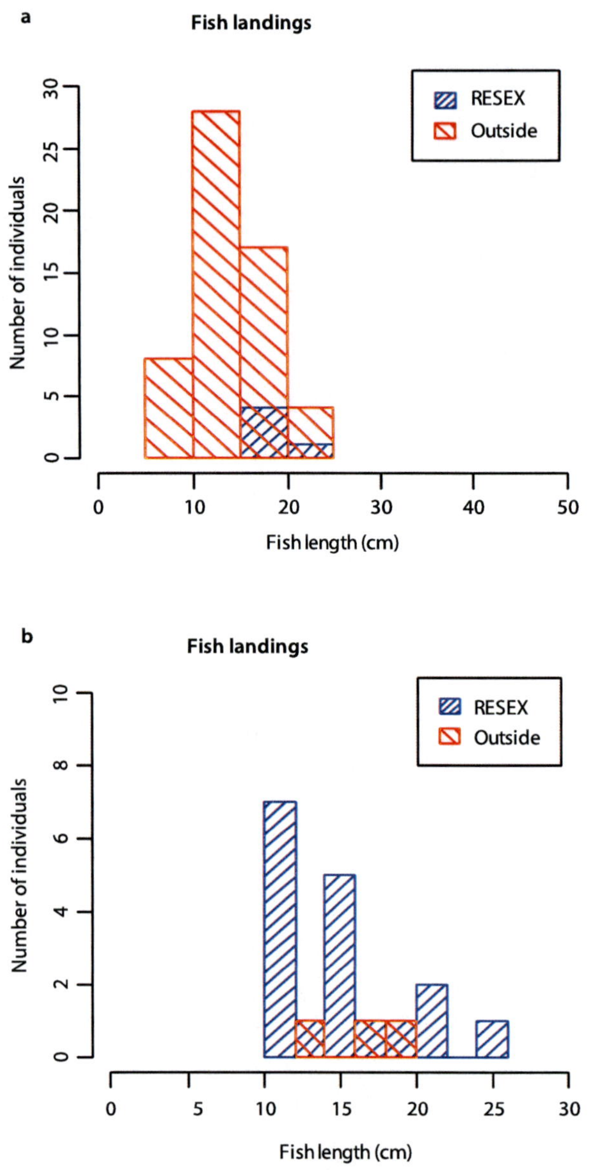

Fig. 6.63 Size distribution of barus or bararuás (*Uaru amphiacanthoides*) caught in small-scale fisheries in 16 communities inside and outside the Extractive Reserves (RESEX) in the rivers **a** Negro, N Resex = 5, N outside = 57 and **b** Tapajós, N Resex = 18, N outside = 3

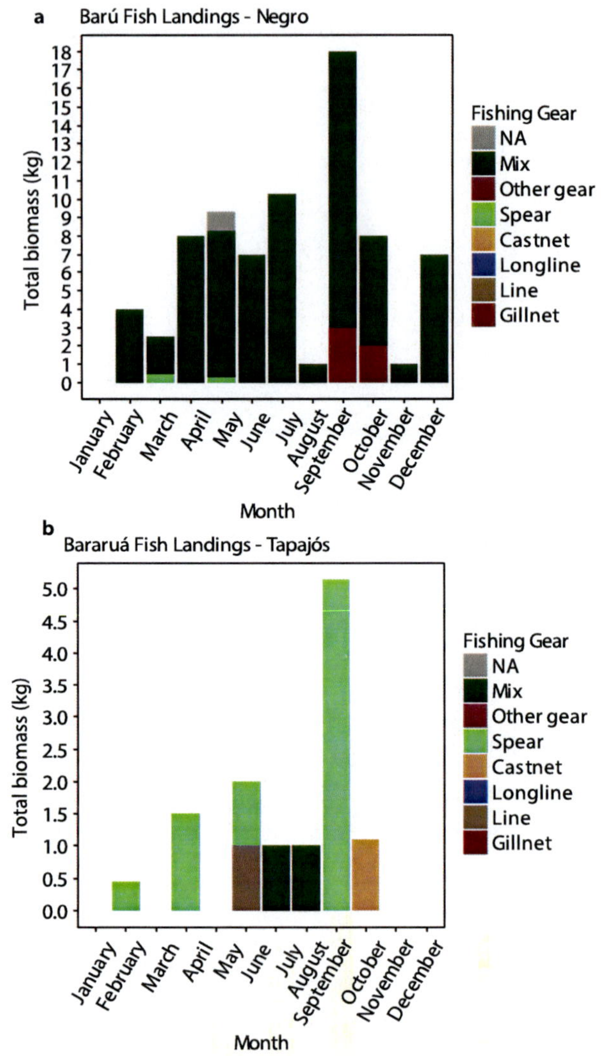

Fig. 6.64 Seasonal variation of biomass of barus or bararuás (*Uaru amphiacanthoides*) caught per fishing gear in small-scale fisheries in 16 communities in the rivers **a** Negro and **b** Tapajós. NA: fishing gear not informed; Mix: more than one gear was used

6.2 Order Perciformes *sedis mutabilis*

6.2.1 Family Sciaenidae

6.2.1.1 Summary of Biology and Ecology (Literature)

Diversity: Family composed of 80 genera and 294 valid species.

Habitat: members of the Scianidae family can be found in the coastal waters, estuaries, and freshwaters of temperate and tropical regions of the world. Six genera are restricted to freshwaters and occur occasionally in estuaries: *Aplodinotus, Boesemania, Pachypops, Pachyurus, Petilipinnis*, and *Plagioscion*; the latter four genera (~ 14 species) occur in South America and are widespread in the Magdalena, Orinoco, São Francisco, Paraná, Guianas, and Amazon river basins.

Ecology and feeding: fishes of this family are benthopelagic fishes with sedentary behavior, nocturnal/crepuscular habits, and carnivorous/piscivorous behavior, feeding mainly on fish, shrimps, and small invertebrates.

Morphology: their body size varies from 5 to 80 cm. The dorsal fin is long with a spiky section with 6 to 13 spines and 18 to 44 soft rays. The lateral line extends to the caudal fin; the cephalic region has well-developed lateral line channels; snout and the mentonian region with conspicuous pores.

Reproduction: the reproduction occurs in periods of high rainfall and high temperature with partitioned spawning and peaks during the dry season. During this period, males of some species produce peculiar sounds ("snores") that are audible outside of the water.

Economic importance: this family has high importance for commercial and artisanal fisheries, especially the species of the genus *Plagioscion*, which usually compose most of the biomass of the fish caught on fisheries in several Amazonian regions and in other regions where this fish has been introduced, such as in reservoirs in the Southeastern Brazil.

References: Santos et al. (2006), Silvano and Begossi (2001), Eschmeyer and Fong (2011), Hallwass et al. (2011, 2013), Ribeiro et al. (2016), Hallwass and Silvano (2016).

Fish species: *Pachypops fourcroi* **(Lacépède 1802)**

Common name(s): corvina, pescada

Size of fish sampled (standard length, both rivers): mean = 11.7 cm, median = 10.3 cm

Fisheries: corvinas (*Pachypops fourcroi* and *Pachyurus junki*) are seldom exploited by fisheries in both rivers. Biomass caught is in ◘ Table 3.3 (▶ Chap. 3).

◘ **Fig. 6.65** *Pachypops fourcroi* sampled in the Negro River. (Photo: Renato Silvano)

6

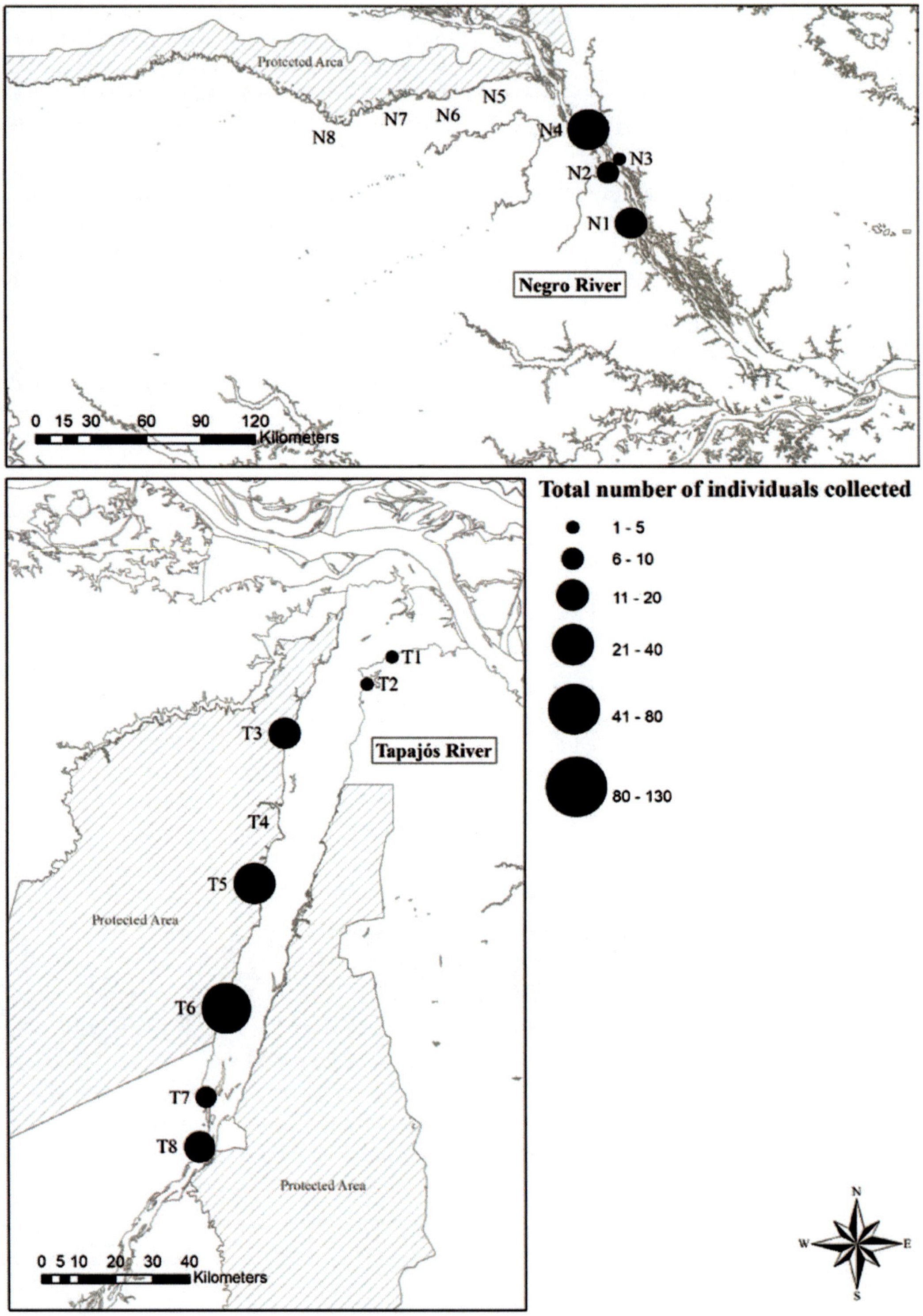

Fig. 6.66 Distribution of individuals of *Pachypops fourcroi* sampled in the Negro and Tapajós rivers

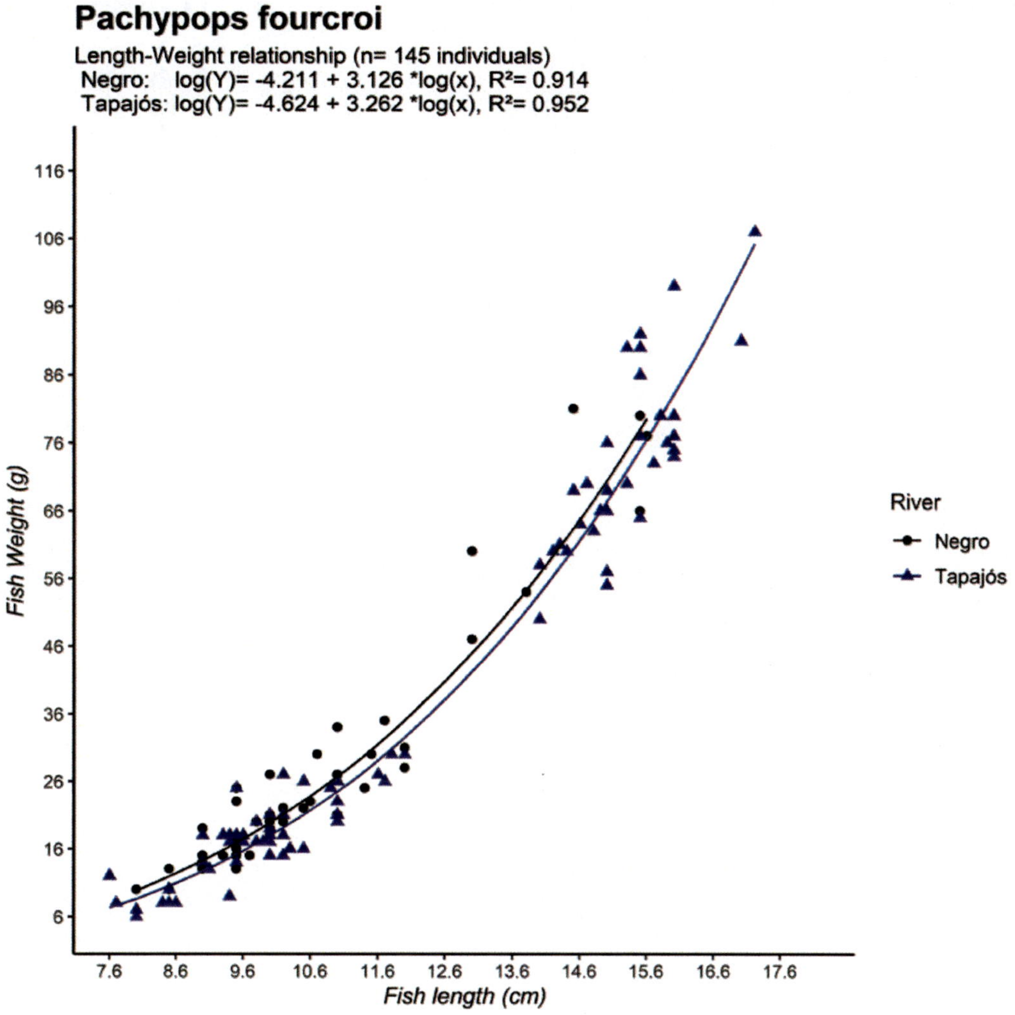

Fig. 6.67 Length-weight relationship for *Pachypops fourcroi* sampled in Tapajós and Negro rivers

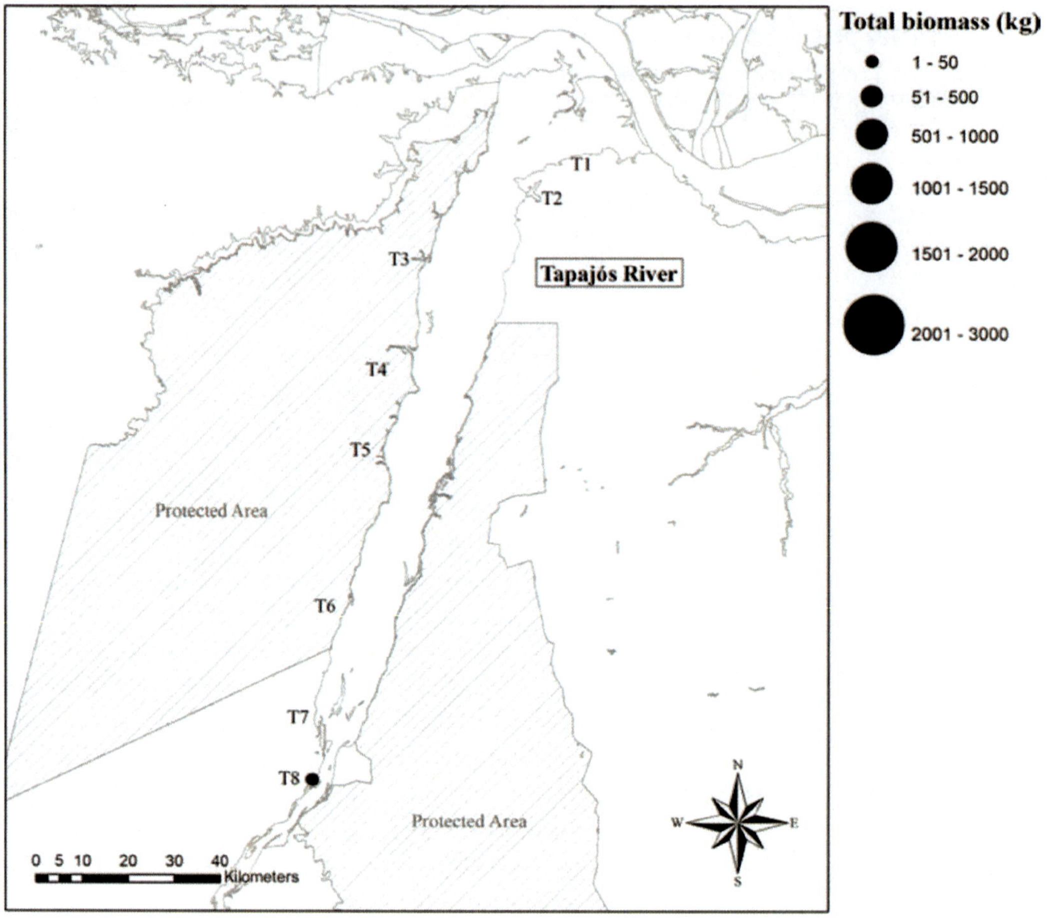

■ **Fig. 6.68** Distribution of biomass of corvinas (*Pachypops fourcroi* and *Pachyurus junki*) caught in small-scale fisheries in the Tapajós River

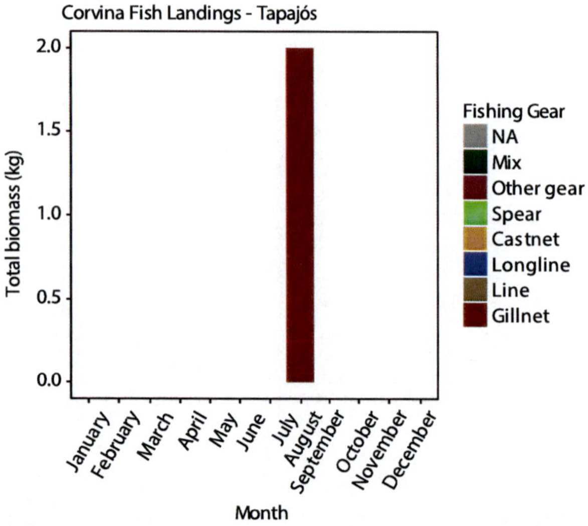

■ **Fig. 6.69** Seasonal variation of biomass of corvinas (*Pachypops fourcroi* and *Pachyurus junki*) caught per fishing gear in small-scale fisheries in eight communities in the Tapajós River. NA: fishing gear not informed; Mix: more than one gear was used

Fish species: ***Pachyurus junki*** **Soares & Casatti 2000**

Common name(s): corvina, corvina-pintada, pescada

Size of fish sampled (standard length, both rivers): min = 15 cm, max = 21.3 cm, mean = 18.1 cm, median = 18 cm

Fisheries: corvinas (*Pachypops fourcroi* and *Pachyurus junki*) are seldom exploited by fisheries in both rivers; see data on corvina fisheries in *Pachypops fourcroi*.

Fig. 6.70 *Pachyurus junki* sampled in the Negro River. (Photo: Renato Silvano)

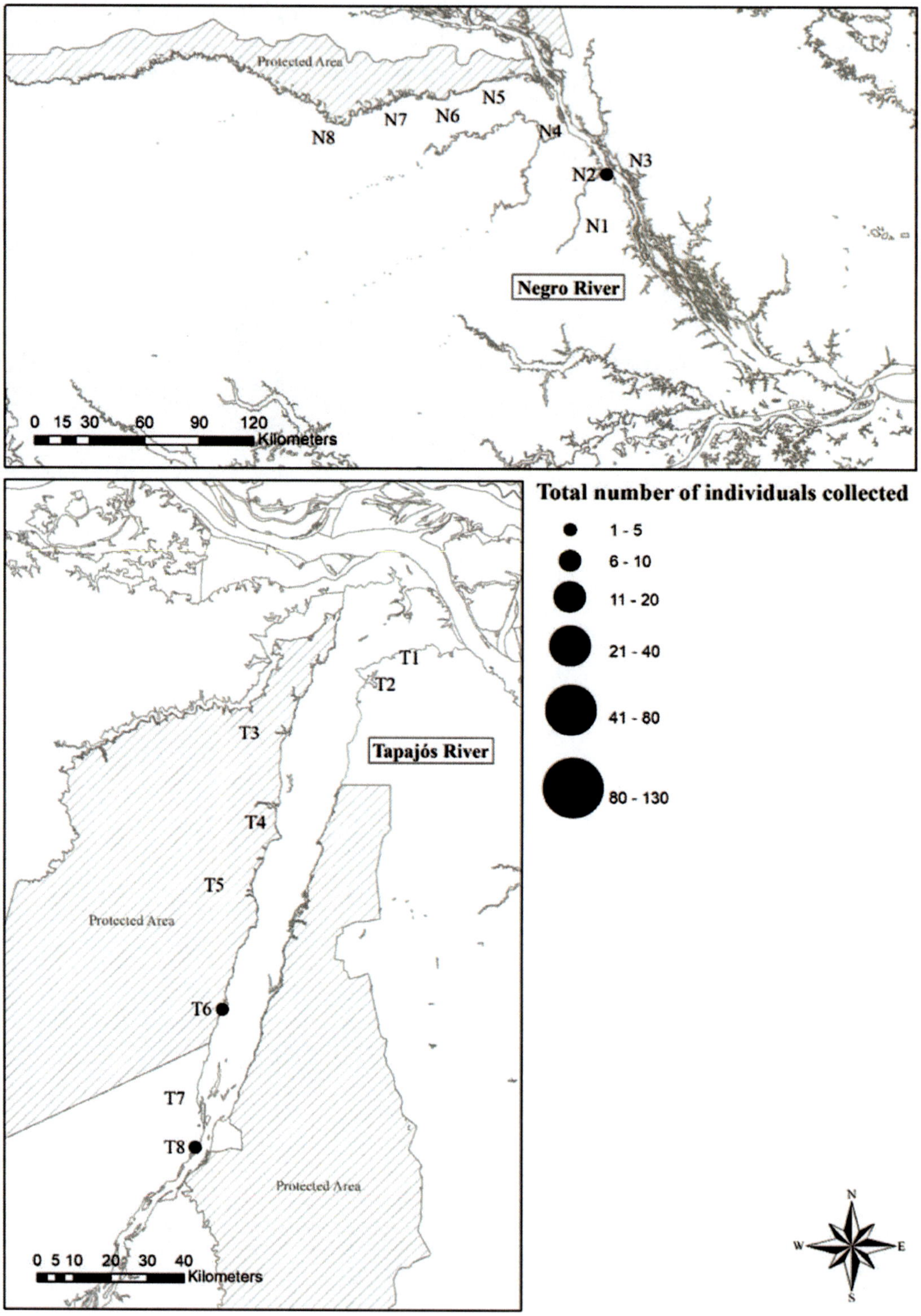

Fig. 6.71 Distribution of individuals of *Pachyurus junki* sampled in the Negro and Tapajós rivers

Fish species: *Plagioscion squamosissimus* (Heckel 1840)

Common name(s): pescada

Size of fish sampled (standard length, both rivers): mean = 19.8 cm, median = 19 cm

Fisheries: pescada (*Plagioscion squamosissimus*) is one of the most exploited fish in the Tapajós River, being also caught in the Negro River; this fish is important for food and commerce, especially in the Tapajós. Biomass caught is in ◘ Table 3.3 (► Chap. 3).

◘ **Fig. 6.72** *Plagioscion squamosissimus* sampled in the Negro River. (Photo: Renato Silvano)

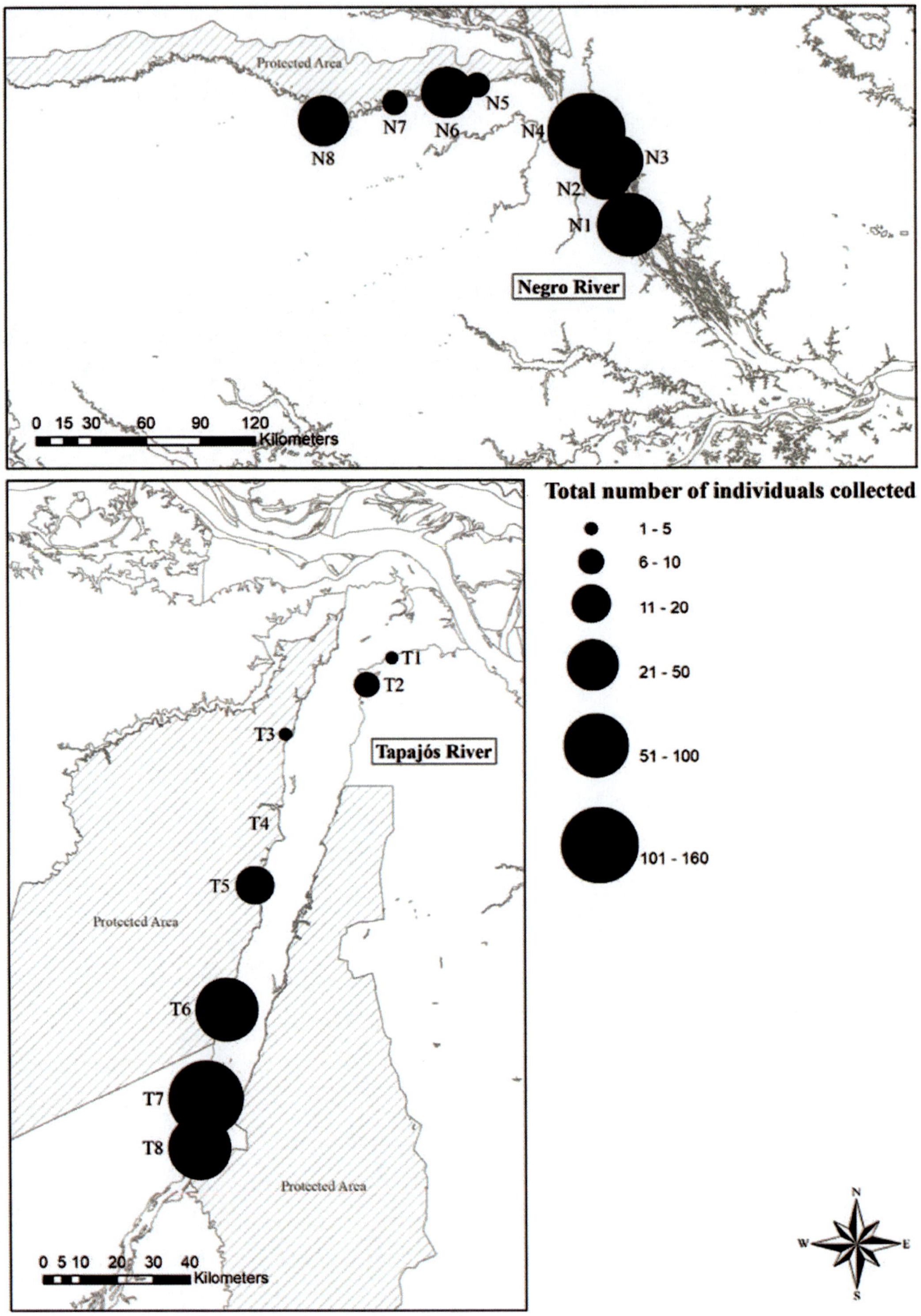

Fig. 6.73 Distribution of individuals of *Plagioscion squamosissimus* sampled in the Negro and Tapajós rivers

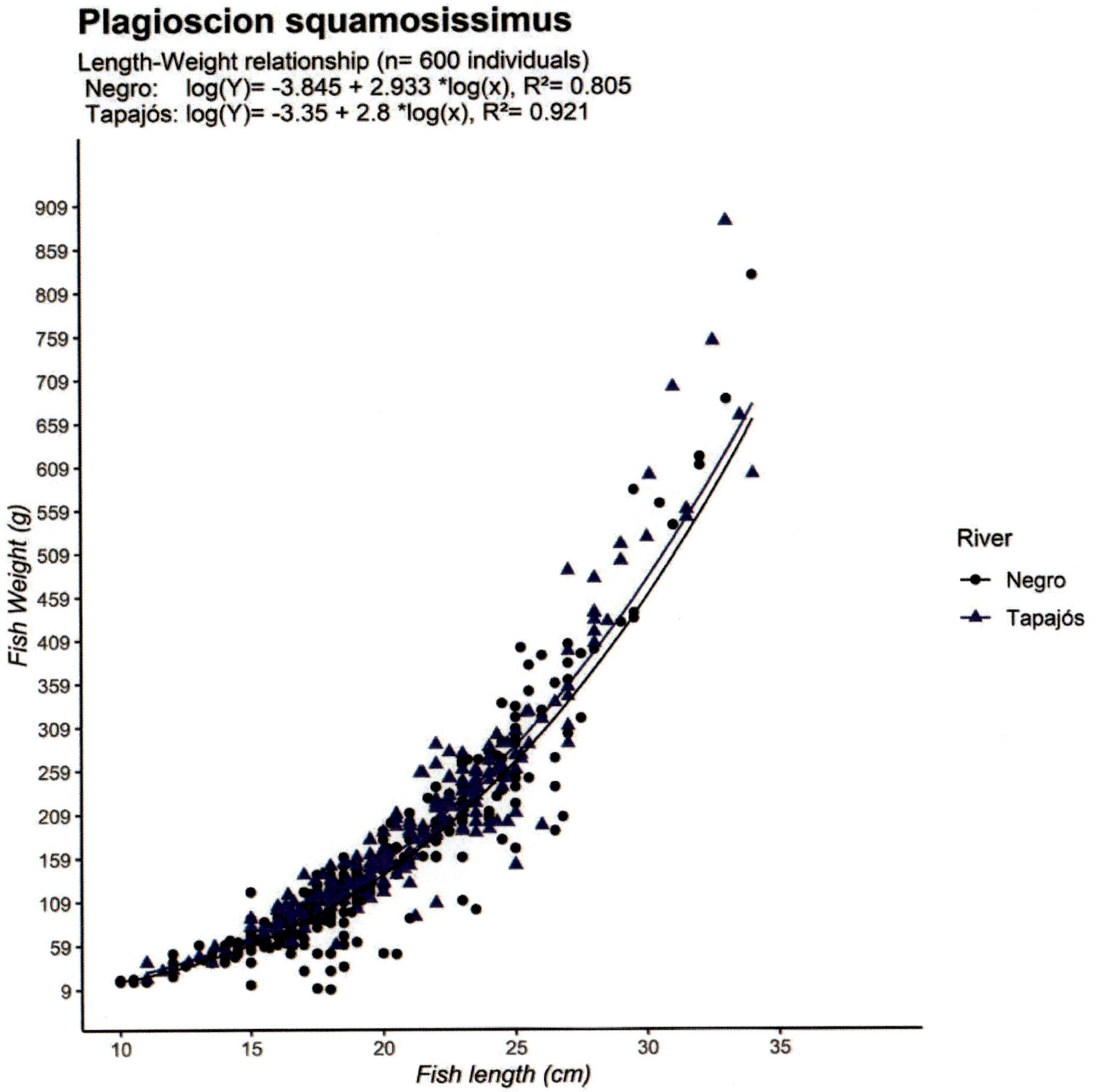

Fig. 6.74 Length-weight relationship for *Plagioscion squamosissimus* sampled in Tapajós and Negro rivers

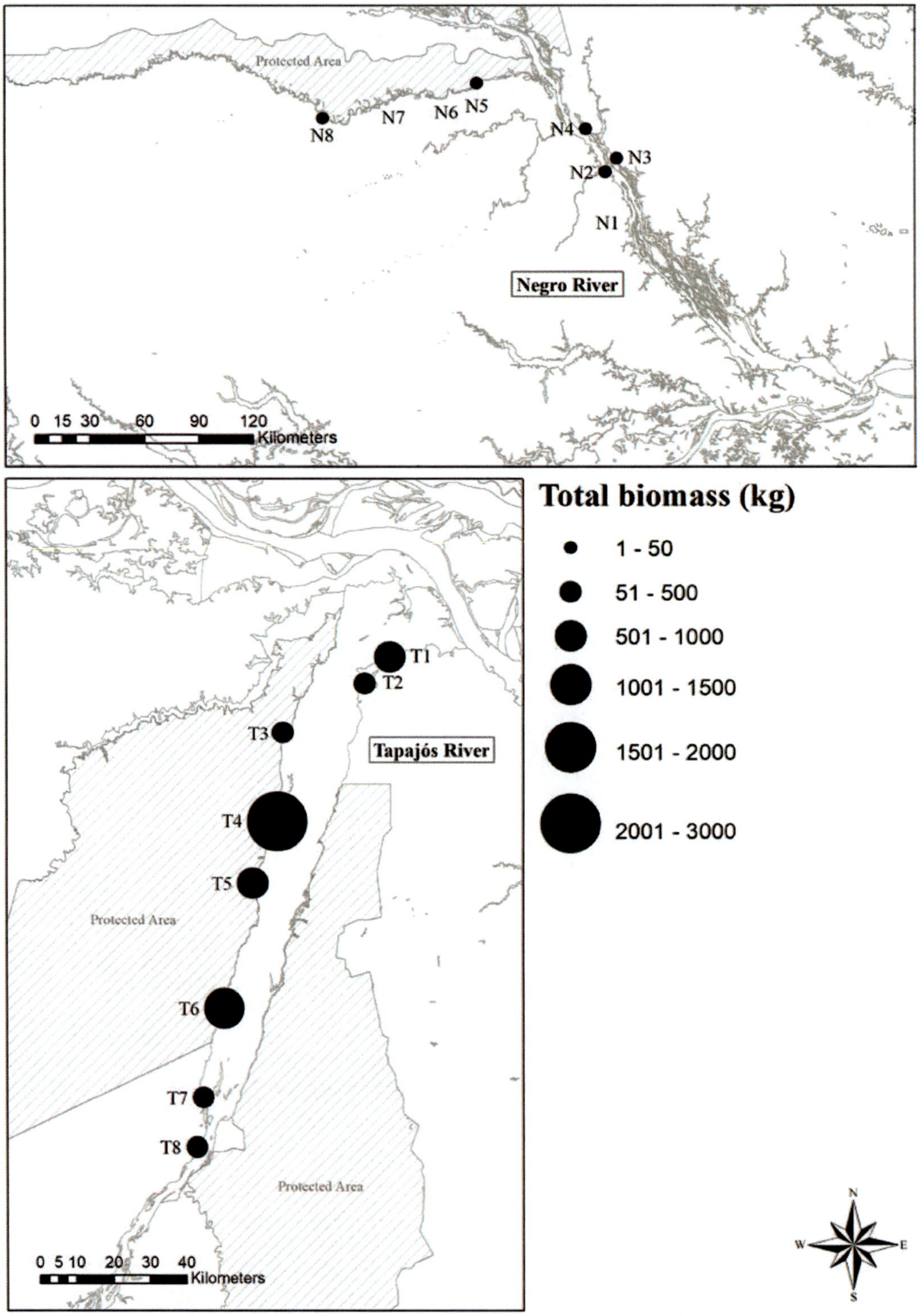

Fig. 6.75 Distribution of biomass of pescada (*Plagioscion squamosissimus*) caught in small-scale fisheries in the Negro and Tapajós rivers

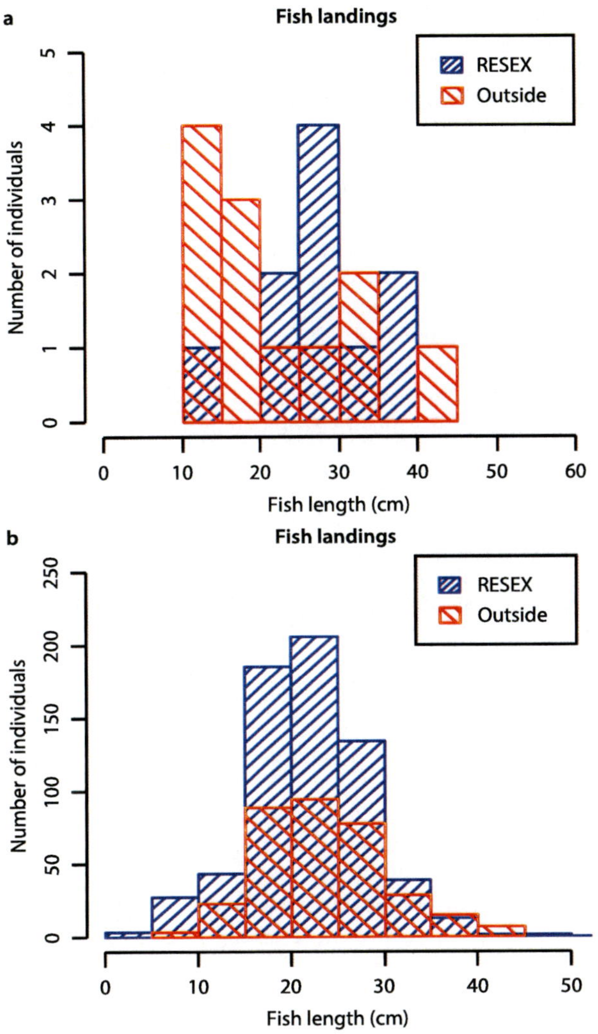

Fig. 6.76 Size distribution of pescada (*Plagioscion squamosissimus*) caught in small-scale fisheries in 16 communities inside and outside the Extractive Reserves (RESEX) in the rivers **a** Negro, N Resex = 10, N outside = 12 and **b** Tapajós, N Resex = 653, N outside = 335

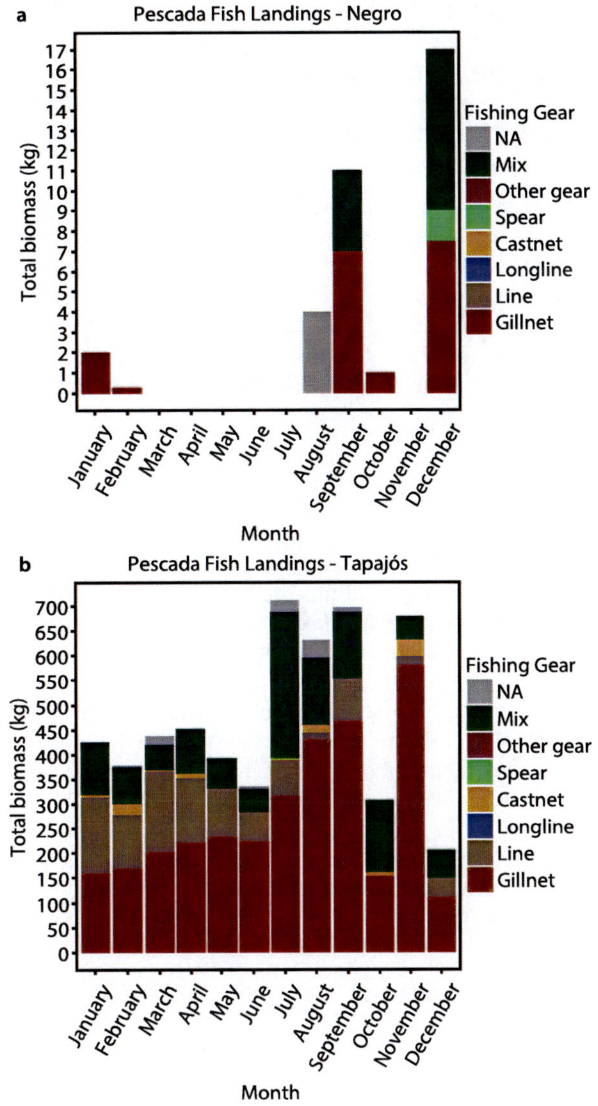

Fig. 6.77 Seasonal variation of biomass of pescada (*Plagioscion squamosissimus*) caught per fishing gear in small-scale fisheries in 16 communities in the **a** Negro and **b** Tapajós rivers. NA: fishing gear not informed; Mix: more than one gear was used

6.3 Order Beloniformes

6.3.1 Family Belonidae

6.3.1.1 Summary of Biology and Ecology (Literature)

Diversity: family composed of 40 species and subspecies distributed in 10 genera. The Belonidae family is found in tropical and subtropical regions of the Pacific, Indian, and Atlantic oceans. Freshwater species are found in basins of South and Central America, Australia, and Asia. The genera *Potamorrhaphis* and *Pseudotylosurus*, from the clade of the New World, are found in rivers of South America.

6

Habitat: no information, but morphology indicates that it occurs near the surface and is a pelagic swimmer.

Morphology: fish with elongated or anguilliform body shape. Dorsal and anal fins are located in the posterior part of the body. Pelvic fins have six rays. The snout is elongated and slim with a mouth containing large conical teeth. After retching, the individual pass through several changes in terms of premaxilla length. Some species have a hemiramphid stage, which consists of a short maxilla and a long dentary. Belonids are variable in terms of body size: some species measure few centimeters while others may reach up to 2 meters. The anguilliform natatorial behavior found in this family is unique among pelagic fishes.

Ecology and feeding: most species are ambush piscivores, except *Potamorrhaphis*, which is insectivore, and *Belonion* that is a plankton feeder. The species that are ambush piscivores attack a prey a single time. In the moments that precede the attack, the fish positions itself orthogonally to the prey, curves the body, and then bursts in the prey direction.

Reproduction: these fishes are oviparous with external fertilization. Eggs are large, spherical, and filamentous. The right gonad is reduced or even loss in some species, particularly in females.

Economic Importance: these fishes usually have no commercial value and are not exploited in Brazilian fisheries.

References: Nichols and Breder (1928), Collette (1982), Collette et al. (1984), Goulding and Carvalho (1984), Collette (2000, 2003), Lovejoy (2000), Collette and Banford (2001), Lovejoy and Collette (2001), Liao (2002), Potter and Motta (2004), Sant'Anna (2011), Sant'Anna et al. (2012), Carpenter and De Angelis (2016).

Fish species: ***Pseudotylosurus microps* (Günther 1866)**

Common name(s): peixe-agulha

Size of fish sampled (standard length, both rivers): mean = 37.2 cm, median = 38 cm

Fisheries: not relevant to fisheries in both rivers.

Fig. 6.78 *Pseudotylosurus microps* sampled in the Tapajós River. (Photo: Renato Silvano)

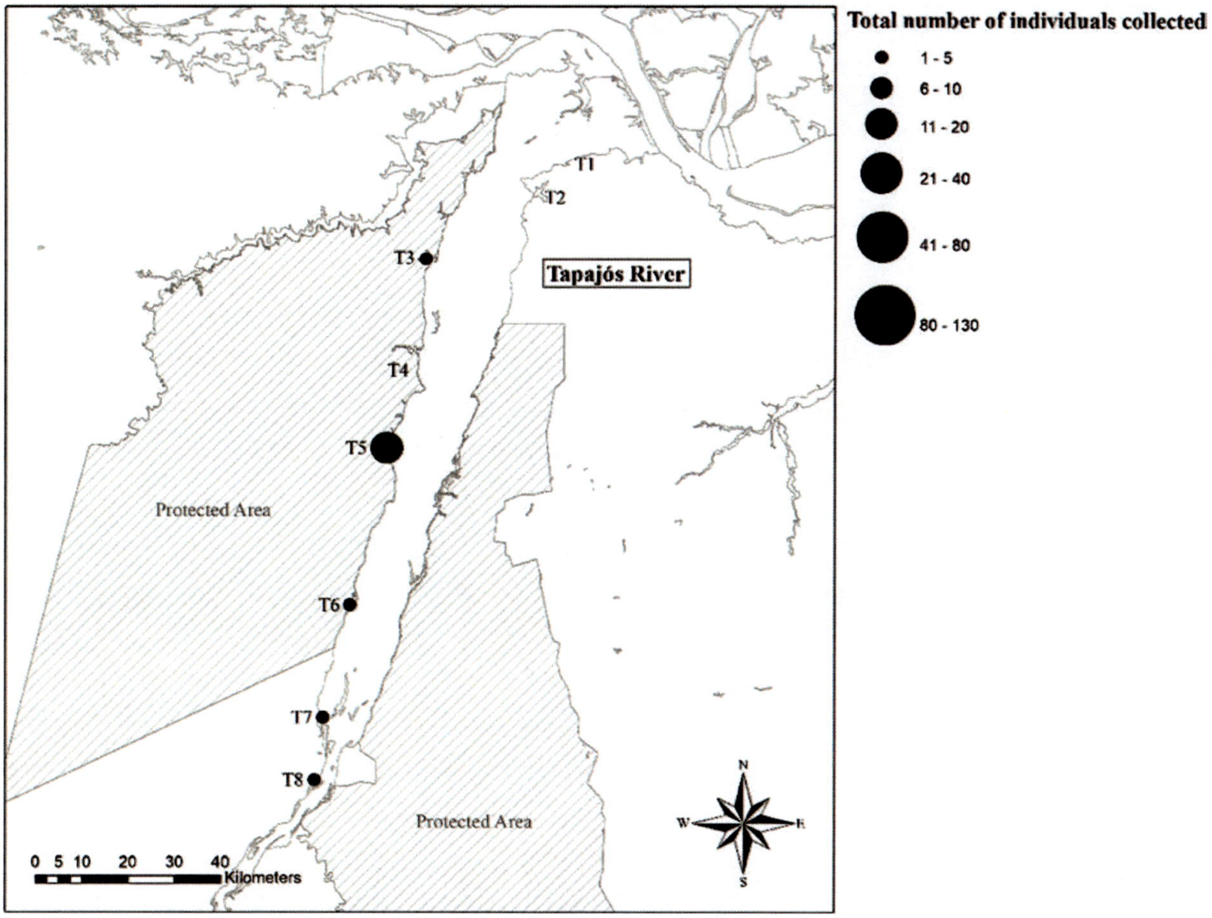

Fig. 6.79 Distribution of individuals of *Pseudotylosurus microps* sampled in the Tapajós River

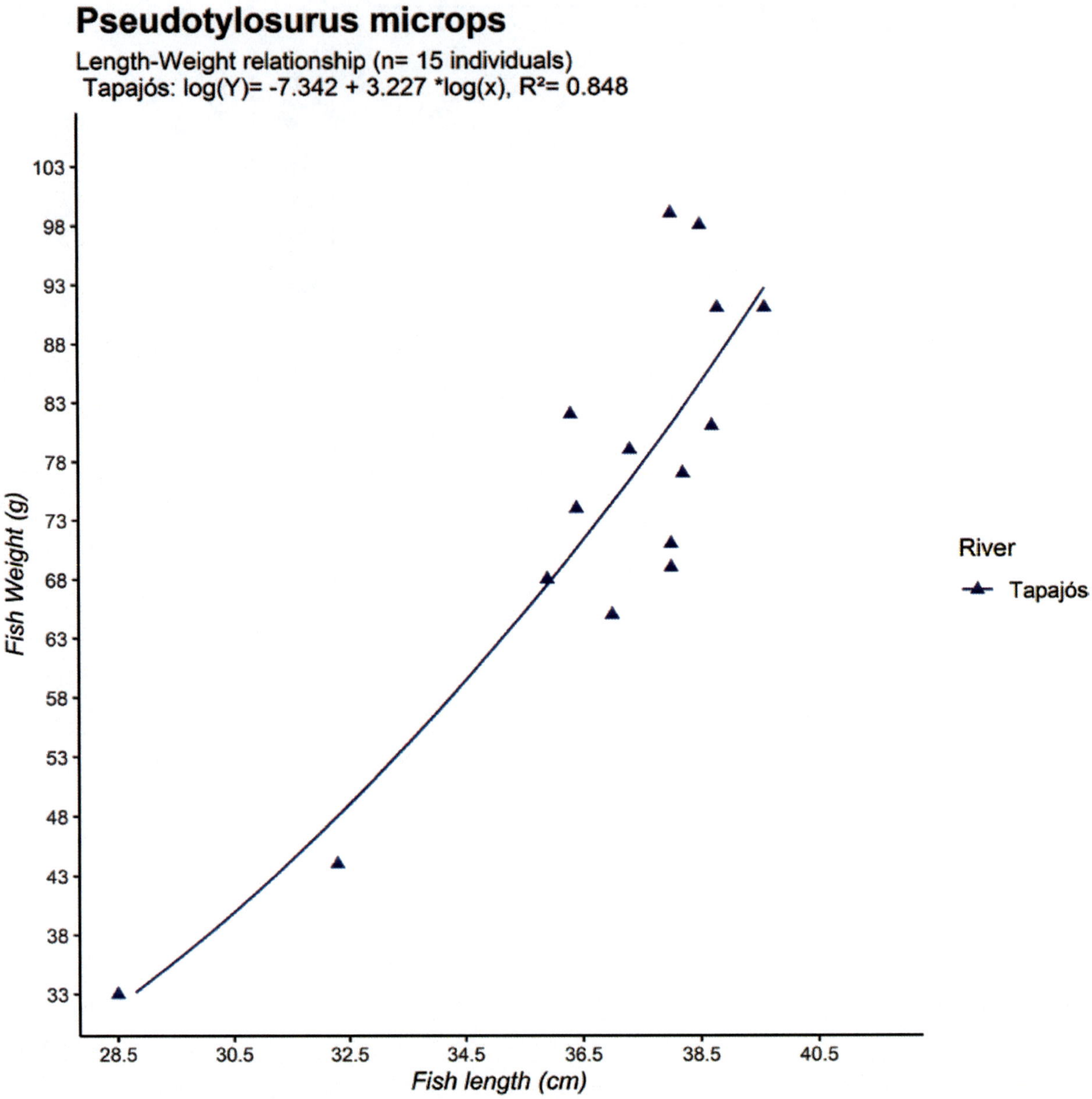

6

Fig. 6.80 Length-weight relationship for *Pseudotylosurus microps* sampled in the Tapajós River

6.4 Order Clupeiformes

6.4.1 Family Pristigasteridae

6.4.1.1 Summary of Biology and Ecology (Literature)

Diversity: this family encompasses 38 valid species distributed in nine genera. Five species are known to occur in the Amazon region, including three genera: *Pellona, Ilisha*, and *Pristigaster*.

Habitat: fishes of this family occur in all tropical oceans and in freshwaters of South America and Southeast Asia. The species *Pellona flavipinnis* possesses the largest geographic distribution among the freshwater species, occurring from the north and northeast until the middle-south of the cisandine South America. These fishes usually occur in pelagic zones of rivers and mouths of lakes.

Ecology and feeding: fishes of this family have an invertivore or piscivore habit when adults. They present a voracious and agile behavior and form large schools.

Morphology: members of this family have medium to large body sizes, presence of a series of abdominal shields forming a pronounced keel, and a long anal fin with more than 30 rays. Their mouth is typically prognata and the posterior end of the pectoral fins usually reach or pass the base of the pelvic fins. Besides that, pristigasterids do not possess lateral lines and adipose fin.

Reproduction: there is evidence that the species *Pellona castelnaeana* have protandric hermaphroditism.

Economic importance: these fishes, especially the larger species of the genus *Pellona*, are relevant for commercial and small-scale fisheries in Brazil and in the Amazon. Furthermore, due to their voracious and agile behavior, some species of this family are also appreciated by recreational fishers.

References: Whitehead (1985), Munroe et al. (1999), Melo (2001), de Pinna and Di Dario (2003), Ikeziri et al. (2008), Ribeiro et al. (2016), Hallwass and Silvano (2016).

Fish species: *Pellona castelnaeana* Valenciennes 1847

Common name(s): sarda, sarda-amarela, apapá, sardinhão

Size of fish sampled (standard length, both rivers): mean = 28 cm, median = 26 cm

Fisheries: The sarda (*Pellona* spp.) is relevant for fisheries in the Tapajós River, for consumption and commerce, being only sporadically caught in fisheries in the Negro River. Biomass caught is in ◘ Table 3.3 (▶ Chap. 3).

◘ **Fig. 6.81** *Pellona castelnaeana* sampled in the Tapajós River. (Photo: Renato Silvano)

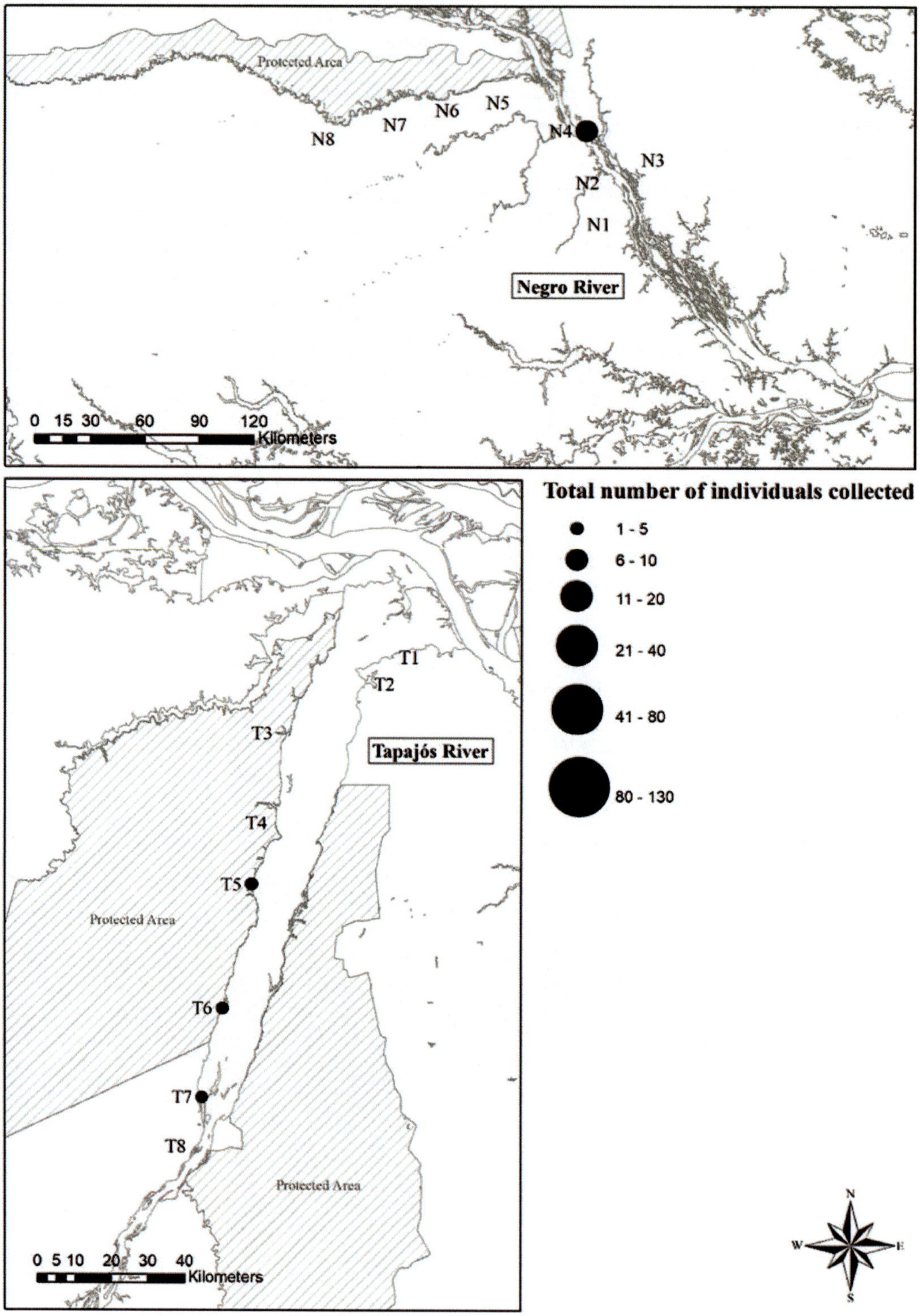

Fig. 6.82 Distribution of individuals of *Pellona castelnaeana* sampled in the Negro and Tapajós rivers

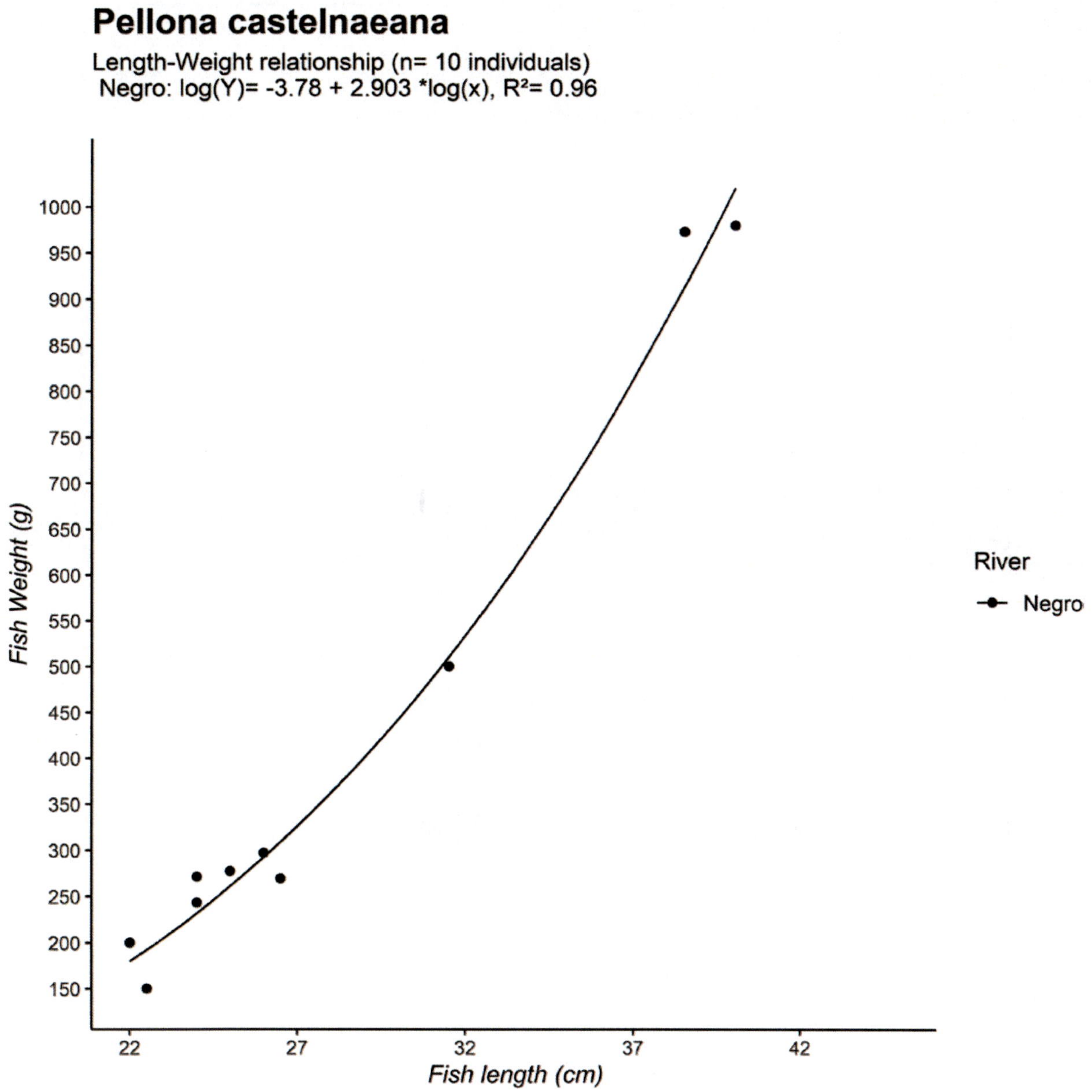

Fig. 6.83 Length-weight relationship for *Pellona castelnaeana* sampled in Negro River

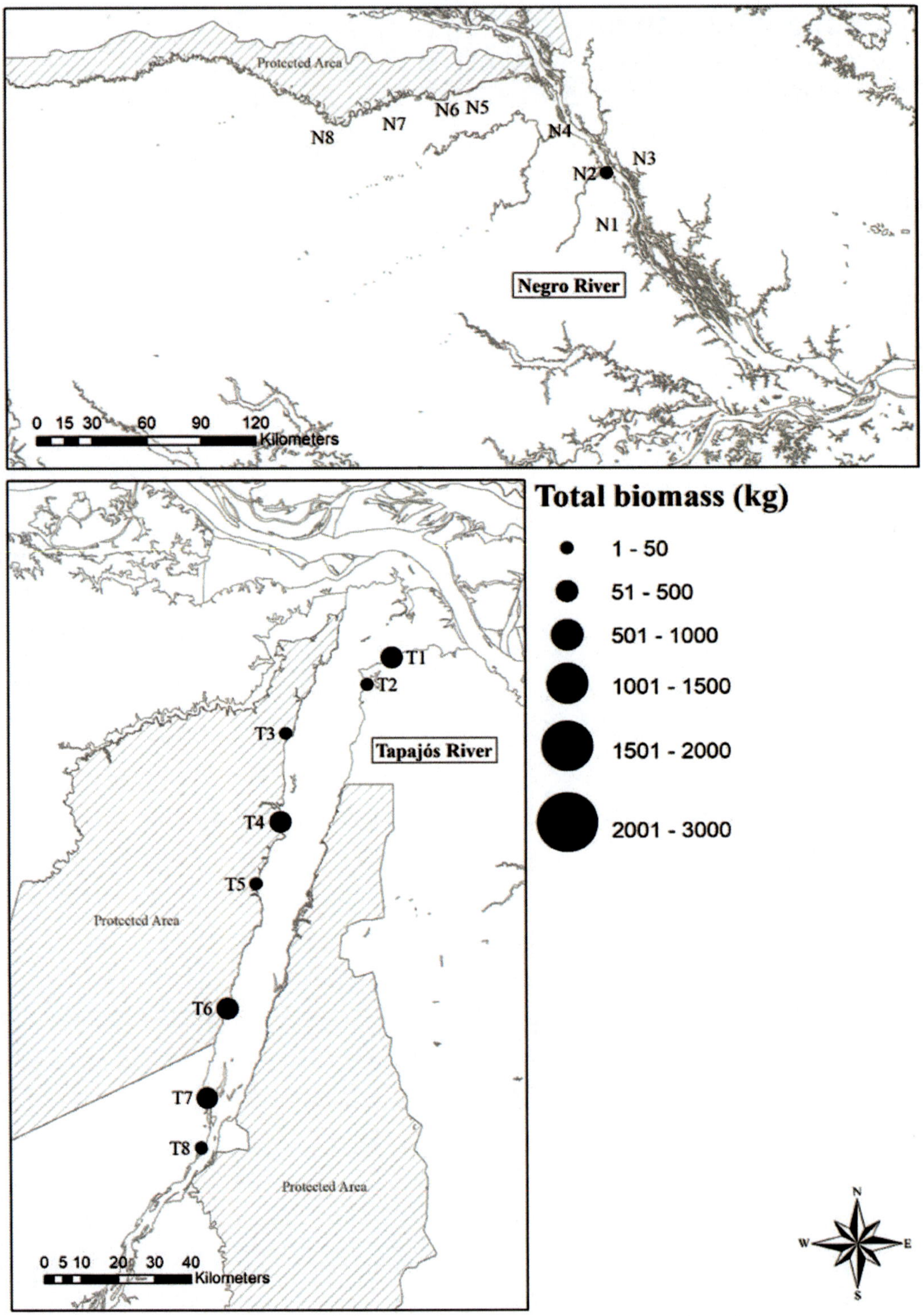

Fig. 6.84 Distribution of biomass of sarda (*Pellona* spp.) caught in small-scale fisheries in the Negro and Tapajós rivers

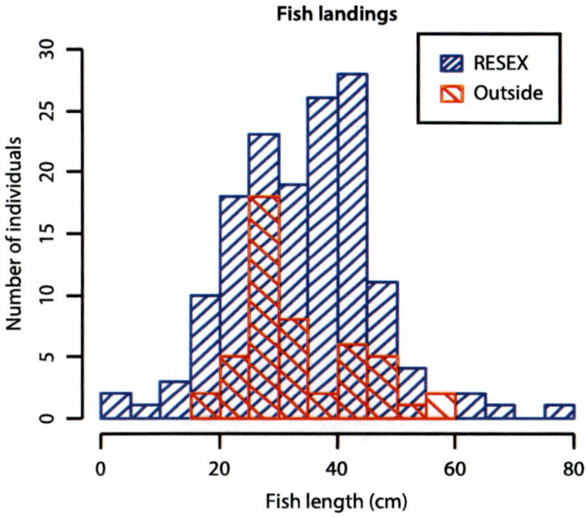

Fig. 6.85 Size distribution of sarda (*Pellona* spp.) caught in small-scale fisheries in eight communities inside and outside the Extractive Reserves (RESEX) in the Tapajós River, N Resex = 149, N outside = 49

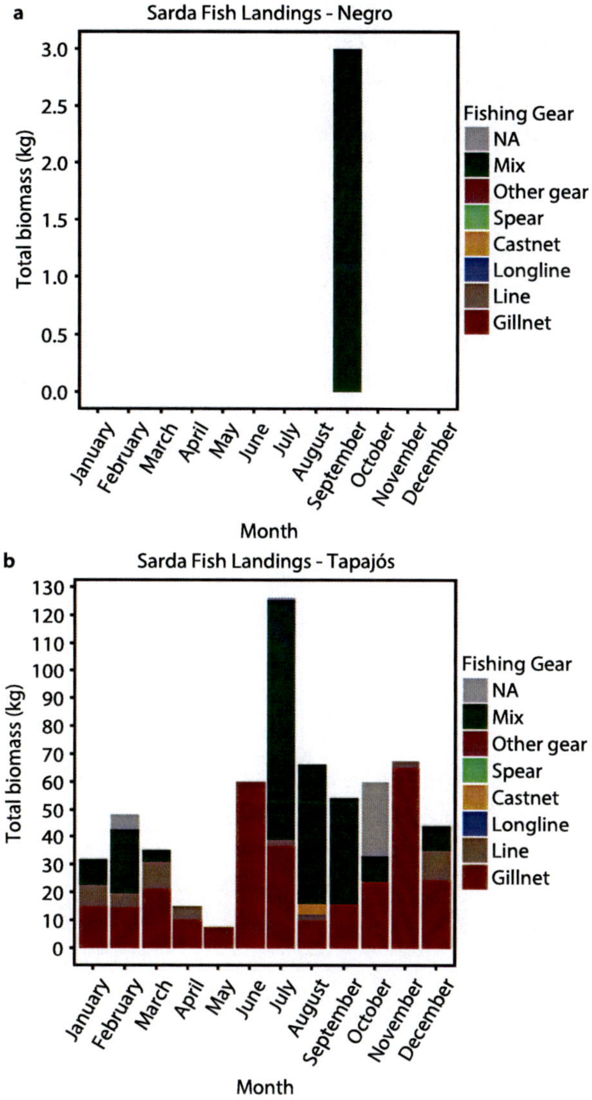

Fig. 6.86 Seasonal variation of biomass of sarda (*Pellona* spp.) caught per fishing gear in small-scale fisheries in 16 communities in the rivers **a** Negro and **b** Tapajós. NA: fishing gear not informed; Mix: more than one gear was used

Fish species: *Pellona flavipinnis* **(Valenciennes 1837)**

Common name(s): sarda, sarda-branca, apapá, sardinhão

Size of fish sampled (standard length, both rivers): min. = 10 cm, max. = 37.3 cm, mean = 17.1 cm, median = 17 cm

Fisheries: The sarda (*Pellona* spp.) is relevant for fisheries in the Tapajós River, for consumption and commerce, being only sporadically caught in fisheries in the Negro River; see *Pellona castelnaeana* for fisheries data on sarda.

Fig. 6.87 *Pellona flavipinnis* sampled in the Tapajós River. (Photo: Renato Silvano)

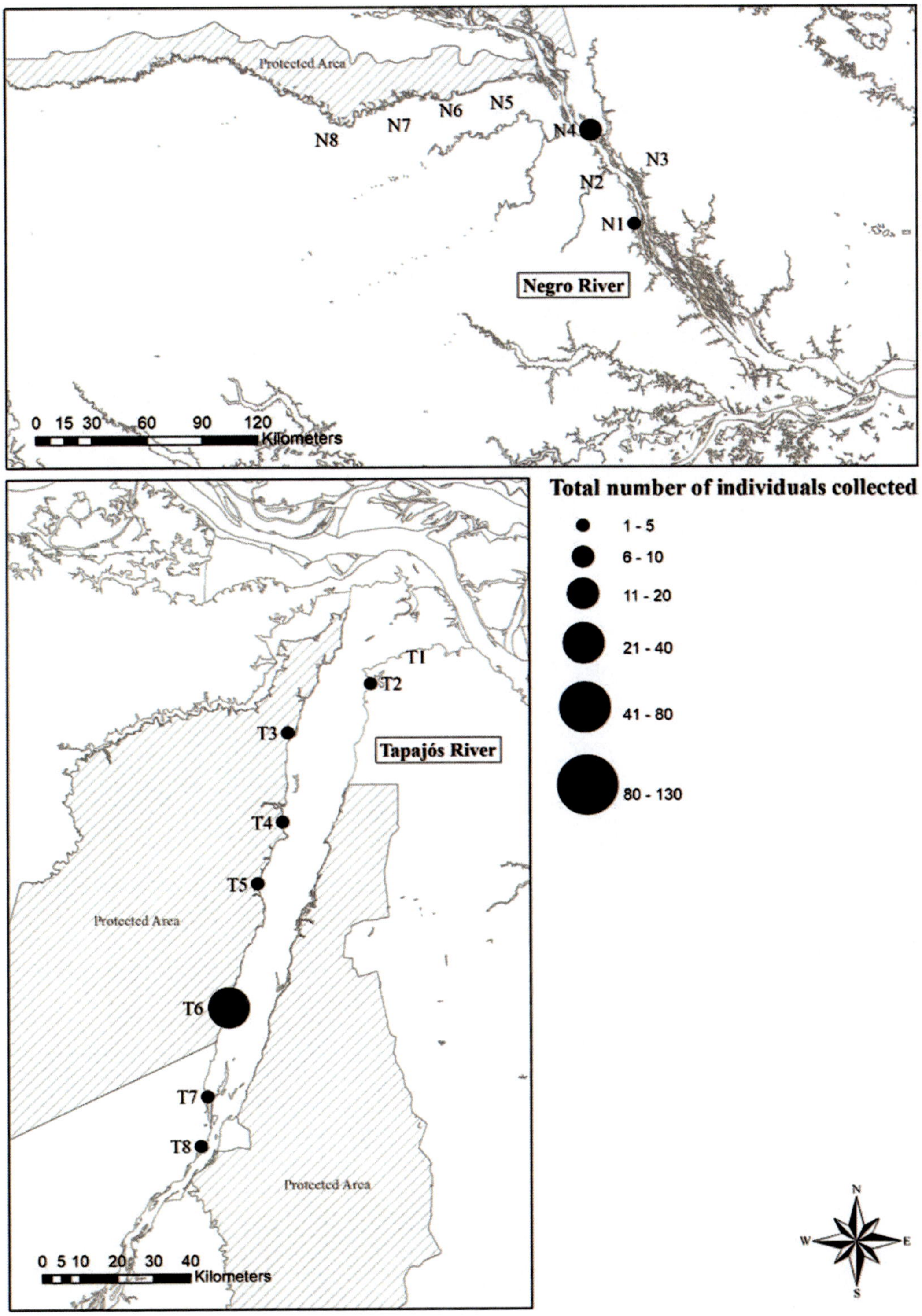

■ **Fig. 6.88** Distribution of individuals of *Pellona flavipinnis* sampled in the Negro and Tapajós rivers

6

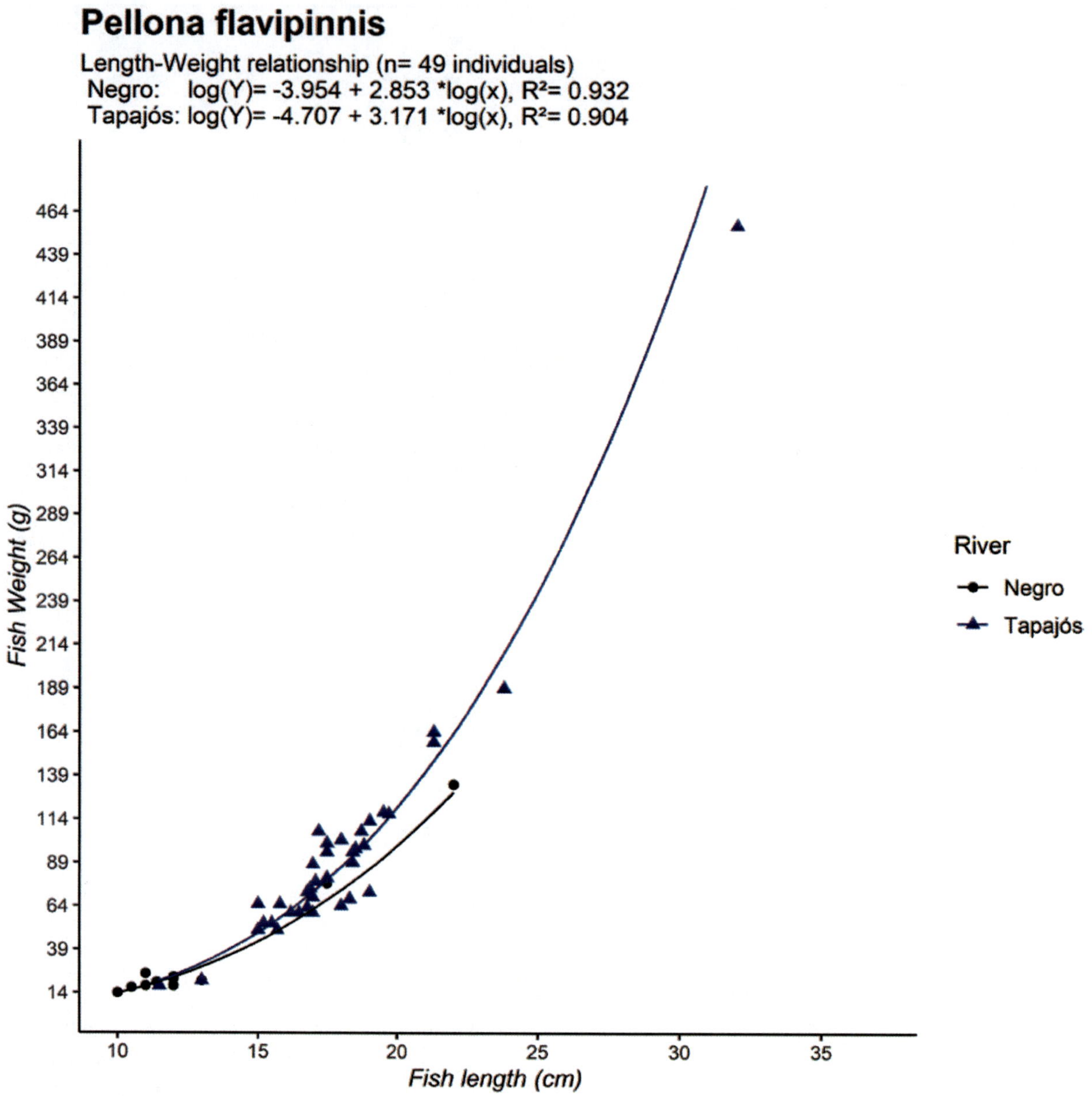

Fig. 6.89 Length-weight relationship for *Pellona flavipinnis* sampled in Tapajós and Negro rivers

6.4.2 Family Engraulidae

6.4.2.1 Summary of Biology and Ecology (Literature)

Diversity: family with 156 species (16 genera) described, 25 of which are found in Brazil (eight occurring in freshwaters).

Habitat: species of this family can be found in the sea, estuaries, and freshwaters. All freshwater species are found in South America, and *Lycengraulis batesii* is one species of this family that occurs in the Amazon basin.

Ecology and feeding: most species of this family are planktivorous, but some, such as *Lycengraulis batesii*, feeds on macroinvertebrates and even on other fishes.

Morphology: most species are small- or medium-sized fish (maximum size ranging from 15 to 20 cm). Fishes of this family can be identified by their enlarged snout and mandibular articulation posterior to the eyes. Their body is covered by medium-sized scales that are easily detached from their bodies. The lateral line is absent and fins do not possess spines.

Reproduction: no information.

Economic importance: although important in marine and coastal fisheries, fishes of this family have no relevance to freshwater fisheries in the Brazilian Amazon.

References: Loeb (2009), Queiroz et al. (2013), Ribeiro et al. (2016), Ohara et al. (2017), Eschmeyer (2019).

Fish species: *Anchovia surinamensis* (Bleeker 1865)

Common name(s): xaréu, sardinha

Size of fish sampled (standard length, both rivers): min. = 7.1 cm, max. = 18 cm, mean = 9.5 cm, median = 8.5 cm

Fisheries: small size, usually not relevant to fisheries.

Fig. 6.90 *Anchovia surinamensis* sampled in the Tapajós River. (Photo: Renato Silvano)

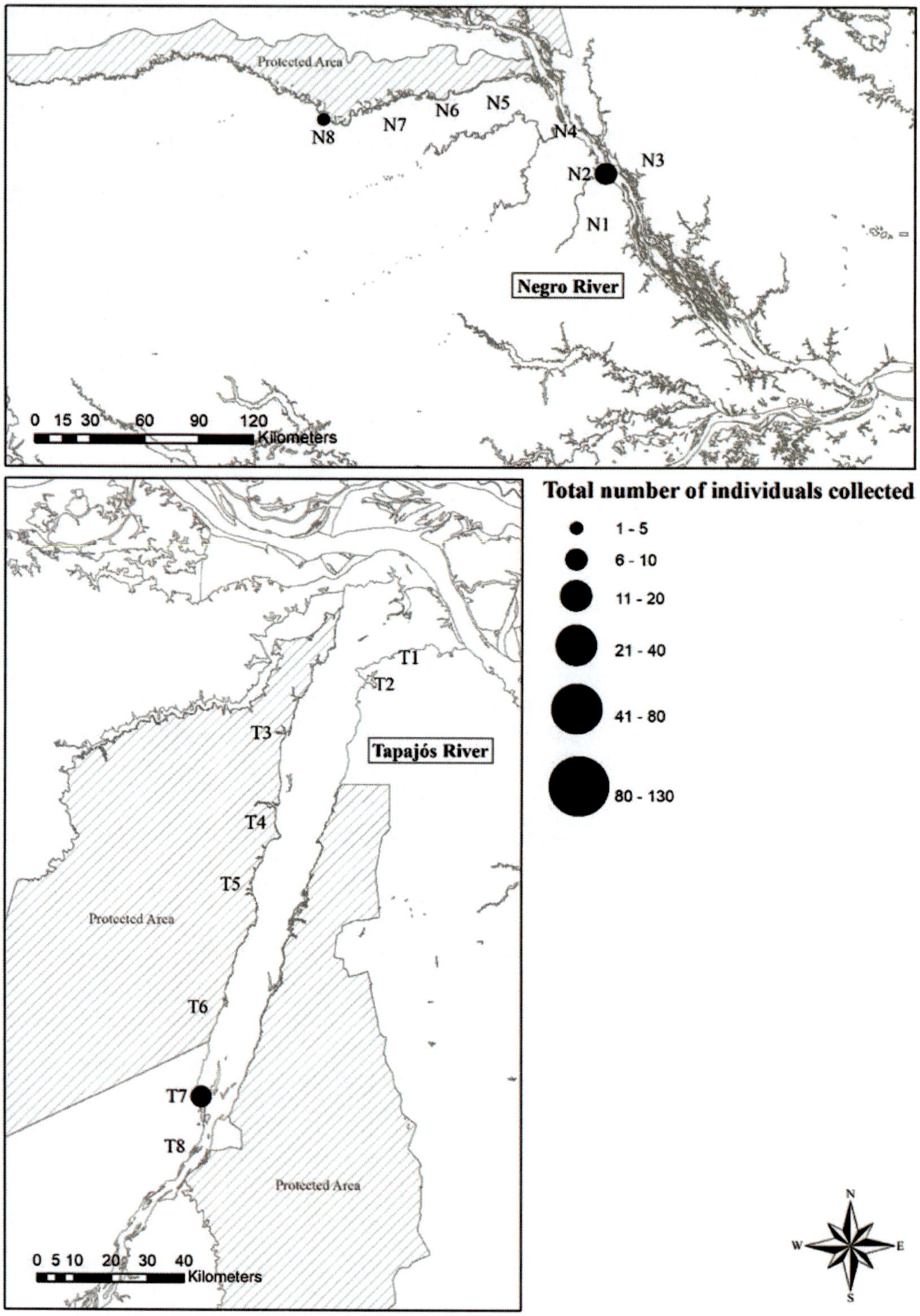

Fig. 6.91 Distribution of individuals of *Anchovia surinamensis* sampled in the Negro and Tapajós rivers

Fish species: ***Lycengraulis batesii*** **(Günther 1868)**

Common name(s): xaréu, xaréu-comprido, sardinha

Size of fish sampled (standard length, both rivers): mean = 14.2 cm, median = 15.8 cm

Fisheries: small size, usually not relevant to fisheries.

Fig. 6.92 *Lycengraulis batesii* sampled in the Tapajós River. (Photo: Renato Silvano)

6

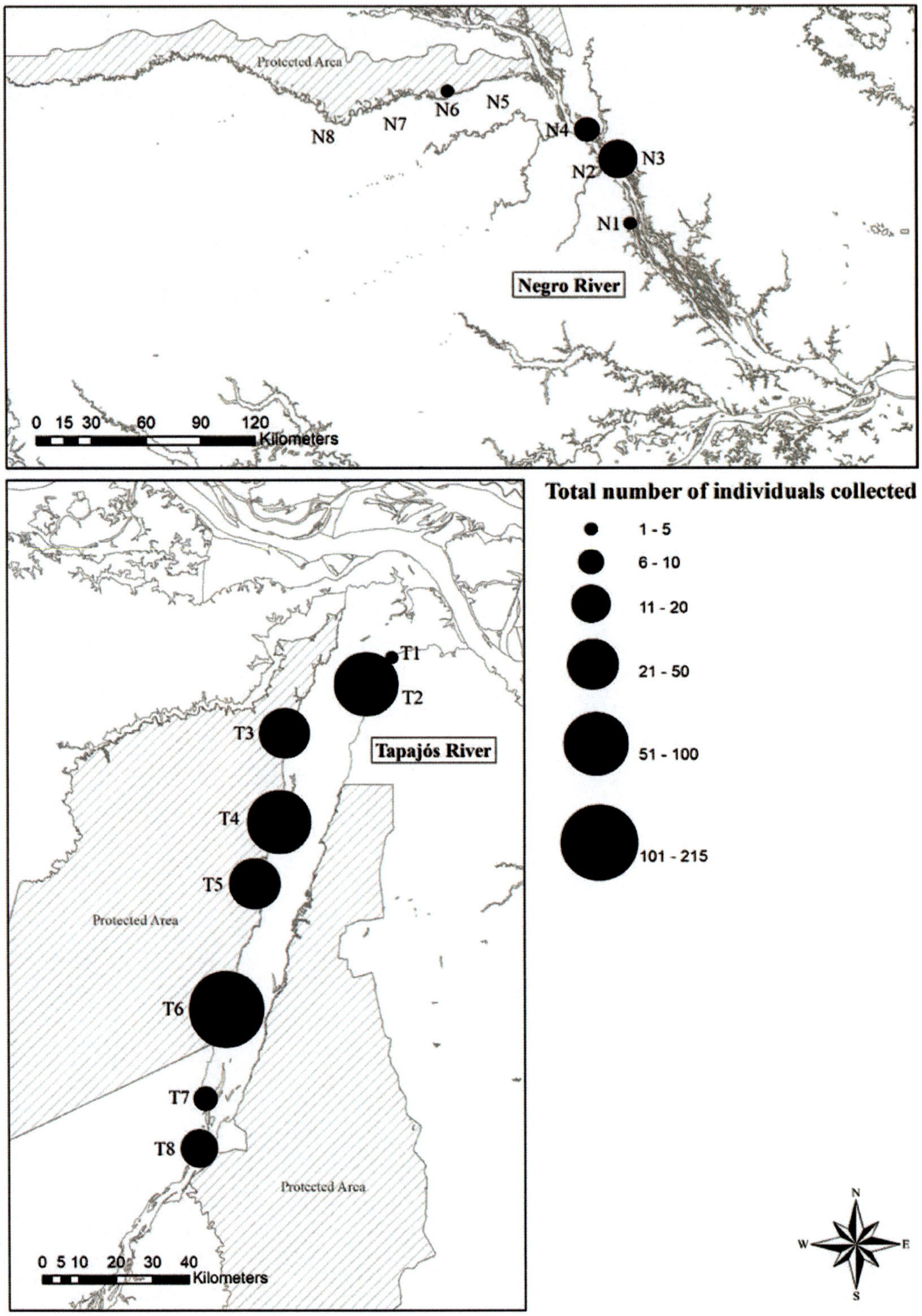

Fig. 6.93 Distribution of individuals of *Lycengraulis batesii* sampled in the Negro and Tapajós rivers

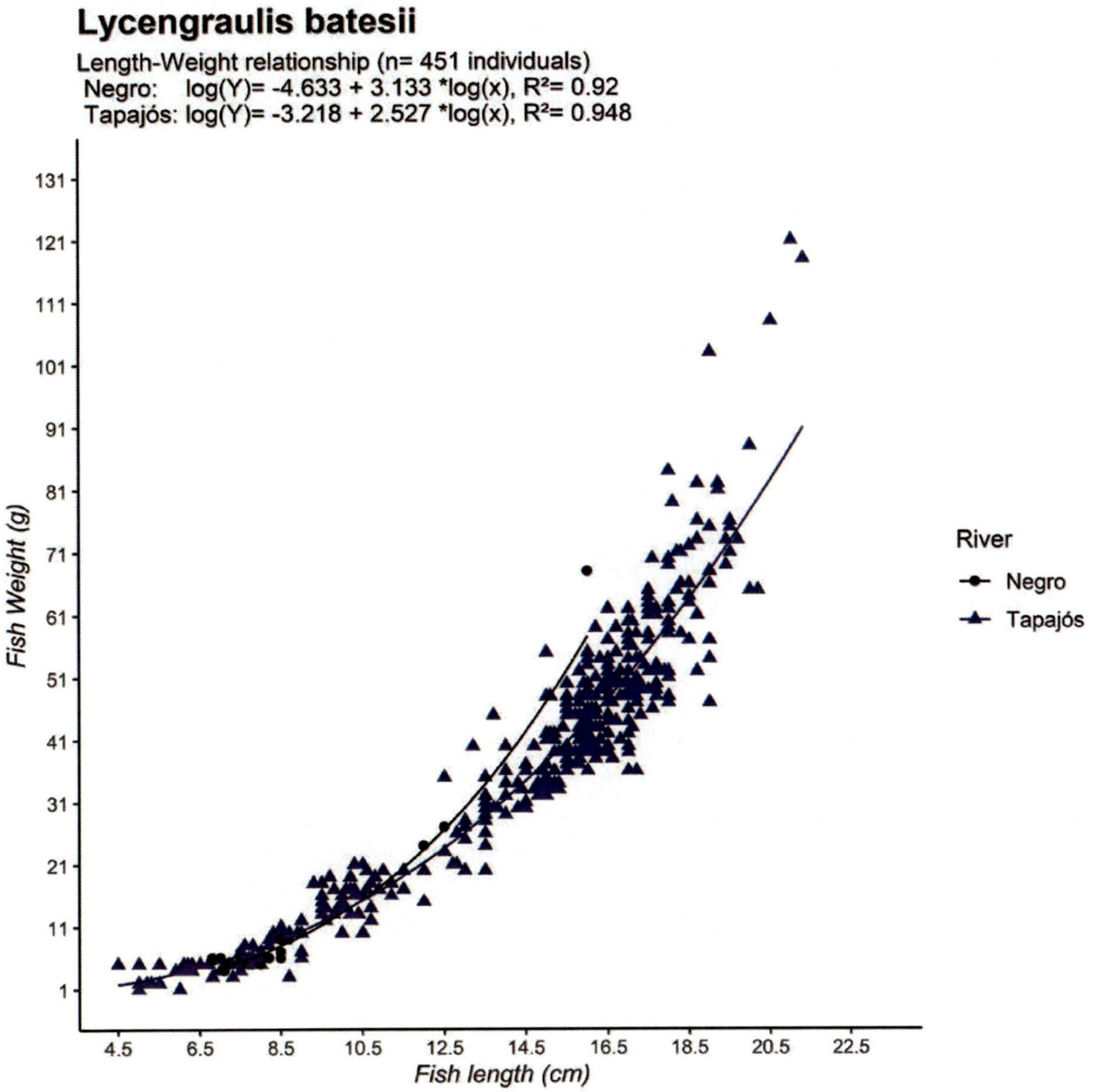

Fig. 6.94 Length-weight relationship for *Lycengraulis batesii* sampled in Tapajós and Negro rivers

6.5 Order Osteoglossiformes

6.5.1 Family Osteoglossidae

6.5.1.1 Summary of Biology and Ecology (Literature)

6

Diversity: these fishes are commonly known as arowanas and comprise a family of six species divided into two genera: *Osteoglossum* and *Scleropages*.

Habitat: fishes of the genus *Osteoglossum* are found in South America (Amazon and Orinoco basins), while fishes of the genus *Scleropages* are restricted to Asia and Australia. They are surface dwellers of rivers and lakes.

Ecology and feeding: these fishes are carnivorous that feed near the water surface, preying on both aquatic and terrestrial animals. They are excellent jumpers and may feed on animals, including other vertebrates, that are sitting on branches close to the water. Some observations indicate that the South American fishes from the genus *Osteoglossum* can jump up to 2 meters out of the water to catch their prey.

Morphology: fishes of this family have a bony tongue, mandibular barbells, highly compressed abdomen, and large and imbricate scales. These are large body size fishes that can reach up to 1 meter of length and 3 kg of weight. Their bodies are elongated with well-developed pectoral fins. Dorsal and anal fins are located in the posterior part of the body.

Reproduction: species of this family have parental care with mouth brooding. Fecundity is usually low and eggs are large (10–18 mm). Period of reproduction may vary, but it is usually during rising and high-water periods. Females lack the right ovary. The species *Osteoglossum bicirrhosum*, which is common in the Brazilian Amazon, has parental care, and males may cooperate in group parental caring, on which some males may take care of broods from other male. The mating system is usually monogamy (one male mating with one female), but may include opportunistic males.

Economic importance: some species of the genus *Osteoglossum* are regularly fished in riverine communities, mainly for food, but sometimes also for commerce, especially in the Negro River. These fishes are popular aquarium fishes, especially in the Asian market. The aquarium trade business was responsible for a sharp decline in the natural population of species of *Scleropages* in Asia. To supply the growing demand for the now rare Asian members of this family, populations of *Osteoglossum bicirrhosum* in South America started to be intensively commercialized in the 1990s for aquarium trade. Currently, *O. bicirrhosum* is considered threatened in Colombia and Peru, and its commercialization for ornamental purpose is not allowed in Brazil.

References: Queiroz (2008), Verba et al. (2014), Duponchelle et al. (2015), Eschmeyer (2019), Froese and Pauly (2019).

Fish species: *Osteoglossum bicirrhosum* (Cuvier 1829)

Common name(s): aruanã, macaco-da-água

Size of fish sampled (standard length, both rivers): mean = 40.7 cm, median = 45.5 cm

Fisheries: The aruanã (*Osteoglossum bicirrhosum*) is relevant for fisheries in the Negro River, and it is occasionally caught in the Tapajós River. Biomass caught is in ◘ Table 3.3 (▶ Chap. 3).

◘ **Fig. 6.95** *Osteoglossum bicirrhosum* sampled in the Negro River. (Photo: Renato Silvano)

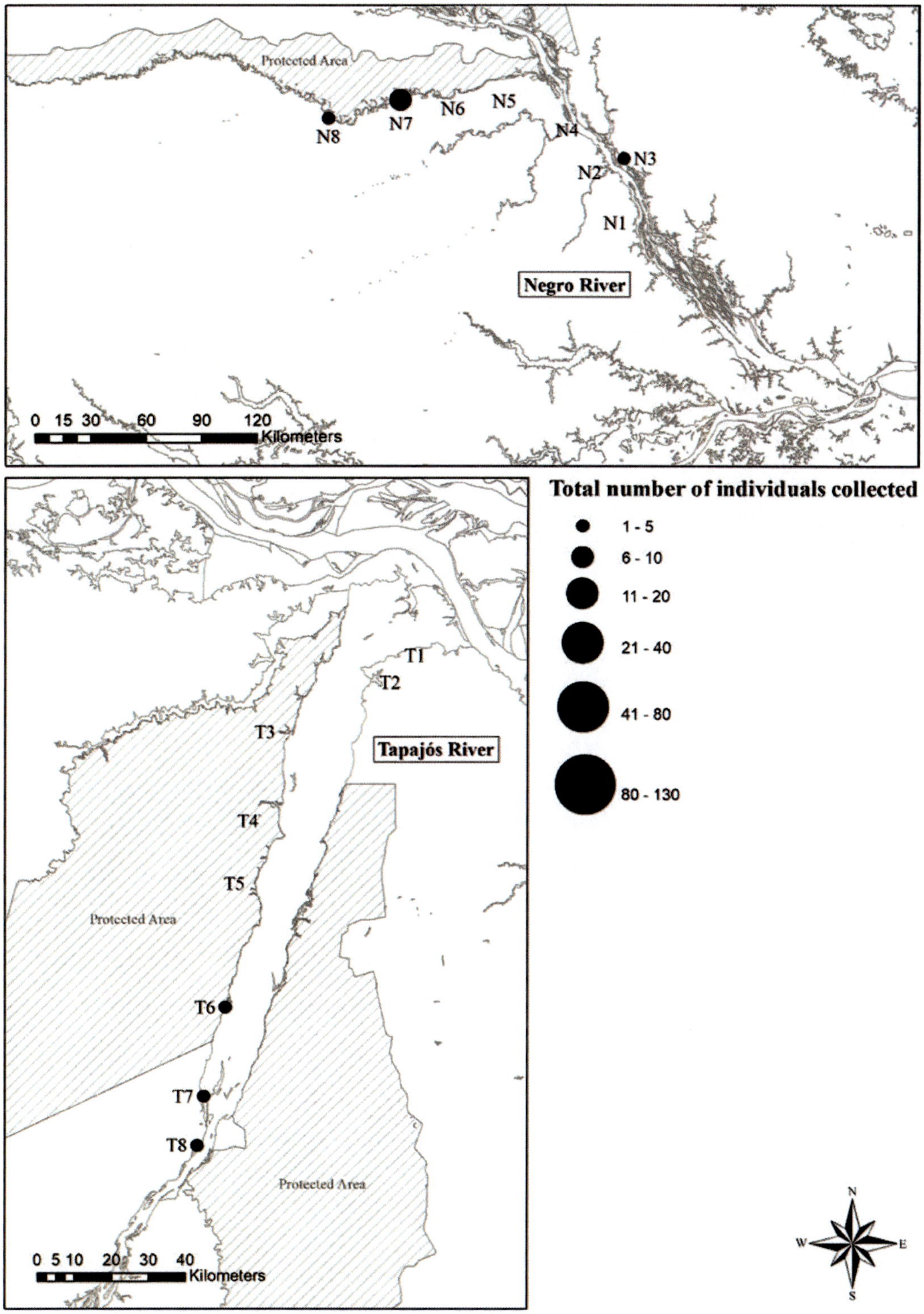

Fig. 6.96 Distribution of individuals of *Osteoglossum bicirrhosum* sampled in the Negro and Tapajós rivers

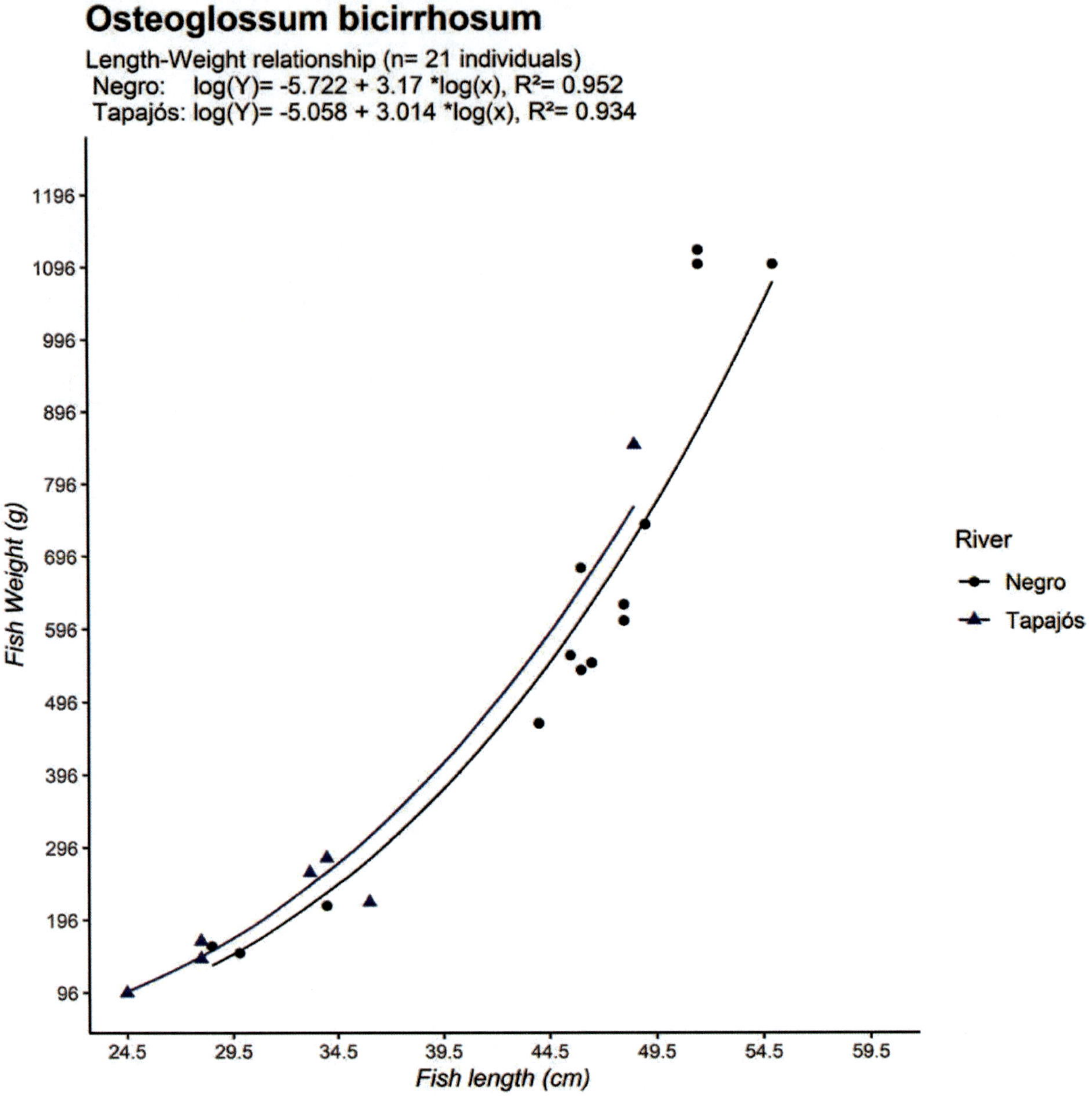

6

Fig. 6.97 Length-weight relationship for *Osteoglossum bicirrhosum* sampled in Tapajós and Negro rivers

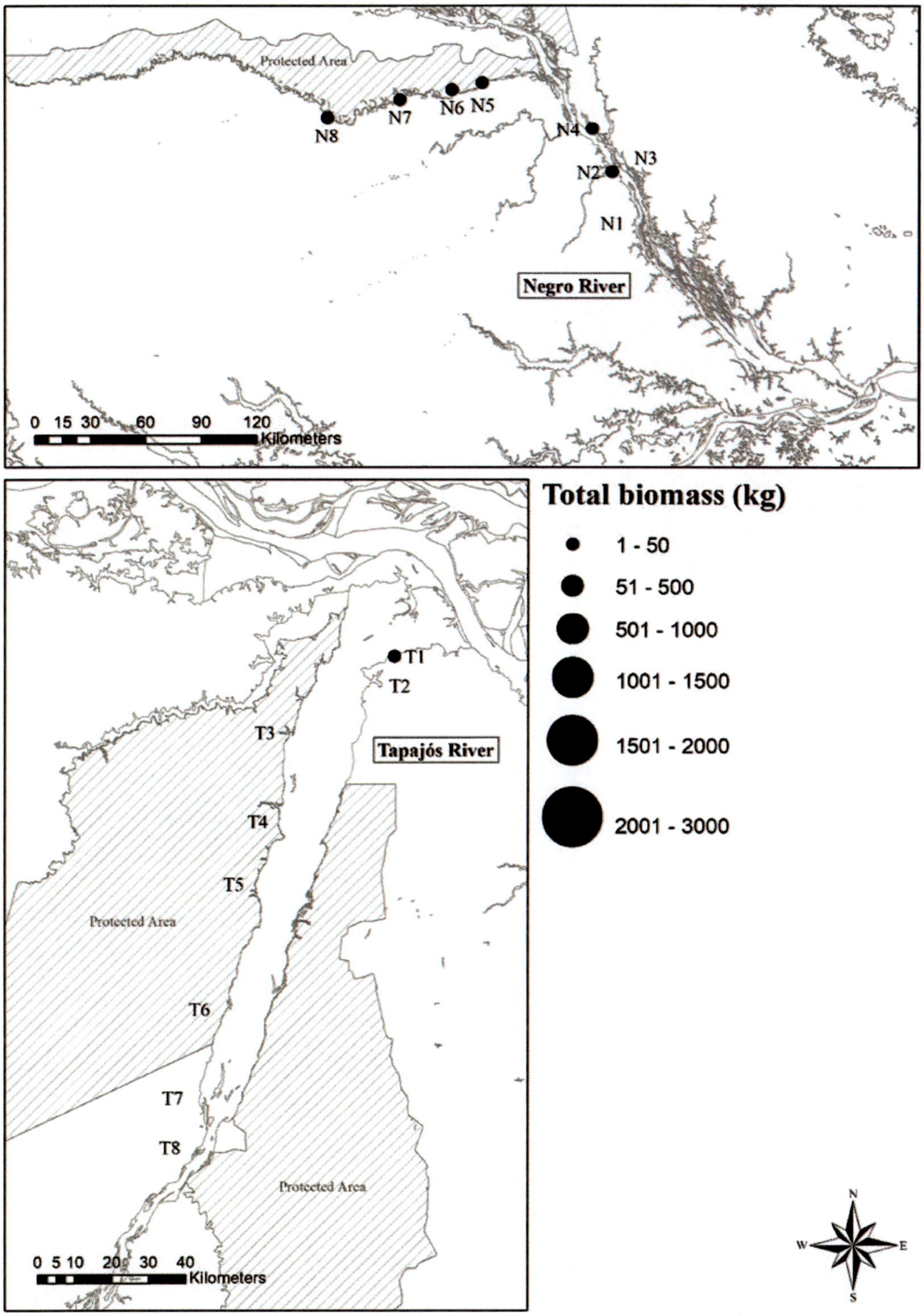

Fig. 6.98 Distribution of biomass of aruanã (*O. bicirrhosum*) caught in small-scale fisheries in the Negro and Tapajós rivers

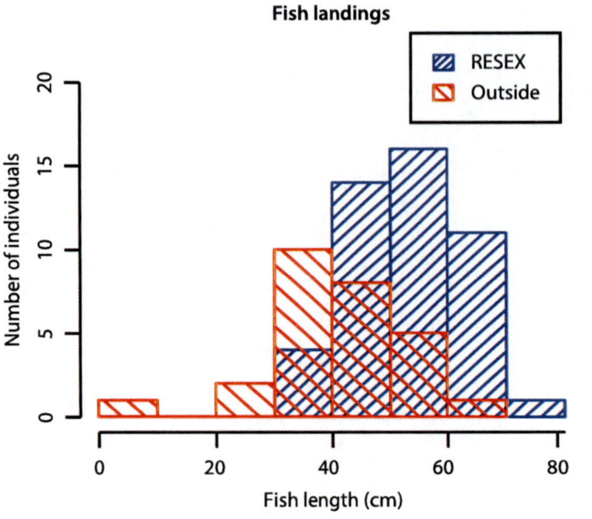

Fig. 6.99 Size distribution of the aruanã (*O. bicirrhosum*) caught in small-scale fisheries in eight communities inside and outside the Extractive Reserves (RESEX) in the Negro River, N Resex = 46, N outside = 27

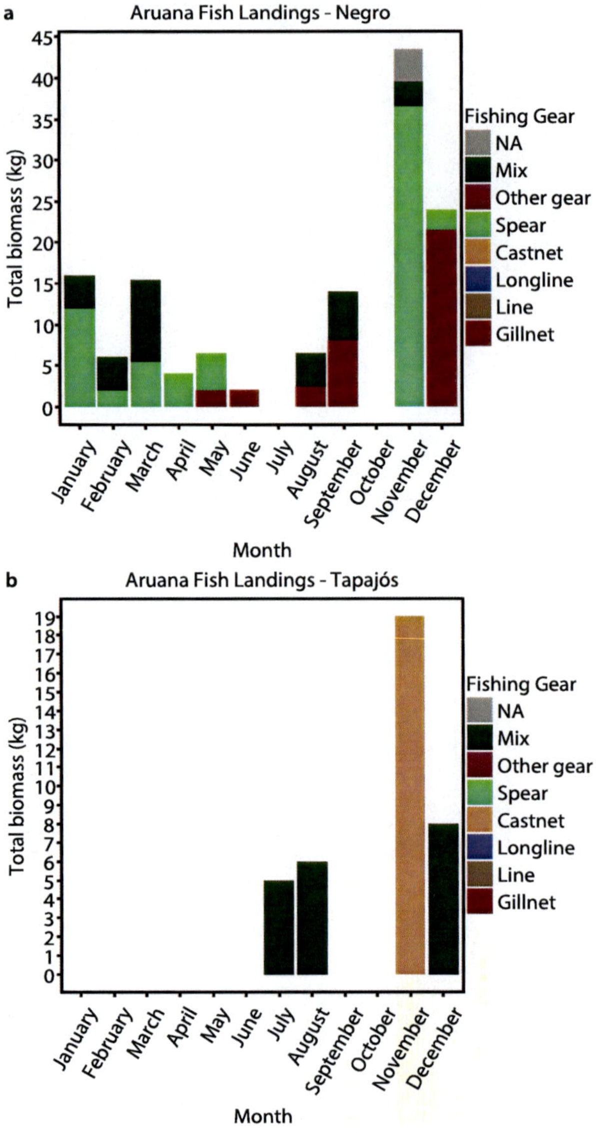

Fig. 6.100 Seasonal variation of biomass of the aruanã (*O. bicirrhosum*) caught per fishing gear in small-scale fisheries in 16 communities in the rivers **a** Negro and **b** Tapajós. NA: fishing gear not informed; Mix: more than one gear was used

6.6 Order Carangiformes

6.6.1 Family Achiridae

6.6.1.1 Summary of Biology and Ecology (Literature)

Diversity: family composed of 35 valid species, 13 of which occur exclusively in freshwaters and five in salt waters (marine and estuaries). All the remaining species are tolerant to both freshwater and salt water, occurring in rivers, lakes, estuaries, and coastal zones.

Habitat: species of this family can be found in the Americas, but freshwater species are restricted to South America.

Ecology and feeding: American members of this family are carnivorous. While larvae and juveniles feed on zooplankton, mainly Cladocera and copepods, adults feed mainly on fish and benthic invertebrates.

Morphology: larvae and juveniles have a symmetrical body plan, and, as they grow up, their left eye migrates to the right side of the body, i.e., the adult becomes asymmetrical. The body is compressed laterally, the inferior lip of the right side of the body is hypertrophied and contains a series of fimbriae (dermic filiform/laminar process with or without ramifications), and the right-side nostril is formed by an elongated longitudinal slit covered by inferior lip when the mouth is closed. Species of this family possess villiform teeth organized in a patch on the blind side of their jaws.

Reproduction: partial spawning in the Summer and Fall. Females tend to be larger than males.

Economic importance: although marine species are usually exploited by commercial fisheries, in the Amazon the freshwater species of this family have no commercial value and are not exploited by commercial nor small-scale fisheries.

References: Nelson (2006), Oliveira and Fávaro (2010), Lima et al. (2013), Ramos (2013), Ribeiro et al. (2016), Ohara et al. (2017).

Fish species: ***Hypoclinemus mentalis*** **(Günther 1862)**

Common name(s): aramaçá, suia

Size of fish sampled (standard length, both rivers): mean = 16 cm, median = 17 cm

Fisheries: non-relevant for fisheries in both rivers

Fig. 6.101 *Hypoclinemus mentalis* sampled in the Tapajós River. (Photo: Renato Silvano)

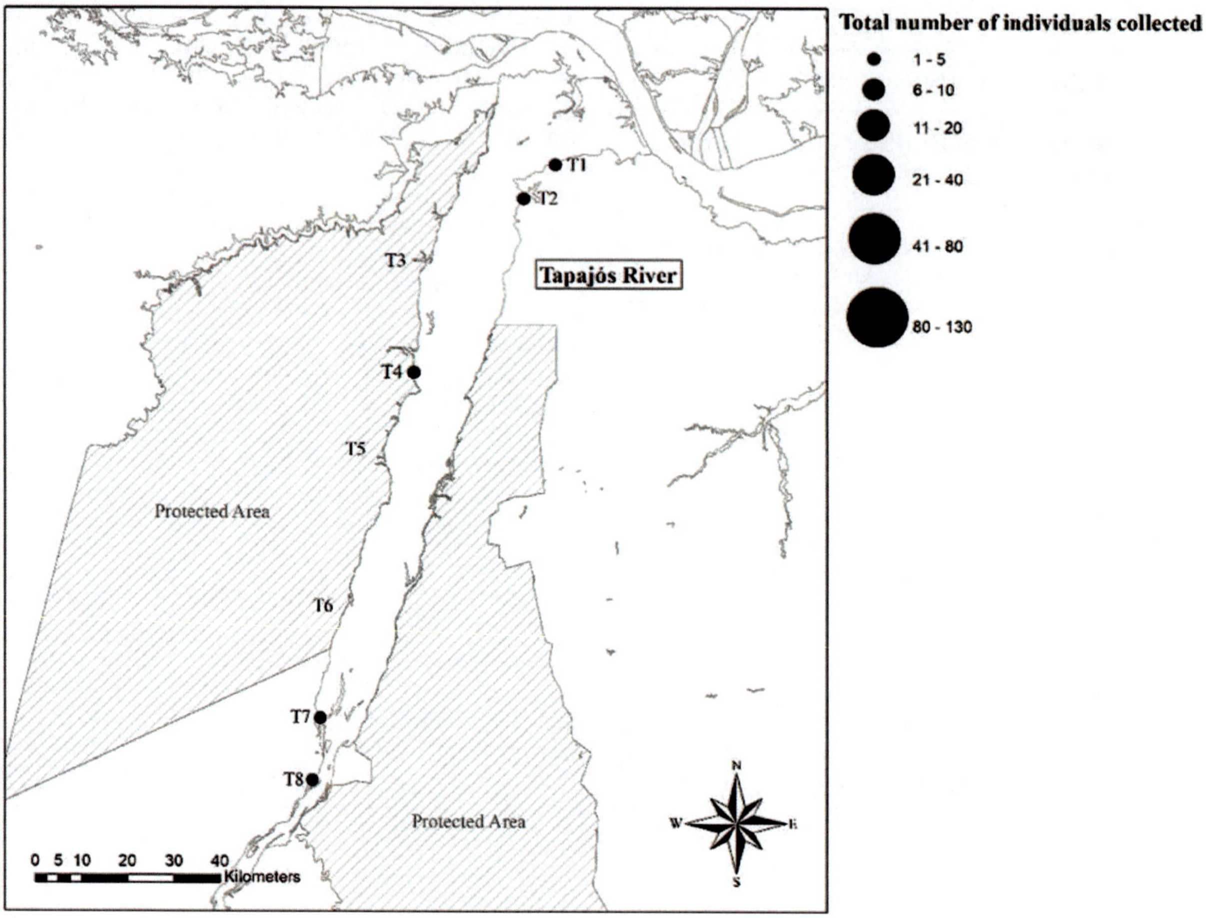

Fig. 6.102 Distribution of individuals of *Hypoclinemus mentalis* sampled in the Tapajós River

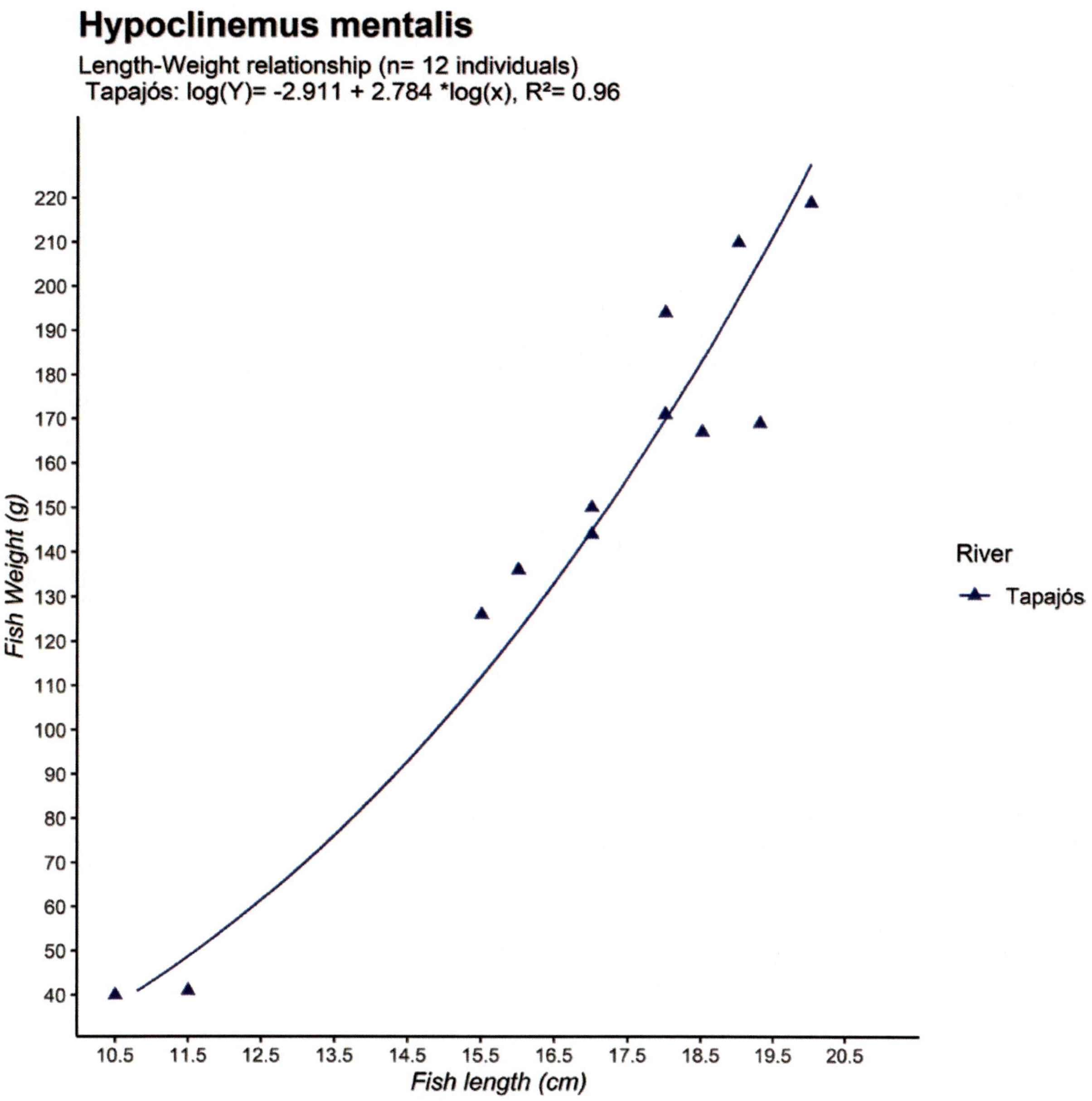

Fig. 6.103 Length-weight relationship for *Hypoclinemus mentalis* sampled in the Tapajós River

6.7 Order Myliobatiformes

6.7.1 Family Potamotrygonidae

6.7.1.1 Summary of Biology and Ecology (Literature)

Diversity: family composed of 32 species and four genera: *Paratrygon* (1 species), *Heliotrygon* (two), *Plesiotrygon* (two) and the more diverse *Potamotrygon* (27).

6

Habitat: fishes from the family Potamotrygonidae (freshwater stingrays) usually inhabit shallow areas with sandy or muddy bottom, where they are usually found revolving the substrate in search for prey or digging it to hide from predators.

Ecology and feeding: fishes from this family are nocturnal foragers. Their feeding consists mainly on insect larvae, but some species can also consume fish, crustaceans, and mollusks. Because they usually dig and revolve the substrate searching for food, fishes of this family may be involved in feeding interactions with other fish and shrimp, which opportunistically follow the stingrays to feed in prey removed from the substrate. At least one species of *Potamotrygon* shows social learning and can learn feeding tactics through observation of the behavior of conspecifics.

Morphology: fishes of this family (river stingrays) have dorsoventrally compressed bodies forming a disc shape that can vary from 25 to up to 100 cm in diameter. The dorsal surface of their bodies is covered with tiny spines and tubercles. One to four caudal spines are present with venom glands attached. Some species also possess a line of non-venomous spines on the dorsal portion of their tails.

Reproduction: fishes of this family are viviparous with internal fecundation.

Economic importance: usually not appreciated as food neither regularly exploited by small-scale fisheries in the Brazilian Amazon. Although some people may eat them, fishes of this family tend to be considered as food taboos and to be avoided in the Amazon. Nevertheless, these fishes may have medicinal use, and, due to the presence of a venomous sting, they are feared by riverine populations. Most accidents happen after these animals are stepped accidentally in beaches with low water visibility.

References: Thorson et al. (1983), Begossi and Braga (1992), Carvalho et al. (2003), Begossi et al. (2004), Almeida et al. (2005), Silva and Uieda (2007), Carvalho and Lovejoy (2011), Carvalho and Ragno (2011), El-Deir et al. (2012), Garrone and Carvalho (2011), Garrone and Uieda (2012), Garrone et al. (2014), Thonhauser et al. (2013), Fontanelle and de Carvalho (2017).

Fish species: ***Potamotrygon humerosa*** **Garman 1913**

Common name(s): arraia

Size of fish sampled (standard length, both rivers): min. = 33 cm, max. = 44.5 cm, mean and median = 38.8 cm

Fisheries: arraias (*Potamotrygon* spp.) are rarely exploited in fisheries in both rivers, but some species may be exploited for aquarium trade. Biomass caught is in ◘ Table 3.3 (▶ Chap. 3).

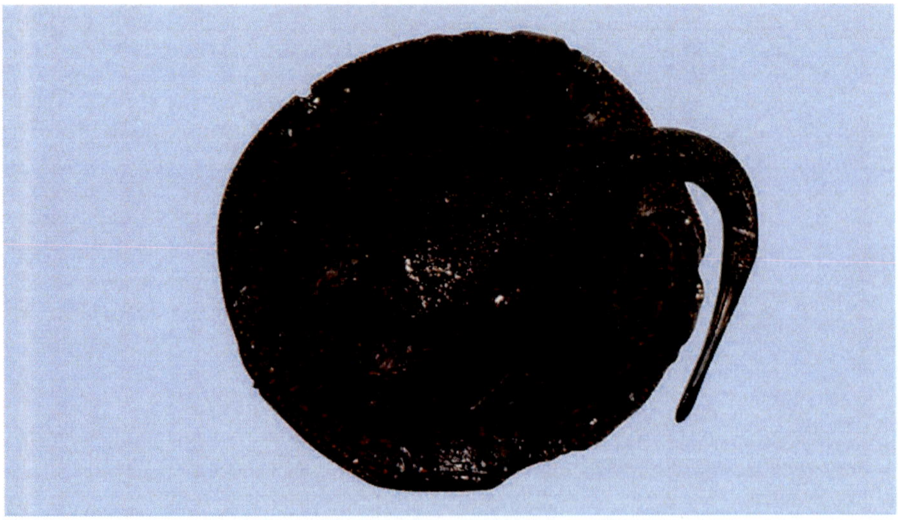

◘ **Fig. 6.104** *Potamotrygon humerosa* sampled in the Tapajós River. (Photo: Renato Silvano)

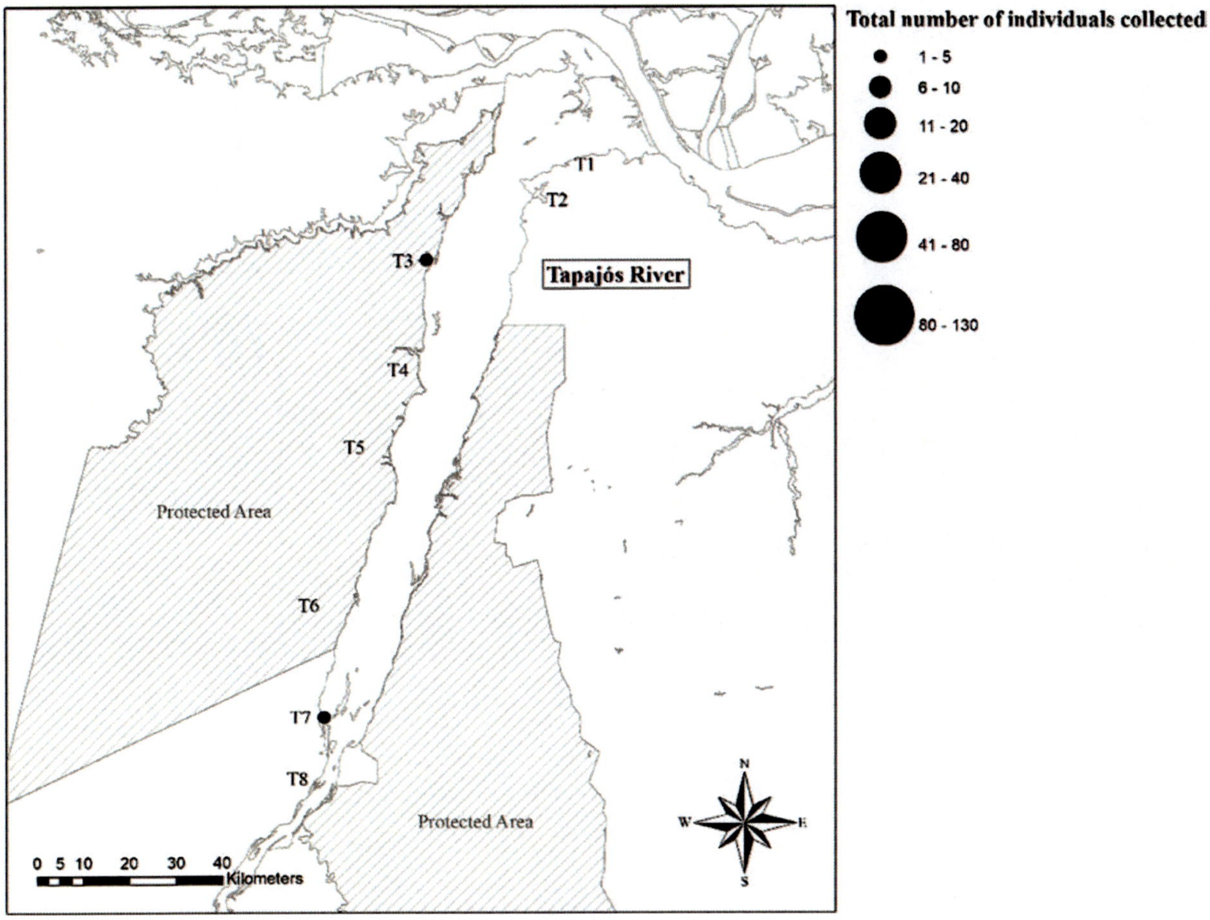

Fig. 6.105 Distribution of individuals of *Potamotrygon humerosa* sampled in the Tapajós River

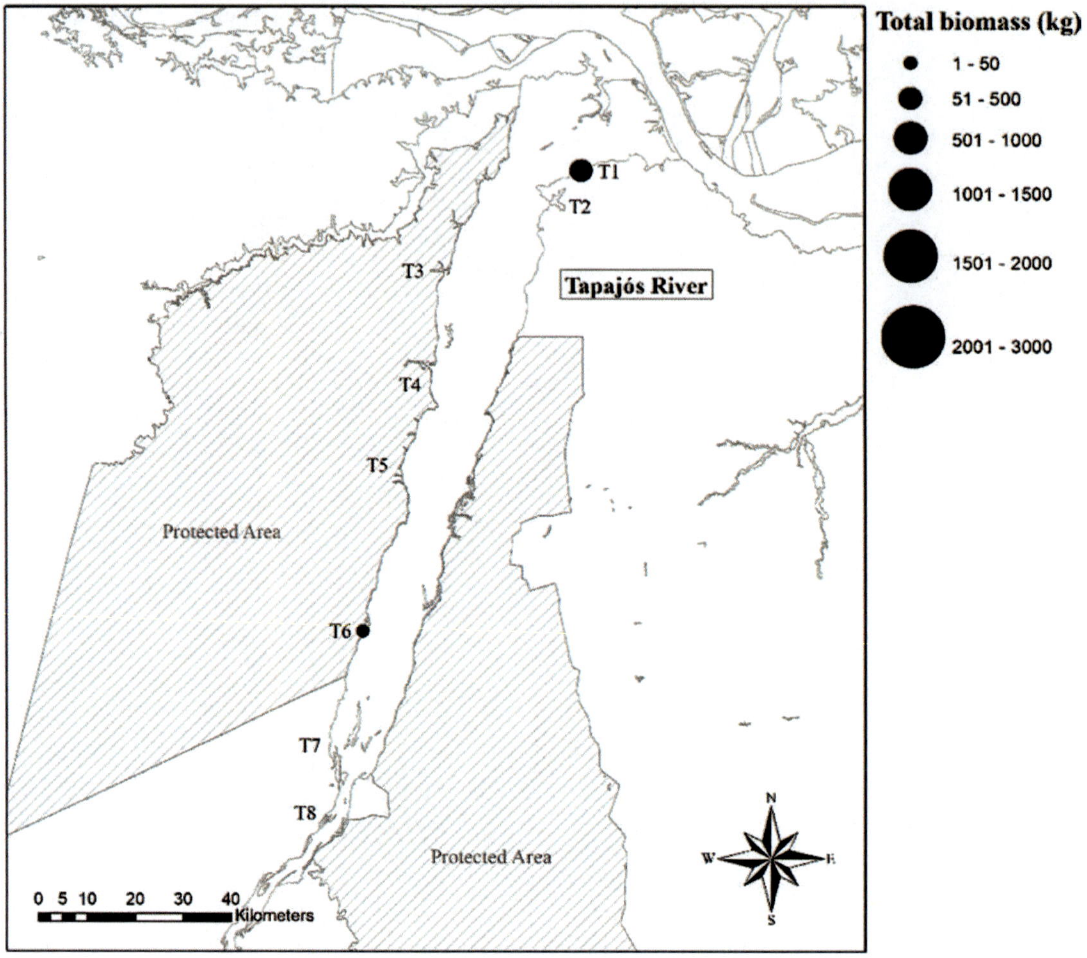

Fig. 6.106 Distribution of biomass of arraias (*Potamotrygon* spp.) caught in small-scale fisheries in the Tapajós River

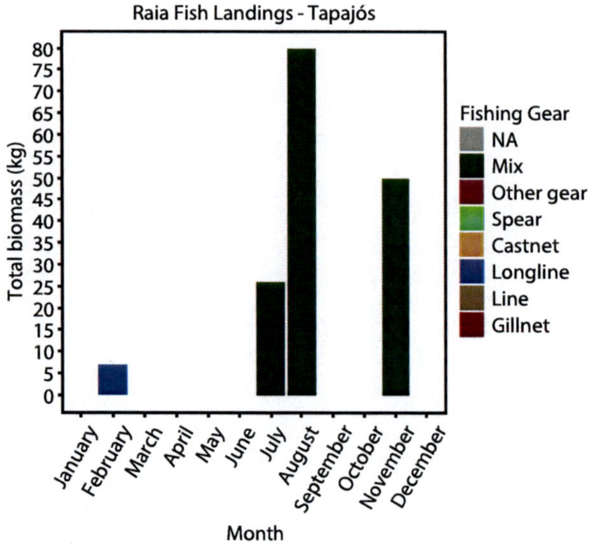

Fig. 6.107 Seasonal variation of biomass of arraias (*Potamotrygon* spp.) caught per fishing gear in small-scale fisheries in eight communities in the Tapajós River. NA: fishing gear not informed; Mix: more than one gear was used

Fish species: *Potamotrygon motoro* **(Müller & Henle 1841)**

Common name(s): arraia, arraia-preta

Size of fish sampled (standard length, both rivers): min. = 24 cm, max. = 51 cm, mean = 37.2 cm, median = 36.5 cm

Fisheries: arraias (*Potamotrygon* spp.) are rarely exploited in fisheries in both rivers, but some species may be exploited for aquarium trade; see *Potamotrygon humerosa* for data on arraia fisheries.

Fig. 6.108 *Potamotrygon motoro* sampled in the Tapajós River. (Photo: Renato Silvano)

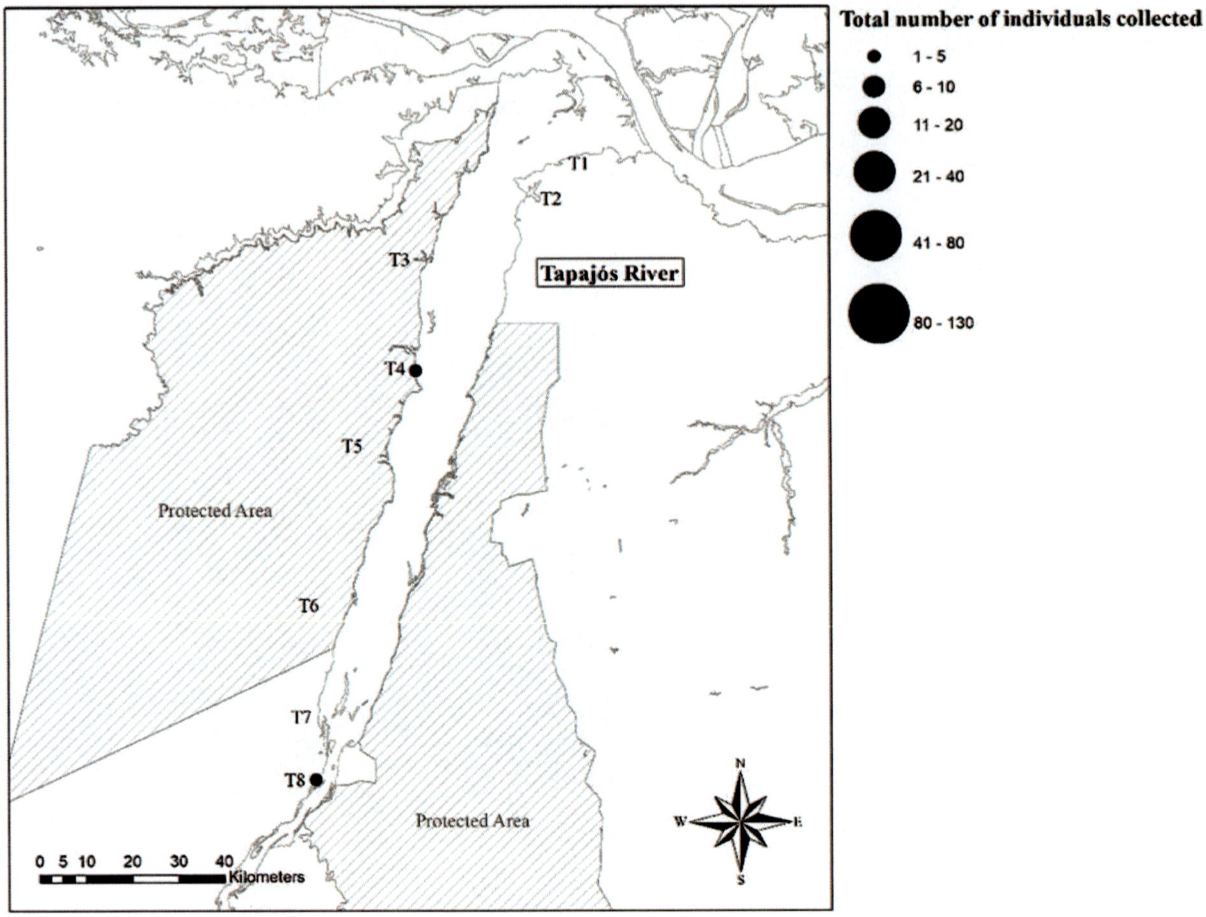

Fig. 6.109 Distribution of individuals of *Potamotrygon motoro* sampled in the Tapajós River

Fish species: *Potamotrygon orbignyi* (Castelnau 1855)

Common name(s): arraia, arraia-tataá

Size of fish sampled (standard length, both rivers): min. = 22 cm, max. = 55 cm, mean = 31.5 cm, median = 26.5 cm

Fisheries: arraias (*Potamotrygon* spp.) are rarely exploited in fisheries in both rivers, but some species may be exploited for aquarium trade; see *Potamotrygon humerosa* for data on arraia fisheries.

Fig. 6.110 *Potamotrygon orbignyi* sampled in the Tapajós River. (Photo: Renato Silvano)

6

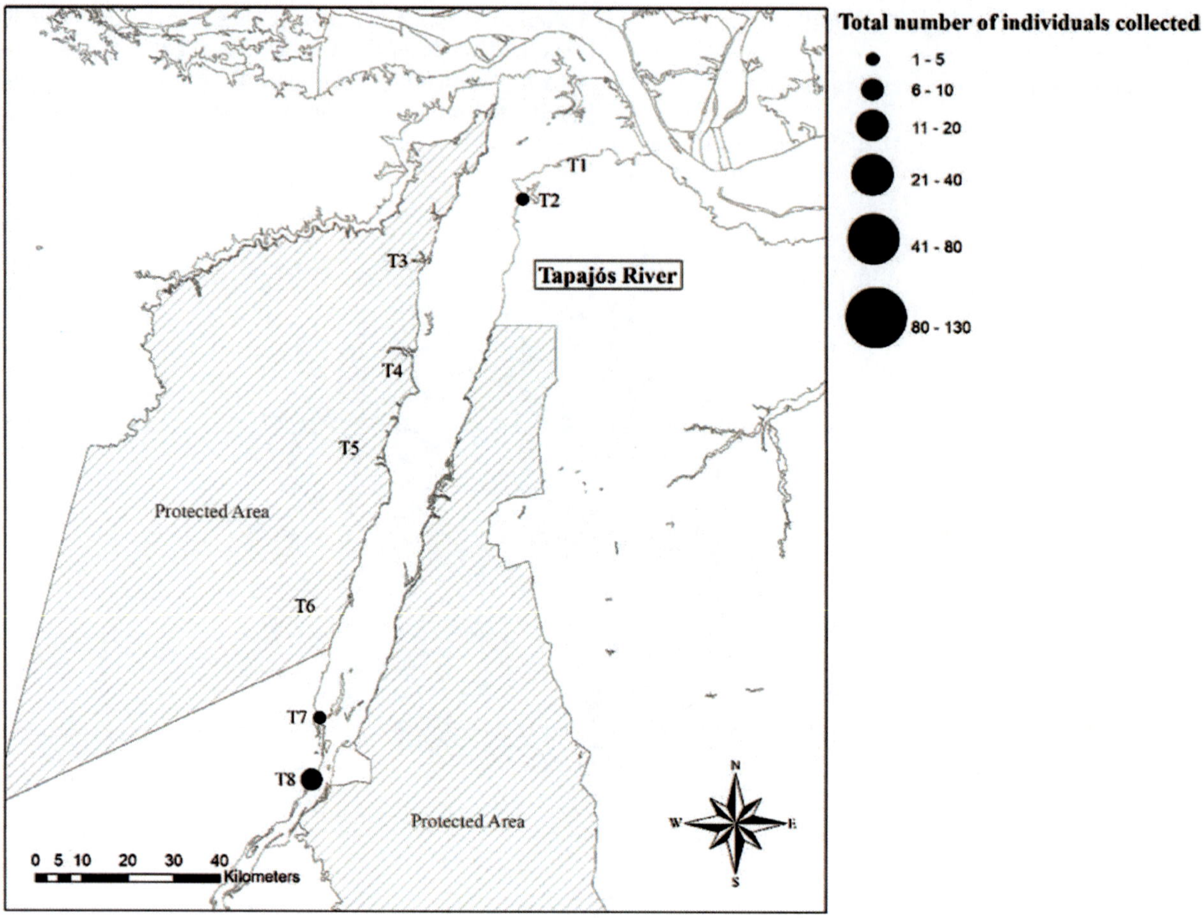

Fig. 6.111 Distribution of individuals of *Potamotrygon orbignyi* sampled in the Tapajós River

6.8 Order Tetraodontiformes

6.8.1 Family Tetraodontidae

6.8.1.1 Summary of Biology and Ecology (Literature)

Diversity: this family is composed of 192 valid species and 29 genera.

Habitat: these fishes occur in tropical and subtropical areas of the Pacific, Indian, and Atlantic Ocean. Most species are marine or estuarine fishes. Only one genus (*Colomesus*) is known to occur in the Amazon basin.

Ecology and feeding: some species are generalists and omnivorous, while others are specialized on certain types of invertebrates or algae. *Colomesus asellus*, a species found in Amazonian rivers, had a diverse diet based on benthic prey, such as aquatic insects (especially Ephemeroptera nymphs). The diet of this fish seems to be weekly affected by seasonal and ontogenetic factors. They present changes in the stomach, which allow them to inflate their body as a defensive mechanism, accompanied by strident sounds produced by pharyngeal teeth. They are very colorful fish, but their flesh is usually highly toxic, as well as their gonads during the breeding period.

Morphology: the species of this group are recognized as having four teeth, two in the jaw and two in the maxilla, as indicated by the name of the family. They have a naked body, sometimes with a series of tiny shields in the ventral area. The caudal fin is slightly rounded and the pelvic fin is absent. Freshwater species are usually much smaller than marine species.

Reproduction: in many species, males court the females for mating. Spawning usually occurs near the surface. Eggs are spherical and buoyant. Some species lay their adhering eggs on algae and other floating structures. During the spawning period, males can defend small territories. Males are also known to guard and blow water over the eggs to maintain them healthy and secure until their hatching.

Economic importance: members of this family are not important for commercial nor to artisanal fisheries in Amazon. Puffers contain a deadly neurotoxin poison known as tetrodotoxin. If eaten without proper care, a person may die in less than 24 hours (fatality rate is ~60%). You can still eat the meat of some puffers from the genus *Takifugu* in licensed restaurants (usually in Japan), but fugu, as popularly known, is expensive.

References: Shipp (1974), Géry (1977), Berra (2001), Galvis et al. (2006, 2007), Hisoshi et al. (2015), Bartolette et al. (2018), Eschmeyer (2019), Froese and Pauly (2019).

Fish species: ***Colomesus tocantinensis*** **Amaral, Brito, Silva & Carvalho 2013**

Common name(s): baiacu, maiacu

Size of fish sampled (standard length, both rivers): 7 cm

Fisheries: Not relevant to fisheries in both rivers.

Fig. 6.112 *Colomesus tocantinensis* sampled in the Tapajós River. (Photo: Renato Silvano)

6

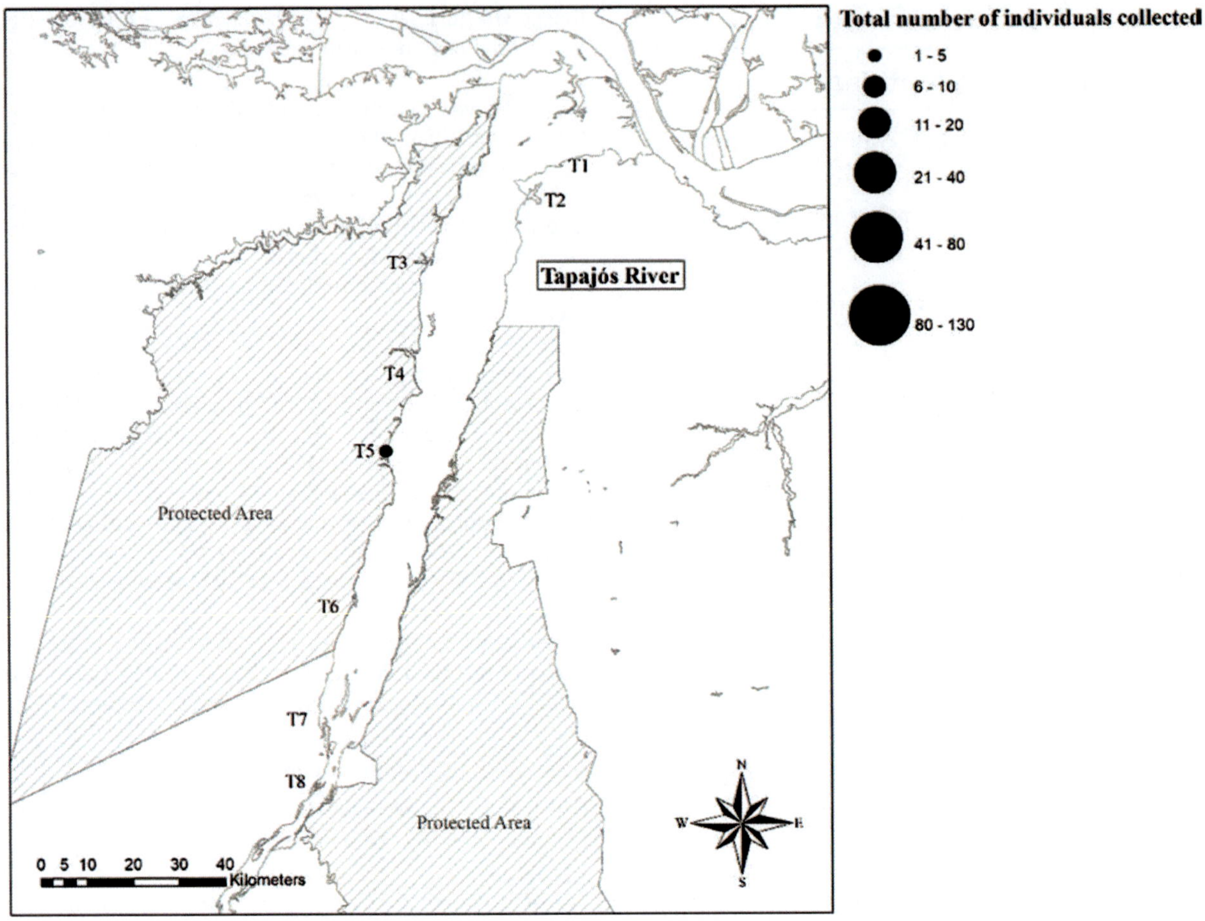

Fig. 6.113 Distribution of individuals of *Colomesus tocantinensis* sampled in the Tapajós River

References

Almeida PC, De Araújo MLG, De Almeida MP (2005) Reproductive aspects of freshwater stingrays (Chondrichthyes: Patamotrygonidae) in the Brazilian Amazon Basin. J Northwest Atl Fish Sci 35:165–171

Bartolette R, Rosa DCO, Beserra DA, Soares BE, Albrecht MP, Brito MFG (2018) Seasonal and ontogenetic diet patterns of the freshwater pufferfish *Colomesus asellus* (Müller & Troschel, 1849) in the upper-middleTocantins River. Acta Scientiarum 40:e35282

Begossi A, Francisco Manoel de Souza Braga (1992) "Food Taboos and Folk Medicine among Fishermen from the Tocantins River (Brazil)." Amazoniana: Limnologia et Oecologia Regionalis Systematis Fluminis Amazonas 12(1):101–118

Begossi A, Natalia H, Rossano MR (2004) "Food Chain and the Reasons for Fish Food Taboos among Amazonian and Atlantic Forest Fishers (Brazil)." Ecological Applications 14(5):1334–1343

Berra (2001) Freshwater fish distribution. Academic Press, New York

Bower LM, Winemiller KO (2019) Fish assemblage convergence along stream environmental gradients: an intercontinental analysis. Ecography. https://doi.org/10.1111/ecog.04690

Carpenter KE, De Angelis N (2016) The living marine resources of the Eastern Central Atlantic. Volume 3: Bony fishes part 1 (Elopiformes to Scorpaeniformes). FAO species identification guide for fishery purposes. FAO, Rome, pp 1511–2342

Carvalho MR, Lovejoy NR (2011) Morphology and phylogenetic relationships of a remarkable new genus and two new species of Neotropical freshwater stingrays from the Amazon basin (Condrichthyes: Potamotrygonidae). Zootaxa 2776:13–48

Carvalho MR, Ragno MP (2011) An unusual, dwarf new species of Neotropical freshwater stingray, *Plesiotrygon nana* sp. nov., from the upper and mid Amazon basin: the second species of *Plesiotrygon* (Chondrichthyes: Potamotrygonidae). Papéis Avulsos de Zoologia, Museu de Zoologia da Universidade de São Paulo 51(7):101–138

Carvalho MR, Lovejoy N, Rosa RS (2003) Potamotrygonidae (River stingrays). In: Reis RE, Malabarba LR, Ferraris C (eds) Checklist of the freshwater fishes of South and Central America. EDIPUCRS, Porto Alegre, pp 22–28

Collette BB (1982) South American freshwater needlefishes of the genus *Potamorrhaphis* (Beloniformes: Belonidae). Proc Biol Soc Wash 95:714–747

Collette BB (2000) Family Belonidae. In: FAO Species Identification. Guide for Fishery Purposes. The Living Marine Resources of the Western Central Pacific. By: Carpenter KE Niem, VH Rome. FAO. 4:2151–2161

Collette BB, Banford HM (2001) Status of the eastern Pacific agujon needlefish *Tylosurus pacificus* (Steindachner, 1876) (Beloniformes: Belonidae). Rev Biol Trop 49:51–57

Collette BB (2003) Family Belonidae Bonaparte 1832 – needlefishes. Calif Acad Sci Annotated Checklists Fish 16:23

Collette BB, McGowen GE, Parin NV, Mito S (1984) Beloniformes: development and relationships. In: Moser HG et al (eds) Ontogeny and systematics of fishes. Amer. Soc. Ichthyol. Herp. Spec. Publ, pp 335–354

de Pinna MCC, Di Dario F (2003) Family Pristigasteridae (Pristigasterids). In: Reis RE, Malabarba LR, Ferraris C (eds) Check list of freshwater fishes of South and Central America. EDIPUCRS, Porto Alegre, pp 43–45

Duponchelle F, Arce AR, Waty A, Vasquez G, Renno JF, Chu-Koo F, Davila CG, Vargas G, Tello S, Ortiz A, Pinedo R, Vasquez RM, Rodriguez JN (2015) Variations in reproductive strategy of the silver Arowana, *Osteoglossum bicirrhosum* Cuvier, 1829 from four sub basins of the Peruvian Amazon. J Appl Ichthyol 31:19–30

El-Deir ACA et al (2012) Ichthyofauna used in traditional medicine in Brazil. Evid Based Complement Alternat Med:1–16

Eschmeyer WN, FONG JD (2011) "Pisces. In: Zhang, Z.-Q. (Ed.) Animal biodiversity: An outline of higher-level classification and survey of taxonomic richness." Zootaxa 3148(1):26–38

Eschmeyer WN (2019) Catalog of Fishes. California Academy of Sciences. Available at: research.calacademy.org/research/ichthyology/catalog/fishcatmain.asp. Accessed July 2019

Fontenelle JP, de Carvalho MR (2017) Systematic revision of the *Potamotrygon scobina* Garman, 1913 species-complex (Chondrichthyes: Myliobatiformes: Potamotrygonidae), with the description of three new freshwater stingray species from Brazil and comments on their distribution and biogeography. Zootaxa 4310(1):1–63

Fricke R, Eschmeyer W, Fong JD (2019) Species by family/subfamily in the catalog of fishes. Califoria Academy of Sciences. Available at: http://researcharchive.calacademy.org/research/ichthyology/catalog/SpeciesByFamily.asp. Accessed July 2019

Froese R, Pauly D (2018) Family Cichlidae, Cichlids. Available at: http://www.fishbase.org/summary/FamilySummary.php?ID=349. Accessed July 2019

Froese R, Pauly D (2019) FishBase. World Wide Web electronic publication. Available at: www.fishbase.org. Accessed July 2019

Galvis G, Mojica JI, Duque SR, Castellanos C, Duarte PS, Arce M, Gutiérrez A, Jiménez LF, Santos M, Vejarano S, Arbeláez F, Prieto E, Leiva M (2006) Peces del medio Amazonas, Región de Leticia. Serie de Guías Tropicales de Campo N°5. Panamericana, Formas e Impresos, Bogotá

Galvis G, Duarte PS, Salazar LMM, Pinto YL, Cortés ÂG, Castaño ML, Castillo CC (2007) Peces de la Amazonía colombiana con énfasis en especies de interés ornamental. Ramon Lopez Editorial, Bogotá

Garrone DN, Carvalho LN (2011) Nuclear-follower foraging associations among Characiformes fishes and Potamotrygonidae rays in clean waters environments of Teles Pires and Xingu rivers basins, Midwest Brazil. Biota Neotropica 11:359–362

Garrone DN, Uieda VS (2012) Activity and habitat use of two species of stingrays (Myliobatiformes: Potamotrygonidae) in the upper Paraná River basin, Southeastern Brazil. Neotrop Ichthyol 10(1):81–88

Garrone DN, Gadig OBF, Zuanon J, Carvalho LN (2014) Cleaning interactions between shrimps (Palaemonidae) and freshwater stingrays (Potamotrygonidae) in the Paraná River, Southeastern Brazil. Ichthyol Explor Fres 24:379–384

Géry J (1977) Characoids of the world, vol 672. T.F.H Publications, Neptune City

Goulding M, Carvalho ML (1984) Ecology of amazonian needlefishes (Belonidae). Revista Brasileira de Zoologia 2(3): 99–111

Hallwass G, Silvano RAM (2016) Patterns of selectiveness in the Amazonian freshwater fisheries: implications for management. J Environ Plan Manag 59:1537–1559

Hallwass G, Lopes PF, Juras AA, Silvano RAM (2011) Fishing effort and catch composition of urban market and rural villages in Brazilian Amazon. Environ Manag 47:188–200

Hallwass G, Lopes PF, Juras AA, Silvano RAM (2013) Behavioral and environmental influences on fishing rewards and the outcomes of alternative management scenarios for large tropical rivers. J Environ Manage 128:274–282

Hisoshi K, Yoji O, Kimiaki I, Akihiko I (2015) Spawning behavior and paternal egg care in a circular structure constructed by pufferfish, *Torquigener albomaculosus* (Pisces: Tetraodontidae). Bull Mar Sci 91:33–43

Ikeziri AASL, Queiroz LJ, Doria CRC, Fávaro LF, Araújo TR, Villara GT (2008) Estrutura populacional e abundância do apapá-amarelo, *Pellona castelnaeana* (Valenciennes, 1847) (Clupeiformes, Pristigasteridae), na Reserva Extrativista do rio Cautário, Rondônia. Revista Brasileira de Zoociências 10(1):41–50

Keppeler FW, Hallwass G, Silvano RAM (2017) Influence of protected areas on fish assemblages and fisheries in a large tropical river. Oryx 51:268–279

Kullander SO, Ferreira EJG (2006) A review of the South American cichlid genus *Cichla*, with descriptions of nine new species (Teleostei: Cichlidae). Ichthyological Exploration of Freshwaters 17(4):289–398

Liao JC (2002) Swimming in needlefish (Belonidae): anguilliform locomotion with fins. J Exp Biol 205:2875–2884

Lima AFD et al (2013) Mudanças morfológicas no trato digestório e composição da dieta de larvas e juvenis do linguado *Catathyridium jenynsii* no reservatório de Itaipu, rio Paraná, Brasil. Iheringia Série Zoologia 103:214–221

Loeb MV (2009) Revisão taxonômica das espécies do gênero *Anchoviella* Fowler, 1911 (Clupeiformes, Engraulidae) das bacias Amazônica e do São Francisco. Instituto de Biociências, Universidade de São Paulo, São Paulo, p 97

Lovejoy NR (2000) Reinterpreting recapitulation: Systematics of needlefishes and their allies (Teleostei:Beloniformes). Evolution 54(4):1349–1362

Lovejoy NR, Collette BB (2001) Phylogenetic relationships of New World needlefishes (Teleostei: Belonidae) and the biogeography of transitions between marine and freshwater habitats. Copeia 2001(2):324–338

Melo ALA (2001) Relações filogenéticas das sardinhas do gênero *Pellona* Valenciennes (1847) (Clupeomorpha: Pellonidae), com revisão taxonômica das espécies sul-americanas. Dissertação de mestrado, Departamento de Biologia, Universidade de São Paulo, Ribeirão Preto. pp. 103

Munroe TA, Wongratana T, Nizinski MS (1999) Family Pristigasteridae. In: Carpenter KE, Niem VH (eds) FAO species identification guide for fishery purposes. The living marine resources of the Western Central Pacific. Volume 3. Batoid fishes, chimaeras and bony fishes part 1 (Elopidae to Linophrynidae). FAO, Rome, pp 1754–1770

Nelson JS, Grande TC, Wilson MVH (2006) Fishes of the world. Wiley, Hoboken

Nichols JT, Breder CM (1928) An annotated list of the Synentognathi with remarks on their development and relationships. Zoologica 8:1–423

Ohara WM, Lima FCT, Salvador GN, Andrade MC (2017) Peixes do Rio Teles Pires: diversidade e guia de identificação. Gráfica Amazonas e Editora Ltda–EPP, Aparecida de Goiânia

Oliveira ECD, Fávaro LF (2010) Reproduction of the flatfish *Achirus lineatus* (Pleuronectiformes: Achiridae) in Paranaguá Bay, state of Paraná, a subtropical region of Brazil. Zoologia (Curitiba) 27:523–532

Porter HT, Motta PJ (2004) A comparison of strike and prey capture kinematics of three species of piscivorous fishes: Florida gar (*Lepisosteus platyrhincus*), redfin needlefish (*Strongylura notata*), and great barracuda (*Sphyraena barracuda*). Mar Biol 145:989–1000

Queiroz HL (2008) Investimento parental e reprodução do Aruanã branco, *Osteoglossum bicirrhosum*, na Reserva Mamirauá. In: Queiroz HL, Camargo M (eds) Biologia, Conservação e Manejo dos Aruanãs na Amazônia Brasileira. Instituto de Desemvolvimento Sustentavel de Mamirauá, Téfé, pp 119–132

Queiroz LJ, Vilara GT, Ohara WM, Pires THS, Zuanon J, Doria CR (2013) Peixes do rio madeira. Editora Dialeto, São Paulo

Ramos RTDC (2013) Achiridae. In: Queiroz LJ, Vilara GT, Ohara WM, THS P, Zuanon J, Doria CR (eds) Peixes do rio Madeira, vol 3. Dialeto, São Paulo, pp 396–401

Ribeiro AC, Vilara GT, Filho JAL, Doria CRC (2016) Ecologia e Biologia de Peixes do rio Madeira. Editora EDUFRO, Porto Velho

Sant'Anna VB (2011) Filogenia dos peixes-agulha da família Belonidae (Atherinomorphae: Beloniformes). Tese de doutorado, Pontifícia Universidade Católica do Rio Grande do Sul, Porto Alegre, Rio Grande do Sul. pp. 280

Sant'Anna VB, Delapieve MLS, Reis RE (2012) A new species of *Potamorrhaphis* (Beloniformes: Belonidae) from the Amazon Basin. Copeia 2012:663–669

Santos G, Ferreira E, Zuanon J (2006) Peixes comerciais de Manaus. Editora PróVarzea, Ibama, Manaus

Shipp R (1974) The pufferfishes (Tetraodontidae) of the Atlantic Ocean. Publications Gulf coast research museum, pp. 4–163

Silva TB, Uieda VS (2007) Preliminary data on the feeding habits of the freshwater stingrays *Potamotrygon falkneri* and *Potamotrygon motoro* (Potamotrygonidae) from the Upper Paraná River basin, Brazil. Biota Neotropica 7(1):221–226

Silvano RAM, Begossi A (2001) Seasonal dynamics of fishery at the Piracicaba River (Brazil). Fish Res 51:69–86

Thonhauser KE, Gutnick T, Byrne RA, Kral K, Burghardt GM, Kuba MJ (2013) Social learning in Cartilaginous fish (stingrays *Potamotrygon falkneri*). Anim Cogn 16:927–932

Thorson TB, Brooks, Mayes MA (1983) The evolution of freshwater adaptation in stingrays. Natl Geogr Res Rep 15:663–694

Turner GF (2007) Adaptive radiation of cichlid fish. Curr Biol 17:827–831

Verba JT, Neto JGR, Zuanon J, Farias I (2014) Evidence of multiple paternity and cooperative parental care in the so called monogamous silver arowana *Osteoglossum bicirrhosum* (Osteoglossiformes: Osteoglossidae). Neotrop Ichthyol 12:145–151

Whitehead PJP (1985) FAO species catalogue. Vol 7. Clupeoid fishes of the World (suborder Clupeoidei). An annotated and illustrated catalogue of the herrings, sardines, pilchards, sprats, anchovies and wolfherrings. Part 1 — Chirocentridae, Clupeidae and Pristigasteridae. FAO Fish Synop 7:1–303

Willis SC, Macrander J, Farias IP, Ortí G (2012) Simultaneous Delimitation of Species and Quantification of Interspecific Hybridization in Amazonian Peacock Cichlids (Genus *Cichla*) Using Multi-Locus Data. BMC Evolutionary Biology 12, no. 1: 96. https://doi.org/10.1186/1471-2148-12-96

Index

C

D

E

F

G

H

I

J

K

L

M

N

O

P

R

S